T0192403

# Design and Analysis of
# Non-Inferiority Trials

# Chapman & Hall/CRC Biostatistics Series

Editor-in-Chief

**Shein-Chung Chow, Ph.D.**
Professor
Department of Biostatistics and Bioinformatics
Duke University School of Medicine
Durham, North Carolina

Series Editors

**Byron Jones**
Senior Director
Statistical Research and Consulting Centre
(IPC 193)
Pfizer Global Research and Development
Sandwich, Kent, U.K.

**Jen-pei Liu**
Professor
Division of Biometry
Department of Agronomy
National Taiwan University
Taipei, Taiwan

**Karl E. Peace**
Georgia Cancer Coalition
Distinguished Cancer Scholar
Senior Research Scientist and
Professor of Biostatistics
Jiann-Ping Hsu College of Public Health
Georgia Southern University
Statesboro, Georgia

**Bruce W. Turnbull**
Professor
School of Operations Research
and Industrial Engineering
Cornell University
Ithaca, New York

# Chapman & Hall/CRC Biostatistics Series

Chapman & Hall/CRC Biostatistics Series

# Design and Analysis of Non-Inferiority Trials

**Mark D. Rothmann**
**Brian L. Wiens**
**Ivan S. F. Chan**

CRC Press
Taylor & Francis Group
Boca Raton   London   New York

CRC Press is an imprint of the
Taylor & Francis Group, an **informa** business

A CHAPMAN & HALL BOOK

Chapman & Hall/CRC
Taylor & Francis Group
6000 Broken Sound Parkway NW, Suite 300
Boca Raton, FL 33487-2742

First issued in paperback 2020

© 2012 by Taylor and Francis Group, LLC
Chapman & Hall/CRC is an imprint of Taylor & Francis Group, an Informa business

No claim to original U.S. Government works

ISBN-13: 978-0-367-57691-2 (pbk)
ISBN-13: 978-1-58488-804-8 (hbk)

### Library of Congress Cataloging-in-Publication Data

Rothmann, Mark D.
  Design and analysis of non-inferiority trials / Mark D. Rothmann, Brian L. Wiens, Ivan S.F. Chan.
       p. ; cm. -- (Chapman & Hall/CRC biostatistics series)
  Includes bibliographical references and index.
  ISBN 978-1-58488-804-8 (hardback : alk. paper)
  1. Drugs--Testing. 2. Experimental design. 3. Therapeutics, Experimental. I. Wiens, Brian L. II. Chan, Ivan S. F. III. Title. IV. Series: Chapman & Hall/CRC biostatistics series.
       [DNLM: 1. Clinical Trials as Topic. 2. Research Design. 3. Therapies, Investigational. QV 771]

RM301.27.R68 2011
615.5'80724--dc22                                                              2011005377

**Visit the Taylor & Francis Web site at**
**http://www.taylorandfrancis.com**

**and the CRC Press Web site at**
**http://www.crcpress.com**

*To Shiowjen, Marilyn, and Lotus*

# Contents

# *Preface*

In recent years there has been frequent use of non-inferiority trial designs to establish the efficacy of an experimental agent. There has also been a proliferation of research articles on the design and analysis of non-inferiority studies. Points to Consider documents involving non-inferiority trials have been issued by the European Medicines Agency, and there is a draft guidance on non-inferiority trials that has been issued by the U.S. Food and Drug Administration. A typical non-inferiority trial randomizes subjects to an experimental regimen or to a standard of care, which is often referred to as an "active" control. A non-inferiority trial places a limit on the amount an experimental therapy is allowed to be inferior to a standard of care to still be considered worthwhile. This limit or non-inferiority margin should be selected so that a loss of efficacy of less than the margin relative to the standard of care implies that the experimental therapy has efficacy (relative to a placebo) and its efficacy is not unacceptably worse than the standard of care. A new treatment that offers a better safety profile or a more preferable method of administration compared to standard treatment may be beneficial even if somewhat less effective than standard treatment. There have been many non-inferiority clinical trials in various medical areas, including thrombolytic, oncology, cardiorenal, and anti-infective drugs, vaccines, and medical devices.

*Design and Analysis of Non-Inferiority Trials* is not intended as a substitute for regulatory guidances on non-inferiority trials, but as a complement to such guidances. This text provides a comprehensive discussion on the purpose and issues involved in non-inferiority trials and will assist the reader in designing a non-inferiority trial and in assessing the quality of non-inferiority comparisons done in practice.

*Design and Analysis of Non-Inferiority Trials* is intended for statisticians and nonstatisticians involved in drug development. Although some sections are technical and written for an audience of statisticians, most of the book is nontechnical and written to be easily understood by a broad audience without any prior knowledge of non-inferiority clinical trials. Additionally, every chapter begins with a nontechnical introduction.

We have strived to provide a thorough discussion on the most important aspects involved in the design and analysis of non-inferiority trials. The first two chapters discuss the history of non-inferiority trials and the design and conduct considerations for a non-inferiority trial. A first step in designing a non-inferiority trial is evaluating the previous effect of the selected active control treatment. Chapters 3 and 4 cover the strength of evidence of an efficacy finding and evaluating the effect size of a treatment. The active control therapy is identified based on knowledge of its performance in previous trials, not independent of the results of those previous trials. Thus, additional efforts are required to understand the effect size of the active control. Chapter 5 presents the two main analysis methods frequently used in non-inferiority trials, their variations, and their properties. Chapter 6 discusses the gold standard non-inferiority design that additionally includes a placebo group. Chapters 7 through 10 cover a variety of individual issues of non-inferiority trials, including multiple comparisons, missing data, analysis population, the use of safety margins, the internal consistency of non-inferiority inference, the use of surrogate endpoints, trial monitoring, and equivalence trials. Chapters 11 through 13 provide specific issues and analysis methods when the data are binary, continuous, and time to event, respectively. *Design and Analysis of Non-Inferiority Trials* can be read fully in the order presented. Various chapters can be comprehended or used as reference directly without reading previous chapters. A reader with little prior exposure to non-inferiority trials should start with Chapters 1 through 6 in the order presented, and cover the remaining material as needed. We have also included a discussion on $p$ values, confidence intervals, and frequentist and Bayesian analyses in the appendix.

We appreciate the assistance of all the reviewers of this book and the book proposal for their careful, insightful review. We are also indebted to so many at Taylor & Francis Publishing, most notably David Grubbs for his guidance and patience.

We thank all of those who have provided discussions and interactions on non-inferiority trials, including David Brown, Kevin Carroll, Gang Chen, George Chi, Ralph D'Agostino, Susan Ellenberg, Thomas Fleming, Paul Flyer, Thomas Hammerstrom, Dieter Hauschke, Rob Hemmings, David Henry, Jim Hung, Qi Jiang, Armin Koch, John Lawrence, Ning Li, Kathryn Odem-Davis, Robert O'Neill, Stephen Snapinn, Greg Soon, Robert Temple, Ram Tiwari, Yi Tsong, Hsiao-Hui Tsou, Thamban Valappil, and Sue Jane Wang. We are particularly grateful to Dr. Ellenberg for providing slides on the history of non-inferiority trials.

We are grateful for the support and encouragement provided by our families. Our deepest gratitude to our wives, Shiowjen (for MR), Marilyn (for BW), and Lotus (for IC), for their patience and support during the writing of this book.

# 1

## What Is a Non-Inferiority Trial?

### 1.1 Definition of Non-Inferiority

If the effect of an experimental therapy on an endpoint is either better than or not too much worse than the effect of a control therapy on that same endpoint, the experimental therapy's effect is said to be *noninferior* to the effect of the control therapy. It is common to say that the experimental therapy is noninferior to the control therapy. A clinical trial comparing an experimental arm with a control arm to determine whether the experimental arm is noninferior to the control arm is often called a "non-inferiority trial." The phrase "at least equivalent as" has also been used instead of "noninferior." We will use "noninferior" throughout this book.

The term "non-inferiority" is well established but can be misleading. A non-inferiority trial places a limit on the amount the experimental therapy is allowed to be inferior to the control therapy and still be beneficial or efficacious. If the experimental therapy had an effect that was "not inferior" to the effect of the control therapy, then the effect of the experimental therapy would either equal or be greater than the effect of the control therapy.

The acceptable amount for which the experimental therapy may be worse than the control therapy and still be noninferior to the control therapy is called the "non-inferiority margin." The margin may represent a threshold for the difference in effects or for the relative size of the effects. In practice, a non-inferiority margin is rarely a priori and usually subjective. Typically, the non-inferiority margin is based on a systematic review of the effect of the control therapy so that the experimental therapy would be regarded as beneficial and/or efficacious should the effect of the experimental therapy be greater than the difference in the effect of the control therapy and the non-inferiority margin.

A non-inferiority comparison may involve an efficacy endpoint, a safety endpoint, or some other clinical endpoint. For many indications where there is an effective standard therapy, it is unethical to deny patients the use of effective therapy. It is therefore not possible to conduct placebo-controlled trials in such settings. A non-inferiority trial with an effective standard therapy as an active control may be necessary to establish the efficacy of the experimental therapy. In these situations, the goals of the non-inferiority trial are to demonstrate (1) that the experimental therapy has any efficacy relative to a placebo or some other reference therapy and (2) that the efficacy of the experimental therapy is not unacceptably worse than that of the active control. A non-inferiority margin is selected for the purpose of satisfying one or both of these goals. An efficacy margin would be selected to show indirectly that the experimental therapy is more effective than a placebo. The determination of any efficacy involves ruling out a difference between the active control and experimental therapies at least the size of the efficacy margin. The size of the efficacy margin should be less than or equal to the effect of the active control therapy in the setting of the non-inferiority trial. The clinical margin is the amount of unacceptable loss of the effect of the active control that needs to be ruled out to declare that the experimental therapy is noninferior to the active control therapy. The clinical margin must be less than or equal to the efficacy margin. The term "non-inferiority margin" will be applied generically to both the efficacy and clinical margins.

While it may not seem worthwhile at first, a new treatment that has not been demonstrated to have better efficacy than a standard therapy can still provide advantages such as superior safety and tolerability. Selective serotonin reuptake inhibitor (SSRI) antidepressants, new "atypical" antipsychotics, nonsedating antihistamines, and treatments for hypertension such as diuretics and reserpine are not more effective than the drugs that preceded them, but they are much better tolerated and, in many cases, clearly safer.[1] New therapies may provide alternatives for people who do not respond to or cannot tolerate available therapies.[2] For therapeutic indications where the future of patient care appears to involve the addition of future therapeutics to the then standard of care, a more tolerable therapeutic with less toxicity may form a better building block than a current standard therapy that is either frequently not tolerated or very toxic. New therapies may also provide methods of administration or treatment schedules that are better preferred by patients. As with generic drugs, allowing new therapies as alternatives to existing ones may also reduce prices because of the resulting competition in the market. Therefore, it is desirable to be able to establish the safety and effectiveness of such treatments and to know how to conduct a meaningful non-inferiority trial.

Other cases that involve non-inferiority trials include experimental therapies that treat or prevent toxicities caused by anticancer therapies where there are concerns involving "tumor/disease protection" or "tumor promotion." For example, a therapy that protects the body from detrimental effects of chemotherapy may also protect the tumors from the action of chemotherapeutic drugs. Experimental therapies that aim to prevent toxicities by promoting the increase of "good cells" may also promote the increase of "bad cells." In both cases, it is important to study whether the use of these experimental therapies alters the benefits or the likelihood of obtaining benefits from the anticancer therapy. For an adequate improvement in the rate of a symptom or adverse event, there may be an acceptable amount of loss of the anticancer therapy's benefits. In such cases, the non-inferiority margin may be related to the amount of proven benefit of the experimental therapy.

## 1.2 Reasons for Non-Inferiority Trials

A randomized, double-blind placebo-controlled trial is the gold standard in the determination of efficacy and the risk–benefit profile of an investigational drug. Such trials provide direct evidence on the effectiveness and safety of the experimental therapy. A high-quality study conduct will result in an unbiased comparison of the study arms. Therefore, any demonstrated difference in outcomes can be attributed to the difference in the therapies used for each arm. The randomization fairly assigns treatments to subjects. This fairness of randomization along with the high quality of study conduct allows for a valid statistical comparison of the study arms.

The existence of blinding is important in achieving a high-quality study conduct. *Double blinding* means that the subject, the subject's investigator, and anyone else associated with the day-to-day conduct of the study do not know to which arm the subject was assigned. This would require that for the entire duration for which information is gathered on a subject, neither the subject nor the investigator is aware to which treatment arm the subject was allocated. In theory, double blinding allows for an equitable distribution across arms of internal and external influences, biases, and errors. For some clinical trials, the unique toxicities or side effects of a drug will break the blind for those who experience them. For a truly double-blind trial, the likelihood of protocol compliance, the general behavior of the subjects and the investigators that influence subject outcomes, the accuracy of subjects' and investigators' measurement, and the ability to obtain such measurements should not systematically favor one study arm over another. Use of a placebo or placebos, such as double dummy blinding in which subjects each receive two treatments (one of which is a placebo), may be necessary to implement a double blind.

There is disagreement on when a placebo-controlled trial would be unethical. Freedman[3] provides a principle of clinical equipoise that requires sound, professional disagreement as to the preferred treatment at the start of the randomized, clinical trial. Freedman[4] provides five conditions, any of which justifies the use of a placebo control: (1) no standard treatment exists; (2) standard treatment is not better than placebo; (3) standard treatment is a placebo (or no treatment); (4) new evidence has shown uncertainty of the risk–benefit profile of the standard treatment; and (5) effective treatment is not readily available due to cost or supply issues.

Placebo-controlled trials are ethical when there are no permanent adverse consequences with delaying treatment and patients are fully informed about the alternatives.[2] Escape clauses could be also included in the protocol.[5] Placebo-controlled trials may also be ethical when no therapy has an obvious, favorable benefit–risk profile (e.g., the only existing therapy may be efficacious but so toxic that patients may refuse therapy).

The use of a placebo control may be unethical when there is an available therapy that has been shown to prevent serious harm, prevent irreversible morbidity, or extend life, or there is an available therapy that has become a benchmark, or one of many benchmarks, by having shown such outstanding benefit with a very favorable benefit–risk profile that denying patients that therapy would be regarded as a safety risk or causing a patient harm. In the latter instance, an investigational drug would need to have better than minimal efficacy to even be considered a possible alternative therapy.

If subjects believe that it is better for them to receive an available therapy that has not demonstrated efficacy than to receive a placebo, recruitment to a placebo-controlled trial may be so difficult as to make the conduct of a placebo-controlled trial impossible. In such a case, an active-controlled trial should be considered. Since the active control has not demonstrated efficacy, the experimental therapy would need to demonstrate superior efficacy over the control therapy.

Subjects may desire a non-inferiority comparison when a new innovative therapy has a small efficacy advantage, no efficacy advantage, or slightly less efficacy than a standard therapy with some notable advantages of the standard therapy. Possible advantages may include a different or easier form of administration, reduced toxicities or side effects, or a more preferable safety profile. Examples where a new product may be preferred even when its efficacy is not greater than that of standard therapy include a new anti-infective product that produces no resistant bacteria and a synthetic respiratory distress product that may be less risky than an animal-derived product.[6]

Additionally, a study of applications studying a different regimen, schedule, or formulation of a therapy that has already demonstrated efficacy and a favorable risk–benefit assessment against placebo may also qualify for a non-inferiority comparison with a currently used version of that same therapy. For example, when a therapeutic protein is "pegylated," its half-life increases. Thus, a pegylated therapeutic protein requires less frequent dosing than its nonpegylated counterpart. To illustrate, consider both pegfilgrastim (pegylated-filgrastim) and filgrastim, which are both approved for the reduction of the incidence of febrile neutropenia induced by cytotoxic chemotherapy. Pegfilgrastim is administered subcutaneously once per chemotherapy cycle, whereas filgrastim is administered as multiple daily injections. This great reduction in the frequency of administration may be preferred by patients, provided there is no important loss in efficacy and no increase in risks.

A guideline from the European Medicines Agency[7] provides situations where a non-inferiority trial may be conducted for registration instead of a superiority trial over placebo. These situations include "Applications based on essential similarity in areas where bioequivalence studies are not possible, e.g., modified release products or topical preparations; products with a potential safety advantage over the standard might require an efficacy comparison to the standard to allow a risk–benefit assessment to be made; cases where a direct comparison against the active comparator is needed to help assess risk/benefit; cases where no important loss of efficacy compared to the active comparator would be acceptable; disease areas where the use of a placebo arm is not possible and an active control trial is used to demonstrate the efficacy of the test product."

The importance of the active control effect is the motivation for choosing a non-inferiority trial as the basis of demonstrating the effectiveness of the experimental therapy. Therefore, it may be unacceptable for the experimental therapy to have a much less effect than the active control. The meaning or purpose of a non-inferiority comparison varies from trial to trial. The following provide different cases for a non-inferiority comparison.

Case 1: A placebo-controlled trial would be unethical or impossible to conduct in practice, and the standard requirement is the demonstration of efficacy against a placebo. In these cases, the purpose of a non-inferiority trial is to provide substantial, indirect evidence that the experimental therapy has better efficacy than a placebo. This is done by providing substantial, direct evidence from clinical trials that the experimental therapy's efficacy is not much worse than the efficacy of an active therapy by some prespecified amount no larger than the effect of the active control therapy against placebo in the setting of the non-inferiority trial. The effect of the active control versus placebo is based on historical trials that evaluated the effect of the active control to placebo (or some other reference therapy) with proper adjustment if necessary to apply this estimated effect to the setting of the non-inferiority trial.

Case 2: In indications where there is a standard therapy that has such a large effect, some minimal efficacy may be necessary for the experimental therapy to have a favorable, worthwhile benefit–risk profile. In these cases, the purpose of a non-inferiority trial is to provide substantial, indirect evidence that the experimental therapy has better efficacy than a placebo by some minimal amount, or that the experimental therapy is not unacceptably worse than the active control therapy. This is done by providing substantial, direct evidence from clinical trials that the experimental therapy's efficacy is not much worse than the efficacy of an active control therapy by some minimal prespecified amount that is comfortably smaller than the difference between the effect of the active control therapy and placebo.

Case 3: There are cases where it may be necessary to demonstrate that an experimental therapy has efficacy either similar to or better than a standard of care. This may be the case when the experimental therapy is in the same "drug class" as the standard therapy. It would be necessary to demonstrate that the experimental therapy has efficacy either better than or not too much worse than the standard therapy.

Case 4: The experimental regimen replaces a drug in a standard regimen of multiple drugs with the experimental drug. For the experimental regimen to be considered as an alternative to the standard regimen, it may be necessary for the experimental regimen to have better efficacy than every drug, drug combination, and regimen for which that standard combination is superior. If each component of the drug combination for the experimental arm is regarded as "active," it may also be necessary for the experimental combination to have more efficacy than any subset of the drugs in that drug combination. When a new standard of care demonstrates improved survival over the previous standard of care, it may (or may not) be unethical to give patients that previous standard of care for that indication or line of therapy. Thus, it is important that any therapy being considered for use has sufficient efficacy to be considered an ethical therapy for the studied indication.

Case 5: The purpose of an experimental drug is to reduce the chance of toxicities or side effects to patients caused by a standard therapy. It is important to study whether the experimental drug interferes with the effectiveness of the standard therapy; that is, to study the amount of efficacy of the standard therapy that may be lost by additionally providing patients with the experimental therapy. The standard therapy with the experimental drug may be worthwhile to patients relative to the standard therapy alone if the standard therapy plus the experimental drug has less toxicities or side effects than the standard therapy alone, despite having a little less efficacy than the standard therapy alone. However, a lower dose (or less frequent use) of the standard therapy may also have less toxicities or side effects than the regular dose of the standard therapy. While a trial comparing the regular dose of the standard therapy with and without the experimental drug provides efficacy and safety data on the two regimens, it may not (depending on the results of the trial) provide evidence of the necessity of the experimental drug, unless the dose–response relationship on efficacy and safety is known for the standard therapy.

Additionally, non-inferiority comparisons are done in safety studies to rule out any unacceptable increase in the risk of some adverse events when deciding to use the experimental therapy.

As a global goal of drug development is the advancement of patient care, the use of non-inferiority trials to evaluate experimental therapies that are not expected to provide important efficacy or safety advantages over standard therapy ("me too drugs") has been criticized.[8,9] This is particularly problematic when nonrigorous margins are used, potentially leading to a "biocreep," in which an inferior therapy is used as the control therapy for the next generation of non-inferiority trials.

There may not be an appropriate choice for the active comparator for a non-inferiority trial of efficacy even when it may seem that a non-inferiority trial is the appropriate choice. Per the International Conference of Harmonization (ICH) E9 Guidance[10]: "A suitable active comparator could be a widely used therapy whose efficacy in the relevant indication has been clearly established and quantified in well-designed and well-documented superiority trials and which can be reliably expected to have similar efficacy in the contemplated active control trial." Per the ICH-E10 guidelines,[11] an active control can be used in a non-inferiority trial when its effect is (1) of substantial magnitude compared to placebo or some other reference therapy, (2) precisely estimated, and (3) relevant to the setting of the non-inferiority trial. Due to the importance of the effect of the active control (the motivation for conducting an active control trial), the non-inferiority margin should be sufficiently small so that demonstrating non-inferiority leads to the conclusion that the experimental therapy preserves a substantial fraction of the active control effect and that the use of the experimental therapy instead of the active control therapy will not result in a clinically meaningful loss of effectiveness.[12]

For an active-controlled clinical trial, the efficacy requirements are less rigorous in a non-inferiority comparison (less needs to be statistically ruled out) than in a superiority comparison. That is, when compared with an effective standard therapy, it is easier to demonstrate that the experimental therapy has noninferior efficacy than to demonstrate that it has superior efficacy. Thus, when the efficacy of an experimental therapy must be determined against an effective standard therapy, non-inferiority may be preferred as the main objective instead of superiority. When the experimental therapy has a small efficacy advantage over the standard therapy, a superiority trial having the standard therapy as the control therapy would require a large number of subjects to be adequately powered (e.g., at least 80% power). When the experimental therapy has no efficacy advantage over the standard therapy, it is impossible to design an adequately powered superiority trial that has the standard therapy as the control therapy.

## 1.3 Different Types of Comparisons

There are various ways whereby an endpoint may compare between an experimental arm and a control arm, as summarized in Table 1.1. Notice that when the experimental arm is superior to the control arm, the experimental

**TABLE 1.1**

Description of Each Type of Comparison

| Type of Comparison | Description |
| --- | --- |
| Inferiority | The experimental arm is worse than the control arm. |
| Equivalence | The absolute difference between the experimental arm and the control arm is smaller than a prespecified margin. |
| Non-inferiority | The experimental arm is either better than the control arm or the experimental arm is inferior to the control arm by less than some prespecified margin. |
| Superiority | The experimental arm is better than the control arm. |
| Difference | The study arms are not equal. Either the experimental arm is worse than the control arm or the experimental arm is better than the control arm. |

arm is also noninferior to the control arm. When the equivalence margin corresponds to the non-inferiority margin and the experimental arm is "equivalent" to the control arm, then the treatment arm is also noninferior to the control arm.

Whether a specific relation can be concluded between the control and experimental arms is often reduced to comparing a confidence interval for the difference in effects with either zero, a non-inferiority margin of $\delta$, or equivalence limits of $\pm\delta$ (for some $\delta > 0$). For a prespecified confidence level, a confidence interval does not contain those cases that have been ruled out by the data. For a confidence level of $100\,(1-\alpha)\%$, the method for determining the confidence interval is such that before observing the data, there was a $100\,(1-\alpha)\%$ chance (or greater chance) that the confidence interval will capture the true value. As such, about 95% of all 95% confidence intervals capture the true value of the parameter that is being estimated.

In Figure 1.1, since confidence interval A contains only negative values for the difference in the effects between the control and experimental therapies (C–E), the experimental therapy is concluded to be superior to the control

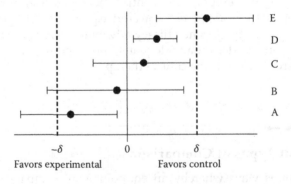

**FIGURE 1.1**

Relationship between different types of conclusions.

therapy. Because confidence interval B contains only values less than $\delta$, the experimental therapy is concluded to be noninferior to the control therapy with respect to the margin $\delta$. However, as confidence interval B contains both positive and negative values for the difference in the effects, the experimental therapy cannot be concluded to be superior or inferior to the control therapy. As confidence interval C contains only values between $-\delta$ and $\delta$, reflecting a small absolute difference in the effects of the experimental and control therapies, the experimental therapy and control therapy are concluded to be "equivalent" or similar with respect to the limits $\pm\delta$. Since confidence interval D contains only positive values for the difference in the effects that are smaller than $\delta$, the experimental therapy is concluded to be inferior and noninferior to the control therapy. This would mean that the experimental therapy is less effective than the control therapy, but not with unacceptably worse efficacy. Since confidence interval E contains only positive values with some of those values larger than $\delta$, the experimental therapy is inferior to the control therapy and cannot be concluded to be noninferior to the control therapy.

The type of comparison that can be done depends on the type or scale of the data. Data may be qualitative or quantitative. Qualitative data may be *nominal* or *ordinal*. The scale is nominal when subjects' outcomes are organized into unordered categories (e.g., gender, type of disease). The scale is ordinal when subjects' outcomes are organized into ordered categories. Quantitative data may have an interval or ratio scale. The scale is interval when differences have meaning (e.g., time of day and temperature in degrees Celsius). Two different scenarios having equal differences display the same meaning. The scale is ratio when ratios or quotients have meaning (e.g., time to complete a task, survival time, temperature in degrees kelvin). Data having a ratio scale have a *meaningful zero*.

When the data have a nominal scale, the relevant parameters are the actual relative frequencies or probabilities for each category. Since the categories are unordered, comparisons between study arms of the distributions for such measurements involve comparing for each category the similarity of the respective relative frequencies. That the distributions are different or that distributions are similar (an "equivalence" type of inference) are the only possible type of inferences. Non-inferiority, superiority, and inferiority inferences require that there is an order to the possible values. One measure of the similarity of two distributions of nominal measurements is the sum over all categories of the smaller relative frequencies between the two arms. For all other scales of measurements, any type of inference (e.g., equivalence, non-inferiority or superiority) can be made.

For data having an ordinal scale, additional relevant parameters would include the actual cumulative relative frequencies or cumulative probabilities for each category. For a given category, its cumulative relative frequency is the relative frequency of observations that either fall into that category or any category having less value. For data that have an interval or ratio scale, parameters of interest include means, medians, specific percentiles, variances, or standard deviations of the distribution. For data having an interval scale, comparisons between study arms of the same type of parameter would involve examining differences in the respective parameters. For data having a ratio scale, comparisons between study arms of the same type of parameter may involve examining differences or quotients in the respective parameters.

## 1.4 A History of Non-Inferiority Trials

The Kefauver–Harris amendment of the Food and Drug Cosmetic Act of 1962 requires that the safety and efficacy of an investigational drug be evaluated on the basis of "evidence consisting of adequate and well-controlled investigations." The gold standard has been evidence from at least two placebo-controlled trials. Placebo-controlled trials provide direct evidence of safety and effectiveness. As effective and highly effective therapies became approved and available, the ethical use and feasibility of placebo-controlled trials became questioned and consideration was given to active-controlled trials.

Makuch and Simon[13] discussed the requirements for evaluating that a less intensive treatment is as good as a more intensive one, including ruling out a difference of a prespecified size (i.e., a margin). There was little discussion on how to choose this margin or in relating the margin with previous results involving the active control effect. Lasagna[14] criticized the use of active-controlled trials for a purpose other than superiority of the experimental therapy, "In the absence of placebo controls, one does not know if the 'inferior' new medicine has any efficacy at all, and 'equivalent' performance may reflect simply a patient population that cannot distinguish between two active treatments that differ considerably from each other, or between active drug and placebo."

Blackwelder[15] provided the statistical hypotheses as one-sided hypotheses with a prespecified margin and the use of confidence intervals for testing for non-inferiority. Blackwelder noted that the choice of the margin may depend on the endpoint, with a more conservative margin being used for a mortality endpoint. Temple[16] discussed the complexities in evaluating the similarity of outcomes between an investigational drug and an active control. Temple proposed that any active-controlled trial aiming to show similarity provide historical results demonstrating that the active control was regularly shown superior to placebo, design the active-controlled study as similarly as possible to the previous studies involving the active control, and estimate the level of response in the experimental arm that would exceed the response expected by placebo.

Fleming[17] examined the case of mitoxantrone in metastatic breast cancer that was presented to the U.S. Food and Drug Administration (FDA) Oncologic Advisory Committee in 1986. The sponsor concluded from four clinical trials comparing mitoxantrone to doxorubicin that the efficacies of the two drugs are comparable and that mitoxantrone is less toxic.[17] Arguments for comparability seem to depend on nonsignificant results for comparisons on overall survival. However, the lack of significant evidence of a difference is not evidence of similarity. Additionally, using nonsignificant results for a determination of equivalence would lead to small, sloppily conducted trials that may increase the chance of failing to detect a difference.

For the determination of efficacy in an active-controlled trial, Fleming[17] emphasized the need for using confidence intervals along with having a reliable estimate of the active control effect. Three of the four mitoxantrone studies had confidence intervals for the overall survival hazard ratio that could not rule out an increase in the instantaneous risk of death of at least 75% by using mitoxantrone instead of doxorubicin. This, accompanied with the lack of historical evidence on an effect or effect size for doxorubicin, makes it impossible to legitimately conclude that mitoxantrone has efficacy (relative to placebo) on overall survival.

According to Fleming,[18] there are three key components of information for an active-controlled trial: (1) the confidence interval relating the effect of the experimental therapy to the active control therapy; (2) an assessment of the difference between the experimental and control therapies in side effects, toxicity, cost, etc.; and (3) a reliable estimate of the effect of the control therapy in the setting of the non-inferiority trial.

In 1992, the Division of Anti-Infective Drug Products of the FDA published a Points to Consider document on the Clinical Development and Labeling of Anti-Infective Drug Products,[19] which established a standard non-inferiority procedure for active-controlled trials of anti-infective therapies. The procedure used a random margin based on the larger of the observed cure rates among the treatment arms. The margin for the difference in cure rates ranged from 10% to 20% and was decreasing in the larger observed cure rate. The susceptibility to a biocreep led the FDA to revise the non-inferiority criterion and use a more conservative approach.[5] In 2007, the FDA issued a draft guidance[20] on the use of non-inferiority products to support approval of antibacterial products. The guidance states that it is unlikely for the indications of acute bacterial sinusitis, acute bacterial exacerbation of chronic bronchitis, and acute bacterial otitis media that data will support a non-inferiority comparison. In the case of acute bacterial sinusitis, there has not been a consistent reliable estimate of the efficacy of a standard therapy against placebo. Per the Code of Federal Regulations (21 CFR 314.126), adequate evidence should be provided to support a proposed non-inferiority margin.

In 1998, the ICH issued a guidance[10] on the statistical principles in clinical trials. This guidance included discussions on non-inferiority trials. An experimental therapy could show effectiveness relative to a standard therapy if only clinically acceptable differences were contained in the confidence interval for the differences in the treatment effects. Two years later the ICH issued a guidance[11] on the choice of the control group in a clinical trial. This guidance discussed various appropriate control groups and the situations in which a given concurrent control would be possible in determining effectiveness. The related trial designs and conduct along with the strengths and weaknesses of each type of control group were also discussed.

There has been a great desire to claim equivalence due to the failure to show a difference. Greene et al.[21] evaluated 88 reports of distinct studies from the literature published between 1992 and 1996 that investigated or concluded equivalence. They found that for 67% (59 of 88) of the studies, equivalence was claimed after a test for superiority yielded a nonsignificant result.

There has also been a desire to switch to a non-inferiority analysis after a failed test for superiority. In 2000, the European Agency for the Evaluation of Medicinal Products released a Points to Consider document on switching between superiority and non-inferiority.[22] One of the conditions for switching from superiority to non-inferiority is the prespecification and justification of a non-inferiority margin. A later guideline from the European Agency for the Evaluation of Medicinal Products was issued in 2005 on the choice of the non-inferiority margin. This document considered both two-arm non-inferiority trials and concurrent placebo-controlled three-arm non-inferiority trials. The usual primary focus of a non-inferiority trial was identified as the relative efficacy of the experimental and control therapies, and not simply that the experimental therapy has an effect. The appropriate choice of margin must provide assurance that the experimental therapy is effective and not substantially inferior to the control therapy. The effect of the experimental therapy should be of a clinically relevant size (greater than zero).

As non-inferiority trials may be more complex in their design and conduct than superiority trials, and there are either inaccurate conclusions of non-inferiority or equivalence in many publications or other unspecified determinations of non-inferiority, Piaggio et al.,[8] for the CONSORT Group, provided a recommendation to authors on how to report the design, conduct, and results of non-inferiority and equivalence randomized trials in publications. Some items on their checklist included providing a rationale for choosing a non-inferiority or equivalence design, providing the results from trials used to base the active control effect, relevant changes in patient characteristics compared with the previous trials that evaluated the active control effect, noting and justifying any differences in outcome measures from the previous trials of the active control including changes in timing of assessment, and interpretation of the results that accounts for sources of potential bias or imprecision.

Henanff et al.[23] reviewed the quality of reporting on non-inferiority and equivalence trials in the literature. Among 116 non-inferiority reports published in 2003 or 2004, only 24 provided justification on the choice of the margin, of which 11 reports provided statistical reasoning in the choice of the margin. In reporting a positive finding of non-inferiority, Henanff et al.[23] recommend language such as "treatment A is not inferior to treatment B with regard to the prespecified margin $\delta$."

The FDA issued a draft guidance on non-inferiority clinical trials in early 2010. This guidance[24] attempts to identify the steps and assumptions made in a non-inferiority trial and suggests approaches to account for the uncertainties. The aim is to reduce the possibility of drawing false-positive conclusions from the non-inferiority trial.

## References

1. Rothmann, M. et al., Design and analysis of non-inferiority mortality trials in oncology, *Stat. Med.*, 22, 239–264, 2003.
2. Ellenberg, S.S. and Temple, R., Placebo controlled trials and active-control trials in the evaluation of new treatments. Part 2: Practical issues and specific cases, *Ann. Intern. Med.*, 133, 464–470, 2000.
3. Freedman, B., Equipoise and the ethics of clinical research, *N. Engl. J. Med.*, 317, 141–145, 1987.
4. Freedman, B., Placebo-controlled trials and the logic of clinical purpose, *IRB: Rev. Hum. Subj. Res.*, 12, 1–6, 1990.
5. D'Agostino, R.B., Massaro, J.M., and Sullivan, L.M., Non-inferiority trials: Design concepts and issues—The encounters of academic consultants in statistics. *Stat. Med.*, 22, 169–186, 2003.
6. Ebutt, A.F. and Frith, L., Practical issues in equivalency trials, *Stat. Med.*, 17, 1691–1701, 1998.
7. Committee for Medicinal Products for Human Use (CHMP), *Guideline on the Choice of the Non-inferiority Margin*, EMA, London, 2005, at http://www.ema .europa.eu/ema/pages/includes/document/open_document.jsp?webContent Id=WC500003636.
8. Piaggio, G. et al., Reporting of non-inferiority and equivalence randomized trials: An extension of the CONSORT statement, *JAMA*, 295, 1152–1160, 2006.
9. Fleming, T.R., Current issues in non-inferiority trials, *Stat. Med.*, 27, 317–332, 2008.
10. International Conference on Harmonization of Technical Requirements for Registration of Pharmaceuticals for Human Use (ICH) E9: Statistical Principles for Clinical Trials, 1998, at http://www.ich.org/cache/compo/475-272-1 .html#E4.
11. International Conference on Harmonization of Technical Requirements for Registration of Pharmaceuticals for Human Use (ICH) E-10: Guidance on Choice of Control Group in Clinical Trials, 2000, at http://www.ich.org/cache/ compo/475-272-1.html#E4.
12. Fleming, T.R. and Powers, J.H., Issues in non-inferiority trials: The evidence in community-acquired pneumonia, *Clin. Infect. Dis.*, 47, S108–120, 2008.

13. Makuch, R. and Simon, R., Sample size requirements for evaluating a conservative therapy. *Cancer Treat. Rep.*, 62, 1037–1040, 1978.

14. Lasagna, L., Placebos and controlled trials under attack, *Eur. J. Clin. Pharmacol.*, 15, 373–374, 1979.

15. Blackwelder, W.C., Proving the null hypothesis in clinical trials, *Control. Clin. Trials*, 3, 345–353, 1982.

16. Temple, R., Government viewpoint of clinical trials, *Drug Inf. J.*, 16, 10–17, 1982.

17. Fleming, T.R., Treatment evaluation in active control studies, *Cancer Treat. Rep.*, 71, 1061–1065, 1987.

18. Fleming, T.R., Evaluation of active control trials in acquired immune deficiency syndrome, *J. AIDS*, 3, 82–87, 1990.

19. U.S. Food and Drug Administration Division of Anti-Infective Drug Products Advisory Committee Meeting Transcript, February 19–20, 2002, at http://www.fda.gov/ohrms/dockets/ac/cder02.htm#Anti-Infective.

20. U.S. Food and Drug Administration. Guidance for Industry Antibacterial Drug Products: Use of Non-inferiority Studies to Support Approval (draft guidance), October 2007.

21. Greene, W.L., Concato, J., and Feinstein A.R., Claims of equivalence in medical research: Are they supported by the evidence?, *Ann. Intern. Med.*, 132, 715–722, 2000.

22. Committee for Proprietary Medicinal Products. *Points to Consider on Switching between Superiority and Non-inferiority*, EMA, London, 2005, at http://www.ema.europa.eu/ema/pages/includes/document/open_document.jsp?webContentId=WC500003658.

23. Henanff, A.L. et al., Quality of reporting of non-inferiority and equivalence randomized trials, *JAMA*, 295, 1147–1151, 2006.

24. U.S. Food and Drug Administration, Guidance for Industry: Non-inferiority Clinical Trials (draft guidance), March 2010.

# 2

## Non-Inferiority Trial Considerations

### 2.1 Introduction

The gold standard in evaluating the safety and efficacy of an experimental agent is a placebo-controlled trial that is designed and conducted so that no or little bias is introduced in the comparison of study arms. It is also necessary for the clinical trial to have *assay sensitivity*—the ability to distinguish an effective therapy from an ineffective therapy. The experimental conditions of the clinical trial should also be such that the results are externally valid.

Poor study conduct will either introduce a bias, favoring one treatment over another, or obscure treatment differences. Obscuring treatment differences makes it more difficult to show that one study arm is better than another. However, for an active-controlled trial, obscuring treatment differences will make it easier to conclude both equivalence and non-inferiority when the experimental therapy is not notably better than the active control. Additionally, since the active control effect size is often assumed before conducting the non-inferiority trial, poor study conduct can reduce the active control effect in the setting of the non-inferiority trial, making it more difficult to distinguish whether an experimental therapy is effective. For a non-inferiority comparison, it is important that the selection of the non-inferiority margin and the effect of the control arm for the current trial are such that a demonstration of noninferior efficacy by the experimental arm compared with the control arm, along with an appropriate, fair study conduct, will imply that the experimental therapy is effective and not unacceptably worse than the active control.

In this chapter we will discuss external validity, assay sensitivity, the steps and issues in designing a non-inferiority trial, including the setting of the non-inferiority margin, the analysis population, and the sizing of a non-inferiority trial. The last section of this chapter briefly discusses the early history and experience of non-inferiority studies in anti-infective products.

## 2.2 External Validity and Assay Sensitivity

Clinical trials that lack external validity, or for which the external validity of the results is suspect, have produced results that cannot be applied to the future use of the medicine to patients. For many indications, for inferences to have external validity, it may be desirable for the control arm to reflect how all or many patients are currently being treated and the experimental arm should as much as possible reflect how a similar group of patients will be treated if the experimental drug becomes approved and is used. Thus, if the experimental drug is not currently available to patients having the disease of interest, it should not be made available to subjects in the control arm. Whatever concurrent or subsequent therapies that patients receive in practice should be made available to subjects in the clinical trial. Whenever there are important differences between a clinical trial and practice due to conduct, medical practice, or patient populations, the inferences from the clinical trial may not apply outside of the trial.

A clinical trial has assay sensitivity if it has the ability to distinguish an effective treatment from a less effective or ineffective treatment. When comparing an experimental therapy with a placebo, a positive finding implies the presence of assay sensitivity. In an active-controlled trial, this is not the case. For a superiority comparison on a given endpoint, it is important that the control arm does not have a negative effect on the endpoint. Then the demonstration of superior efficacy of the experimental arm to the control arm, along with an appropriate, fair study conduct, implies that the experimental therapy has positive efficacy. Poor study conduct or providing the experimental therapy to control subjects (when the experimental therapy is not available in medical practice) can make the outcomes similar between treatment arms and reduce assay sensitivity.

Failure to demonstrate superiority of an experimental therapy to placebo can either be due to the experimental therapy being ineffective or the trial lacking assay sensitivity. Additional use of an active control arm (i.e., a three-arm trial) can assist in determining whether the study has assay sensitivity. If the active control is demonstrated to be superior to placebo, then the trial has assay sensitivity. If neither the active control nor the experimental therapy is demonstrated to be superior to placebo, then the trial may be lacking assay sensitivity.

The non-inferiority margin is the minimum amount by which the efficacy of the experimental therapy can be less, but not unacceptably worse, than that of the active control. For a non-inferiority trial, assay sensitivity is the ability of the trial to distinguish whether the efficacy of the experimental therapy is unacceptably worse than the active control when the experimental therapy is noninferior to the active control. The absence of a placebo arm in the non-inferiority trial makes it difficult to assess whether the clinical trial conditions modified the active control effect and impaired the ability of the trial to have assay sensitivity. The assay sensitivity of a non-inferiority trial depends on the active control having an effect that is at least the size of the non-inferiority margin or the assumed effect of the active control used to determine the non-inferiority margin.[1] As such, the determination of assay sensitivity relies on the external results of previous trials evaluating the effect of the active control.

More will be discussed on external validity and assay sensitivity in Section 2.3 as they relate to issues involving the design and conduct of a non-inferiority trial.

## 2.3 Critical Steps and Issues

According to International Conference on Harmonization (ICH) E-10 guidelines,[2] the *historical evidence of sensitivity to drug effects* and *appropriate trial conduct* may be used to determine whether a non-inferiority trial has assay sensitivity. For the studied disease in past clinical trials, an evaluation should be made on whether similarly designed trials using the control therapy (or related therapy) regularly had assay sensitivity and on how sensitive or variable were the therapeutic effect and the conduct of the trials. Ideally, the control therapy regularly shows a consistent positive effect with respect to a placebo or some other reference therapy in trials that were similarly designed with acceptable study conducts that are analogous to the expected conduct of the non-inferiority trial.

The non-inferiority trial may require a similar design as trials used to study the effect of the control therapy. The conduct of the non-inferiority trial must not undermine its ability to distinguish whether an experimental therapy has any effectiveness or an acceptable effectiveness. Only after the non-inferiority trial is completed can the conduct of that trial be assessed. An evaluation should be made on features of the study conduct of the non-inferiority trial that may have altered the effect of the control arm from its historical effect or may have altered that reference against which the effect of the control arm is defined. The actual patient population represented by the subjects who entered the non-inferiority trial and the usage of concomitant therapies should not be dissimilar to the patient populations represented by the historical studies. If they are different, the difference will either need to be accounted for in determining the non-inferiority margin or a non-inferiority comparison may not even be appropriate (i.e., superiority would need to be shown).

According to ICH-E10,[2] there are four critical steps in the design and conduct of non-inferiority trials:

1. Determining that historical evidence of sensitivity to drug effects exists
2. Designing a trial

3. Setting a margin

4. Conducting the trial

### 2.3.1  Historical Evidence of Sensitivity to Drug Effects

There is no standard way of determining whether there is sensitivity to the effect of the control therapy on the basis of the results of previous trials. Such is determined on a case-by-case basis and such determinations may involve some subjectivity. It is important to understand whether the control arm will have an effect on the non-inferiority trial and how quantifiable that effect is. Ideally, there would be

1) Consistent results from multiple trials on the estimation of the effect of the control therapy on that endpoint of interest or that such an effect was so consistently large that its specific magnitude was not relevant to establish the non-inferiority criterion.

2) The design and conduct of the non-inferiority trial makes the results of the previous trials with the control therapy relevant to the current practice of medicine for the disease studied.

3) The design and the conduct of the non-inferiority trial make its results and the results of the non-inferiority comparison externally valid for the current care of patients.

There is historical sensitivity to drug effects when previously conducted clinical trials using the active control therapy consistently demonstrated that the active control was effective (i.e., superior to placebo). For the non-inferiority trial, this can provide assurance that the active control will be effective and may make it easier to evaluate the potential size of the effect of the active control and ultimately easier to set the non-inferiority margin. The non-inferiority margin should be based on the historically estimated effect, the variability in that estimate, the variability across trials in the active control effect, and any effect modification that has occurred since the previous trials as it relates to the setting of the non-inferiority trial. This evaluation is simplest when the design and conduct of the non-inferiority trial is similar to those of the previous studies.

When the outcome is very different for patients treated with an effective therapy versus untreated patients, the sensitivity to drug effects is more obvious. For such indications, it may be typical to have large effects for therapies versus placebo—for example, the effects on short-term cure rates in many infectious diseases.

In cases where there has been an inconsistency among past trials as to whether a therapy is effective, there will probably be no minimal effect that the control therapy can be expected to have in a non-inferiority trial. Therefore, a non-inferiority comparison would not be possible. According to ICH E10,[2] indications where this has been a concern include depression, anxiety, dementia, angina, symptomatic congestive heart failure, seasonal allergies, and symptomatic gastroesophageal reflux disease. For these indications, therapies have been shown effective in multiple well-controlled trials. However, because of the lack of sensitivity to having even a minimal effect size, a non-inferiority margin cannot be established for which the effectiveness of the experimental arm could be inferred from a non-inferiority comparison. In such situations, it would be necessary to conduct trials for the purpose of demonstrating superiority to either a placebo or a standard therapy.

## 2.3.2 Designing a Trial

For a comparative clinical trial, the choice of the control arm is very influential in the overall design of a trial and the potential conclusions that may be drawn from the trial. As mentioned in ICH-E10,[2] the choice of the control arm affects "the inferences that can be drawn from the trial; the ethical acceptability of the trial; the degree to which bias in conducting and analyzing the study can be minimized; the types of subjects that can be recruited and the pace of recruitment; the kind of endpoints that can be studied; the public and scientific credibility of the results; the acceptability of the results by regulatory authorities; and many other features of the study, its conduct, and its interpretation." The ICH-E10 guidelines[2] further discuss the advantages and disadvantages of various choices for a control arm. Potential choices for a control therapy may include no treatment or background therapy only, a placebo or a placebo plus background therapy, a different dose or regimen of the test drug, a different drug that may be used in practice for the disease being studied, or a different drug plus a placebo as in a "double dummy" design where a placebo will also be provided with the experimental therapy. Thus, the subjects in the control group will receive no treatment, a placebo, or an active therapy. The choice of control therapy may depend on whether there are any known, available effective therapies; whether it is ethical to give a placebo; the therapeutic practice for the disease studied; and whether it is necessary to demonstrate that the experimental therapy has any efficacy or that the experimental therapy has adequate efficacy.

In certain situations it may be desirable to have multiple control arms—for example, an active control arm and a placebo control arm. When possible, a three-arm randomized trial with a non-inferiority comparison can provide advantages to a two-arm non-inferiority trial. Three-arm randomized trials, having both standard therapy and placebo control arms, allow for the simultaneous assessment of the experimental therapy having superior efficacy versus placebo and noninferior efficacy versus the standard therapy. The non-inferiority comparison will not require any use of historical trials to evaluate the efficacy of the standard therapy since the efficacy of the standard therapy can be assessed against the placebo arm within the non-inferiority trial. Because only direct comparisons are needed to evaluate the effectiveness of the experimental therapy, there will be fewer issues involving the sensitivity of the trial to establish that the experimental therapy has efficacy or adequate efficacy. For example, there are no similar issues, as with a two-arm non-inferiority trial, about whether the results of the previous trials are transferable to the non-inferiority trial. Additionally, a three-arm trial allows for the control of both the precision of the estimated effect of the experimental therapy versus placebo and the precision of the estimated difference between the experimental and active control therapies. This is usually an advantage as the precision of the historical effect of the active control therapy is what it is, possibly leading to imprecise indirect estimates of the effect of the experimental therapy relative to placebo. If the precision of the historically estimated effect of the active control is very low, this historical estimation may not be useful in designing a two-arm, active-controlled non-inferiority trial.

Recall that for an experimental drug that treats or prevents toxicity caused by another drug, it may be important to study whether the use of the experimental drug alters the benefits or likelihood of benefit from the original therapy. Since a reduction of the dosage of the original drug should produce less toxicity, for the experimental drug to be useful, it is important that the use of the experimental drug with the studied dosage of original drug have a benefit–toxicity profile that is as good as or better than the benefit–toxicity profile of any given reduction of dosage of the original drug. Without knowing the dose–response relationship of the original drug on benefit and toxicity, the demonstration of less toxicity and noninferior benefit of adding the experimental drug to the standard dosage of the original drug may not be sufficient to show that the experimental drug is absolutely necessary.

### 2.3.3 Selecting the Margin

We will briefly discuss issues surrounding the selection of the non-inferiority margin. Further discussion and details are provided in Chapter 4 on evaluating the active control effect and in Chapter 5 on analysis methods.

The non-inferiority margin is that loss in the control therapy's effect on an endpoint that needs to be ruled out from using the experimental therapy instead of the control therapy. Any loss in the control therapy's effect greater than the non-inferiority margin is regarded as unacceptable. For efficacy comparisons, the non-inferiority margin will necessarily be smaller than the effect size of the control therapy. A quality non-inferiority inference depends largely on the ability to estimate or quantify the effect of the active control therapy within the setting of the non-inferiority trial. This is usually done in the absence of a placebo arm in the non-inferiority trial and on the basis of the results from past clinical trials evaluating the effect of the active control.

Wiens[3] summarized the various definitions of the non-inferiority margin in regulatory documents on the basis of one or more of the following criteria:

1. A value that is small enough to conclude an effect of the test treatment compared with placebo

2. The smallest value that would represent a clinically meaningful difference or the largest value that would represent a clinically meaningless difference

3. A value that is small compared to background variability or has other good statistical properties

Wiens[3] discussed three considerations for choosing a non-inferiority margin/criterion: comparison with putative placebo, the clinical importance of the active control effect, and statistical considerations.

The basic requirement for demonstrating efficacy is substantial evidence provided in an unbiased comparison of an experimental therapy to a placebo. When there is no placebo arm in the non-inferiority trial, a putative placebo strategy attempts to indirectly compare the experimental therapy to a placebo on the basis of the past comparisons of the active control to placebo and the comparison of the experimental therapy to the active control therapy in the non-inferiority trial. The inferences about the effect of the active control therapy in the setting of the non-inferiority may be based on a meta-analysis of previous trials using the control therapy with adjustments due to any design or conduct differences between those previous trials and the non-inferiority trial. The potential impact of any planned or expected design and/or conduct difference between the non-inferiority trial and the previous trials should be considered upfront when establishing the non-inferiority criterion.

According to ICH-E10,[2] the non-inferiority margin chosen for the planned trial "... cannot be greater than the *smallest effect size that the active drug would be reliably expected to have* compared with placebo in the setting of the planned trial." In that way, a conclusion that the relative effects of the experimental arm and the active control arm differ by less than that margin is tantamount to concluding that the experimental therapy is more effective than placebo.

A lower limit of a confidence interval (CI) for the active control effect based on a meta-analysis of past studies evaluating the active control effect or a fraction of that lower limit has been used as a surrogate for the effect of the active control therapy in the non-inferiority trial. Then, the test for non-inferiority compares the upper limit of a CI for the difference of effects between the experimental arm and the control arm with the non-inferiority margin. If the upper limit of the CI is smaller than the non-inferiority margin, the experimental arm is concluded to be noninferior to the active control arm with respect to that margin.

The assumption that the historical estimation of the effect of the control therapy is unbiased for the setting of the non-inferiority trial has been called the *constancy assumption*.

Because the evaluation of the active control effect is based on past studies, it is often unclear whether these estimated effects apply to the non-inferiority trial setting. Even when the historical studies show a fairly constant active control effect, there may be factors that would alter the effect of the active control in the setting of the non-inferiority trial. The non-inferiority trial may be conducted in subjects with less responsive or more resistant disease; subjects may now have access to better supportive care or different concomitant interventions that may attenuate the active control effect, or there may be lower adherence in the non-inferiority trial.[4] The definitions of the primary endpoint and/or how the primary endpoint is measured may also vary across studies. If it is believed that the effect size of the active control therapy has diminished or otherwise will be smaller in the non-inferiority trial than in the previous trials, the estimated effect of the control therapy should be reduced when applied to the setting of the non-inferiority trial. If the active control effect is smaller in the non-inferiority trial than in the historical trials and this is not accounted for in the analysis, the assay sensitivity of the trial will be low and there will be an increased risk of claiming an ineffective therapy as effective. The non-inferiority margin is often conservatively chosen because of concerns that the effect of the standard therapy may have diminished. As stated in the ICH-E10 guidance[2]: "The determination of the margin in a non-inferiority trial is based on both statistical reasoning and clinical judgment, and should reflect uncertainties in the evidence on which the choice is based, and should be suitably conservative."

As the effect of the active control depends on external experience, the non-inferiority comparison is an across-trials comparison. As such, formal cause-and-effect conclusions cannot be made from across-trials comparison without either making assumptions or providing evidence or arguments that the conditions and conduct of the current trial and previous trial are exchangeable, or the results are so marked that the lack of such exchangeability is not impactful.

Although many of these across-trials issues are shared with historically controlled trials, other issues are different. Essentially, historically controlled studies compare subject outcomes where subjects were not randomized. Unaddressed imbalances between groups on known and unknown prognostic factors can invalidate a historical comparison. Differences between the historical trials used to evaluate the effect of the active control and the non-inferiority trial in factors associated with the size of the active control effect (effect modifiers) that are not accounted for in the analysis can invalidate a non-inferiority comparison. More on effect modification is discussed in Chapter 4 on evaluating the active control effect.

The non-inferiority margin should account for effect modifiers and also for biases in the estimation of the active control effect. Biases in the estimation of the historical effect of the active control can arise owing to selection biases in choosing the historical studies and regression to the mean bias in identification of the active control. If the historical trials were found through a literature search, there may be a publication bias. If studies having unfavorable or less favorable results were not published, and thus not included in the estimation of the active control therapy's effect, the historical active control effect will be overestimated. Furthermore, the active control is likely selected on the basis of outcome (i.e., positive results from previous trials) and thus the estimated active control effect will be biased and greater than the true effect size.

Additionally, in the absence of the ability to estimate the between-trial variability of the effect of the active control therapy, some additional variability may need to be added to the variance of the estimator of the control therapy's historical effect to account for potential unknown factors that influence the effect of the active control. This would be particularly true if there were only one or two previous studies that could be used to estimate the effect of the control therapy and the disease of interest has a history of therapies having between-trial variability in their effects.

A constant or slightly varying effect size across studies for the control therapy is more important when the effect size is always small or moderate. The planned non-inferiority trial may still have assay sensitivity in demonstrating that the experimental arm has adequate efficacy when there are inconsistent, but all large, demonstrated effect sizes across studies. When large effects have been demonstrated across studies, there may be little statistical uncertainty in the choice of the acceptable amount of loss of a control therapy's effect that an experimental therapy can have for the experimental therapy to be noninferior.

The U.S. Food and Drug Administration (FDA) draft guidance[1] discusses an efficacy margin (M1), used in evaluating whether the experimental therapy has any efficacy, and a clinical margin (M2), used to evaluate whether the experimental therapy has unacceptably less efficacy than the active control. The reason for considering a clinical margin that is smaller than the efficacy margin is attributable to the importance of the effect of the active control therapy. The importance of the active control effect is often the reason why a placebo-controlled trial cannot be conducted.

As noted in an FDA advisory committee meeting for antibiotic drugs,[5] for some diseases (e.g., pneumonia), the reasons that make it unethical to do a placebo-controlled trial are the same reasons attributed to the unwillingness to have an experimental therapy that is much less effective than the standard therapy. For clinical trials, in such diseases, it may therefore be worthwhile to consider how much less efficacious a new therapy could be compared with an existing therapy when choosing the non-inferiority margin. Such margins may be based on clinical practice guidelines, patient opinion, other sources, and/or sound reasoning.

For endpoint of mortality or irreversible morbidity, it may be more difficult or impossible to define any margin that is clinically acceptable. However, if the available therapies include intolerable side effects or onerous administration, patients may be willing to give up a modest amount of efficacy for a better overall quality of treatment experience. There is subjectivity in the selection of such a margin. Different physicians and different patients may have different views on what would constitute a clinically important loss in efficacy. An appropriate non-inferiority margin may also depend on the yet-to-be-determined toxicities and/or additional benefits of the experimental therapy.

In many instances, numerical evaluations or statistical considerations may need to be considered when establishing a non-inferiority margin or criterion.

For continuous data, the choice of the non-inferiority margin for a difference in means may depend on the amount of variability (the standard deviation) of the outcomes on the control therapy. This allows for an evaluation of whether the distributions of the outcomes between the experimental and the control arms are likely to have significant overlap.

For a difference in proportions, inverting the margin can provide a *number needed to treat* to prevent one event. With a margin of 2 percentage points, treating at least 50 subjects will be required to observe one extra event; with a margin of 20 percentage points, treating only five subjects will result in one extra event.

The choice of analysis method for the comparison in the non-inferiority trial and the comparisons in the historical trials may also influence the validity of transferring a historical estimate of the effect of the control therapy to the non-inferiority trial. There may be better validity when all trials (the non-inferiority and historical trials used to evaluate the effect of the control therapy) adjust for the same meaningful covariates. This may be particularly true for modeling log-odds and log-hazard ratios, as in those cases in which the parameter being estimated from an adjusted analysis is different from the parameter being estimated from an unadjusted analysis.

Additionally, one choice of a metric may have more stable estimated effects for the active control therapy. If a product is able to protect 50% of the subjects who would otherwise acquire the disease, the active control effect as defined by a relative risk will be more stable than as defined by a difference in proportions.

Lenient margins introduce the risk that a treatment having little benefit will be endorsed after two or three generations of non-inferiority trials.[6] A *biocreep* is said to have occurred when slightly inferior therapy becomes the control therapy for the next generation of non-inferiority trials. This can occur by using a therapy whose efficacy was established from a non-inferiority trial as the control therapy in future non-inferiority trials. Unless great care is taken in designing the non-inferiority trials, continuing such a process can potentially lead to control therapies that are not any better than placebo and to clinical trials lacking assay sensitivity. Indications where there have been concerns about biocreep include mild or common infections.[7]

The potential for biocreep can be greatly reduced by using as the control therapy the therapy (or one of the therapies) with the greatest demonstrated effect.[5,8]

Fleming[6] proposed that a clinical margin that takes into consideration the perspective of the patient be determined by a team of clinical and statistical researchers—that is, how much clinical benefit would a patient be willing to exchange for greater ease of administration or less risk of adverse events.

Probably the most common choice of a non-inferiority margin is half of the lower limit of the 95% CI of the effect of the control therapy based on a meta-analysis of historical studies comparing that therapy with placebo. Different individuals have agreed on using such a margin but have disagreed on its interpretation. Some have viewed this margin as acceptable only for indirectly concluding that the experimental treatment is better than placebo— that is, the lower limit of the 95% CI of the effect of the control therapy is used as an "estimate" of the historical effect of the control therapy and is then decreased by 50% to apply it to the setting of the non-inferiority trial (see Snapinn[9]). Others have viewed such a non-inferiority margin as using the lower limit of the 95% CI of the effect of the control therapy as a conservative estimate of the control therapy's effect for the non-inferiority, and it is required that the experimental therapy retain 50% of the effect of the control therapy.[10] In both perspectives, how conservative such an approach for selecting a margin changes from case to case and is independent of the concerns on how transferable the estimates based on historical trials are to the non-inferiority trial.

Synthesis test procedures have also been used in testing non-inferiority. Typically, the results from the active-controlled trial and the results from estimating the historical effect of the control therapy are integrated through a normalized test statistic. The goal is to demonstrate that the experimental therapy retains a fraction of the control therapy's effect greater than a pre-specified fraction of the control therapy's effect. Examples and discussion on particular synthesis methods can be found in the papers of Hasselblad and Kong,[11] Holmgren,[12] Simon,[13] and Rothmann et al.[10] The procedures used by Rothmann et al.,[10] Hasselblad and Kong,[11] and Holmgren,[12] are designed to maintain a desired type I error rate when the estimation of the effect of the control therapy is unbiased for the setting of the non-inferiority trial. For a synthesis test procedure, Wang, Hung, and Tsong[14] examined how the type I error rate changes in various cases when the historical estimation of the control effect is used and the constancy assumption is false.

*Efficacy and Clinical Margins.* As stated in the FDA draft guidance,[1] "Determining the NI margin is the single greatest challenge in the design, conduct, and interpretation of NI trials." We discussed in Section 2.3 much of the issue involved in selecting a non-inferiority margin. Temple and Ellenberg[15] described three possible margins, M0, M1, and M2. M0 (or just zero) is the margin used when the active control is not regularly superior to placebo. M1 is the efficacy margin used to determine whether the experimental therapy has any efficacy and M2 is the clinical margin used to evaluate whether the experimental therapy has unacceptably less efficacy than the active control.

The efficacy margin has been regarded as the assumed active control effect size for the non-inferiority trial.[1] The value for M1 is often based on previous trials evaluating the active control effect with appropriate adjustments due to factors (effect modifiers) that may lead to a different effect size for the active control in the setting of the non-inferiority trial. Since the quality of the non-inferiority trial cannot be assessed beforehand,[1] "the size of M1 cannot be entirely specified until the NI study is complete." As concluding that an ineffective therapy is effective comes with a great cost, there is a tendency to conservatively choose the assumed effect of the active control and the corresponding clinical margin.[1]

Clinical judgment is used in determining M2, which cannot be greater than M1. M2 is often determined by taking a fraction of M1. This particular fraction, the retention fraction, depends on the importance of the endpoint and the size of the active control effect. The importance of the active control effect is the motivation for choosing a non-inferiority trial as the basis of demonstrating the effectiveness of the experimental therapy. Therefore, it may be unacceptable for the experimental therapy to have an effect much less than the active control. Situations that influence the retention fraction are provided in the FDA draft guidance.[1] If the active control has a large effect in reducing the mortality rate, retaining a large fraction of that effect will be desirable. If it is known that the experimental therapy is associated with a lower incidence of serious adverse events or is more tolerable for patients, the retention fraction may be lowered.

Statistical hypotheses involving such an M1 or M2 are surrogate hypotheses. The intention or hope is that ruling out that the experimental therapy has an effect that is less than the effect of the active control by M1 or more will imply that the experimental therapy is effective. Likewise, the intention is that ruling out that the experimental therapy has an effect that is less than the effect of the active control by M2 will imply that the experimental therapy does not have unacceptably worse efficacy than the active control therapy. When M2 is much smaller than M1, ruling out a difference in effects between the experimental therapy and active control of M2 should provide persuasive evidence that the experimental therapy is effective.

Assurance that the active control will have an effect at least the size of M1 in the setting of the non-inferiority trial is the "single most critical determination" in planning the non-inferiority trial.[1] Whether the non-inferiority trial will have assay sensitivity is based on whether the effect of the active control will be at least M1 in the setting of the non-inferiority trial, and the quality of the design and conduct of the non-inferiority trial.

The FDA draft guidance[1] prefers basing M1 on the lower limit of a high-percentage confidence for the active control effect (e.g., a 95% CI) from a meta-analysis of clinical trials that evaluates the effect of the active control. In ruling out that the experimental therapy is unacceptably worse than the active control, the synthesis method with an appropriate retention fraction can be considered.[1]

This choice of the lower limit of the 95% CI for the active control effect for M1 is intended as a conservative estimate or assumed value of the active control effect in the setting of the non-inferiority trial. There are situations provided in Chapters 4 and 5 where the regression to the mean bias will be so great that such a choice of M1 will not be conservative. A motivation for selecting the lower limit of the 95% CI for the active control effect for M1 is that it may protect against any decrease in the effect of the active control since the previous trials. However, as the issue in the departure of constancy is one of bias, not the precision of the estimate, Fleming[16] cautions against using the precision of the historical estimate to make adjustments but instead recommends addressing departures from constancy by adjusting the estimate of the active control effect. When there is only one previous study that evaluated the active control effect, and thus between-trial variability cannot be evaluated, the FDA draft guidance[1] recommends as a "potential cautious approach" defining M1 as the lower limit of the 99% CI for the active control effect.

When the experimental therapy is pharmacologically similar to the active control therapy, a single trial with possibly a less conservative choice of the non-inferiority margin may be appropriate.[1] The rationale is that pharmacologically similar products are expected to have a similar performance. Other possibilities in the FDA draft guidance[1] where one study with a less conservative choice of a margin may be appropriate include: when similar activity has been demonstrated between the experimental and active control therapies on a "very persuasive biomarker," the experimental therapy has already been shown to be effective in a closely related clinical setting, or when the experimental therapy has been shown effective in "distinct but related populations" (e.g., pediatric versus adult populations for the same disease).

### 2.3.4 Study Conduct and Analysis Populations

The conduct of the non-inferiority trial should allow for inferences that have external validity. Differences between the conduct of the historical trials and conduct that is needed for the non-inferiority trial to have external validity should be addressed, if possible, when determining the non-inferiority margin. When better efficacy than placebo is the goal of the active-controlled trial, the demonstrated efficacy of the experimental arm should be such that it is highly unlikely to have observed such efficacy or greater efficacy if the experimental therapy was a placebo. In other words, the study conduct should not introduce bias that would increase the risk of claiming an ineffective therapy as effective.

Poor or inappropriate study conduct can undermine the anticipated assay sensitivity of a non-inferiority trial. The specific results of previous trials are often instrumental to the non-inferiority margin to the point where it is necessary for the design and conduct of the non-inferiority trial to be similar to those previous trials. The ICH-E10 guidelines[2] provide some factors in the conduct of a study that could reduce the difference in the effects of an experimental therapy and a control therapy that should not be related to the treatment effects and therefore reduce the assay sensitivity of the trial. These factors include (from ICH-E10[2])

1. Poor compliance with therapy
2. Poor responsiveness of the enrolled study population to drug effects
3. Use of concomitant nonprotocol medication or other treatment that interferes with the test drug or that reduces the extent of the potential response
4. An enrolled population that tends to improve spontaneously, leaving no room for further drug-induced improvement
5. Poorly applied diagnostic criteria (subjects lacking the disease to be studied)
6. Biased assessment of endpoint because of knowledge that all subjects are receiving a potentially active drug—for example, a tendency to read blood pressure responses as normalized, potentially reducing the difference between test drug and control

For an equivalence trial, the aim is to determine whether the effects of therapies on an endpoint are fairly similar. Any undesirable design feature or undesirable study conduct that tends to make the observed effects of the study arms on some endpoint closer to each other without influencing the corresponding standard errors would increase the chance that a statistical procedure will arrive at the conclusion of "equivalence." For a non-inferiority comparison, such factors that make the observed effects on an endpoint closer without influencing the corresponding standard errors would increase (decrease) the chance of statistically concluding non-inferiority when the experimental arm is inferior (superior) to the control arm. Poor study conduct that obscures treatment differences makes it more difficult to show a difference (i.e., to show superiority) but can make it easier to show non-inferiority.

Randomization fairly assigns subjects to study arms. It is this fairness of randomization that allows for a valid statistical comparison of the study arms when the analysis is based on the "as-randomized" groups. The as-randomized population is referred to as the intent-to-treat (ITT) population. When the experimental therapy is inferior to the control therapy, the use of the ITT population for the non-inferiority analysis may not be conservative when poor study conduct obscures differences. Non-inferiority analysis on the basis of a per protocol (PP) subject population have also been performed in the hope of reducing or eliminating biases that tend to make the subject outcomes more similar between the study arms. It is important that the results and/or conclusions of these analyses be similar. Any difference in the results of the analyses may be indicative of influential, poor study quality. Too much missing data may potentially introduce a large bias and invalidate both analyses, even if the results are similar.

For a non-inferiority comparison, an ITT analysis need not be more conservative than a PP analysis. For non-inferiority comparisons of anti-infective products, in most studies evaluated by Brittain and Lin,[17] the PP analysis was less conservative than the ITT analysis. In fact, sloppiness due to poor study conduct may introduce a bias that favors a particular treatment arm. Study 1 of Rothmann et al.[18] in an advanced cancer setting had a high percentage of subjects prematurely censored for progression-free survival. On the basis of poorer prognosis for overall survival among subjects prematurely censored for progression-free survival compared with those still under observation for a progression-free survival event on the experimental arm, this premature censoring appears to be highly informative, whereas the premature censoring on the control arm does not appear to be informative. More on analysis populations is discussed in Chapter 8.

Because biases can also occur in subtle, unknown ways, the robustness of the results and primary conclusions should be evaluated[19]—that is, how sensitive the conclusions are to the limitations of the data and the unverifiable assumptions made. Open-label trials may be particularly vulnerable to bias. The limitations of the data will likely not be known until they are analyzed. It is important to keep missing data to a minimum. Proper sensitivity analyses are important in addressing data limitations and the potential impact of missing data. Sensitivity analyses should be prespecified to the extent possible. While sensitivity analyses are recommended, the use of sensitivity analyses is not a substitute for poor trial conduct or poor adherence to protocol, and does not rescue the results of a poor-quality clinical trial.

It is thus important that the conduct of the non-inferiority trial be of high quality so as to not compromise the non-inferiority comparison by either obscuring differences in the effects of the study arms on the endpoint of interest, or being so dissimilar to the study conduct of those previous trials whose results were used to establish the non-inferiority criterion so as to make the non-inferiority margin irrelevant.

The comparison of interest with the greatest real-world relevance is that between a control arm of a standard therapy along with best medical management and an experimental arm consisting of the experimental therapy along with best medical management. This compares how all or many patients are currently being treated with how the same group of patients could be treated if the experimental drug becomes approved for that indication. Influences or biases that interfere with having an unbiased comparison reduce the assay sensitivity of the trial. However, this comparison of interest may not require that all aspects of the trial conduct be equal between arms. Differences in the tolerability and effectiveness of different study therapies may result in the subjects in one study arm complying more frequently in taking their study therapy than subjects in other study arms. This unevenness in taking the assigned study therapy is an outcome of being on different arms and not a bias to any comparison that will be made. The analysis should not adjust for such unevenness. The potential subsequent therapies that are used and their distribution of usage may naturally be different between study arms, and such would be expected for that comparison of interest. If the control therapy is available for subsequent use in practice, or would be available for subsequent use if the experimental therapy is approved (as part of its best medical management), then it may be natural for the control therapy to be made available to subjects on the experimental arm for subsequent use. Although this feature may make it more difficult to show that the experimental therapy has better efficacy than the active control therapy, it can make it easier to show non-inferiority or equivalence. Delayed use of the control therapy may be noninferior to immediate use of the control therapy. In which case, an experimental therapy being noninferior to the control therapy where many subjects cross-in to the control therapy may not distinguish the experimental therapy from a placebo. Therefore, unless it was true for the historical studies used to establish the non-inferiority margin, allowing the control therapy to be available to the subjects in the experimental arm can obscure a noninferiority comparison and the determination of effectiveness of the experimental therapy. If it is unethical to deny the control therapy for later use to subjects on the experimental arm, either the non-inferiority margin would need to account for this cross-in to the control therapy or a superiority comparison may need to be required. In most instances, previous studies evaluating the effect of that standard therapy (the active control therapy) would probably not have subjects on the placebo arm later use the standard therapy. This makes it difficult to evaluate the effect of the active control therapy in the setting of the non-inferiority trial where cross-in to the control therapy may be allowed.

The study conduct of the non-inferiority trial cannot be evaluated until the trial has ended. It is only at that time an assessment or reassessment can be made as to the transferability of the results of previous trials that had been used to establish the non-inferiority margin. If the non-inferiority margin was based on previous trials and the conduct of the non-inferiority trial is not consistent with the required conduct, a reevaluation of the non-inferiority margin may be necessary.

## 2.4 Sizing a Study

Table 2.1 summarizes sample size formulas for designing superiority trials, non-inferiority trials with a fixed margin, and equivalence trials based on

**TABLE 2.1**

Sample Size Formulas

| Type of Comparison | Superiority[a] | Non-Inferiority with a Fixed Margin[b] | Equivalence[c] |
|---|---|---|---|
| Alternative hypothesis | $H_a: \Delta > 0$ | $H_a: \Delta > -\delta$ | $H_a: -\delta < \Delta < \delta$ |
| Required total sample size | $n \geq 4\left(\dfrac{Z_\alpha + Z_\beta}{\Delta_a/\sigma}\right)^2$ | $n \geq 4\left(\dfrac{Z_\alpha + Z_\beta}{(\Delta_a + \delta)/\sigma}\right)^2$ | $n$ satisfies $1-\beta \geq \Phi\left(\dfrac{\delta - \Delta_a}{2\sigma/\sqrt{n}} - Z_\alpha\right) - \Phi\left(\dfrac{-\delta - \Delta_a}{2\sigma/\sqrt{n}} - Z_\alpha\right)$ |
| Multiplier for a randomization ratio other than 1:1 | | $\dfrac{1}{4\pi(1-\pi)}$ | |

[a] Superiority is concluded if the lower bound of the two-sided 95% CI for Δ is greater than zero.

[b] Non-inferiority is concluded if the lower bound of the two-sided 95% CI for Δ is greater than −δ.

[c] Equivalence is concluded if the two-sided 90% CI for Δ lies within the interval (−δ,δ).

continuous data. It is assumed that outcomes with larger values are more desirable. Here,

$\Delta$ is the true difference in the effects of the experimental arm and the control arm $(E - C)$.

$\Delta_a$ is the assumed difference in the effects of the experimental arm and the control arm chosen to size the study.

$\delta$ is the non-inferiority (equivalence) margin.

$\alpha$ is the significance level.

$1 - \beta$ is the power (the probability of making the respective conclusion of superiority, non-inferiority, or equivalence) at $\Delta_a$.

$\sigma^2$ is the common population variance of the values for each study arm.

$z_\gamma$ is the $100(1 - \gamma)$th percentile of a standard normal distribution.

$\pi$ is the proportion of patients randomized to the control arm.

For time-to-event endpoints where effect sizes are measured with a log-hazard ratio, formulas for the required number of events are obtained by replacing $\sigma$ with the numeral 1. Note that the sample size formula for a superiority trial is just the sample size for a non-inferiority trial when $\delta = 0$ (or when $\delta \to 0^+$). Note also that

a) For $\delta > 0$, the same $\alpha$, $\beta$, and $\sigma$, and the same alterative $\Delta_a$, the required sample size is smaller for a non-inferiority trial than for a superiority trial.

b) For $\delta > 0$ and the same $\alpha$, $\beta$, and $\sigma$, the sample size for a superiority trial powered at the alterative $\Delta_a$ equals the sample size for a non-inferiority trial powered at the alternative $\Delta_a - \delta$.

c) For both superiority and non-inferiority trials, the required sample size decreases as $\Delta_a$ increases within the alternative hypothesis.

d) For an equivalence trial, the required sample size decreases as $|\Delta_a - 0|$ decreases.

It is very common to size an equivalence trial at the alternative where the experimental therapy and the control therapy have the same effect. This alternative provides the maximum power among all alternatives and thus leads to the smallest sample size for a fixed power. Since it is not likely that the experimental and control therapies have exactly the same effect, the equivalence trial will have less power than desired. It is recommended to size an equivalence trial at some small difference between the experimental and control therapies to ensure that the trial is adequately powered.

It is perhaps unfortunate that the sample size for a non-inferiority trial is often powered at the alternative where the experimental arm and the control arm have the same effect. This may be consistent with the thought that "non-inferiority" is "one-sided equivalence." When the active control has a small effect (i.e., when the non-inferiority margin is small), a non-inferiority trial powered at no difference in the effects of the experimental and control therapies will generally require a rather large sample size. When the control therapy has a very small effect versus a placebo, sizing a trial at the alternative where the experimental arm and the control arm have the same effect means that the trial is being powered at an alternative where the experimental arm has a very small effect versus a placebo. A placebo-controlled clinical trial with such an experimental arm having a small effect relative to placebo would require a large study size to be adequately powered to demonstrate superiority. When the non-inferiority margin is much smaller than the true effect of the control therapy versus placebo, a large sample size may be needed.

As the active control therapy may represent the therapy having the best effect among all therapies previously evaluated for a disease, it may also be unrealistic to assume that the next arbitrary product for that disease is equal in effect to the active control therapy. In these instances, it may be reasonable to size the non-inferiority trial on the basis of an assumed difference in effects where the experimental therapy is less effective than the active control therapy.

In an active-controlled, substitution trial, a misconception frequently arises about whether a non-inferiority trial or a superiority trial would require more subjects. When comparing an experimental therapy with an active control therapy, lesser values need to be statistically ruled out by the CI for a non-inferiority comparison than for a superiority comparison. Thus, for a fixed power, a larger sample size is required for a superiority comparison than for a non-inferiority comparison of the same two treatment arms. The misconception arises from comparing sample size calculations based on different assumed differences in effects. For the superiority analysis, the calculated sample size has adequate power when the experimental arm has greater efficacy than the control arm by some meaningful amount. For the non-inferiority analysis, the calculated sample size has adequate power when the experimental arm and the control arm have the same effect. The sample sizes should be compared on the basis of a single assumed difference in the effects of the experimental and active control therapies. When comparing an experimental therapy to an active control therapy for the same assumed difference in effects and power, a non-inferiority comparison requires a smaller sample size than a superiority comparison.

There is also a misconception that the more efficacious the control therapy, the easier it is for an experimental therapy (E) to demonstrate non-inferiority. Suppose that, in designing a non-inferiority trial, there are two candidates that may be chosen as the active comparator of the trial, C1 and C2, where C2 is more effective than C1 (C2 > C1). It is easier for an experimental therapy to demonstrate superiority (more probable or requires a smaller size) against a slightly efficacious control therapy than against a very efficacious control therapy. Thus, it would be easier to demonstrate that E is superior to C1 than to demonstrate that E is superior to C2 in a randomized, comparative trial. It is also easier to demonstrate that E is noninferior to C1 than to demonstrate that E is noninferior to C2 in a randomized, comparative trial. We will discuss this further in an example later in this section.

Suppose the non-inferiority comparisons of maintaining greater than some fixed proportion of the control effect were powered at the same assumed difference in effects—for example, that the mean for E is greater than the mean for C2 by 10. For the same power, a smaller sample size would be needed for the non-inferiority comparison with C1 than would be needed for the non-inferiority comparison with C2. This can easily be shown algebraically. It is true that a non-inferiority trial of E versus C1 powered at E = C1 + 10 will require a larger sample size than a non-inferiority trial of E versus C2 powered at E = C2 + 10. However, E = C1 + 10 and E = C2 + 10 are different assumptions regarding the effect of E. In general, the less efficacious the control therapy is, the easier it is for an experimental therapy to demonstrate non-inferiority (superiority).

To illustrate some ideas from this section, we will compare the sample sizes of various possible designs of a two-arm trial having a one-to-one randomization where all of the effects between the experimental therapy, control therapy, and a placebo are known. It is understood that these effects are unknown in practice. Here, the effects are assumed to be known so that comparisons will isolate the effect of a particular design feature on the overall sample size.

$\Delta_{E-P}$, $\Delta_{C-P}$, and $\Delta_{E-C}$ denote the true difference in the effects between the test therapy and placebo, between the control therapy and placebo, and between the test therapy and control therapy, respectively. As before, $\delta_a$ is the alternative used to power the study and $\delta$ is the non-inferiority margin. Table 2.2 gives three different cases of direct or indirect superiority comparisons of the test therapy versus placebo. In each case, $\Delta_{E-P} = 10$ and the overall sample size is 378.

**TABLE 2.2**

Sample Sizes for Direct or Indirect Superiority Comparisons of Test Therapy versus Placebo

| $\alpha, \beta, \sigma$ | $\alpha = 0.025, \beta = 0.10,$ $\sigma = 30$ | $\alpha = 0.025, \beta = 0.10,$ $\sigma = 30$ | $\alpha = 0.025, \beta = 0.10,$ $\sigma = 30$ |
|---|---|---|---|
| Type of trial | Superiority of E vs. P | Non-inferiority of E vs. C | Non-inferiority of E vs. C |
| $\Delta_a$ and $\delta$ | $\Delta_a = 10 = \Delta_{E-P},$ $\delta = 0$ | $\Delta_a = 0 = \Delta_{E-C},$ $\delta = 10 = \Delta_{C-P}$ | $\Delta_a = 5 = \Delta_{E-C},$ $\delta = 5 = \Delta_{C-P}$ |
| N | $N = 378$ | $N = 378$ | $N = 378$ |

**TABLE 2.3**

Sample Sizes for Non-Inferiority Comparisons of Test Therapy versus Control Therapy where Greater Than 50% Retention of Control Therapy's Effect Is Required

| $\alpha, \beta, \sigma$ | $\alpha = 0.025, \beta = 0.10,$ $\sigma = 30$ | $\alpha = 0.025, \beta = 0.10,$ $\sigma = 30$ | $\alpha = 0.025, \beta = 0.10,$ $\sigma = 30$ |
|---|---|---|---|
| Type of trial | Non-inferiority of E vs. C | Non-inferiority of E vs. C | Non-inferiority of E vs. C |
| $\Delta_a$ | $\Delta_a = 0 = \Delta_{E-C}$ | $\Delta_a = 5 = \Delta_{E-C}$ | $\Delta_a = 8 = \Delta_{E-C}$ |
| $\delta$ | $\delta = 5 = 0.5 \times \Delta_{C-P}$ | $\delta = 2.5 = 0.5 \times \Delta_{C-P}$ | $\delta = 1 = 0.5 \times \Delta_{C-P}$ |
| $N$ | $N = 1513$ | $N = 673$ | $N = 467$ |
| $N/378$ | 4 | 1.78 | 1.23 |

Table 2.3 gives three different cases of non-inferiority comparisons of the test therapy versus the control therapy. In each case, $\Delta_{E-P} = 10$ and greater than 50% retention of the control therapy's effect is required. Notice that the overall sample size decreases as the control therapy becomes less efficacious. The overall sample sizes for the first, second, and third cases in Table 2.3 are 300%, 78%, and 23%, respectively, greater than the common sample size in Table 2.2. Additionally, Table 2.3 illustrates that, for a fixed effect of an experimental therapy versus placebo (here, $\Delta_{E-P} = 10$), it becomes easier to demonstrate non-inferiority (here, with respect to 50% retention) as the active control therapy becomes less effective. Note that for a superiority comparison with $\beta = 0.10$, for the second case in Table 2.3, the needed sample size is 1513 when $\Delta_a = 5$, 2.25 times the sample size for the non-inferiority trial. For a superiority comparison with $\beta = 0.10$, for the third case in Table 2.3, the needed sample size is 591 when $\Delta_a = 8$, 1.27 times the sample size for the non-inferiority trial. Again, whatever is the true difference in effects between the experimental and active control arms, the power will be greater in demonstrating non-inferiority than the power in demonstrating superiority.

The necessary sample size also depends on the choice of the endpoint, the metric chosen to compare the outcomes between study arms, and the method of analysis. When there is more than one choice for the clinical endpoint, choosing that endpoint for which there is less (relative) variability can lead to a smaller non-inferiority trial.[20] Furthermore, for binary outcomes, an odds ratio or relative risk may be a more stable measure of the effectiveness of the active control therapy than a difference in proportions. The use of an odds ratio in a non-inferiority trial where the mortality rates are fairly low (comfortably below 50%) across risk groups may encourage the enrolment of subjects with greater risk of mortality.[4] A study conducted only in subjects with greater risk of mortality would be smaller than a study conducted only in subjects with less risk of mortality. This provides an incentive to study more seriously ill subjects, to both reduce trial size and to provide robust evidence for safety and effectiveness in that group where a drug can have the greatest impact in lowering the mortality rate. Additionally, choosing adjusted analyses over prognostic factors can also lead to smaller sample sizes.[20]

## 2.5 Example of Anti-Infectives

Perhaps more so than any other therapeutic area, drug development of anti-infective products has relied on non-inferiority trials. There have been many issues about relying on such designs, including the choice of an appropriate margin, the potential for biocreep, and the impact of study conduct on the ITT and PP populations.

Clinical trials for the registration of antibiotic therapies have almost exclusively been active-controlled, non-inferiority trials. Owing to highly effective therapies and the serious consequences in providing a placebo to a patient, the use of placebo-controlled trials is rare. In 1992, the Division of Anti-Infective Drug Products of the FDA published a Points to Consider document on the Clinical Development and Labeling of Anti-Infective Drug Products,[21] which included the establishment of a standard non-inferiority test procedure for efficacy for active control trials of anti-infective therapies. The procedure did not use a prespecified or a priori margin, but rather a random margin that was based on the larger of the observed cure rates in the experimental and control arms from the non-inferiority trial. The margin was 10% if the larger cure rate was greater than 90%, 15% if the larger cure rate is between 80% and 90%, and 20% if otherwise. Non-inferiority would be concluded if the lower limit of the two-sided 95% CI for the experimental versus control difference in the cure rates is greater than the "observed" margin. This type of approach to post hoc selection of a non-inferiority margin has been called the "step function" approach.

For $\hat{p}_{max}$ equal to the larger observed cure rate, the observed margin, $\hat{\delta}$, is expressed as

$$\hat{\delta} = \begin{cases} 0.10, & \text{if } \hat{p}_{max} > 0.90 \\ 0.15, & \text{if } 0.80 < \hat{p}_{max} \leq 0.90 \\ 0.20, & \text{if } \hat{p}_{max} \leq 0.80 \end{cases}$$

Because the margin was based on the larger cure rate, the margin used at the time of analysis may be different from the anticipated margin at the time of study design. If the anticipated control cure rate was 81%, the anticipated margin is 15%. If the observed cure rate for the control arm is 79%, with a lower observed cure rate for the experimental arm, a 20% margin would be used. This type of "margin selection" seems counterintuitive. If the control therapy is better than the experimental therapy, the smaller its apparent effect (the smaller its cure rate), the larger the non-inferiority margin. Since the variance in the estimates becomes larger as the true cure rate approaches 0.5, it appears that this type of "margin selection" was based on maintaining a desired power for the trial for a sample size or a given range of sample sizes.

Consider the following hypothetical example with results summarized in Table 2.4. Three hundred subjects are randomly divided among the experimental and control arms. This sample size will provide at least 80% power for a margin of 0.15 when there is a common cure rate between 80% and 90%. Suppose 112 (75%) patients on the experimental arm and 121 (81%) patients on the control arm are cured. Then the lower limit of the two-sided 95% CI is less than −0.15, and thus non-inferiority is not demonstrated. However, had the control cure rate been 2% lower and the experimental cure rate been 6% lower, non-inferiority would have been demonstrated; in fact, the results would have statistically demonstrated that the experimental arm has an inferior cure rate when compared with the control arm. Thus, a small reduction in the observed cure rate of the control arm with a more dramatic reduction in the cure rate of the experimental arm changed the conclusion from the experimental arm not demonstrating noninferior efficacy to the control arm to one in which the experimental arm has demonstrated noninferior (and inferior) efficacy.

There is no transitivity to conclusions of non-inferiority as there is for conclusions of superiority (inferiority). Table 2.5 illustrates a three-arm trial where, with respect to cure rates, arm C would be considered noninferior to arm B and arm B would be considered noninferior to arm A, but arm C would not be considered noninferior to arm A. In fact, in this example, the cure rate for arm C is demonstrated to be inferior to the cure rate for arm A.

In relation to the lack of transitivity, there is a real possibility for a biocreep for such trials. These trials were often designed to have at least 80% power when the cure rates between the arms are equal. This allows for the experimental arm to have a lower observed cure rate than the control arm and still demonstrate noninferior efficacy. In fact, these trials have sufficient power when the experimental arm has even a moderately lower cure rate. Additionally, since drugs with even much lower cure rates than current

**TABLE 2.4**

Possible Cases in Deciding whether Non-Inferiority Has Been Demonstrated

| Case | Sample Size per Arm | Experimental Arm Number Cured (Rate) | Control Arm Number Cured (Rate) | 95% CI for the Difference in Cure Rates | Margin |
|---|---|---|---|---|---|
| 1 | 150 | 112 (0.75) | 121 (0.81) | (−0.154, 0.034) | −0.15 |
| 2 | 150 | 103 (0.69) | 118 (0.79) | (−0.199, −0.001) | −0.20 |

**TABLE 2.5**

Example Showing Lack of Transitivity of a Non-Inferiority Conclusion

| Cure Rate | | | 95% CI for the Difference in Cure Rates and Margin | | |
|---|---|---|---|---|---|
| Arm A | Arm B | Arm C | B vs. A | C vs. B | C vs. A |
| 122/150 | 115/150 | 103/150 | (−0.139, 0.045) | (−0.180, 0.020) | (−0.224, −0.030) |
| (0.81) | (0.77) | (0.69) | Margin = −0.15 | Margin = −0.20 | Margin = −0.20 |

standards could be approved with wider margins up to 20%, a biocreep may occur with the future use of such drugs as the control therapy for the next generation of non-inferiority trials. Table 2.6 illustrates an example of biocreep for two-arm studies where the number of patients per arm is 150, where the observed cure rates would satisfy a non-inferiority conclusion when the observed results of each therapy are duplicated when that therapy is used as the control therapy in a future trial. We see that when the experimental therapy in an anti-infective trial is used as the control therapy in a future non-inferiority trial, there is a potential for continually concluding non-inferiority (and inferiority) while the new experimental therapy has an observed cure rate between 8% and 10% lower than its comparator.

Additionally, such standard test procedures do not differentiate between different indications. For diseases with high mortality or where approved therapies add little to the "cure" rate, a drug having a too much lower cure rate than standard therapy may not be acceptable, particularly when there is a small chance of curing a patient with subsequent therapy.

Upon further review of the step-function recommendation, the FDA in February 2001 stated on its Web site that the step function method is no longer in use and that they are developing a detailed guidance on selecting an appropriate non-inferiority margin.[22] The FDA encouraged the discussion on the selection of a margin with drug sponsors before the initiation of the

**TABLE 2.6**

Potential Change in Observed Cure Rates When Experimental Arm for Each Study Is Control Arm of the Next Study

| Therapy | Sample Size per Arm | Experimental Arm Number Cured (Rate) | Control Arm Number Cured (Rate) | 95% CI for Difference in Cure Rates | Margin |
|---|---|---|---|---|---|
| A | 150 | 118 (0.79) | — | — | — |
| B | 150 | 103 (0.69) | 118 (0.79) | (−0.199, −0.001) | 0.20 |
| C | 150 | 90 (0.60) | 103 (0.69) | (−0.195, −0.021) | 0.20 |
| D | 150 | 78 (0.52) | 90 (0.60) | (−0.192, −0.032) | 0.20 |
| E | 150 | 66 (0.44) | 78 (0.52) | (−0.193, −0.033) | 0.20 |
| F | 150 | 53 (0.35) | 66 (0.44) | (−0.197, −0.024) | 0.20 |
| G | 150 | 39 (0.26) | 53 (0.35) | (−0.197, −0.010) | 0.20 |

clinical trials. In February 2002 the FDA held an advisory committee meeting on the selection of the non-inferiority margin for antibiotic drugs.[5] The outcomes from the meeting[8] included

- An agreement that infectious disease indications are different and that it is impractical to have a common statistical requirement across such indications.
- Addressing the potential for biocreep by using appropriate comparator agents (e.g., that comparator with the largest demonstrated effect) for which there is assurance of a sufficient effect. When not possible, the use of three-arm clinical trials having a placebo-arm with early withdrawal should be considered.

According to D'Agostino, Massaro, and Sullivan,[23] from February 2001 to February 2002, a 10% margin has been a fairly popular choice for the non-inferiority margin regardless of the indication. A non-inferiority trial should have a prespecified non-inferiority margin. The FDA draft guidance on the use of non-inferiority products to support approval of antibacterial products[24] states that it is unlikely for the indications of acute bacterial sinusitis, acute bacterial exacerbation of chronic bronchitis, and acute bacterial otitis media that historical data will support the determination of a non-inferiority margin or a non-inferiority comparison. In November 2008, the FDA held an advisory committee meeting on the selection of the non-inferiority margin for anti-infective drugs.[25] The committee recommended that non-inferiority trials are acceptable in complicated skin and skin structure infections (SSSI). There was general agreement that a margin of 10% may be appropriate in most settings, although the particular margin would depend on the non-inferiority trial setting and the patient population. The great difficulty in establishing a non-inferiority margin is the evaluation of the effect of an active control therapy. Available data date back many decades and there is a concern that antimicrobials lose effectiveness over time. The committee recommended against using non-inferiority trials in uncomplicated SSSI. Superiority trials may be needed against a standard therapy.

Another main issue in the evaluation of antibiotic non-inferiority trials has been the role of the ITT and PP populations in the assessment of the efficacy of the experimental therapy. The list of possible criteria for excluding a patient from the PP population varies from trial to trial. Brittain and Lin[17] provided some fairly common criteria. These fairly common criteria include:

- Insufficient compliance with the assigned drug
- No assessment at the primary endpoint visit, unless previously classified as a failure
- Use of concomitant antibiotics effective against target infection
- Failure to meet baseline eligibility criteria

Some PP populations exclude subjects who die before the primary endpoint assessment where the cause of death is not regarded as related to the infection. Alternatively, these patients have been treated as nonfailures in the PP population and treated as failures in the ITT population. True ITT analyses where all subjects are followed to the endpoint or the end of study rarely occur due to missing outcomes. Patients with missing outcomes are often treated as failures in the cure rate analyses.

Brittain and Lin[17] compared the PP and ITT analyses from 20 trials that were presented to the FDA Anti-Infective Drug Products Advisory Committee between October 1999 and January 2003. Each trial studied a specific infection. The characteristic of the trials and the results are summarized as follows:

- The overall sample sizes ranged from 20 to 819 with a median of 400.
- The percentage of patients in the ITT population that were excluded from the PP population ranged from 2% to 43% with a median of 22%.
- The estimated treatment effect was more favorable for the experimental therapy in the ITT analysis for 13 of the 20 trials.
- The 95% CI for the difference in cure rates was wider for the ITT analysis for 12 of the 20 trials.
- The absolute differences in the treatment effect between the PP and ITT analyses ranged from 0.03% to 18.9% (the trial with the overall sample size of 20) with a median of 1.3%. The second largest absolute difference was 4.8%.

## References

1. U.S. Food and Drug Administration, Guidance for industry: Non-inferiority clinical trials (draft guidance), March 2010.
2. International Conference on Harmonization of Technical Requirements for Registration of Pharmaceuticals for Human Use (ICH) E10: Guidance on choice of control group in clinical trials, 2000, at http://www.ich.org/cache/compo/475-272-1.html#E4.
3. Wiens, B., Choosing an equivalence limit for non-inferiority or equivalence studies, *Control. Clin. Trials*, 1–14, 2002.
4. Fleming, T.R. and Powers, J.H., Issues in non-inferiority trials: The evidence in community-acquired pneumonia, *Clin. Infect. Dis.*, 47, S108–S120, 2008.
5. U.S. Food and Drug Administration Division of Anti-Infective Drug Products Advisory Committee meeting transcript, February 19–20, 2002, at http://www.fda.gov/ohrms/dockets/ac/cder02.htm#Anti-Infective.
6. Fleming, T.R., Design and interpretation of equivalence trials, *Am. Heart J.*, 139, S171–S176, 2000.

7. Tally, F., Challenges in antibacterial drug development, presented at the U.S. Food and Drug Administration Division of Anti-Infective Drug Products Advisory Committee meeting, meeting transcript, February 19–20, 2002, at http://www.fda.gov/ohrms/dockets/ac/02/slides/3837s1_06_Tally%20.ppt.

8. Shlaes, D.M., Reply, *Clin. Infect. Dis.*, 35, 216–217, 2002.

9. Snapinn, S., Alternatives for discounting in the analysis of non-inferiority trials, *J. Biopharm. Stat.*, 14, 263–273, 2004.

10. Rothmann, M. et al., Design and analysis of non-inferiority mortality trials in oncology, *Stat. Med.*, 22, 239–264, 2003.

11. Hasselblad, V. and Kong, D.F., Statistical methods for comparison to placebo in active-control trials, *Drug Inf. J.*, 35, 435–449, 2001.

12. Holmgren, E.B., Establishing equivalence by showing that a specified percentage of the effect of the active control over placebo is maintained, *J. Biopharm. Stat.*, 9, 651–659, 1999.

13. Simon, R., Bayesian design and analysis of active control clinical trials, *Biometrics*, 55, 484–487, 1999.

14. Wang, S.-J., Hung, H.M.J., and Tsong, Y., Utility and pitfalls of some statistical methods in active controlled clinical trials, *Control. Clin. Trials*, 23, 15–28, 2002.

15. Temple, R. and Ellenberg, S.S., Placebo-controlled trials and active-controlled trials in the evaluation of new treatments, Part 1: Ethical and scientific issues, *Ann. Intern. Med.*, 133, 455–463, 2000.

16. Fleming, T.R., Current issues in non-inferiority trials, *Stat. Med.*, 27, 317–332, 2008.

17. Brittain, E. and Lin, D., A comparison of intent-to-treat and per-protocol results in antibiotic non-inferiority trials, *Stat. Med.*, 24, 1–10, 2005.

18. Rothmann, M. et al., Examining the extent and impact of missing data in oncology clinical trials, *ASA Biopharm. Sec. Proc.*, 4014–4019, 2009.

19. International Conference on Harmonization of Technical Requirements for Registration of Pharmaceuticals for Human Use (ICH), E9: Statistical principles for clinical trials, 1998, at http://www.ich.org/cache/compo/475-272-1.html#E4.

20. Koch, G.G., Comments on 'current issues in non-inferiority trials' by Thomas R. Fleming, *Stat. Med.*, 27, 333–342, 2008.

21. U.S. Food and Drug Administration, Division of Anti-Infective Drug Products, Clinical Development and Labeling of Anti-Infective Drug Products, *Points-to-Consider*, U.S. Food and Drug Administration, Washington, DC, 1992.

22. U.S. Food and Drug Administration, Division of Anti-Infective Drug Products, *Points to Consider: Clinical Development and Labeling of Anti-infective Drug Products*, Disclaimer of 1992 Points to Consider document, U.S. Food and Drug Administration, Washington, DC, 2001.

23. D'Agostino, R.B. Massaro, J.M., and Sullivan, L.M., Non-inferiority trials: Design concepts and issues—The encounters of academic consultants in statistics, *Stat. Med.*, 22, 169–186, 2003.

24. U.S. Food and Drug Administration, Guidance for industry antibacterial drug products: use of non-inferiority studies to support approval (draft guidance), October 2007.

25. U.S. Food and Drug Administration Division of Anti-Infective Drug Products Advisory Committee meeting transcript, November 19–20, 2008, at http://www.fda.gov/ohrms/dockets/ac/cder08.html#AntiInfective.

# 3

# *Strength of Evidence and Reproducibility*

## 3.1 Introduction

It is important to evaluate from the evidence in the data whether an observed finding is real and can be reproduced in any similar or different relevant setting. In evaluating the strength of the evidence in the data, the Kefauver–Harris amendment of the Food and Drug Cosmetic Act of 1962 defines "substantial evidence" as "evidence consisting of adequate and well-controlled investigations."[1] According to Huque,[2] the U.S. Food and Drug Administration (FDA) has interpreted this as "the need to conduct at least two adequate and well-controlled studies, each convincing on its own, as evidence of efficacy of a new treatment of a given disease." There are conditions where data from a single adequate and well-controlled trial can be considered to constitute substantial evidence.[3] An interpretation of this is data from a large, adequate, and well-controlled multicenter study with a sufficiently small $p$-value that is internally consistent and clinically meaningful.[2]

Studying how the results of a clinical trial or an experiment would change if it were repeated under the exact same conditions (i.e., the same time in history with the same clinical investigators and potential patient pool, etc.) is a study of the *variability* of the results. Studying how the results of a clinical trial or an experiment would change if an additional study (or studies) were performed under a different environment (e.g., using different clinical investigators or a different potential pool of patients) is a study of the *reproducibility* of the results. Understanding the variability of the results from a clinical trial is necessary but usually not sufficient in understanding the reproducibility of the results.

For some indications, it may often be reasonable that if the results of a given clinical trial are quite marked and internally consistent—having a large estimated effect size that is many times greater than its corresponding standard error—a statistically significant result will be reproduced if an additional clinical trial were conducted having the same number of subjects as the earlier trial. The only way to truly know would be to conduct the additional clinical trial.

In Section 3.2, we will evaluate the strength of evidence both based on *p*-values (probabilities of observing as strong or stronger evidence supporting effectiveness than actually observed when the therapy is ineffective) and based on Bayesian probabilities that a therapy is ineffective given the data. The former starts with a hypothesis and under assumptions determines a probability based on the data (a form of inductive reasoning), whereas the latter starts with data and under assumptions determines a probability of a hypothesis (a form of deductive reasoning). In Section 3.32, we examine the reproducibility of a finding in an additional, identically designed and conducted clinical trial.

## 3.2 Strength of Evidence

### 3.2.1 Overall Type I Error

The gold standard is achieving a statistically significant favorable result at a one-sided significance level of 0.025 in each of the two trials. When two clinical trials are simultaneously designed to try to achieve this standard, the overall type I error probability of getting false-positive conclusions from both trials is $0.000625 = (0.025)^2$. This provides a rationale for designing a single clinical trial to achieve favorable statistical significance at a one-sided 0.000625 level. However, if that single trial achieves statistical significance at the 0.025 level, but not at the 0.000625 level (i.e., $0.000625 < p < 0.025$), a second clinical trial may be designed to achieve favorable statistical significance at the one-sided 0.025 level. This "one after the other" approach has an overall one-sided type I error probability of approximately $0.001234 \approx 0.000625 + (0.025 - 0.000625)(0.025)$. In essence, the type I error probability is greater than 0.000625 because the results of the first trial are not abandoned by failing to achieve statistical significance at the 0.000625 level. For the "one after the other" approach to have an overall one-sided type I error probability of 0.000625, a single trial would need to have a one-sided $p$-value $< 0.000315 \approx (0.0178)^2$ or have $0.000315 < p < 0.0178$ accompanied with a $p$-value $< 0.0178$ from an additional trial.

### 3.2.2 Bayesian False-Positive Rate

How convincing the evidence that an observed effect is real depends on both the size of the *p*-value and the general likelihood that a given product is effective for the studied indication. When 95% of all investigational products are effective for an indication, observing a one-sided *p*-value of 0.02 will be convincing of a real effect. However, when 5% of all investigational products are effective for an indication, observing a one-sided *p*-value of 0.02 may be suggestive but not convincing that an effect is real.

Given the results from the trial (i.e., "given the data"), the probability that the experimental therapy is or is not effective can be determined on the basis of some prior probability that a random experimental therapy is effective. This differs from a *p*-value, which is a probability involving the likelihood of observing the actual data (or data that would provide stronger evidence) given that the null hypothesis is true. The *p*-value does not consider the likelihood that the null hypothesis is true. Suppose 5% of investigated agents for a given indication are truly effective (and meaningfully so). Additionally, when an agent is effective, there is 80% power to achieve a one-sided *p*-value of less than 0.025. When an agent is ineffective, there is a 2.5% probability of achieving a one-sided *p*-value less than 0.025. For a typical 100 cases, Table 3.1 gives the number of cases for each combination of whether the investigated agent is truly effective and whether the observed one-sided *p*-value is less than 0.025. From Table 3.1, in 2.4 out of 6.4 cases (37.5%) where the one-sided *p*-value is less than 0.025, the agent was truly ineffective. Thus, when 5% of the investigated agents for a given indication are truly effective, simply achieving a one-sided *p*-value of less than 0.025 from a single clinical trial may be suggestive, but far from convincing, evidence of effectiveness. A similar example as that provided here can be found in the paper by Fleming.[4] We will refer to the posterior probability that the experimental agent is truly ineffective, given that the experimental therapy has been concluded as effective, as the *Bayesian false-positive rate*.

Suppose that two simultaneously conducted clinical trials are done per investigational agent. An investigational agent is concluded as effective when each of the two studies has a one-sided *p*-value less than 0.025. As before, when an agent is effective, there is 80% power to achieve a one-sided *p*-value less than 0.025 within a single clinical trial. For a typical 100 cases, Table 3.2 gives the number of cases for each combination of whether the investigated agent is truly effective and whether the agent is concluded as effective by achieving one-sided *p*-values less than 0.025 in both studies. From Table 3.2, in 3.2 out of 3.26 cases (≈98.2%) where the conclusion was "effective," the agent was truly effective. Therefore, the Bayesian false-positive rate is about 1.8%. Thus, when 5% of the investigated agents for a given indication are truly effective, achieving a one-sided *p*-value less than 0.025 from each of two clinical trials is fairly convincing evidence of effectiveness.

**TABLE 3.1**

Number of Cases for Which the Agent Is Effective According to Observed *p*-value

| One-Sided *p*-value | Truth | | |
| --- | --- | --- | --- |
| | Effective | Ineffective | |
| <0.025 | 4 | 2.4 | 6.4 |
| ≥0.025 | 1 | 92.6 | 93.6 |
| | 5 | 95 | 100 |

**TABLE 3.2**

Number of Cases for Which the Agent is Effective According to
Conclusion Drawn from Two Studies

|                          | Truth     |             |        |
| ------------------------ | --------- | ----------- | ------ |
| Conclusion of "Effective" | Effective | Ineffective |        |
| Yes                      | 3.2       | 0.06        | 3.26   |
| No                       | 1.8       | 94.94       | 96.74  |
|                          | 5         | 95          | 100    |

Formulas can be derived for determining the Bayesian false-positive and Bayesian false-negative rates. Let $\eta$ denote the probability that a random agent is truly effective, and let $1 - \beta$ denote the power at the assumed effect and $\alpha/2$ denote the one-sided significance level. When an agent is not effective, we will assume the effect is zero. The Bayesian false-positive rate, $\alpha^*$, that an experimental therapy is ineffective given a favorable test results (i.e., a conclusion of effectiveness) equals

$$\alpha^* = \frac{\alpha(1-\eta)/2}{\alpha(1-\eta)/2+(1-\beta)\eta} \tag{3.1}$$

The Bayesian false-positive rate increases as the power or trial size decreases. Thus, a group of small studies would have a larger Bayesian false-positive rate than an analogous group of large studies. The Bayesian false-positive rate also increases as the probability that a random agent is truly effective decreases or as the significance level increases. The Bayesian false-negative rate, $\beta^*$, that an experimental therapy is effective given a nonfavorable test results equals

$$\beta^* = \frac{\beta\eta}{\beta\eta+(1-\alpha/2)(1-\eta)}$$

The Bayesian false-negative rate increases as the power or trial size decreases. Thus, a group of small studies would have a larger Bayesian false-negative rate than an analogous group of large studies. The Bayesian false-negative rate also increases as the probability that a random agent is truly effective increases or as the significance level increases (and the power remains unchanged). Our notation of $\alpha^*$ and $\beta^*$ for the Bayesian false-positive and Bayesian false-negative rates is the reverse of the notation by Lee and Zelen.[5]

Lee and Zelen[5] examined 87 studies conducted by the Eastern Cooperative Oncology Group. Most studies used a one-sided significance level of 0.05 and sized for 80–90% power. Twenty-five of those studies had "significant outcomes." On the basis of a model that considers only no effect and the assumed effect to size the study as the possibilities for the true effect, it was deduced that the true fraction of studies with effective experimental therapies was between 0.28 and 0.32. Moreover, "on average," 3 of the 25 studies (12%) having significant outcomes are expected to have false-positive conclusions and 4–10% of the 62 nonpositive studies are expected to be false-negative conclusions.

For fixed power and probability that a random trial uses an effective experimental agent, the one-sided significance level ($\alpha/2$) for a single trial or overall level for two trials can be determined so as to lead to a desired Bayesian false-positive rate. For fixed $\beta$, $\eta$, and $\alpha^*$,

$$\alpha/2 = \frac{\alpha^*(1-\beta)\eta}{(1-\alpha^*)(1-\eta)} = \left(\frac{\alpha^*}{1-\alpha^*}\right)\left(\frac{\eta}{1-\eta}\right)(1-\beta) \tag{3.2}$$

The required significance level equals the product of the odds of a false-positive result, the odds a random study has an effective experimental agent, and the power at the assumed effect.

From Equation 3.2, for $\alpha^* = 0.025$ and $1 - \beta = 0.9$, $\alpha/2 \approx (0.0231)(\eta/(1 - \eta))$. When $\eta > 0.52$, $\alpha/2 > \alpha^* = 0.025$. For $\alpha^* = 0.025$, $0.01$ and $0.000625$ and $1 - \beta = 0.9$, Table 3.3 gives the value of $\alpha/2$ for various $\eta$ values. When the probability that a random agent is truly effective is 0.2, a single-study significance level of 0.0058 leads to a Bayesian false-positive rate of 0.025, whereas a single-study significance level of 0.00014 leads to a Bayesian false-positive rate of 0.000625.

### 3.2.3 Relative Evidence between the Null and Alternative Hypotheses

For a placebo-controlled trial (a non-inferiority trial), the null effect (null difference in effects) is less than the assumed effect (assumed difference in effects) to size the trial by $(z_{\alpha/2} + z_\beta)s$, where $s$ is standard error for estimated effect (estimated difference in effects). For a placebo-controlled trial, when the estimated treatment effect is less than $(z_{\alpha/2} + z_\beta)s/2$, the results of the trial support no effect more than the assumed effect. For 90% power and a one-sided significance level of 0.025, this means that a one-sided $p$-value greater

**TABLE 3.3**

One-Study Significance Levels Leading to Given Bayesian False-Positive Rates

| $\eta$ | $\alpha^* = 0.025$ | $\alpha^* = 0.01$ | $\alpha^* = 0.000625$ |
|---|---|---|---|
| 0.2 | 0.0058 | 0.0023 | 0.00014 |
| 0.3 | 0.0099 | 0.0039 | 0.00024 |
| 0.4 | 0.0154 | 0.0061 | 0.00038 |
| 0.5 | 0.0231 | 0.0091 | 0.00056 |
| 0.6 | 0.0346 | 0.0136 | 0.00084 |

than 0.053 supports no effect more than the assumed effect. For a non-inferiority trial, an estimated difference in effects between the experimental and active control arms of $-\delta/2$ supports the null difference of $-\delta$ more than the alternative of no difference in effects. When the study has 90% power at no difference with a one-sided significance level of 0.025, an estimated difference in effects of $-\delta/2$ corresponds to a one-sided $p$-value of 0.053.

*Bayes Factor.* Goodman[6] proposes the use of the Bayes factor as a measure of the strength of evidence instead of a $p$-value. The greater the data support the alternative hypothesis relative to the null hypothesis, the more likely the alternative hypothesis is true. For testing two simple hypotheses, we have that

$$\begin{matrix} \text{Posterior odds} \\ \text{of null hypothesis} \end{matrix} = \begin{matrix} \text{Prior odds} \\ \text{of null hypothesis} \end{matrix} \times \text{Bayes factor} \qquad (3.3)$$

where the Bayes factor equals the probability of the data given the null hypothesis/probability of the data given the alternative hypothesis.

For testing the simple hypotheses $H_o{:}\theta = \theta_o$ versus $H_a{:}\theta = \theta_a$, Equation 3.3 can be expressed as

$$\frac{g\left(\theta_o|\mathbf{x}\right)}{g\left(\theta_a|\mathbf{x}\right)} = \frac{h\left(\theta_o\right)}{h\left(\theta_a\right)} \frac{f\left(\mathbf{x}|\theta_o\right)}{f\left(\mathbf{x}|\theta_a\right)}$$

The Bayes factor, $f(\mathbf{x}|\theta_o)/f(\mathbf{x}|\theta_a)$, is the Neyman–Pearson likelihood ratio.

For testing a simple hypothesis against a composite hypothesis (i.e., $H_o{:}\theta = \theta_o$ versus $H_a{:}\theta \in \Theta_a$), Goodman proposes a minimum Bayes factor defined as

$$\text{Minimum Bayes factor} = \frac{f(\mathbf{x}|\theta_o)}{\sup_{\theta \in \Theta_a} f(\mathbf{x}|\theta)} \qquad (3.4)$$

which is also the generalized likelihood ratio that is often used as a frequentist test statistic. In practice, the supremum in the denominator of Equation 3.4 occurs at the maximum likelihood estimate of $\theta$.

In many applications where the maximum likelihood estimator has an approximate normal distribution, the minimum Bayes factor is approximated by $\exp(-z^2/2)$, where $z$ is the number of standard errors the maximum likelihood estimate is different from $\theta_o$.[6] Goodman evaluated the strength of evidence for various "small" $p$-values and prior odds that the null hypothesis is true. On the basis of a fairly pessimistic prior that the alternative hypothesis is true, Goodman regarded a one-sided $p$-value of 0.05 as providing moderate evidence (at best) against the null hypothesis, a one-sided $p$-value of 0.001–0.01 as at best moderate to strong evidence against the null hypothesis, and a one-sided $p$-value of less than 0.001 as strong to very strong evidence. Data leading to a *p*-value less than 0.001 yields posterior odds that the null hypothesis is true that is less than 1/216 of the prior odds that the null hypothesis is true.[6]

When testing with two composite hypotheses (i.e., $H_o{:}\theta \in \Theta_o$ vs. $H_a{:}\theta \in \Theta_a$), a natural extension chooses the generalized likelihood ratio as that Bayes factor when determining the posterior odds that the null hypothesis is true. For two composite hypotheses, Goodman[6] proposes having the selected Bayes factor based on a weight function. For a nonnegative function *w* defined on the parameter space, the weight-based Bayes factor is given by

$$\frac{\int_{\Theta_o} w(\theta)f(\theta|\mathbf{x})d\theta \,/\, \int_{\Theta_o} w(\theta)d\theta}{\int_{\Theta_a} w(\theta)f(\theta|\mathbf{x})d\theta \,/\, \int_{\Theta_a} w(\theta)d\theta}$$

A weight function can also used when testing a simple hypothesis against a composite hypothesis. When the weight function is the prior density function *h*, the posterior odds that the null hypothesis is true is given by

$$\frac{\int_{\Theta_o} h(\theta)f(\theta|\mathbf{x})d\theta}{\int_{\Theta_a} h(\theta)f(\theta|\mathbf{x})d\theta}$$

which is the posterior odds that the null hypothesis is true on the basis of the posterior distribution for $\theta$.

For fixed prior odds that the null hypothesis is true, Goodman[6] notes that the weights or prior densities for the possibilities in the alternative hypothesis can be distributed to focus on whether the true difference or effect is meaningful. When the observed effect is small and not meaningful, such a weight-based Bayes factor would account for this and lead to unimpressive posterior odds that the null hypothesis is true.

In practice, it may be better or more appropriate for the prior odds of the null hypotheses to be based on a typical or random therapy for that indication, not on the prior belief involving the given experimental therapy. This leads to a consistent criterion across all studies in that indication. Different decisions from studies involving different experimental therapies would be based on the differences in the study results.

### 3.2.4 Additional Considerations

Demonstrating superiority to a highly effective active control would probably be substantial evidence of (any) effectiveness as this would mean that the experimental therapy has a great effect versus a theoretical placebo. Likewise, ruling out even a small inferior difference when the non-inferiority margin is quite large would mean that the experimental therapy has a great effect versus a theoretical placebo and minimally retains a large fraction of the active control effect. A conclusion that the experimental therapy is non-inferior to the active control therapy has as a "cushion" the distance between the confidence interval from the non-inferiority trial and the non-inferiority margin along with the corresponding confidence coefficient.

One-study and two-study approaches for the determination of efficacy in a placebo-controlled trial have been compared,[2,7,8] including non-inferiority trials.[8] For powering at no difference in the effects of the experimental and active control therapies, Koch[8] noted that one study using a margin of $0.7\delta$ with a one-sided significance level of 0.01 and 90% power at equal effectiveness would require essentially the same total sample size from two studies having 95% power at equal effectiveness, each with a margin of $\delta$ and a one-sided significance level of 0.025. In the two-study approach, the overall power in demonstrating non-inferiority in both trials with a margin of $\delta$ is approximately 90.25%. Additionally, for powering at no difference in the effects of the experimental and active control therapies, one study using a margin of $0.802\delta$ with a one-sided significance level of 0.000625 with 81% power would require essentially the same sample size as two studies having 90% power each with a margin of $\delta$ and a one-sided significance level of 0.025.

## 3.3 Reproducibility

It is important that a finding in one laboratory by one investigator can be reproduced in another laboratory by a different investigator. A finding that fails to be reproduced when tried at different laboratories by different investigators may not be of great consequence and may have been a fluke. Likewise, it is important to know whether a positive finding from a clinical trial can be reproduced from an independent clinical trial having different subjects and different investigators. If a positive finding from a given clinical trial fails to be reproduced by other conducted clinical trials, the finding will lack external validity. Hung and O'Neill[9] investigated the distribution for the $p$-value under the alternative hypothesis and the likelihood of reproducing a positive result in an identical, second trial when the true effect is the observed effect from the first trial for the second trial. When the observed one-sided $p$-value in the first trial is 0.025, there is a 50% probability of achieving a one-sided $p$-value less than 0.025 in the second trial when the true effect is the observed effect from the first trial. When the observed one-sided $p$-value in the first trial is 0.000625, there is a 90% probability of achieving a one-sided *p*-value less than 0.025 in the second trial when the true effect is the observed effect from the first trial.

In practice, the reproducibility of a positive finding from a clinical trial need not require clinical trials of identical designs. Separate positive findings from clinical trials involving different stages of the same disease may support each other and represent reproducibility of a positive finding for that disease. Similarly, positive findings from a clinical trial using subjects who were previously treated and also from a clinical trial using subjects who were previously untreated of the disease may support each other and represent reproducibility of a positive finding.

Reproducibility is also important for a non-inferiority efficacy claim. However, there may be differing views on what reproducibility should mean for a non-inferiority inference,[10] which depends jointly on both the comparison from the non-inferiority trial and the historical experience of the active control. Conceptually, repeating the entire non-inferiority inference can be considered as jointly repeating both the historical experience of the active control and the non-inferiority trial.[11] Alternatively, the reproducibility of a non-inferiority inference can be viewed by separately assessing the reproducibility in the estimated active control effect across the previous trials and the reproducibility in the difference in effects between the active control and the experimental therapy across multiple non-inferiority trials. When the non-inferiority trial is based on an assumed active control effect size and that effect size or a larger effect size is "regularly reproduced" across trials studying the active control, the testing of non-inferiority will generally be associated with a rather small false-positive rate ($\alpha^*$) for a conclusion that the experimental therapy is effective and constitute substantial evidence of any efficacy, provided that the active control effect is at least the size of the assumed effect.

A consistent, reproduced conclusion of efficacy across trials not only increases the likelihood that the finding of efficacy is real but also can justify that a model used to estimate the active control effect may approximately hold. Before observing the results of any study, the estimated treatment effect or treatment difference is unbiased. However, as the decision to evaluate the effect of a selected active control is dependent on the already observed effects, the retrospective estimation of the active control effect is biased. When a finding of efficacy across studies has reproducibility, this bias should be small.

For indications where there is only one effective standard therapy that can be difficult to tolerate, a second clinical trial comparing the experimental therapy with a placebo can use subjects that do not tolerate the standard therapy. A demonstration of effectiveness for that trial may involve demonstrating superior efficacy to the placebo or some other therapy. In some instances, the dose–response relationship of an experimental therapy may provide supportive information on the efficacy of the experimental therapy.

### 3.3.1 Correlation across Identical Non-Inferiority Trials

The non-inferiority analyses from two different trials are positively corrected (not independent) when the same active control is used with the same estimation of the active control effect.[11] Therefore, the likelihood of two false-positive conclusions may be greater for two identical non-inferiority trials than from two identical superiority trials. For the standard synthesis method under the constancy assumption, the probability of getting $x$ false positives out of $n$ identical non-inferiority trials was provided and assessed by Lawrence.[11] The greater the required fraction of the active control effect that needs to be retained, the smaller the correlation between the test statistics (i.e., the tests) of two non-inferiority trials that use the same historical studies to evaluate the active control effect.[12] Additionally, for two non-inferiority trials using the same estimation of the active control effect, Tsong, Zhang, and Levenson[13] investigated the correlation of a two confidence interval-based test statistic for means.

It should be noted that if the likelihood of conducting a non-inferiority trial is greater when the active control effect is overestimated, the collective, average type I error rate across non-inferiority trials will be greater than the desired or anticipated rate.

### 3.3.2 Predicting Future and Hypothetical Outcomes

For a real or hypothetical clinical trial, Bayesian and frequentist prediction methods can be used to assess the likelihood of a favorable outcome. This can involve predicting whether an ongoing trial will produce a favorable outcome; whether a clinical trial repeated under the exact same conditions or different conditions produces a favorable outcome; or whether a hypothetical, standard placebo-controlled trial would demonstrate that experimental therapy is superior to placebo.

*Predictive Distributions.* The predictive distribution for a future observation is based on the prior distribution for the parameter(s), the distribution for an observation given a fixed value of the parameter, and the data that have presently been observed. The predictive density (or mass function) for a random variable $X_{n+1}$ can be determined from the posterior distribution of the parameter and the density for $X_{n+1}$ (conditioned on the value of the parameter). That is, the predictive density for $X_{n+1}$ is given by

$$f^*(x_{n+1}) = \int_{\Omega} f(x_{n+1}|\theta)g(\theta|x_1, x_2, \ldots, x_n)\,d\theta \tag{3.5}$$

For example, consider a Jeffreys prior (a beta distribution with both parameters equal to 0.5) for the probability that a random study subject will respond to therapy. Suppose that 9 of 20 patients have responded to therapy. Solely on the basis of these data, the predictive probability that a future random study patient will respond to therapy is 19/42 ($\approx$0.452). This value (19/42) was found by evaluating $f^*(1) = \int_0^1 u \cdot \frac{1}{B(9.5, 11.5)} u^{9.5}(1-u)^{11.5}\, du$, where $B(9.5, 11.5) = \int_0^1 u^{9.5}(1-u)^{11.5}\, du$. Thus, with Bernoulli (dichotomous) data, the posterior mean is the predictive probability that a future random observation will be a success. Additionally, the predictive distribution for the number of the next $m$ subjects that will respond to therapy will be a binomial distribution with parameters $m$ trials and a probability of success of 19/42.

Example 3.1 illustrates determining the predictive probability of a favorable outcome from a second identical trial when the first clinical trial has a favorable outcome.

## Example 3.1

Suppose a randomized, controlled clinical trial was performed, having a one-to-one randomization between the two study arms. The primary endpoint will be some time-to-event endpoint where the event is undesirable. After 400 events, an analysis yields an experimental versus control hazard ratio of 0.75, which is statistically significant (two-sided $p$-value = 0.004 < 0.05). An approximate posterior distribution for the actual log-hazard ratio, $\theta$, can be determined on the basis of a noninformative prior, an approximate normal distribution for an observed log-hazard ratio based on 400 events, and the actual observed log-hazard ratio of log 0.75. The approximate posterior distribution for $\theta$ is a normal distribution with mean log 0.75 and variance 4/400 = 0.01. For an identical clinical trial with an analysis after 400 events, the log-hazard ratio will be modeled as having a normal distribution with mean $\theta$ and variance 4/400 = 0.01. Using this model and the posterior distribution for $\theta$, the predictive distribution for the observed log-hazard ratio based on 400 events can be found in Equation 3.5. The predictive distribution for the log-hazard ratio from the additional trial reduces to the convolution of two normal distributions (one having mean 0 and variance 0.01 and the other having mean log 0.75 and variance 0.01), which yields a normal distribution with mean = log 0.75 and variance = 0.02. Statistical significance based on 400 events at a one-sided 0.025 level requires the observed log-hazard be less than −0.196 (hazard ratio less than 0.822). Here, the predictive probability of achieving statistical significance based on 400 events equals 0.742.

When it is believed that the true effect in the second trial is the same as the true effect in the first trial, the posterior distribution for the common effect based on the results in the first trial forms the prior distribution for the common effect in the second trial. The predictive probability of achieving statistical significance can also be determined under a model for differing true effects between clinical trials by adding variability into the prior distribution for the common effect in the second trial or from a hierarchical model.

When indirect comparisons are made between the experimental therapy and a placebo (or other reference therapy) through the control therapy, a predictive distribution can also be used to determine the predictive probability that a hypothetical trial comparing the experimental therapy with placebo will result in a statistically significant result on the primary endpoint favoring the experimental arm. This is illustrated in Example 3.2.

### Example 3.2

We consider two non-inferiority trials comparing capecitabine to the Mayo clinic regimen of 5-fluorouracil with leucovorin (5-FU + LV) in first-line metastatic colorectal cancer. This case is presented in greater detail in Chapter 5. Estimates from each trial and a meta-analysis of those trials along with a meta-analysis of the effect of 5-FU + LV relative to 5-fluorouracil (5-FU) can be found in previous studies.[14–16] The endpoint of interest is overall survival—the time from randomization to death.

Posterior distributions for the actual 5-FU + LV versus 5-FU log-hazard ratio for overall survival, $\theta_{C/P}$, and the actual capecitabine versus 5-FU + LV log-hazard ratio for overall survival, $\theta_{E/C}$, can be determined on the basis of noninformative prior distributions and the results from the meta-analysis for the effect of 5-FU + LV and the two active-controlled clinical trials. Assuming that the effects of therapies remain constant across studies, the posterior distribution for the actual capecitabine versus 5-FU log-hazard ratio for overall survival, $\theta_{E/P}$, is determined on the basis of the relation $\theta_{E/P} = \theta_{E/C} + \theta_{C/P}$, where $\theta_{E/C}$ and $\theta_{C/P}$ are independent. Table 3.4 provides the posterior distributions for $\theta_{C/P}$, $\theta_{E/C}$, and $\theta_{E/P}$, and the predictive distribution for the observed overall survival capecitabine versus 5-FU log-hazard ratio, $x_{E/P}$, on the basis of 533 events from the hypothetical trial comparing capecitabine and 5-FU arms. The event number of 533 is chosen since it was the number of events for the overall survival analysis in both capecitabine clinical trials.

For our hypothetical 5-FU–controlled trial, statistical significance at a one-sided 0.025 level favoring capecitabine on overall survival requires an observed log-hazard ratio be less than –0.170 (hazard ratio less than 0.844) on the basis of 533 events. Here, the predictive probability of reaching statistical significance favoring capecitabine equals 0.797. Note that a superiority inference at a one-sided 0.025 significance level based on 533 events would have 79.7% power for a hazard ratio of 0.785.

### TABLE 3.4

Posterior and Predictive Distributions

| Parameter | Posterior Distribution |
| --- | --- |
| $\theta_{C/P}$ | Normal distribution with mean –0.234 and standard deviation 0.0750 |
| $\theta_{E/C}$ | Normal distribution with mean –0.044 and standard deviation 0.0613 |
| $\theta_{E/P}$ | Normal distribution with mean –0.278 and standard deviation 0.0969 |
| **Estimator** | **Predictive Distribution** |
| $x_{E/P}$ | Normal distribution with mean –0.278 and standard deviation 0.1300 |

In these examples, we have started with noninformative prior distributions before any results involving the active control and/or experimental therapy are known. In certain situations, an informative prior distribution may be more appropriate than a noninformative prior for the parameter of interest. Knowledge of the history of the reproducibility of clinical trial results among all therapies used for studied indication and/or the reproducibility of clinical trial results using the experimental therapy in related indications may also be used to establish a prior distribution on the parameter of interest.

*Frequentist Prediction.* When dealing with random quantities $U$ and $V$ that have a bivariate normal distribution, where the value of $U$ is used as the predicted value for $V$ and the expected value of $V - U$ is zero, then the $100(1 - \alpha)\%$ prediction interval for $V$ is given by

$$u \pm z_{\alpha/2} \sigma_{V-U} \tag{3.6}$$

where $u$ is the observed value for $U$, $1 - \Phi(z_{\alpha/2}) = \alpha/2$, and $\sigma_{V-U}$ is the standard deviation of $V - U$. In Example 3.1, $U$ is the estimated log-hazard ratio from the first clinical trial based on 400 events, $V$ is the estimator of the log-hazard ratio for the second clinical trial based on 400 events, and $\sigma_{V-U} = \sqrt{0.02}$. From Equation 3.6 the 95% prediction interval for the observed log-hazard ratio in the second clinical trial, on the basis of 400 events, is (−0.565, −0.010) ((i.e., the 95% prediction interval for the hazard ratio is 0.568, 0.990)). An observed hazard ratio less than 0.822 is needed for statistical significance at a one-sided 0.025 level. The one-sided 74.2% prediction interval for the observed hazard ratio in the second clinical trial is (0, 0.822), an analogous result to the Bayesian predictive probability.

For Example 3.2, prediction limits can be determined for the comparison of capecitabine with 5-FU from a hypothetical 5-FU–controlled trial. Let $U_{C/P}$ and $U_{E/C}$ denote the estimators of the overall survival log-hazard ratio for 5-FU + LV versus 5-FU ("placebo") and capecitabine versus 5-FU + LV, respectively. Let $V$ denote the estimator of the overall survival log-hazard ratio for capecitabine versus 5-FU based on 533 events from the hypothetical trial comparing capecitabine and 5-FU arms. We will assume that $E(V) = E(U_{C/P} + U_{E/C})$ and use normal distributions for the sampling distributions. The 95% prediction interval for $V$ is (−0.533, −0.023). The corresponding 95% prediction interval for the hazard ratio based on 533 events is (0.587, 0.977). This interval includes 0.844, the threshold needed for the observed hazard ratio for achieving statistical significance, and larger possibilities for the observed hazard ratio. A one-sided 79.7% prediction interval for the log-hazard ratio based on 533 events is (−∞, −0.170), the analog to the result of the Bayesian predictive analysis.

## References

1. U.S. Food and Drug Administration. Statement regarding the demonstration of effectiveness of human drug products and devices. *Federal Register*, 60, Docket No. 9500230, 39180–39181, August 1, 1995.

2. Huque, M.F., Commentaries on statistical consideration of the strategy for demonstrating clinical evidence of effectiveness—one larger vs two smaller pivotal studies by Z. Shun, E. Chi, S. Durrleman and L. Fisher, *Stat. Med.*, 24, 1639–1651, 2005.

3. U.S. Food and Drug Administration, Guidance for industry: Providing clinical evidence of effectiveness for human drug and biological products, 1998, at http://www.fda.gov/downloads/Drugs/GuidanceComplianceRegulatory Information/Guidances/UCM078749.pdf.

4. Fleming, T.R., Clinical trials: Discerning hype from substance, *Ann. Intern. Med.*, 153, 400–406, 2010.

5. Lee, S.J. and Zelen, M., Clinical trials and sample size considerations: Another perspective, *Stat. Sci.*, 15, 95–110, 2000.

6. Goodman, S.N., Toward evidence-based medical statistics 2: The Bayes factor, *Ann. Intern. Med.*, 130, 1005–1013, 1999.

7. Shun, Z. et al., Statistical consideration of the strategy for demonstrating clinical evidence of effectiveness—one larger vs two smaller pivotal studies, *Stat. Med.* 24, 1619–1637, 2005.

8. Koch, G.G., Commentaries on statistical consideration of the strategy for demonstrating clinical evidence of effectiveness—one larger vs two smaller pivotal studies by Z. Shun, E. Chi, S. Durrleman and L. Fisher, *Stat. Med.*, 24, 1639–1651, 2005.

9. Hung, H.M.J. and O'Neill, R.T. Utilities of the $p$-value distribution associated with effect size in clinical trials, *Biometrical J.*, 45, 659–669, 2003.

10. Rothmann, M.D., Issues to consider when constructing a non-inferiority analysis, *ASA Biopharm. Sec. Proc.*, 1–6, 2005.

11. Lawrence, J., Some remarks about the analysis of active control studies, *Biometrical J.*, 47, 616–622, 2005.

12. Tsong, Y. et al., Choice of $\lambda$-margin and dependency of non-inferiority trials, *Stat. Med.* 27, 520–528, 2008.

13. Tsong, Y., Zhang, J., and Levenson, M., Choice of $\delta$ non-inferiority margin and dependency of the non-inferiority trials, *J. Biopharm. Stat.*, 17, 279–288, 2007.

14. FDA Medical-Statistical review for Xeloda (NDA 20-896), dated April 23, 2001.

15. FDA/CDER New and Generic Drug Approvals: Xeloda product labeling, at http://www.fda.gov/cder/foi/label/2003/20896slr012_xeloda_lbl.pdf.

16. Rothmann, M. et al. Design and analysis of non-inferiority mortality trials in oncology, *Stat. Med.*, 22, 239–264, 2003.

# 4

## Evaluating the Active Control Effect

### 4.1 Introduction

According to the U.S. Food and Drug Administration (FDA) Draft Guidance on Non-inferiority Trials,[1] "The first and most critical task in designing an NI study is obtaining the best estimate of the effect of the active control in the NI study (i.e., M1)." The FDA draft guidance on non-inferiority trials[1] provides instances for which a non-inferiority margin can be defined in the absence of controlled clinical trials evaluating the active control effect. The circumstances are similar to those for which historically controlled trials can provide persuasive evidence.[2] For example, there should be a good understanding or estimate of the outcome (e.g., spontaneous cure rate) without treatment, and the outcomes or cure rate for the active control from multiple historical experiences should be substantially different from those seen without treatment (e.g., substantially different spontaneous cure rates). The assumed effect of the active control in the setting of the non-inferiority would be conservatively chosen.

Usually, there are data on the effect of the active control therapy from other clinical trials. It is a daunting task to determine whether the estimated effects of the active control therapy from previous trials apply to the setting of the non-inferiority trial. Differences in patient populations, in the natural history of the disease, and in supportive care are just some of the factors that can alter the effect of a therapy from one clinical trial to another. Additionally, bias may be introduced by identifying the active control therapy after its effect has been estimated. Bias may also be introduced by selective, post hoc determination of which studies to include in the evaluation of the active control effect. How to integrate results across trials and whether the integrated results would apply to a future clinical trial are also key concerns. The potential heterogeneity in the active control effect across trials needs to be considered and investigated. For the setting of the non-inferiority trial, explained heterogeneity should be accounted for in the estimated effect of the active control and unexplained heterogeneity should be accounted for in the corresponding variance. These issues and topics are discussed in this chapter.

## 4.2 Active Control Effect

### 4.2.1 Defining the Active Control Effect

For monotherapy, the effect of the active control therapy is usually defined by the difference in outcomes in a randomized study between an arm of the active control therapy and a placebo arm. When the endpoint is objective and beyond manipulation, it may be appropriate to define the active control effect relative to an observation arm in a randomized study. For combination therapy, when an experimental regimen substitutes the experimental therapy for one or more components in the control regimen, the active control effect would be defined as the difference in outcomes in a randomized study between an arm of the active control regimen and an arm where subjects are given a regimen made up of the active control regimen minus the substituted components (or the experimental regimen minus the experimental therapy).

The non-inferiority criterion should assure that the experimental therapy is effective. However, since active-controlled trials tend to be done because it may be unethical to give subjects a placebo, an experimental therapy simply being better than a placebo may not provide assurance that it would be ethical to give the experimental therapy to subjects in another clinical trial or to patients in practice. For example, it may be unethical to give a slightly effective treatment for a serious or fatal disease when there is a much more effective therapy and there is only one shot at cure or treatment. Also, for combination therapy, the reference (active control regimen minus the substituted components) for defining the active control effect may be a regimen that itself may be unethical to give to subjects.

Standards for what is a safe, effective, and ethical treatment may change over time. When an indication has a history of replacing standards of care with better, more efficacious standards of care, the previous standard of care may become unethical to give. There may be some part of the active control effect that must be retained by the experimental therapy for the experimental therapy to be ethically administrable to patients.

### 4.2.2 Modeling the Active Control Effect

For a two-arm non-inferiority trial, the effect of the active control is assumed not measured or estimated within the non-inferiority trial. By relying on an across-trials inference, two-arm non-inferiority trials can have similar concerns as historical controlled trials. Ideally, there are multiple studies of the active control therapy, providing consistent estimates of the size of an effect that is undoubtedly real and relevant to the setting of the non-inferiority trial.

Modeling the effect of the active control relative to a placebo in the setting of the non-inferiority trial is imperfect. Modeling the active control effect involves examining the effect of the active control from relevant previous trials, adjusting for any potential biases, and understanding and adjusting for any differences between the historical trials and the non-inferiority trial. The between-trial variability that cannot be explained should also be considered when modeling the uncertainty of an estimate of the active control effect. When there are no historical studies providing relevant information on the effect of the control therapy, a two-arm non-inferiority trial cannot be done. When there are relevant randomized comparative studies, it may be possible to assess the effect of the control therapy. In assessing the size of the active control effect in the setting of the non-inferiority trial, relevant information from previous trials needs to be considered, including the consistency of the size of the estimated effects, consistency of any effect, and similarities and differences in the designs of the trials (e.g., differences in patient populations, concurrent therapies, and subsequent therapies).

If there are concerns that the effect of the control therapy has diminished by some fraction, $\varepsilon$, then the estimated control effect can also be reduced by this fraction.

When there is one historical, randomized trial comparing the active control with placebo, between-trial variability cannot be assessed. It is also difficult to quantify the between-trial variability with just two historical trials. The potential between-trial variability should be considered, particularly when that disease or indication has a history of inconsistent estimated effects across clinical trials investigating the effects of the same therapy.

"Likes" should be combined with "likes." For example, it may be inappropriate to combine the results of observational studies with blinded, placebo-controlled studies. Therefore, in a meta-analysis of a collection of studies, it may be necessary to first divide the overall collection of studies into subsets where, within each subset, the studies are fairly homogeneous on the most important design and conduct features relative to the treatment effect and its estimation. Then a meta-analysis is done for each subset of the studies. The use of multiple definitions of the endpoint, differences in how the endpoint is measured or how frequently the endpoint is monitored, differences in the amount of follow-up on the endpoint, or meaningfully different patient populations may be the basis for dividing the overall collection of studies into subsets.

### 4.2.3 Extrapolating to the Non-Inferiority Trial

Across-trials non-inferiority comparisons need to account for effect modifiers and biases in the estimation of the active control effect. Biases in the estimation of the historical effect of the active control can arise owing to publication or other selection biases in choosing the historical studies and regression to the mean bias in that the identification of the active control was based on the positive results of some or all of the selected studies.

In establishing the non-inferiority margin or criterion, it is important to understand what effect the active control may have in the setting of the non-inferiority trial. Without a placebo arm in the non-inferiority trial, the effect of the active control cannot be directly assessed. An assessment of the effect that the active control has had in previous clinical trials and an understanding of the differences between the previous trials and the non-inferiority trial that may affect the effect of the active control are used to judge the effect size that the active control may have in the non-inferiority trial. Failure to account for biases in the estimation of the historical active control effect and any effect modification will lead to a biased estimation of the active control effect in the setting of the non-inferiority trial. Additionally, since the selection of the active control is likely outcome based (based on positive results of previous trials), the estimation of the active control effect will be biased with a tendency to produce an estimate greater than the true effect size. The studies selected to estimate the size of the historical effect of the active control may also be outcome dependent as would occur when there is publication bias in the available studies that can be found in a systematic review.

Factors that may change the effect of the active control from one study to the next study include:

1. Changes in the distribution of a covariate where the covariate is associated with the size of the treatment effect

2. Changes in supportive care

3. Improvements in diagnosing the disease and the true prognosis of patients

4. Differences in the dose or regimen of the active control

5. Differences in the patient population that would volunteer for an active control trial and the patient population that would volunteer for a placebo-controlled trial

6. Differences in the timing of the measurement of the endpoint and, for time-to-event endpoints, differences in the censoring distributions when there are nonproportional hazards

7. Differences in "cross-over" to the active control therapy

8. Differences in the level of adherence and the amount and impact of missing data

*Analog to Nonrandomized Comparisons.* A comparison of two groups involves a comparison of the patient-level outcomes. A randomized, controlled trial comparing two groups fairly allocates treatments to subjects, which allows for valid comparisons. Nonrandomized comparisons that include observational and historical studies do not fairly allocate subjects to groups and do not control for known and unknown prognostic factors. Unaddressed imbalances between groups can invalidate a comparison of the groups. In many instances, conclusions based on marked differences from nonrandomized comparisons have been proven false on the basis of the evidence from later randomized, controlled trials. This includes a suggested 40–50% reduction in the incidence of coronary heart disease for postmenopausal hormone therapy from observational studies that was later disproven by a Women's Health Initiative clinical trial that found an elevated incidence of coronary heart disease for postmenopausal hormone therapy.[3]

To validate a nonrandomized comparison, the analysis would need to adjust not only for different distributions in known prognostic factors but also for biases due to potential differences in unknown prognostic factors, the lack of fair assignment of patients to groups, and differences in external factors (e.g., environmental/external differences due to differences in the calendar time for which data are obtained for each group as in a historical controlled study).

While the lack of balance of known and unknown prognostic factors can invalidate a nonrandomized comparison, differences between the historical trials used to evaluate the effect of the active control and the non-inferiority trial in the distribution of covariates that are associated with the effect size of the active control and/or the existence of other effect modifiers can invalidate an across-trial comparison when unaccounted for in the non-inferiority analysis.[4] The non-inferiority analysis in a two-arm trial involves an across-trials comparison of the estimated active control effect in the historical trials with the estimated difference in effects between the experimental therapy and the active control in the non-inferiority trial. The validity of the non-inferiority analysis depends on whether the true effect of the active control in the setting of the non-inferiority trial is different from the true historical effect of the active control and, if so, how different. It is thus important that the relative frequencies of effect size-related covariates are similar in the non-inferiority trial as in the historical trials and that the effects attributed to such covariates has not changed (e.g., due to improvements in supportive care or in monitoring or diagnosing the disease) from the historical trials to the non-inferiority trial. A non-inferiority analysis should adjust for differences in known and potentially unknown covariates that are effect modifiers between the historical trials and the non-inferiority trial, as well as how the effect of the active control may have changed and for any biases in the estimation of the historical effect of the active control.

*What Does the Constancy Assumption Mean?* When the true effect of the active control varies across previous trials (i.e., the active control effect is not constant), what does the constancy assumption mean? One interpretation is that the active control effect in the non-inferiority trial equals the global mean active control effect across studies (i.e., the location parameter in the random-effects model).[5] Another interpretation is that the true active control effect in the non-inferiority trial, $\gamma_{k+1}$, has the same distribution (i.e., $N(\gamma, \tau^2)$) as the true effects in the previous trials. In either case, the random-effects meta-analysis estimator of the global mean effect across studies, $\hat{\gamma}$, is "unbiased" for the effect of the active control in the non-inferiority trial (i.e., $E(\hat{\gamma}) = E(\hat{\gamma}_i) = \gamma$). The difference is the variance that is attributed to $\hat{\gamma}$. Since $E(\hat{\gamma} - \hat{\gamma}_{k+1})^2 = E(\hat{\gamma} - \gamma)^2 + E(\gamma_{k+1} - \gamma)^2$, the variance for the second case is larger. While the estimator is unbiased for the active control effect in both cases, the modeling of the uncertainty of the estimator and its sampling distribution is different. In general, the constancy assumption is more than just having an unbiased estimator of the active control effect in the non-inferiority trial, but also correctly modeling or identifying the sampling distribution for the estimator.

The constancy of effect may depend on the chosen metric. The benefit of a therapy used to prevent a disease may depend on the placebo rate of getting the disease. An experimental therapy that prevents occurrence of the disease in one out of two subjects who would have otherwise acquired the disease has an occurrence rate of 25% when the placebo rate is 50% (a difference of 25%). The occurrence rate would be 15% when the placebo rate is 30% (a difference of 15%). How to make adjustments for departures from constancy in the active control effect is often a matter of judgment.[1]

### 4.2.4 Potential Biases and Random Highs

Inferior design features and poor study conduct can introduce bias in the estimated treatment effects. Bias is also introduced when the timing of the definitive analysis is outcome dependent or when the selection of a study used in estimating the effect of the active control depends on the results of the study. The fact that the active control is selected because it had positive results in previous clinical trials also introduces a bias.

Open-label studies of the active control versus observation may not provide unbiased estimates of the active control effect that placebo-controlled trials would provide when the blind is maintained until the endpoint is determined. Additionally, when the endpoint of interest is not the primary endpoint of a study, there may not be quality scrutiny in obtaining the measurements, thereby yielding a biased estimate of the active control effect. For example, for a time-to-event secondary endpoint, it may be deemed less important to follow subjects to the endpoint or the end of study as according to the intent-to-treat principle. Subjects who were lost to follow-up in determining the endpoint may have their times censored when they were lost to follow-up (i.e., at their last evaluation for the endpoint) for the analysis of the endpoint. Such censoring may be informative, thus leading to bias estimates of the effect of the active control versus placebo.

*Publication Bias.* It is crucial to have an appropriate, valid way of estimating the effect of the active control therapy in the non-inferiority trial, which often involves a meta-analysis. Having all available relevant information from past studies on the effect of the active control therapy is crucial.[6-8] It is important that the studies to be included in the meta-analysis are selected in a manner that would avoid or minimize bias. When there are only a few studies, all of which are known, this will be an easier task. In many instances, a literature search is done to obtain known relevant studies that can be used to quantify the effect of the active control therapy. There may be concern of a publication bias, that only the results from studies indicative of a treatment effect will be published or that such studies are more likely to be published (and thus found in a literature search) than the results from those studies that did not indicate a treatment effect. There are various techniques that can assist in recognizing that a publication bias may exist. The techniques tend to assume that the true treatment effect is constant across trials or that the true effect is not dependent on the size of the trial. Hopefully, the recent creation of a clinical trials data bank (i.e., clinicaltrials.gov) will reduce the possibility of publication bias for estimating the effects of many future active controls.

A "funnel plot" is a graphical display often used to evaluate for potential publication bias.[9] Plotted for each study is the pair of the estimated effect and a measure related to the associated variability in the estimate. The greater the variability associated with the estimate (as with smaller studies), the more spread there is in the observed estimates; thus, when there is no publication or sampling bias, a funnel-like shape is expected.

Search strategies that attempt to find all relevant studies and minimize bias should be used. The methods for abstracting estimates and the standard error from summaries of the results should also be considered. For example, it is easy and valid to derive the estimate and corresponding standard deviation from a 95% confidence interval that was based on a normal distribution. The estimated effect would be the midpoint of the confidence interval, whereas the standard deviation would equal the difference of the upper and lower limits of the 95% confidence interval divided by 3.92. When the subjects are monitored indefinitely for an event (e.g., death) and accrued over time, it would probably be inappropriate to use the fraction of subjects who had events in both arms to arrive at an estimate of the hazard ratio.

*The Timing of the Definitive Analysis Is Random.* When a study includes one or more interim analyses, the sample treatment effect is often regarded as the estimated treatment effect when an efficacy boundary is first crossed, or as the estimated treatment effect at the final planned analysis should no efficacy boundary be crossed. For this definition, when the experimental therapy is more effective than the control therapy or placebo, the sample treatment effect is biased, having a mean or expected value greater than the true treatment effect. If the study were replicated over and over, the long-run arithmetic average sample treatment effect would exceed the true-treatment effect. This long-run average gives equal weight to each replication. If each study result was weighted by the amount of the information used at the definitive analysis (analogous to a fixed-effects meta-analysis), then the corresponding long-run weighted average sample treatment effect would converge almost surely to the true treatment effect. On the basis of a sequence of independent, identical clinical trials having at least one interim analysis, the cumulative fixed-effects meta-analysis estimator is asymptotically unbiased (i.e., the bias decreases towards zero) as the number of trials increases, whereas the bias in the arithmetic average is constant in the number of trials. When a given clinical trial necessarily continues to the final analysis, regardless of the results of earlier analyses, the estimated treatment effect at the time of the final planned analysis is an unbiased estimator of the true treatment effect, provided no design or conduct changes occur on the basis of the results of the interim analysis.

When there is zero treatment effect and parallel boundaries are used at the interim analyses, the expected sample treatment effect will be zero. The bias in the observed treatment effect at the time of the definitive analysis is attributable to the randomness of the amount of information at the time of that analysis. The amount of bias will depend on the true effect size, the $\alpha$ allocation, and the timings of the analyses.

Adaptive designs having a sample size reestimation component also have sample treatment effects that are biased with the mean sample effect greater than the true treatment effect when the treatment is effective.

*Random Highs.* Random highs in the estimated effect of the active control therapy in historical studies are a real issue. For example, "data dredging" leads to estimates of treatment effects that tend to overstate the true effect. Situations include selecting a subgroup retrospectively on the basis of the estimated effect seen in that group. The estimates that generate hypotheses for further studies are in themselves conditionally biased, tending to be larger than the true effect size. Likewise, conditional bias estimates can occur when a claim is limited to a subgroup either due to quite positive results seen for that subgroup or for quite negative results seen for the complement subgroup. There are various other scenarios when random highs are likely to be more prevalent, which include the following: when the use of the control therapy in the non-inferiority trial was predicated on the success of one or two trials designed to study that therapy in the indication; the first trial in an indication to yield a favorable statistically significant result after many other trials (possibly based on different therapies) previously failed to do so; the estimated effect is from an interim analysis that resulted in favorable statistical significance or is from a design having a sample size reestimation; and a retrospective or nonprespecified analysis on a demographic or genomic subgroup.

*Statistical Significance Bias.* Studies whose results are responsible for motivating the use of a therapy as an active control introduce some bias or conditional bias into the historical estimation of a treatment effect. Before a trial that will be well conducted and well controlled is started, the estimated treatment effect is unbiased for the true treatment effect of a population that is represented by the subjects in the clinical trial. At the start of the trial, the observed treatment effect will or will not wind up being large enough to achieve statistical significance. Conditional on statistical significance being achieved, the expected or mean sample treatment effect is greater than the true treatment effect. Because active control therapies in a non-inferiority trial are often selected because statistical significance was reached in one or two trials that were designed to study that therapy in the indication, there will be a tendency for the estimated active control effect to be greater than the true effect. It is therefore necessary for the estimated control effect to be either reproduced in multiple trials or be "adjusted" for this conditional bias.

When a drug has demonstrated an effect in a clinical trial (e.g., one-sided $p$-value < 0.025), it is more likely than not that the estimated effect in a second trial of the same design will be smaller than that seen in the first trial. Statistically, consider two normalized test statistics from two separate, identically designed clinical trials, $Z_1$ and $Z_2$, that have standard normal distributions when there is no difference in effects between treatments. For the $i$th trial ($i = 1,2$), a one-sided $p$-value of less than 0.025 is equivalent to $Z_i > 1.96$. It can be shown that $P(Z_1 > Z_2 | Z_1 > 1.96) > 0.5$. In other words, given that the first trial achieved statistical significance, it is more likely that the first trial had a smaller $p$-value (and also a larger estimated effect) than the second trial.

Let $\sigma$ denote the standard error for the estimated treatment effect. Consider a study having $100\gamma\%$ power at the actual treatment effect $\mu$, which is based on a large sample normalized test statistic with a one-sided significance level of $\alpha$. When statistical significance has been achieved ($p$-value < $\alpha$), the conditional expected or mean estimated treatment effect is approximately $\mu + g_U(\gamma)\sigma$, where $g_U(\gamma) = [\phi(\Phi^{-1}(\gamma))/\gamma]$ and $\phi$ and $\Phi$ are the density and distribution functions for a standard normal distribution, respectively. Note that for a one-sided significance level of $\alpha$, $\gamma = \Phi(\mu/\sigma - z_\alpha)$. When statistical significance is not reached, the conditional expected or mean estimated treatment effect is approximately $\mu - g_L(\gamma)\sigma$, where $g_L(\gamma) = [\phi(\Phi^{-1}(\gamma))/(1 - \gamma)]$. Because of the symmetry of $\phi$ about zero, $g_U(\gamma) = g_L(1 - \gamma)$. Table 4.1 provides values of $g_U(\gamma)$ for various $\gamma$ values. For 90% power at the actual treatment effect $\mu$, based on a large sample normalized test, the conditional expected or mean estimated treatment effect given that statistical significance has been reached is approximately $\mu + 0.195\sigma$. Thus, if the same study having 90% power was repeated over and over where only the estimated effects from those replications having a one-sided $p$-value of < 0.025 are retained, the long-run average

**TABLE 4.1**

Number of Standard Error Bias in Achieving Statistical Significance at a One-Sided Level by Power

| $\gamma$ | $g_U(\gamma)$ | $\gamma$ | $g_U(\gamma)$ | $\gamma$ | $g_U(\gamma)$ |
|---|---|---|---|---|---|
| 0.05 | 2.06 | 0.30 | 1.16 | 0.75 | 0.42 |
| 0.10 | 1.75 | 0.40 | 0.97 | 0.80 | 0.35 |
| 0.15 | 1.55 | 0.50 | 0.80 | 0.85 | 0.27 |
| 0.20 | 1.40 | 0.60 | 0.64 | 0.90 | 0.195 |
| 0.25 | 1.27 | 0.70 | 0.50 | 0.95 | 0.11 |

among the retained estimated effects would be approximately $\mu + 0.195\sigma$. For a case where the true effect size is marginal, Example 4.1 illustrates the relationship between the true effect size and the mean observed effect size when statistical significance is achieved.

### Example 4.1

Consider a time-to-event endpoint compared between two arms after 400 events in a placebo-controlled clinical trial having a 1:1 randomization. Suppose the true experimental to placebo hazard ratio is 0.894, which provides 20% power to achieve statistical significance at a one-sided $\alpha$ of 0.025 (which occurs when the observed experimental to placebo hazard ratio is less than 0.822). Given that statistical significance is achieved, the mean for the observed experimental to placebo log-hazard ratio is $-0.252$ ($=\ln 0.894 - 1.40 \times 0.1$) from Table 4.1, which corresponds to a hazard ratio of 0.777. In cases where the true power is 20%, the typical observed experimental to placebo hazard ratio when statistical significance is achieved will be 13% less than the true value.

*Maximum and Regression to the Mean Biases.* In baseball, the best rookie player in each league receives the Rookie of the Year (ROY) award. The performance of the winners in their next seasons tends to be worse than in their rookie year. This is often referred to as the "sophomore jinx." The ROY has the maximum performance or outcome among all rookies in their league their first year. Also, among the same group of players for the next year, the ROY cannot do any better than having the maximum performance and can do comparatively worse.

The sophomore jinx occurs because the ROY is identified, not at random, but on the basis of having the maximum performance. The sophomore jinx is an example of "regression to the mean." The bias that occurs from using the rookie outcome of the ROY to project their second (sophomore) year performance or estimate their ability, while ignoring and not adjusting for the fact that the ROY is being identified on the basis of maximum performance/ outcome (not identified at random), is an example of regression to the mean bias. In a non-inferiority trial, the active control is usually identified on the basis of past performance in clinical trials. Often, the active control will be the therapy that performed best (or one therapy among the therapies that performed the best) in previous clinical trials. Therefore, unless a proper adjustment is made, including the outcomes from previous clinical trials that were used to identify the active control will lead to biased estimation of the active control effect with a tendency to overestimate the true effect even when the true effect of the active control remains constant across previous trials and the non-inferiority trial.

Regression to the mean refers to the phenomenon in a simple linear regression when an observed value of an explanatory variable $X$ of $x$ is $k$ standard deviations away from its mean ($\mu_X$), and the expected value for the response variable $Y$ of $\mu_{Y|x}$ is $\rho k$ of its standard deviations away from its mean $(\mu_Y)$, where $-1 < \rho < 1$ is the correlation coefficient. Since $|\rho k| < |k|$, in relative terms of respective standard deviations, $\mu_{Y|x}$ is closer to $\mu_Y$ than $x$ is to $\mu_X$. For example, if $\rho = 0.5$ with standard deviations of $\sigma_X$ and $\sigma_Y$, then when we observe $x = \mu_X + 2\sigma_X$, the corresponding expected value of $Y$ is $\mu_{Y|x} = \mu_Y + \sigma_Y$. For the sophomore jinx, $X$ is the performance in the rookie year and $Y$ is the performance in the second year.

In Statistics, when an outcome or estimate represents a maximum, the outcome or estimate will tend to be greater than the true mean of the underlying distribution with high probability. Thus, in the subject area of clinical trials, when the estimated effect represents a maximum across studies and/or subgroups, it is highly likely that the estimated effect is greater than the true effect. Conditional bias is also introduced when the selection of historical studies used to estimate the effect of the active control is outcome dependent. For example, limiting the selected studies to a narrow indication where a study achieves statistical significance and ignoring the results from related indications will lead to a bias and an exaggerated estimate of the active control effect and potentially inflate the type I error rate for the non-inferiority trial. The maximum of a random sample tends to be larger than the mean of the underlying distribution (i.e., larger than the true effect). The bias of the maximum in estimating the underlying mean increases as the number of studies increases. When an observation represents a maximum, it should not be evaluated as if it were an isolated, random observation.

Consider an investigational agent, $A$, being studied for a first-line metastatic cancer in three large, equally sized, randomized clinical trials. Each clinical trial compared the addition of agent $A$ to a different standard chemotherapy regimen with that standard chemotherapy regimen alone. The three clinical trials used a different background standard chemotherapy regimen ($X1$, $X2$, and $X3$). Suppose that the only trial that demonstrated improved overall survival when agent $A$ is added is the trial that used $X1$ as the background chemotherapy. Now, a sponsor wants to study the addition of the experimental agent $B$ in a non-inferiority trial that compares $X1$ plus $B$ with $X1$ plus $A$. As the observed effect of adding $A$ to $X1$ represents the maximum observed effect across three trials, the observed effect of adding $A$ to $X1$ probably overestimates the true effect. Therefore, if the estimation of the effect of the active control, $A$, only considers the previous trial that used $X1$ as the background chemotherapy and ignores the fact that the observed effect represents a maximum observed effect, the true effect of adding $A$ to $X1$ will tend to be overestimated. This may then lead to an inappropriately large non-inferiority margin and an increase in the likelihood of concluding that an ineffective experimental therapy is effective. If only the results from the clinical trial using $X1$ as the background therapy are used, the estimated effect in that trial needs to be interpreted and modeled as representing a maximum observed effect. It is important to note that when improved survival is not demonstrated, it does not mean that improved survival was ruled out. If the other two trials had slightly favorable observed effects, their failure to demonstrate a survival improvement does not mean that the effect of adding agent $A$ to chemotherapy is heterogeneous across background chemotherapies. The observed effects across the studies may still be consistent with homogeneous effects. Knowledge of the observed effects from the other two studies is needed to correctly interpret the results from the study using X1 as the background chemotherapy.

Similar situations would also arise when an investigational agent is studied in multiple lines of an advanced or metastatic cancer, when an investigational agent is studied in separate trials in different disease settings, or when the chosen dose for the active control in the non-inferiority trial is the dose with the greatest estimated effect and only data on that dose is used to estimate the effect of the active control. Treating a better or best finding as coming from an isolated trial will tend to overstate the true effect. Treating a better or best finding as a maximum or the relevant upper order statistic of a sample when the effects are homogeneous will be correct when the effects are homogeneous and will be conservative when the effects are heterogeneous. However, when the effects are homogeneous, the most reliable estimate of the common effect integrates the estimated effects from all trials.

*Dealing with Maximum Bias.* For a random sample, the observed maximum is not an appropriate estimator of the common true mean or treatment effect. When assumptions are made on the shape of the underlying distribution and/or the shape of the distribution of the maximum, the observed maximum can be used to make inferences on the common true mean or treatment effect.

Let $X_1, \ldots, X_k$ be a random sample from a distribution with underlying distribution function $H$. Let $X_{(k)}$ denote the maximum of $X_1, \ldots, X_k$, and let $H_{(k)}$ denote its distribution function. Then for $-\infty < t < \infty$, $H_{(k)}(t) = (H(t))^k$. The quantiles/percentiles for $X_{(k)}$ are given by $H_{(k)}^{-1}(\gamma) = H^{-1}(\gamma^{1/k})$ for $0 < \gamma < 1$. The mean and variance for $X_{(k)}$ are given as $\mu_{X_{(k)}} = \int_0^1 H^{-1}(x^{1/k})dx$ and $\sigma^2_{X_{(k)}} = \int_0^1 (H^{-1}(x^{1/k}) - \mu_{X_{(k)}})^2 dx$, respectively.

Let $Z_1, \ldots, Z_k$ be a random sample from a standard normal distribution with distribution function denoted by $\Phi$. Let $Z_{(k)}$ denote the maximum of $Z_1, \ldots, Z_k$, and let $\Phi_{(k)}$ denote its distribution function. The quantiles/percentiles for $Z_{(k)}$ are given by

$$\Phi_{(k)}^{-1}(\gamma) = \Phi^{-1}(\gamma^{1/k}) \tag{4.1}$$

for $0 < \gamma < 1$. The mean and variance for $Z_{(k)}$ are given, respectively, as

$$\mu_{Z_{(k)}} = \int_0^1 \Phi^{-1}(x^{1/k})dx \tag{4.2}$$

and

$$\sigma_{Z_{(k)}}^2 = \int_0^1 \left( \Phi^{-1}(x^{1/k}) - \mu_{Z_{(k)}} \right)^2 dx \qquad (4.3)$$

Table 4.2 provides the mean, standard deviation, and various percentiles based on Equations 4.1 through 4.3.

When the underlying distribution for $X_1, \ldots, X_k$ is a normal distribution with mean $\mu$ and standard deviation $\sigma$, $X_{(k)}$ is equal in distribution to $\mu + Z_{(k)}\sigma$. The behavior of the minimum treatment effect is analogous to that of the maximum treatment effect, with the roles of the treatment arms reversed. Example 4.2 illustrates using the distribution of the maximum in constructing a confidence interval for the true common treatment effect for a time-to-event endpoint.

## Example 4.2

Suppose that there are five randomized, placebo-controlled clinical trials evaluating an experimental therapy with each trial based on a one-to-one randomization. For each study, as in Example 4.1, the same time-to-event endpoint will be compared after 400 events where the true experimental versus placebo hazard ratio is 0.894 (which provides 20% power to achieve statistical significance at a one-sided $\alpha$ of 0.025) in every study. Then, using Table 4.2, the maximum observed treatment effect is represented by the minimum observed experimental versus placebo log-hazard ratio, which has mean $-0.228$ (= ln $0.894 - 1.16 \times 2/\sqrt{400}$), which corresponds to a hazard ratio of 0.796. The standard deviation for the minimum log-hazard ratio is 0.067. The median minimum experimental versus placebo hazard ratio is 0.799 (= ln $0.894 - 1.13 \times 2/\sqrt{400}$). Using the 2.5th and 97.5th percentiles in Table 4.2, an equal-tailed 95% prediction interval for the minimum hazard ratio is 0.691, 0.899.

**TABLE 4.2**

Means, Standard Deviations, and Various Percentiles for the Maximum of a Random Sample from a Standard Normal Distribution

| $K$ | $\mu_{Z_{(k)}}$ | $\sigma_{Z_{(k)}}$ | Percentiles | | | | |
|---|---|---|---|---|---|---|---|
| | | | 2.5th | 25th | 50th | 75th | 97.5th |
| 2 | 0.56 | 0.83 | −1.00 | 0 | 0.54 | 1.11 | 2.24 |
| 3 | 0.85 | 0.75 | −0.55 | 0.33 | 0.82 | 1.33 | 2.39 |
| 5 | 1.16 | 0.67 | −0.05 | 0.70 | 1.13 | 1.59 | 2.57 |
| 10 | 1.54 | 0.59 | 0.50 | 1.13 | 1.50 | 1.91 | 2.80 |
| 25 | 1.97 | 0.51 | 1.09 | 1.61 | 1.92 | 2.28 | 3.09 |

Suppose instead that the true common experimental versus placebo hazard ratio is unknown and that only the best (minimum) observed hazard ratio is considered. If the minimum observed hazard ratio is 0.75, then, based solely on that, a 95% equal-tailed confidence interval for the true common experimental versus placebo hazard ratio is 0.746, 0.970. This confidence interval is based on the 2.5th and 97.5th percentiles in Table 4.2 and the relation $X_{(5)} = \mu + Z_{(5)}\sigma$ in distribution, which is applied to the placebo versus the experimental log-hazard ratio.

Once the estimates have been observed along with their respective order, the confidence coefficients for the confidence intervals change. The confidence coefficients will depend on the distributions (or conditional distributions) of the order statistics. In Example 4.2, where there is a random sample of five estimated effects, the confidence coefficient for the error symmetric 95% confidence interval for that individual study that had the maximum (minimum) estimated effect is now an error asymmetric 88.1% confidence interval when the order of the estimated effects across all five studies is considered. The confidence coefficient for the 95% confidence interval for that study having the second largest (second smallest) estimated effect is 99.4% when the order of the estimated effects is considered. The confidence coefficient for the 95% confidence interval for that study having the median estimated effect is 99.97% when the order of the estimated effects is considered. For a random sample of estimated effects, the confidence coefficient for the 95% confidence interval for the individual study that had the maximum (median) estimated effect decreases (increases) toward zero (one) as the number of studies increases.

*Simultaneous Confidence Bounds.* Fairly analogous to having the inference based on a maximum is requiring simultaneous one-sided confidence intervals to maintain a desired overall coverage. For $k$ studies and a probability of $1 - \alpha$ that every one-sided confidence interval will capture the respective true effect, the common confidence coefficient for each confidence interval is $(1 - \alpha)^{1/k}$. When the estimated effects across studies is a random sample (e.g., the studies are identical in design and conduct), the largest (smallest) of the one-sided simultaneous lower (upper) confidence bounds each with confidence coefficient $(1 - \alpha)^{1/k}$ equals the lower confidence bound of coefficient $1 - \alpha$ based solely on the maximum (minimum) observed effect. For example, when $k = 5$ and $\alpha = 0.025$, the confidence coefficient for each confidence interval is 0.995 ($=0.975^{0.2}$). Note that the formula for determining the common confidence coefficient for each confidence interval is the same as the formula for relating the $(1 - \alpha)$th quantile for the maximum to the $[(1 - \alpha)^{1/k}]$th quantile of the underlying distribution.

It is fairly common to use for the non-inferiority trial the lower limit of a 95% confidence interval for the true active control effect (usually from a meta-analysis) as a surrogate or substitute for the unknown true effect of the active control. When only the result from the study that produced the largest estimated effect among the $k$ studies is considered, it seems a reasonable analog to base the surrogate or substitute for the unknown true effect of the active control for the non-inferiority trial as the $(0.975)^{1/k} \times 100\%$ lower confidence bound calculated solely from that study.

More extensive modeling based on order statistics can also be done. For example, suppose it is believed that there may be a specific number of small studies that did not get published because of unfavorable results for the treated arm. A model can be applied to the results from the known small studies that assumes that those known results represent better-order statistics from some samples of independent observations. Two approaches used in Example 4.3 are based on the maximum of a sample of estimated effects that are not a random sample. In Example 4.3 we consider various ways of integrating the available information from two studies on the overall survival effect of docetaxel in second-line non-small cell lung cancer (NSCLC).

## Example 4.3

The JMEI trial studied the use of pemetrexed against the active control of docetaxel at a dose of 75 mg/m$^2$ (D75) with subjects in second-line NSCLC. A non-inferiority claim for pemetrexed versus docetaxel on overall survival was sought.[10] Thus, it would be necessary to understand the effect of docetaxel on overall survival in second-line NSCLC. There have been several clinical trials studying the effects of docetaxel in NSCLC and other cancers. For the sake of this example, only two studies of docetaxel in second-line NSCLC (TAX 317 and TAX 320) will be considered. For the TAX 320 study, 373 subjects were randomized to either 100 mg/m$^2$ docetaxel (D100), D75, or a control therapy (vinorelbine or ifosfamide, V/I). There is little evidence that vinorelbine or ifosfamide extends life in a second-line setting of NSCLC. For the TAX 317 study, 100 subjects were randomized to D100 or best supportive care (BSC) in phase A of the study, and 104 subjects were randomized to D75 or BSC in phase B of the study.

How the results are modeled or integrated will have a great impact on the estimation of the relevant effect of D75. When an approach is selected retrospectively and dependent on the trial results, it will produce biased estimates. Prespecification of an approach before the conduct of the TAX 320 and TAX 317 studies (or independent of their results) would be necessary to produce unbiased estimates. Some possible approaches are listed below.

1. A naïve approach that uses only the results from phase B of the TAX 317 study.
2. There is no strong enough evidence from the TAX 320 study to rule out that the effects are equal between the docetaxel regimens. Therefore estimation of the active control effect based on the assumption that the effects of the docetaxel regimens are equal and constant across studies can be considered.
   a. Use only the results from the TAX 317 study.
   b. Use results from both studies treating the control arms of vinorelbine or ifosfamide, and BSC as exchangeable.
3. An approach that integrates the results in the TAX 320 study of the comparison of D100 with D75, with the separate comparisons of each phase of docetaxel to BSC from the TAX 317 study. The effects of each docetaxel

regimen are allowed to be different and are assumed to be constant across studies.

4. An approach used during an Oncology Drugs Advisory Committee meeting[10] treated the results from phase B of TAX 317 as isolated results (i.e., as the sole results available for estimating the effect of docetaxel). However, the results from phase B of TAX 317 are not isolated results.

   a. Therefore, another approach uses only the results from phase B of TAX 317, but treats the estimate and the corresponding variability as representing the maximum observed effect across all observed effects.

   b. Further condition on the maximum observed effect being the observed effect in phase B of TAX 317.

For each approach, Table 4.3 summarizes the estimated hazard ratio, the corresponding 95% confidence interval for the true D75 versus BSC hazard ratio, and the one-sided $p$-value for testing that D75 is superior to BSC. For approach 1, the estimate of the D75 versus BSC hazard ratio from TAX 317 is 0.56 with the corresponding 95% confidence interval of 0.35–0.88.[11] From the confidence interval, the standard error for the log-hazard ratio estimator is approximately 0.235 and the one-sided $p$-value for superiority of D75 versus BSC is approximated as 0.007.

From the overall survival results provided in the Statistical review of NDA 20449/S11 for TAX 317,[12] with data cutoff date of April 12, 1999, the observed D100 versus BSC hazard ratio is either 0.96 or 1.04 = 1/0.96 (using the $p$-value and the number of events for each group) and the corresponding standard error for the log-hazard ratio estimator is 0.221 ($=\sqrt{1/40+1/42}$). For this example, we will use 0.96 as the observed hazard ratio. For approach 2a, applying a fixed-effects meta-analysis to the independent comparison of phases A and B of TAX 317 leads to an estimated D75/D100 versus BSC hazard ratio of 0.743 (= exp([ln 0.56/(0.235)² + ln 0.96/(0.221)²]/[1/(0.235)² + 1/(0.221)²])) and the corresponding standard error for the log-hazard ratio estimator of 0.160 ($=\sqrt{1/(0.235)^2+1/(0.221)^2}$).

For the TAX 320 study, there were 104, 97, and 110 deaths in the D100, D75, and V/I treatment groups, respectively.[12] The D75 versus V/I hazard ratio is provided in the product label for Taxotere[11] as 0.82, and the D100 versus V/I hazard ratio is determined to be either 0.99 or 1.01 = 1/0.99. For this example, we will use

**TABLE 4.3**

Estimates of D75 versus BSC Hazard Ratio by Approach

| | D75 vs. BSC Hazard Ratio | | One-Sided |
|---|---|---|---|
| Approach | Estimate | 95% Confidence Interval | $p$-Value |
| 1 | 0.56 | (0.35, 0.88) | 0.007 |
| 2a | 0.743 | (0.543, 1.018) | 0.032 |
| 2b | 0.842 | (0.698, 1.015) | 0.035 |
| 3 | 0.655 | (0.466, 0.921) | 0.007 |
| 4a | 0.675 | (0.524, 0.938) | 0.011 |
| 4b | 0.704 | (0.536, 0.985) | 0.021 |

0.99 as the observed hazard ratio. The geometric mean of the two hazard ratio estimates is 0.901, which will be used as the combined estimate of the D75/D100 versus V/I hazard ratio. The estimated standard error for the combined log-hazard ratio estimator is 0.119 (= $0.5 \times \sqrt{1/104 + 1/97 + 4/110}$). When the combined results for TAX 320 and TAX 317 (determined for approach 2a) are integrated by a fixed-effects meta-analysis, the estimated D75/D100 versus V/I/BSC hazard ratio is 0.842 (= exp([ln 0.743/(0.160)² + ln 0.901/(0.119)²]/[1/(0.160)²+1/(0.119)²])), with a corresponding estimated standard error for the log-hazard ratio estimator of 0.095 (= $\sqrt{1/(0.160)^2 + 1/(0.119)^2}$).

For approach 3, the D75 versus D100 hazard ratio from TAX 320 is determined as 0.828 (=0.82/0.99), with a corresponding standard error for the log-hazard ratio of 0.141 (= $\sqrt{1/104 + 1/97}$). This result is combined with the results of the D100 versus BSC, yielding an estimated hazard ratio of 0.795 and a corresponding standard error for the log-hazard ratio estimator of 0.262. This indirect comparison is now integrated with the direct comparison of D75 versus BSC, yielding an overall D75 versus BSC hazard ratio of 0.655 with a corresponding standard error for the log-hazard ratio estimator of 0.175. This estimate of the D75 versus BSC hazard ratio is the maximum likelihood estimate under the model where the log-hazard ratio estimators have independent normal distributions with respective standard deviations equal to the estimated standard errors and the true log-hazard ratios of D75 versus BSC, BSC versus D100, and D100 versus D75 are required to sum to zero.

Let $\beta$ denote the common true D75/D100 versus BSC/V/I log-hazard ratio across studies. Let $\hat{\beta}_1$, $\hat{\beta}_2$, $\hat{\beta}_3$, and $\hat{\beta}_4$ denote the log-hazard ratios of D100 versus V/I, D75 versus V/I in TAX 320, D100 versus BSC, and D75 versus BSC in TAX 317, respectively. For approaches 4a and 4b, the distribution of the deviation between the minimum observed log-hazard ratio (maximum observed effect) was studied by simulations. This deviation equals the minimum deviation across estimates/estimators (i.e., $\min\{\hat{\beta}_1, \hat{\beta}_2, \hat{\beta}_3, \hat{\beta}_4\} - \beta = \min\{\hat{\beta}_1 - \beta, \hat{\beta}_2 - \beta, \hat{\beta}_3 - \beta, \hat{\beta}_4 - \beta\}$). Let $(Z_1, Z_2)$, $Z_3$, $Z_4$ be independent, where each $Z_i$ has a standard normal distribution and $(Z_1, Z_2)$ has a bivariate normal joint distribution with a correlation of 0.5. For the comparison of D100 versus V/I and D75 versus V/I in TAX 320, the respective deviations $\hat{\beta}_1 - \beta$ and $\hat{\beta}_2 - \beta$ are modeled as $0.138Z_1$ and $0.138Z_2$ (0.138 represents the average of the two estimated standard errors). For the comparisons of D100 versus BSC and D75 versus BSC in TAX 317, the respective deviations $\hat{\beta}_3 -$ and $\hat{\beta}_4 - \beta$ are modeled as $0.221Z_3$ and $0.235Z_4$.

On the basis of 100,000 replications, the simulated mean minimum deviation is −0.188 with simulated 2.5th and 97.5th percentiles of −0.515 and 0.067, respectively. On the basis of retaining only the maximum observed effect of a hazard ratio of 0.56, this leads to the estimated common hazard ratio of 0.675 (= exp(ln 0.56 + 0.188)) and limits of the corresponding 95% confidence interval of 0.524 (= exp(ln 0.56 − 0.067)) and 0.938 (= exp(ln 0.56 + 0.515)).

In 31,899 of the 100,000 replications, $\hat{\beta}_4 = \min\{\hat{\beta}_1, \hat{\beta}_2, \hat{\beta}_3, \hat{\beta}_4\}$. Conditioning on $\hat{\beta}_4 = \min\{\hat{\beta}_1, \hat{\beta}_2, \hat{\beta}_3, \hat{\beta}_4\}$, the simulated mean minimum deviation is −0.229 with simulated 2.5th and 97.5th percentiles of −0.565 and 0.043, respectively. On the basis of retaining only the maximum observed effect of a hazard ratio of 0.56 and that this maximum observed effect came from phase B of TAX 317, this leads to the estimated common hazard ratio of 0.704 (= exp(ln 0.56 + 0.229)) and limits of the corresponding 95% confidence interval of 0.536 (= exp(ln 0.56 − 0.043)) and 0.985 (= exp(ln 0.56 + 0.565)).

From Table 4.3, the results vary across approaches with the estimated effects ranging from a D75 versus BSC hazard ratio of 0.56–0.842. The upper limits of the 95% confidence intervals range from 0.88 to 1.018. Additionally, the more information used in an approach or the more restrictive the assumptions, the narrower the 95% confidence interval. Two approaches failed to achieve statistical significance at a one-sided 0.025 level. All in all, the results do not provide substantial evidence that the true D75 versus BSC hazard ratio for overall survival is less than 1.

## 4.3 Meta-Analysis Methods

Meta-analysis was defined by Glass[13] as "the statistical analysis of a large collection of analysis results from individual studies for the purpose of integrating the findings." Some reasons for performing a meta-analysis include the following: to obtain a more precise estimate for the entire population or subgroup, to evaluate the secondary endpoint when the power is inadequate in any given trial, to more reliably understand the dose–response relationship, and to qualify the relationship between a surrogate endpoint and a clinical benefit endpoint. For non-inferiority trials, meta-analyses have been used to estimate the historical effect of the active control therapy. The goal for a non-inferiority trial is estimating the effect of the active control in the setting of the non-inferiority trial along with appropriately modeling the corresponding variability/uncertainty. This includes determining the appropriate shape for the sampling distribution of the estimated effect of the active control. For a random-effects model involving the mean of continuous data where the underlying variances within the studies are unknown, Follmann and Proschan[14] showed that, in many cases, the sampling distribution for the estimator for the global mean across studies should be modeled with the appropriate $t$ distribution.

There may be many differences among the clinical trials considered in the meta-analysis. Some of the studies may have bias in the estimation of the effect of the endpoint of interest. The lack of blinding can introduce bias. The primary objective of the studies may differ, potentially leading to different scrutiny in obtaining measurements on the endpoint of interest for the meta-analysis.

The extent of monitoring and follow-up for an endpoint may be different between a primary and a secondary endpoint. In investigating the follow-up for overall survival in 10 studies in advanced solid tumor cancers, Rothmann et al.[15] found that the percentage of loss to follow-up for overall survival ranged from 1% to 5% for the six studies having overall survival as the primary endpoint and ranged from 10% to 12% for the four studies where overall survival was a secondary endpoint. Poor follow-up and monitoring may introduce bias in the estimated effect.

We will discuss some meta-analytic methods. These include fixed-effects meta-analyses, which assume a common effect across clinical trials; random-effects meta-analyses, which assume varying effects across studies; and "covariate-adjusted" approach, which estimates the treatment effect on the basis of a particular distribution for effect modifiers. Meta-analytic models used to estimate the historical active control effect ignore that the active control was identified on the basis of one, some, or all the studies used in the meta-analysis. This will lead to a (regression to the mean) biased estimation of the active control effect with a tendency to overestimate the effect. The presence of reproduced estimated effects across studies can provide assurance that this bias is small and unimportant and that the modeling assumptions may approximately hold.

### 4.3.1 Fixed Effects Meta-Analysis

Consider that there are $k$ studies comparing the active control with placebo. Let $\gamma_i$ denote the true active control effect for the $i$th study. For a fixed-effects meta-analysis, it is assumed that $\gamma_1 = \gamma_2 = \gamma_k = \gamma$. The standard model for the sample estimators $\hat{\gamma}_1, \hat{\gamma}_2, \dots, \hat{\gamma}_k$ is given by $\hat{\gamma}_i = \gamma + \varepsilon_i$, where $\varepsilon_1, \varepsilon_2, \dots, \varepsilon_k$ are independent and $\varepsilon_i \sim N(0, \sigma_i^2)$. The maximum likelihood estimator of $\gamma$ is given by $\hat{\gamma}_{\text{FE}} = \sum_{i=1}^{k} (1/\sigma_i^2)\hat{\gamma}_i / \sum_{i=1}^{k} (1/\sigma_i^2)$. The variance of $\hat{\gamma}_{\text{FE}}$ is $1 / \sum_{i=1}^{k} (1/\sigma_i^2)$.

Commonly, the true variances of the within-study sample effects are not known, but are estimated. Let $s_i^2$ denote the estimated variance of $\hat{\gamma}_i$ for $i = 1, \dots, k$. Then the standard fixed-effects estimator of $\gamma$ is given by

$$\hat{\gamma}_{\text{FE}} = \sum_{i=1}^{k} (1/s_i^2)\hat{\gamma}_i / \sum_{i=1}^{k} (1/s_i^2)$$

and the corresponding estimated variance is given by $s^2 = 1 / \sum_{i=1}^{k} (1/s_i^2)$.

The existence of heterogeneity invalidates the assumptions of a fixed-effects meta-analysis. The presence of heterogeneous effects across studies can be tested on the basis of the statistic $Q = \sum_{i=1}^{k} (1/s_i^2)(\hat{\gamma}_i - \hat{\gamma}_{\text{FE}})^2$ (see DerSimonian and Laird[16]). When the effects are homogeneous, $Q$ has an approximate $\chi^2$ distribution with $k - 1$ degrees of freedom. For $0 < \alpha < 1$, let $\chi_{k-1,\alpha}^2$ denote the upper $\alpha$th percentile from a $\chi^2$ distribution with $k - 1$ degrees of freedom. When $Q > \chi_{k-1,\alpha}^2$, a single common effect is rejected. Formally, the conclusion is that there are at least two different values among $\gamma_1, \gamma_2, \dots, \gamma_k$. The formal conclusion is neither that there are $k$ distinct values for $\gamma_1, \gamma_2, \dots, \gamma_k$ nor that $\gamma_1, \gamma_2, \dots, \gamma_k$ are independent and identically distributed random variables. The test of heterogeneity tends to have low power in most practical situations. A formula for determining the power for testing heterogeneity on the basis of the test statistic $Q$ is given by Jackson.[17]

## 4.3.2 Peto's Method

For binary data, a popular fixed-effects meta-analysis uses Peto's odds ratio of a success.[18,19] For a given clinical trial comparing an experimental arm to a control arm, let $d_j$ denote the number of successes in arm $j$ among $n_j$ subjects, $j = C, E$, and let $d = d_C + d_E$ and $n = n_C + n_E$ denote the overall number of subjects. Peto's odds ratio and its distribution are conditioned on the total number of successes, $d$, between the two treatment arms. Then the number of successes in the experimental arm, $d_E$, is random, having a hypergeometric distribution. To determine Peto's odds ratio, first calculate

$$Z = d_E - n_E \times d/n \text{ and } V = (n_E \times n_C \times d \times (n - d))/(n^2 \times (n - 1)) \quad (4.4)$$

Conditioned on the value of $d$, $Z$ represents the difference between the observed number of successes and the expected number of successes for the experimental arm, and $V$ is the conditional variance for both $d_E$ and $Z$. Peto's estimator of the log-odds ratio and odds ratio are given by $\hat{\theta} = Z/V$ and $\exp(Z/V)$. For large samples, the estimator of the log-odds ratio, $\hat{\theta}$, has an approximate normal distribution with a mean true log-odds ratio of $\theta$ and a variance of $1/V$. In the fixed-effects meta-analysis, the log-odds estimates will be weighted by the corresponding values for $V$. Note that the test statistic for testing $H_o: \theta = 0$ against $H_a: \theta \neq 0$ is $Z/\sqrt{V}$, which has an approximate standard normal distribution when $\theta = 0$ for large sample sizes.

Let $Z_i$ and $V_i$ denote the corresponding quantities in Equation 4.4 for the $i$th study, $i = 1,..., k$. Then the meta-analysis estimator of the common log-odds ratio is given by $\hat{\theta} = \sum_{i=1}^{k} V_i \hat{\theta}_i \Big/ \sum_{i=1}^{k} V_i = \sum_{i=1}^{k} Z_i \Big/ \sum_{i=1}^{k} V_i$. The estimated variance of $\hat{\theta}$ is given by $1 \Big/ \sum_{i=1}^{k} V_i$. The estimator $\hat{\theta}$ has the form $\sum_{i=1}^{k} (O_i - E_i) \Big/ \sum_{i=1}^{k} \text{Var}(O_i)$, where $O$ denotes an observed value and $E$ denotes an expected value when $\theta = 0$. Peto's method may have substantial bias when the allocation is not balanced.[20]

The presence of heterogeneous effects across studies can be tested on the basis of the statistic $R = \sum_{i=1}^{k} Z_i^2 \Big/ \sum_{i=1}^{k} V_i - \left( \sum_{i=1}^{k} Z_i \right)^2 \Big/ \sum_{i=1}^{k} V_i$ (see Yusuf et al.[18]). When the effects are homogeneous, $R$ has an approximate $\chi^2$ distribution with $k - 1$ degrees of freedom. For $0 < \alpha < 1$, let $\chi^2_{k-1,\alpha}$ denote the upper $\alpha$th percentile from a $\chi^2$ distribution with $k - 1$ degrees of freedom. When $R > \chi^2_{k-1,\alpha}$, a single common effect is rejected.

The Cochran–Mantel–Haenszel approach for proportions is another type of fixed-effects meta-analysis (see Section 11.6 for details).

### 4.3.3 Random-Effects Meta-Analysis

One way of modeling different effect sizes is with the DerSimonian and Laird random-effects model. For a standard random-effects meta-analysis, $\gamma_1, \gamma_2, \ldots, \gamma_k$ are assumed to be a random sample from a normal distribution having mean $\gamma$ and variance $\tau^2$ (see DerSimonian and Laird[16]). The standard model for the sample estimators $\hat{\gamma}_1, \hat{\gamma}_2, \ldots, \hat{\gamma}_k$ is given by $\hat{\gamma}_i = \gamma + \eta_i + \varepsilon_i$, where $\eta_1, \eta_2, \ldots, \eta_k$ are a random sample from a normal distribution having mean zero and variance $\tau^2$ and $\varepsilon_1, \varepsilon_2, \ldots, \varepsilon_k$ are independent with $\varepsilon_i \sim N(0, \sigma_i^2)$.

The estimator $\hat{\gamma}_i$ "unbiasedly estimates" the realized value for $\gamma_i = \gamma + \eta_i$. If $\sigma_1, \ldots, \sigma_k$ and $\tau^2$ are known, then the maximum likelihood estimator of $\gamma$ is given by $\hat{\gamma}_{RE} = \sum_{i=1}^{k} (\tau^2 + \sigma_i^2)^{-1} \hat{\gamma}_i \Big/ \sum_{i=1}^{k} (\tau^2 + \sigma_i^2)^{-1}$. The variance of $\hat{\gamma}_{RE}$ is $1 \Big/ \sum_{i=1}^{k} (\tau^2 + \sigma_i^2)^{-1}$. In practice, $\sigma_1, \ldots, \sigma_k$ and $\tau^2$ are not known. Then the between-study variance, $\tau^2$, is estimated by

$$\hat{\tau}^2 = \max\left\{ 0, \frac{\sum_{i=1}^{k}(1/s_i^2)(\hat{\gamma}_1 - \hat{\gamma}_0)^2 - (k-1)}{\sum_{i=1}^{k}(1/s_i^2) - \sum_{i=1}^{k}(1/s_i^4)\Big/\sum_{i=1}^{k}(1/s_i^2)} \right\}$$

where $\hat{\gamma}_0 = \sum_{i=1}^{k}(1/s_i^2)\hat{\gamma}_i \Big/ \sum_{i=1}^{k}(1/s_i^2)$. Then the random-effects estimator of $\gamma$ is given by $\hat{\gamma}_{RE} = \sum_{i=1}^{k}(\hat{\tau}^2 + s_i^2)^{-1}\hat{\gamma}_i \Big/ \sum_{i=1}^{k}(\hat{\tau}^2 + s_i^2)^{-1}$. The corresponding estimated variance for $\hat{\gamma}_{RE}$ is $s^2 = 1 \Big/ \sum_{i=1}^{k}(\hat{\tau}^2 + s_i^2)^{-1}$.

For the random-effects model, the units are studies, not subjects. Inference is formally on studies. For the inference to apply at the subject level, either all studies should have the same study size or all individual estimated effects should have the same study standard error. Otherwise, the study standard error should not be correlated (not even spuriously correlated) with the study effect size. It is difficult to evaluate and be certain that the study standard error and study effect size are not correlated.

While the existence of heterogeneity invalidates the assumptions of a fixed-effects meta-analysis, the existence of a correlation between the estimated effects and the within-study standard errors invalidates the assumptions of a random-effects meta-analysis. There are other circumstances that can also invalidate the assumptions of a random-effects meta-analysis.

For these meta-analysis methods, the common or average effect in the models reflects the expected value (or conditional expected value for a random-effects model) of the estimated effects. This is important as study conduct, missing data, or design features can introduce bias in estimating the true study effect.

Differential bias across studies will resemble and contribute to between-trial variability. Heterogeneity in effects should be investigated. The estimated effect of the active control in the setting of the non-inferiority trial should account for explained heterogeneity (effect modification) and bias. The corresponding variance or standard error should account for the unexplained heterogeneity.

Additionally, there are other options to modeling the study effects when the effects are heterogeneous. The within-trial effects can be viewed as fixed, unknown, and not necessarily equal, instead of using a random model. When there is a distinct effect modifier that is either present or absent in a trial, modeling can be done separately for studies where the effect modifier is present and those studies where the effect modifier is absent. This may be the case when each study only involves patients from one of two distinct populations where the responsiveness to therapy is known to be different between the populations.

### 4.3.4 Sampling Distributions

For a fixed-effects meta-analysis, $\gamma$ represents the common effect across studies or the mean effect across subjects. Subjects can be regarded as the units in a fixed-effects meta-analysis. For a fixed-effects meta-analysis, the asymptotic distribution of $\hat{\gamma}_{FE}$ depends on the total number of subjects in each arm across studies going to infinity. In most settings involving a fixed-effects meta-analysis, the overall number of subjects will be large and $(\hat{\gamma}_{FE} - \gamma)/s$ will have an approximate standard normal distribution.

For a random-effects meta-analysis, $\gamma$ represents the mean of the within-study effects or the mean effect across studies. Studies are the units in a random-effects meta-analysis, not subjects. The asymptotic normality of $\hat{\gamma}_{RE}$ depends on the total number of studies, $k$, going to infinity for a random-effects meta-analysis. Typically, the number of studies considered in a random-effects meta-analysis is not large enough for $(\hat{\gamma}_{RE} - \gamma)/s$ to have an approximate standard normal distribution. The distribution for $(\hat{\gamma}_{RE} - \gamma)/s$ under a random-effects model has been investigated by many.[14,21,22]

For constructing confidence intervals for the global study mean effect, $\gamma$, Larholt, Tsiatis, and Gelber[21] utilized Sattherwaite's approximation for the distribution of $Q$ under the restriction that each within-trial variance, $\sigma_i^2$, is much smaller than $\tau^2$ (i.e., $\sigma_i^2 \ll \tau^2$ all $i$). Biggerstaff and Tweedie[23] extended the work of Larholt, Tsiatis, and Gelber[21] to determine the confidence intervals for $\tau^2$ and the alternative confidence intervals for $\gamma$ on the basis of the distribution of

$$\hat{\tau}_{BT}^2 = \sum_{i=1}^{k}\left[(1/s_i^2)(\hat{\gamma}_i - \hat{\gamma}_0)^2 - (k-1)\right]\bigg/\left[\sum_{i=1}^{k}(1/s_i^2) - \sum_{i=1}^{k}(1/s_i^4)\bigg/\sum_{i=1}^{k}(1/s_i^2)\right]$$

that are not restricted to $\sigma_i^2 \ll \tau^2$ all $i$.

For random-effects models, Follmann and Proschan[14] criticized the approach of using a standard normal distribution as the distribution for $\hat{\gamma}_{\mathrm{RE}}$ when $\gamma = 0$. For the cases that they considered, which involved continuous data, the type I error rate was greatly inflated in testing $H_o$: $\gamma = 0$ against $H_a$: $\gamma \neq 0$ when standard normal percentiles were used as the critical values. This included a potential inflation of the type I error rate by 100% from the desired rate when the meta-analysis involves 16 studies. For cases similar to those they considered, Follmann and Proschan recommended using as critical values the appropriate percentiles from a $t$ distribution with $k - 1$ degrees of freedom. They also recommended performing simulations to determine more appropriate critical values.

When $\sigma_1^2, \ldots, \sigma_k^2$ are known, how close the distribution of $(\hat{\gamma}_{\mathrm{RE}} - \gamma)/s$ is to a standard normal distribution or a $t$ distribution with $k - 1$ degrees of freedom depends on both the number of studies and the relative size of $\tau^2$ to $\sigma_1^2, \ldots, \sigma_k^2$. For example, when $\tau^2 > 0$ is unknown and it is known that $\sigma_1^2 = \cdots = \sigma_k^2 = 0$, then $(\hat{\gamma}_{\mathrm{RE}} - \gamma)/s$ will have a $t$ distribution with $k - 1$ degrees of freedom. When $\tau^2 > 0$ is unknown and $\sigma_1^2 = \cdots = \sigma_k^2 = \sigma^2 > 0$ is known, an approximate distribution for $(\hat{\gamma}_{\mathrm{RE}} - \gamma)/s$ will be between a standard normal distribution and a $t$ distribution with $k - 1$ degrees of freedom. The larger $\sigma^2/\tau^2$, the closer the approximate distribution of $(\hat{\gamma}_{\mathrm{RE}} - \gamma)/s$ tends to be to a standard normal distribution.

When $\sigma_1^2, \ldots, \sigma_k^2$ are unknown, and therefore need to be estimated, the approximate distribution for $(\hat{\gamma}_{\mathrm{RE}} - \gamma)/s$ will also depend on how reliably $\sigma_1^2, \ldots, \sigma_k^2$ can be estimated. For example, when a log-hazard ratio is considered, the asymptotic standard deviation can be closely approximated when the analysis is based on a prespecified number of events and the true log-hazard ratio is not far from zero (see Chapter 13). However, when data are from a normal distribution with unknown variances, the estimated standard deviation for the observed treatment difference is more variable relative to the true standard deviation.

When the between-trial variability dominates the within-trial variability, a $t$ distribution with $k - 1$ degrees of freedom or a similar number of degrees of freedom may appropriately model the distribution for $(\hat{\gamma}_{\mathrm{RE}} - \gamma)/s$. A $t$ distribution with $k - 1$ or a similar number of degrees of freedom may also be an appropriate model when $\tau^2 > 0$ is unknown and the estimators of $\sigma_1^2, \ldots, \sigma_k^2$ are fairly variable, as with continuous data. When the within-trial variability dominates the between-trial variability and the within-trial variability is either known or very reliably estimated, a distribution with smaller tails (e.g., a $t$ distribution with degrees of freedom larger or much larger than $k - 1$ or a standard normal distribution) may be appropriate as an approximate null distribution for $(\hat{\gamma}_{\mathrm{RE}} - \gamma)/s$.

Ziegler, Koch, and Victor[22] showed for the log-odds ratio that the test for effectiveness based on a random-effects meta-analysis with standard normal determined critical values (the RE test) can be both conservative and anticonservative. The amount of type I error rate inflation increases as the within-trial variance and the number of studies decreases. For known and equal study-specific variances of $\sigma^2$, they provided an approximate type I error rate for the RE test of

$$2\left\{1 - \Phi\left(z_{\alpha/2}/\sqrt{1+\tau^2/\sigma^2}\right)F_{k-1}\left((k-1)/(1+\tau^2/\sigma^2)\right)\right.$$

$$\left. - \int_{(k-1)(1+\tau^2/\sigma^2)}^{\infty} \Phi\left(z_{\alpha/2}/\sqrt{x/(k-1)}\right)f_{k-1}(x)dx\right\}$$

where $F_{k-1}$ and $f_{k-1}$ are the distribution and density function for a $\chi^2$ distribution with $k-1$ degrees of freedom, respectively. As noted by Ziegler, Koch, and Victor,[22] for a fixed number of studies, the type I error rate is increasing in $\tau^2/\sigma^2$. Thus, for fixed $k$ and $\tau^2$, the type I error rate is decreasing in $\sigma^2$ (increasing in the sample size/the number of events for a time-to-event endpoint). For fixed $\tau^2$ and $\sigma^2$, the type I error rate decreases as the number of studies, $k$, increases. As $\sigma^2 \to 0$, the type I error rate converges to $2(1 - G_{k-1}(Z_{\alpha/2}))$, where $G_{k-1}$ is the distribution function for a $t$ distribution with $k-1$ degrees of freedom.

For selected numbers of studies, Table 4.4 provides the limiting type I error rates as $\sigma^2 \to 0$ for $\alpha/2 = 0.025$. When there are only three studies and the within-trial variability is much less than the between-trial variability, the type I error rate for the superiority test will be about 9.5%. The type I error inflation may be quite large when the number of studies is small.

As $\sigma^2 \to 0$, the form of the asymptotic distribution function, $H_{k-1}$, for the RE test statistic when $\gamma = 0$ is provided in the paper of Ziegler, Koch, and Victor.[22] They proposed using $H_{k-1}^{-1}(1-\alpha/2)$ as the critical value for the RE test statistic when testing for effectiveness and as a multiplier when determining confidence intervals for $\gamma$. From their simulations, the new test either maintains the approximate type I error rate or is conservative.

**TABLE 4.4**

Limiting Type I Error Rates as $\sigma^2 \to 0$ for $\alpha/2 = 0.025$

| Number of Studies | Limiting Type I Error Rate |
| --- | --- |
| 3 | 0.095 |
| 10 | 0.041 |
| 25 | 0.031 |
| 50 | 0.028 |

We recommend performing simulations that use models consistent with the observed $\hat{\gamma}_{RE}$, $\hat{\tau}^2$, and $s_1^2, ..., s_k^2$ to determine an appropriate approximate distribution for $(\hat{\gamma}_{RE} - \gamma)/s$ or appropriate critical values or multipliers to be used in hypotheses testing or the determination of confidence intervals.

### 4.3.5 Concerns of Random-Effects Meta-Analyses

There are various concerns with the application of random-effects models to estimate the active control effect in the setting of the non-inferiority trial. For example, the estimation of the mean effect across studies (global mean) is frequently used to infer the effect of the active control in the non-inferiority trial from either the use of a 95% confidence interval or by the use of the point estimate with its standard error (e.g., in a synthesis test statistic). Such an inference is not on a study-specific estimate of the active control effect but on the mean of such an effect across the trials that were considered in the random-effects meta-analysis. Also, the standard deviation and the width of the 95% confidence interval converges to zero as the number of studies used in the random-effects meta-analysis increases without bound, regardless of the extent of the heterogeneity. When the within-study effects are greatly heterogeneous, using a precisely estimated global or average effect to establish a non-inferiority margin will likely not be appropriate and be accompanied with a great chance of concluding from the non-inferiority trial that an ineffective experimental therapy is effective. Additionally, as will be seen in an example later, the lower limit of a two-sided 95% confidence interval for the global mean may increase as the largest estimated effect is reduced. This would seem counterintuitive and is a result of the inference being based on a global mean, the assumption of between-study effects being normally distributed, and the large confidence coefficient for the 95% confidence interval. This seeming contradiction is of particular importance as the lower limit of the 95% confidence interval for the global mean effect is often used as a surrogate, substitute, or "estimate" of the treatment effect in a non-inferiority trial.

In one case from experience, there were eight previous studies involving the active control with sample sizes ranging from 50 to 1500 subjects. The non-inferiority trial had about 20,000 subjects. The logistics involving a 20,000-subject trial are bound to be quite different (location of sites, types of subjects, study conduct, quality of follow-up) from the logistics of the trials used to estimate the effect of the active control. Estimating the active control effect from trials quite different from the non-inferiority trial is likely to give results not relevant to the non-inferiority trial. In this case, because there are no previous randomized comparative trials involving the active control of comparable size to the non-inferiority trial, using the results of those previous trials to determine the effect of the active control in the setting of the non-inferiority trial may involve notable extrapolation.

In another case, the meta-analysis to estimate the effect of the active control was based on the results of 10 small and 3 large studies (each large study having thousands of subjects). The small studies collectively had a huge estimated treatment effect, whereas the large studies averaged to a slightly negative or adverse estimated effect for the "active" control. The three large studies accounted for more than 95% of the total number of subjects across 13 clinical trials. The random-effects meta-analysis yielded a 95% confidence interval for the average study effect that was entirely on the side of having an effect (entirely positive). However, the fixed-effects meta-analysis yielded no estimated difference between the "active" control and placebo. The stark difference in the results of the two meta-analyses is due to the correlation across studies between the observed effects and the trial size. The smaller trials had much larger observed treatment effects. When the observed effects are correlated with the trial size, a random-effects meta-analysis pulls the estimated effect away from the estimated effect from a fixed-effects meta-analysis toward the observed effects from the smaller studies.

As the non-inferiority trial would likely have been similar in size to the larger clinical trials, it may also be more similar in other aspects to the larger previous trials than the smaller trials, and thus the observed effects from the larger studies may be more relevant. It is not clear why the observed effects would be so different between the small and large studies. There may be differences between the groups of studies in the general health of subjects enrolled or in the quality of medical care. Additionally, rather large studies get published regardless of the results, and the publication of small studies may be affected by whether the results of the study are positive (i.e., a positive small study may be more likely to be published then a negative or nonpositive small study). If among small studies only the positive ones are published, the estimated effect from the meta-analyses will be biased and will overstate the true effect (if any). This bias would be more profound in a random effects meta-analysis than in a fixed effects meta-analysis. Furthermore, heterogeneity in treatment effects across studies may be due to different distributions across studies in important effect modifiers. Example 4.4 illustrates some of the complexities in using the results of meta-analyses to interpret the effect of the active control in the non-inferiority setting.

### Example 4.4

In this example, there are three previous randomized, clinical trials comparing the active control therapy to placebo on a continuous outcome. We will assume that each within-trial estimator of the treatment effect is unbiased and has a normal distribution. Table 4.5 provides the estimated active control effect along with the corresponding standard deviation and 95% confidence interval for each trial and for the fixed-effects and random-effects meta-analyses. Figure 4.1 displays the corresponding 95% confidence intervals.

For a random-effects meta-analysis, Table 4.5 provides the 95% confidence intervals based on both percentiles from a standard normal distribution

**TABLE 4.5**

Trial and Integrated Estimated Effects, Standard Deviations, and 95% Confidence Intervals

| Trial/Analysis | Estimated Effect | Standard Deviation | 95% Confidence Interval |
| --- | --- | --- | --- |
| Trial 1 | 7 | 3 | (1.1, 12.9) |
| Trial 2 | 6 | 2 | (2.1, 9.9) |
| Trial 3 | 37 | 3.5 | (30.2, 43.8) |
| Fixed effects | 12.0 | 1.50 | (9.0, 14.9) |
| Random effects | 16.5 | 9.01 | (−1.2, 34.2)[a]  (−22.3, 55.3)[b] |

[a]  Based on the 2.5th and 97.5th percentiles of a standard normal distribution.
[b]  Based on the 2.5th and 97.5th percentiles of a $t$ distribution with 2 degrees of freedom.

and percentiles from a $t$ distribution with 2 degrees of freedom. The estimated between-trial variance is given by $\hat{\tau}^2 = 235$. As the between-trial variability dominates the within-trial variances (which range from 4 to 12.25), if a random-effects model is the correct model, the 95% confidence interval for the global mean should be based on the percentiles from a $t$ distribution with 2 degrees of freedom rather than based on the percentiles from a standard normal distribution. A real issue is whether a random-effects model is appropriate.

Despite the confidence intervals from the individual trials being consistently positive and comfortably away from zero, the 95% confidence interval for the global mean from a random-effects model contains zero, regardless of the approach. This would suggest that there is uncertainty that the active control is typically effective.

Table 4.6 provides the estimated active control effects and the corresponding standard deviation and 95% confidence interval for each trial and for the fixed-effects meta-analysis had the estimated active control effect from trial 3 been 12 instead of 37. The estimated effects are similar for the fixed-effects and random-

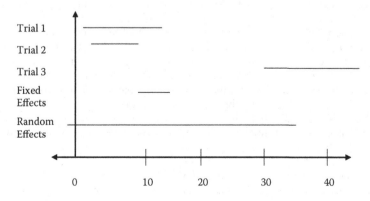

**FIGURE 4.1**
Trial and meta-analysis 95% confidence intervals.

**TABLE 4.6**

Trial and Integrated Estimated Effects, Standard Deviations, and 95% Confidence Intervals

| Trial/Analysis | Estimated Effect | Standard Deviation | 95% Confidence Interval |
| --- | --- | --- | --- |
| Trial 1 | 7 | 3 | (1.1, 12.9) |
| Trial 2 | 6 | 2 | (2.1, 9.9) |
| Trial 3 | 12 | 3.5 | (5.2, 18.8) |
| Fixed effects | 7.4 | 1.50 | (4.4, 10.3) |
| Random effects | 7.5 | 1.62 | (4.3, 10.6)[a]  (0.5, 14.4)[b] |

[a]  Based on the 2.5th and 97.5th percentiles of a standard normal distribution.
[b]  Based on the 2.5th and 97.5th percentiles of a *t* distribution with 2 degrees of freedom.

effects meta-analyses. The estimated between-trial variance is $\hat{\tau}^2 = 0.91$, which is much smaller than all the within-trial variances. Additionally, as there is little difference between the estimated effects and the standard deviation from the two methods, standard normal percentiles would be more appropriate than percentiles from a *t* distribution with 2 degrees of freedom. The actual sampling distribution for $(\hat{\gamma}_{RE} - \gamma)/s$ would probably have slightly fatter tails than a standard normal distribution leading to a slightly wider confidence interval than 4.3–10.6. Simulations can be performed to determine the appropriate multipliers that yield an approximate equal-tailed confidence interval that has roughly a 95% coverage probability.

Note that reducing the estimated effect in trial 3 from 37 to 12 led to an increase in the lower limit of the 95% confidence interval from a random-effects meta-analysis. This is particularly important since the lower limit of a 95% confidence interval for the global mean has been used as a surrogate of the active control effect in a non-inferiority trial. It seems counterintuitive to use a smaller value for the assumed effect of the active control when the third trial has an estimated effect of 37 than when the third trial has an estimated effect of 12.

Here, when the third trial has an estimated effect of 37, it may also be very difficult to defend the use of the lower limit of the 95% confidence interval for the global mean as the effect of the active control when there is great heterogeneity of the estimated effects across trials. The actual treatment effect in trial 3 is clearly bigger than the actual treatment effect in trials 1 and 2. It is important to investigate why the treatment effect in the third study is different so as not to provide too large or too small a non-inferiority margin in the non-inferiority trial. There may be effect modifiers for the active control that are distributed differently across trials. For example, the third trial may have been designed after learning that the active control therapy may only be beneficial and have a large effect in a particular subgroup of subjects. This may lead to the designing of a clinical trial with only subjects belonging to that subgroup randomized between therapy and placebo that would likely be sized at a large effect (thus yielding a greater standard deviation for the estimated effect than in the previous trials). It is important to use the historical studies of the active control therapy to make an inference on the expected or (minimum) likely effect that the active control would have in the non-inferiority trial with respect to the actual subject population, schedule and dose of the active control therapy, and study conduct in the non-inferiority trial. Differences in the bias on the estimated effects, due to differences in study conduct and study design, also contribute to heterogeneity in the estimated effects. This heterogeneity is not properly dealt with by being treated as unexplained variability in treatment effects.

When there is convincing evidence that $\gamma_i$ and $\sigma_i^2$ are correlated, the assumptions for the random effects probably do not hold. That is, the assumption that $\gamma_1, \gamma_2, \ldots, \gamma_k$ are identically distributed is probably false.

The fixed-effects estimator, $\hat{\gamma}_{FE}$, is also an unbiased estimator of $\gamma$ when the random-effects model holds. The variance for $\hat{\gamma}_{FE}$ under the random-effects model is given by $\left( \sum_{i=1}^{k} 1/\sigma_i^2 \right)^{-1} + \tau^2 \left( \sum_{i=1}^{k} 1/\sigma_i^4 \right) \Big/ \left( \sum_{i=1}^{k} 1/\sigma_i^2 \right)^2$.

The variance of $\hat{\gamma}_{FE}$ under a fixed-effects model and the variances for $\hat{\gamma}_{RE}$ and $\hat{\gamma}_{FE}$ under a random-effects model are respectively ordered with

$$\left( \sum_{i=1}^{k} 1/\sigma_i^2 \right)^{-1} \leq \left( \sum_{i=1}^{k} 1/(\sigma_i^2 + \tau^2) \right)^{-1} \leq \left( \sum_{i=1}^{k} 1/\sigma_i^2 \right)^{-1} + \tau^2 \left( \sum_{i=1}^{k} 1/\sigma_i^4 \right) \Big/$$

$\left( \sum_{i=1}^{k} 1/\sigma_i^2 \right)^2$. The closer $\tau^2$ is to zero, the more similar the variances. When the fixed effects and random effects estimated effect sizes are quite different, this may indicate that the assumptions for the random effects model do not hold.

When the effects are heterogeneous, Greenland and Salvan[20] suggest modeling the study differences instead of providing a single estimated effect.

In a case like that given in Table 4.5, where there is enormous heterogeneity of the estimated effects, it is likely that much of that heterogeneity is explainable. Both the fixed-effects and the DerSimonian–Laird random-effects meta-analyses are not appropriate. It is important to investigate the heterogeneity of the estimated effects. The potential bias in the estimates should also be considered. The variability in the estimated effects that can be explained should be used to estimate the active control effect in the setting of the non-inferiority trial along with any further effect modification anticipated in the setting of the non-inferiority trial. The precision of the estimate would be based on the within-trial variances of the estimated effects and the unexplained between-trial variability in the estimated effects of active control therapy.

### 4.3.6 Adjusting over Effect Modifiers

There may be key differences among the historical trials that attribute to differences in the estimated effects across clinical trials. For example, when the active control effect varies greatly across meaningful subgroups and the relative frequencies of these subgroups vary across the studies, there will be heterogeneous overall true effects of the active control across studies. In these instances, an estimate of the active control effect in the setting of the non-inferiority trial will not be relevant if it does not consider the heterogeneity of the effects across meaningful subgroups and the likely relative frequencies of these subgroups in the non-inferiority trial. Zhang[24] considered the problem of estimating the active control effect in the setting of the non-inferiority trial when the effects are heterogeneous across important subgroups or covariates (i.e., effect modifiers) by adjusting the historical effect of the active control according to the distribution of the effect modifiers/covariates within the non-inferiority trial. When the estimated effect, $\hat{\gamma}(\mathbf{x})$, for any given set of covariates, $\mathbf{x}$, can be determined from the previous trial comparing the active control to placebo, then the estimated effect of the active control in the setting of the non-inferiority trial, $\hat{\gamma}_{EM}$, is the average of the $\hat{\gamma}(\mathbf{x})$'s across all subjects in the non-inferiority trial (i.e., $\hat{\gamma}_{EM} = \sum_i \hat{\gamma}(\mathbf{x}_i)/n$).

The standard deviation for the resulting estimator of the active control effect is larger than the corresponding standard deviation for the estimator not based on a covariate adjustment.

In the absence of other biases previously discussed, the unbiased application of this estimated effect to the non-inferiority trial requires the conditional constancy assumption that for any given set of covariates the conditional active control effect is constant across all studies (including the non-inferiority trial) and that all effect modifiers have been accounted for. There should be biological plausibility that a covariate is an effect modifier with preferably reproduced results on the effect size of the active control. Selection of a covariate should not be based on data dredging by selecting an arbitrary covariate that just happens to have differing observed effects across its subgroups.

Such a procedure is most relevant when multiple trials evaluating the active control have demonstrated similar heterogeneous effects. When there are no underlying differences in the effects across subgroups, there will always be some anticipated difference in the estimated effects across subgroups. Observing similar but small differences in the estimated effects within subgroups in two, three, or even four trials may not be strong evidence of heterogeneous effects (or at least meaningful heterogeneous effects that would deserve attention). A conservative approach may select the smaller of the margin (or the more conservative estimation of the active control effect) from an approach that adjusts the estimated active control effect by the relative frequencies of important subgroups and from an approach of homogeneous effects across subgroups.

This problem of heterogeneous effects across subgroups cannot be solved by using adjusted analyses within each study, as such analyses either assume homogeneous effects across the corresponding subgroups, or weights the effects by the corresponding relative frequencies, which would be different across trials (leading to heterogeneous effects across trials).

However, when the effect of the active control varies across important subgroups, a non-inferiority or any efficacy conclusion overall is really a conclusion on an overall or weighted-average result with the weights being the relative frequencies of the important subgroups. A conclusion of non-inferiority or any efficacy for every meaningful subgroup requires individual non-inferiority comparisons for each subgroup. Unless the results are quite marked, it is very difficult to interpret subgroup analyses in a two-arm non-inferiority trial.

## 4.4 Bayesian Meta-Analyses

For a Bayesian fixed-effects meta-analysis, the output is a posterior distribution for the common treatment effect across studies, $\gamma$. As this would be a meta-analysis of the results from possibly all known studies, it may make sense to use a noninformative prior distribution for $\gamma$. When appropriate, the within-study estimators $\hat{\gamma}_1, \hat{\gamma}_2, \ldots, \hat{\gamma}_k$ may be modeled as independent and normally distributed with common mean $\gamma$ and respective variances of $\sigma_1^2, \sigma_2^2, \ldots, \sigma_k^2$. The variances $\sigma_1^2, \sigma_2^2, \ldots, \sigma_k^2$ may be regarded as known or as having some prior distribution. In the known variances case, $\gamma$ has a normal posterior distribution with mean equal to the observed value of $\sum_{i=1}^{k} (1/\sigma_i^2)\hat{\gamma}_i \Big/ \sum_{i=1}^{k} (1/\sigma_i^2)$ and variance equal to $1/\sum_{i=1}^{k} (1/\sigma_i^2)$. When the constancy assumption holds, the derived posterior distribution for the effect of the active control can validly be used in the setting of the non-inferiority trial as the distribution for the effect of the active control. If appropriate the active control effect can be discounted when applied to the setting of the non-inferiority trial.

For a Bayesian analog to a random-effects meta-analysis, the output can be either a posterior distribution for the random within-study treatment effect, $\gamma_{k+1}$, or a posterior distribution for the mean treatment effect across studies (i.e., the global mean), $\gamma$. Let $\psi = \tau^2$. We will consider an improper prior distribution for $(\gamma, \psi)$ whose density depends only on the value of $\psi$ (i.e., $g(\gamma, \psi) = j(\psi)$). Conditional on $(\gamma, \psi)$, the true within-study effects, $\gamma_1, \gamma_2, \ldots, \gamma_k$, are assumed to be a random sample from a normal distribution having mean $\gamma$ and variance $\psi$. Conditional on $\gamma_1, \gamma_2, \ldots, \gamma_k, \hat{\gamma}_1, \hat{\gamma}_2, \ldots, \hat{\gamma}_k$ are independently normally distributed, where $\hat{\gamma}_1$ has a normal distribution with mean $\gamma_i$ and variance $\sigma_i^2$ for $i = 1, \ldots, k$. The variances $\sigma_1^2, \sigma_2^2, \ldots, \sigma_k^2$ may be regarded as known or as having some prior distribution. In the known variances case, the joint posterior distribution for $(\gamma, \psi)$ is determined conditional on the observed values, $x_1, x_2, \ldots, x_k$, of $\hat{\gamma}_1, \hat{\gamma}_2, \ldots, \hat{\gamma}_k$. The joint posterior density will factor into the product of the marginal distribution for $\psi$ and a normal conditional distribution for $\gamma$ given $\psi$ having a mean equal to $\sum_{i=1}^{k} (\sigma_i^2 + \psi)^{-1} x_i \Big/ \sum_{i=1}^{k} (\sigma_i^2 + \psi)^{-1}$

and variance equal to $1/\sum_{i=1}^{k}(\sigma_i^2+\psi)^{-1}$. The density for the marginal distri-

bution for $\psi$ is proportional to $\exp[-(1/2)q(\psi)]\prod_{i=1}^{k}(\sigma_i^2+\psi)^{-1/2}\times j(\psi)$, where

$q(\psi)=\sum_{i=1}^{k}x_i^2(\sigma_i^2+\psi)^{-1}-\left(\sum_{i=1}^{k}x_i(\sigma_i^2+\psi)^{-1}\right)^2/\sum_{i=1}^{k}(\sigma_i^2+\psi)^{-1}$. For simulat-

ing probabilities involving $\gamma$, a random value for $\psi$, $\psi_r$, can be selected from its posterior distribution and then a random value for $\gamma$ can be taken from a

normal distribution having mean equal to $\sum_{i=1}^{k}(\sigma_i^2+\psi_r)^{-1}x_i/\sum_{i=1}^{k}(\sigma_i^2+\psi_r)^{-1}$

and variance equal to $1/\sum_{i=1}^{k}(\sigma_i^2+\psi_r)^{-1}$.

Alternatively, the modeling can use a posterior distribution for $\psi$ based on some prior distribution and the sampling distribution of a chosen estimator of $\psi$, while still letting $\gamma$ given $\psi$ have a normal distribution with mean equal

to $\sum_{i=1}^{k}(\sigma_i^2+\psi)^{-1}x_i/\sum_{i=1}^{k}(\sigma_i^2+\psi)^{-1}$.

Under the assumption that the same model applies for the next trial or the non-inferiority trial (i.e., a form of the constancy assumption), the distribution for the treatment or active control effect in the next study, $\gamma_{k+1}$, is based on the posterior distribution for $(\gamma,\psi)$ and that conditional on $(\gamma,\psi)$, $\gamma_{k+1}$ has a normal distribution with mean $\gamma$ and variance $\psi$. Thus, the distribution for $\gamma_{k+1}$ can be approximated by further taking a random value from the normal distribution with mean equal to the simulated value for $\gamma$ and variance equal to the simulated value for $\psi$. As noted earlier, if appropriate the active control effect can be discounted when applied to the setting of the non-inferiority trial.

In the cases where the variances $\sigma_1^2,\sigma_2^2,\ldots,\sigma_k^2$ are unknown, the joint posterior distribution for $(\gamma_i,\sigma_i^2)$ can be determined for each $i = 1, 2, \ldots, k$. For continuous data, Section 12.2.4 provides an example of a joint posterior distribution for the mean and variance. Approximating the posterior distribution for $\gamma$ involves:

1. Simulating values for $(\gamma_i,\sigma_i^2),\ldots,(\gamma_k,\sigma_k^2)$ from their respective joint posterior distributions.

2. Simulating a value $\psi_r$ from the posterior distribution of $\psi = \tau^2$ based on the simulated values for $(\gamma_1,\sigma_1^2),\ldots,(\gamma_k,\sigma_k^2)$.

3. Simulating a value for $\gamma$, $\gamma_r$, from a normal distribution with mean

$\sum_{i=1}^{k}(\sigma_i^2+\psi_r)^{-1}x_i/\sum_{i=1}^{k}(\sigma_i^2+\psi_r)^{-1}$ and variance $1/\sum_{i=1}^{k}(\sigma_i^2+\psi_r)^{-1}$.

To approximate the distribution for $\gamma_{k+1}$ further, this entails

4. Simulating a value for $\gamma_{k+1}$ from a normal distribution with mean $\gamma_r$ and variance $\psi_r$.

# References

1. U.S. Food and Drug Administration, Guidance for industry: Non-inferiority clinical trials (draft guidance), March 2010.
2. International Conference on Harmonization of Technical Requirements for Registration of Pharmaceuticals for Human Use (ICH) E-10: Guidance on choice of control group in clinical trials, 2000, at http://www.ich.org/cache/compo/475-272-1.html#E4.
3. Prentice, R.L. et al., Combined postmenopausal hormone therapy and cardiovascular disease: Toward resolving the discrepancy between observational studies and the Women's Health Initiative clinical trial, *Am. J. Epidemiol.*, 162, 404–414, 2005.
4. Fleming, T.R. et al., Some essential considerations in the design and conduct of non-inferiority trials, submitted manuscript, 2010.
5. Wang, S.-J., Hung, H.M.J., and Tsong, Y., Utility and pitfalls of some statistical methods in active controlled clinical trials, *Control. Clin. Trials*, 23, 15–28, 2002.
6. Hedges, L.V., Modeling publication selection effects in meta-analysis, *Stat. Sci.*, 7, 246–255, 1992.
7. Dear, K.B. and Begg, C.B., An approach for assessing publication bias prior to performing meta-analysis, *Stat. Sci.*, 7, 237–245, 1992.
8. Sterling, T.D., Rosenbaum, W.L., and Weinkam, J.J., Publication decisions revisited: The effect of the outcome of statistical tests on the decision to publish and vice-versa. *Am. Stat.*, 49, 108–112, 1995.
9. Light, R.J. and Pillemer, D.B., *Summing Up: The Science of Reviewing Research*, Harvard University Press, Boston, MA, 1984.
10. U.S. Food and Drug Administration Oncologic Drugs Advisory Committee meeting, July 27, 2004, transcript, at http://www.fda.gov/ohrms/dockets/ac/04/transcripts/2004-4060T1.pdf.
11. Product label for Taxotere, at http://www.accessdata.fda.gov/drugsatfda_docs/label/2010/020449s059lbl.pdf.
12. Statistical review of NDA 20449/S11 dated December 15, 1999, at http://www.accessdata.fda.gov/drugsatfda_docs/nda/99/20449-S011_TAXOTERE_statr.pdf.
13. Glass, G.V., Primary, secondary and meta-analysis of research, *Educ. Res.*, 5, 3–8, 1976.
14. Follmann, D.A. and Proschan, M.A., Valid inferences in random effects meta-analysis, *Biometrics*, 55, 732–737, 1999.
15. Rothmann, M.D. et al., Missing data in biologic oncology products, *J. Biopharm. Stat.*, 19, 1074–1084, 2009.
16. DerSimonian, R. and Laird, N., Meta-analysis in clinical trials, *Control. Clin. Trials*, 7, 177–188, 1986.
17. Jackson, D., The power of the standard test for the presence of heterogeneity in meta-analysis, *Stat. Med.*, 25, 2688–2699, 2006.
18. Yusuf, S. et al., Beta blockade during and after myocardial infarction: An overview of the randomized trials. *Prog. Cardiovasc. Dis.*, 27, 335–371, 1985.
19. Peto, R., Why do we need systematic overviews of randomised trials? *Stat. Med.*, 6, 233–240, 1987.

20. Greenland, S. and Salvan, A., Bias in the one-step method for pooling study results, *Stat. Med.*, 9, 247–252, 1990.
21. Larholt, K., Tsiatis, A.A., and Gelber, R.D., Variability of coverage probabilities when applying a random effects methodology for meta-analysis, Harvard School of Public Health Department of Biostatistics, unpublished, 1990.
22. Ziegler, S., Koch, A., and Victor, N., Deficits and remedy of the standard random effects methods in meta-analysis, *Methods Inform. Med.*, 40, 148–155, 2001.
23. Biggerstaff, B.J. and Tweedie, R.L., Incorporating variability in estimates of heterogeneity in the random effects model in meta-analysis. *Stat. Med.*, 16, 753–768, 1997.
24. Zhang, Z., Covariate-adjusted putative placebo analysis in active-controlled clinical trials, *Stat. Biopharm. Res.*, 1, 279–290, 2009.

# 5

## *Across-Trials Analysis Methods*

### 5.1 Introduction

A non-inferiority analysis is frequently conducted based on the determination of a non-inferiority margin or threshold. The choice of the margin should depend on prior experience of the estimated effect of the active control in adequate, well-controlled trials, and account for regression-to-the-mean bias, effect modification, and clinical judgment. The non-inferiority margin must be small enough to preclude that a placebo (or a treatment that is no better than placebo on a given endpoint) is noninferior to the active control. Other concerns about the non-inferiority margin might make the margin even smaller, but it should not be larger than the smallest anticipated difference between a placebo and the active control in the setting of the non-inferiority trial.

From experience there are basically two philosophies in constructing a non-inferiority analysis. One philosophy involves making adjustments to the estimation of the active control effect to account for biases, effect modification, and any additional uncertainty, and then use a test procedure that targets a desired type I–like error rate. The other philosophy involves applying a conservative method of analysis (e.g., comparing the most conservative limits of 95% confidence intervals) that includes the results from an unadjusted estimation of the active control effect and from the non-inferiority trial. The hope is that the conservative method will account for any biases in the estimate of the active control effect and any deviation from the constancy assumption.

There will be instances when the non-inferiority analysis will not be based on either philosophy. For example, the clinical judgment of the unacceptable amount of loss of the active control effect is necessarily smaller than determined from either philosophy. Another exception can occur when there is great heterogeneity in the effect of the control in previous studies. If the heterogeneity cannot be explained, the non-inferiority analysis may need to consider this heterogeneity and how small the active control effect may need to be in the non-inferiority trial. If the active control therapy has not regularly shown efficacy in clinical trials, the non-inferiority margin may need to be zero, meaning that the experimental therapy must show superiority to the active control to be deemed effective.

In this chapter, we discuss two confidence interval and synthesis methods for non-inferiority testing in an active-controlled trial. These methods are compared in Section 5.4. Additionally, the type I error rates are also assessed in Section 5.4, including under practical models where the estimation of the active control effect is subject to regression-to-the mean bias. In Section 5.5, we compare the results of two confidence interval and synthesis methods with an example in oncology.

## 5.2 Two Confidence Interval Approaches

### 5.2.1 Hypotheses and Tests

It is tempting to use the point estimate of the historical effect of the active control versus placebo as the true effect of the active control therapy in the non-inferiority trial. Such an inference allows for considerable room for error, as half the time an estimated effect overestimates the true effect. Additionally, by using the point estimate of the historical effect of the active control versus placebo as the margin in the non-inferiority trial, the probability of concluding that an ineffective experimental therapy is effective can exceed 50% even when the constancy assumption holds.[1,2]

To incorporate variability in the point estimate when choosing the margin, the lower bound of a confidence interval for the active control effect has been considered a surrogate for the unknown active control effect. The selected non-inferiority margin may be smaller than this lower bound of the confidence interval when it is desired to show that the experimental therapy has more than some minimal amount of efficacy, or to account for uncertainty in whether the constancy assumption holds. For an endpoint such as mortality or irreversible morbidity, allowing an investigational treatment to be non-inferior if it is only a little better than placebo is unacceptable if an alternative therapy exists that provides much superior benefit. In these cases, it is therefore common to take a fraction, such as one-half, of the lower bound of the confidence interval as the non-inferiority margin. This has been called "preservation of effect" in that it guarantees that some fraction of the effect of the active control is preserved.[3] With such a margin and accounting for any regression-to-the mean bias and effect modification, there is, with high confidence, little chance that an experimental therapy having no effect compared with placebo would be called noninferior to the active control.

Temple and Ellenberg[4] described three possible margins, M0, M1, and M2. M0 (or just zero) is the margin used when the active control is not regularly superior to placebo. M1, the effectiveness margin, is the smallest effect the active control may have versus placebo in the setting of the non-inferiority trial. If the active control has an effect of M1 in the non-inferiority trial, the trial will have assay sensitivity in determining whether an experimental therapy is effective or ineffective provided adequate study conduct. The non-inferiority margin M2 is a fraction of M1 chosen to assure that the experimental therapy retains at least some desired amount of the active control effect. The margins of M1 and M2 are used respectively for two objectives: (1) demonstrating that the experimental therapy is superior to placebo and (2) demonstrating that the experimental therapy is not unacceptably worse than placebo.

M2 has been treated as a fixed margin, despite often being based on or influenced by the estimated active control effect. In this section, we will consider testing involving statistical hypotheses that treat M2 as a fixed value. In Section 5.4 on evaluating error rates, we will treat M1 and M2 as realized values involving the estimated active control effect. We will consider comparing the treatment arms using metrics based on undesirable outcomes. This includes the difference in means where the smaller the value the better, differences in proportions on an undesirable event, the log-relative risk of an undesirable event, and the log-hazard ratio where the longer the time the better. Then the hypotheses of interest are expressed as

$$H_o: \beta_N \geq M2 \text{ vs. } H_a: \beta_N < M2 \qquad (5.1)$$

where $\beta_N$ is the experimental therapy versus the active control therapy (i.e., E–C or E/C) parameter of interest (i.e., the true treatment difference) in the non-inferiority trial. The inequalities in the hypotheses in Equation 5.1 would be reversed for "positive" or desirable outcomes (e.g., adverse cure, prevention, time-to-relief). For these cases, each hypothesis is expressed by multiplying each side of the inequality by –1 and defining the parameter in terms of the active control therapy versus the experimental therapy.

Since M1 and M2 are based on the estimated effect of the active control, they are realizations of random quantities, not a priori. The hypotheses in Equation 5.1 are surrogate hypotheses for whether the experimental therapy is unacceptably worse than the active control in the setting of the non-inferiority trial. The hope is that rejecting $H_o$ in Expression 5.1 and concluding $H_a$ will imply that the experimental therapy is effective, with an effect that is not unacceptably worse than the active control. The null hypothesis in Expression 5.1 is rejected and non-inferiority is concluded when the upper limit of a $100(1 - \alpha)\%$ confidence interval for $\beta_N$ is less than M2. Normal-based confidence intervals are often used, so $H_o$ would be rejected when $\hat{\beta}_N + z_{\alpha/2} s_N < M2$, where $\hat{\beta}_N$ is the estimated value for $\beta_N$ and $s_N$ is the estimated standard deviation for $\hat{\beta}_N$.

Additionally, it is popular to define M1 as the lower limit of a $100(1 - \gamma)\%$ confidence interval for the historical active control effect, $\beta_H$ (expressed in terms of placebo vs. active control). For some $0 \le \lambda_o \le 1$, we would then have M2 = $(1 - \lambda_o) \times$ M1. Such approaches have been referred to as "fixed-margin" approaches. The non-inferiority analysis is then expressed as a two–confidence interval approach. For estimators that are normally or approximately distributed, non-inferiority would be concluded when

$$\hat{\beta}_N + z_{\alpha/2}s_N < (1 - \lambda_o)(\hat{\beta}_H - z_{\gamma/2}s_H) \tag{5.2}$$

Consistent with Hung, Wang, and O'Neill,[5] an "Y–X method" will refer to a two–confidence interval procedure where a two-sided $Y\%$ confidence interval is determined from the non-inferiority trial and the active control effect is based on a two-sided $X\%$ confidence interval. The definitions of $Y$ and $X$ are the reverse in the U.S. Food and Drug Administration (FDA) draft guidance.[6]

Using 95% confidence intervals for both the historical effect of the active control therapy and for the comparison of the experimental and active control therapies in the non-inferiority trial is common. We will refer to this approach as the 95–95 method or approach. This approach has been described as comparing the two statistically worst cases. For the 95–95 approach, Rothmann et al.[1] showed that when the constancy assumption holds, the one-sided type I error rate is between 0.0027 and 0.025 in falsely concluding that an ineffective therapy is effective. Sankoh[7] called the two–confidence interval approach "uniformly ultraconservative" and preferring to use a fraction of the point estimate instead of the lower bound of a confidence interval for the active control effect. Although using a fraction of the lower bound of a confidence interval may be conservative in many situations, it may not be conservative (and certainly not uniformly ultraconservative) in all situations, particularly in indications where regression-to-the-mean bias and/or effect modification are major concerns.

Such a margin (the lower bound of the 95% confidence interval or some fraction thereof) will be conservative when the constancy assumption holds. However, in many cases, the constancy assumption does not hold, or cannot be proven to hold. The use of the lower limit of the 95% confidence interval for the estimated active control effect provides some adjustment for bias and deviation from the constancy. Subjects enrolled in the current study may be fundamentally different from subjects enrolled in the historical study, owing to changes in diagnosis or standards of concomitant care since the historical study was completed; or the disease is fundamentally different (such as infectious diseases, which are known to change over time as they adapt in response to medications); or logistics differ (when a study is run in a different set of geographic sites than the historical comparison used). When the constancy assumption may not hold, choosing a fraction of the lower bound of a confidence interval for the historical treatment effect can provide an allowance for deviation from the constancy assumption.

The width of the confidence interval for the historical effect of the active control will depend on the sample sizes of the historical studies. A large estimated effect for the active control therapy from large studies may produce a confidence interval with a lower bound that is a large effect, and thus require a smaller sample size for the non-inferiority trial to rule out a difference of practical importance. Conversely, a single small study may produce a confidence interval with a lower bound that corresponds to a small effect, even if the point estimate of the active control effect was large, and thus require a large sample size for the non-inferiority trial to rule out an appropriate non-inferiority margin. In such cases, it is tempting, although it may not be possible, to increase the margin because of the lack of precision in the estimate of the historical treatment effect. When warranted, the confidence level for the historical effect of the active control therapy can be adjusted to be higher for a more conservative, smaller margin or be adjusted lower for a more liberal, larger margin. Hauck and Anderson[8] suggested utilizing the lower bound of a confidence interval with a confidence level of 68–90%. The lower confidence level will lead to a larger lower bound, and hence a larger non-inferiority margin.

Fixed-effects and random-effects meta-analyses have been used in determining the confidence interval for the historical active control effect. Between-trial variability in the active control is a concern especially when the heterogeneity in the active control effect cannot be explained. When there is a single study, the heterogeneity in the active control effect cannot be assessed. Also, with the lack of a reproduced effect size, the significant or highly significant result from a single trial may have a large associated regression-to-the-mean bias and thus greatly overstate the true active control effect. The existence of multiple studies that provide consistent estimates of the active control effect gives assurance that the regression-to-the-mean bias is small, and that the meta-analysis reliably estimates the active control effect when the historical effect of the active control applies in the setting of the non-inferiority trial. Concerns involving the use of the estimated active control effect to the setting of the non-inferiority trial may lead to either discounting the estimated active control effect (i.e., discounting the lower limit of the confidence interval for the active control effect) or basing the non-inferiority margin on a larger-level confidence interval for the active control effect.

If the multiple historical comparisons of the active control to placebo provide inconsistent estimates of the active control effect, confidence in a common active control effect decreases. In such a case, the choice of non-inferiority margin should consider the between-trial variability in the active control effect. When a random-effects meta-analysis is used for the estimation of the active control effect, Lawrence[9] proposes using a 95% prediction interval for the active control effect in a next random trial as a replacement in the 95–95 method for the 95% confidence interval for the active control effect.

Example 5.1 summarizes one of the first two confidence interval procedures that involved thrombolytic products.

**Example 5.1**

As life-saving therapeutics became available in thrombolytics, placebo-controlled
trials could not ethically be done. The Cardio-Renal advisory committee in 1992
recommended that a new thrombolytic must retain 50% of the control thrombolytic's effect.[10]

The Assent-2 trial compared the experimental agent tenecteplase to the active
control of activase on 30-day mortality. There are no direct comparisons of activase with placebo. The historical effect of activase on 30-day mortality was evaluated on the basis of a meta-analysis that included a comparison of activase to
streptokinase from the Global Utilization of Streptokinase and t-PA for Occluded
Coronary Arteries (GUSTO) study and four studies comparing streptokinase to
placebo.[10] As there were concerns that the GUSTO study was not blinded, a
weighted meta-analysis was used that gave weight to the GUSTO study, consistent
with constructing a one-sided 99.5% confidence interval.[10] The resulting non-
inferiority threshold was based on half of the lower limit of a one-sided 95%
confidence interval for the historical effect of activase from the meta-analysis. This
approach yielded a non-inferiority threshold for the experimental versus activase
relative risk of 30-day mortality of 1.143 (which corresponds to a non-inferiority
margin on the log-relative risk of 0.134). The Assent-2 trial randomized 16,949
subjects between the tenecteplase and activase arms. For the primary analysis,
the estimated percentages for the 30-day mortality were 6.18% and 6.15% for the
tenecteplase and activase arms, respectively.[11] The relative risk was 1.004, with a
corresponding 90% confidence interval of 0.914–1.104. As 1.104 < 1.143, tenecteplase was deemed noninferior to activase on 30-day mortality.

For an application in second-line non-small cell lung cancer (NSCLC),
Example 5.2 revisits Example 4.3 in Section 4.2.4 and illustrates using a 95%
confidence interval for the historical active control effect to define a margin
or margins. The determined margins are then applied to a later trial of pemetrexed versus docetaxel in second-line NSCLC.

**Example 5.2**

To illustrate the use of the margins just discussed, we revisit Example 4.3, which
considered six approaches for estimating the effect of docetaxel versus best supportive care (BSC) on overall survival in second-line NSCLC. Table 4.3 provided
the estimates and 95% confidence intervals for the docetaxel versus BSC hazard ratios. Here the BSC versus docetaxel log-hazard ratio is the docetaxel effect
parameter, $\beta_H$. For each of the six approaches using Table 4.3, Table 5.1 gives the
95% confidence intervals for $\beta_H$ and the corresponding margins obtained from the
confidence interval where M2 represents 50% of M1 where M1 is the lower limit
of the 95% confidence interval for $\beta_H$. For approaches 2a and 2b, a superiority
comparison to docetaxel would be required for a new investigation agent.

**TABLE 5.1**

95% Confidence Intervals for Docetaxel Effect and Corresponding Margins

| Approach | 95% Confidence Interval for $\beta_H$ | Margins |
|---|---|---|
| 1 | (0.128, 1.050) | M1 = 0.128, M2 = 0.064 |
| 2a | (−0.018, 0.611) | M0 = 0 |
| 2b | (−0.015, 0.360) | M0 = 0 |
| 3 | (0.082, 0.764) | M1 = 0.082, M2 = 0.041 |
| 4a | (0.064, 0.646) | M1 = 0.064, M2 = 0.032 |
| 4b | (0.015, 0.624) | M1 = 0.015, M2 = 0.0075 |

The JMEI trial studied the use of pemetrexed against the active control of doc-etaxel at a dose of 75 mg/m$^2$ with subjects in second-line NSCLC. From the FDA Oncologic Drugs Advisory Committee meeting transcript,[12] the 95% confidence interval for the pemetrexed versus docetaxel hazard ratio in the JMEI study is 0.817–1.204. Taking the natural logarithms of each limit in the confidence interval gives a 95% confidence interval for the pemetrexed versus docetaxel log-hazard ratio, $\beta_N$, of −0.202 to 0.186. The upper limit of the 95% confidence interval for $\beta_N$ of 0.186 exceeds every margin specified in Table 5.1. Thus when a margin is based on a 95% confidence interval for the docetaxel effect versus BSC, the results from the JMEI trial fail to conclude non-inferiority to docetaxel, regardless of the approach used to estimate the docetaxel effect.

*Reducing the Potential for Biocreep.* In general, when possible, the seemingly most effective, available standard of care should be used as the active control in a non-inferiority trial. However, as the selected standard of care is the therapy or regimen that has the best estimated effect, the estimation of the effect of that standard of care will have a regression-to-the-mean bias. This bias should be accounted for by either making an appropriate adjustment to the estimation of its effect or including the estimated effects of all potential candidates for a standard of care into the meta-analysis.

For a given indication, once the first non-inferiority trial has established a criterion for non-inferiority, it may be reasonable that all future non-inferiority trials have the same or more stringent criterion regardless of the active control used in the trial. For example, suppose that a margin of 5 days was used for the duration of an adverse event for the original active control (A) in non-inferiority trials. If another therapy (B) is to be used as an active control in a non-inferiority trial and the 5-day margin to A is still relevant, the non-inferiority margin for B as a control, $\delta$, should be such that it guar-antees that if the experimental therapy (C) is noninferior to B with margin $\delta$, then C is noninferior to A with a margin of 5 days. Suppose B was previously compared with A in randomized trials and the 95% confidence interval for the difference in days of the mean durations was −0.8 to 1.6. Using the phi-losophy of a 95–95 method, a margin of 5.0 − 1.6 = 3.4 days may be justified for B as the control therapy in a non-inferiority trial. In practice, it may also be necessary to make further adjustments due to effect modifiers/deviations from constancy.

This way of establishing margins for the next generation of therapies so that non-inferiority is maintained to the original active control will avoid a biocreep. The use of the same margin for a next-generation active control therapy that is not superior to the original active control is susceptible to a biocreep. If the next-generation therapy demonstrated superiority to the original active control, then the margin to the original active control will probably be no longer relevant.

## 5.3 Synthesis Methods

### 5.3.1 Introduction

A synthesis method is a non-inferiority test procedure whose decision is based on the difference in effects of the active control and experimental arms in the non-inferiority trial and an external estimate of the effect of the active control along with the corresponding standard errors. A synthesis method usually compares a test statistic on the basis of the aforementioned estimates, their standard errors, and a retention fraction, to a critical value. The estimate of the active control effect is based on the results of previous clinical trials studying the active control and may be adjusted to fit the setting of the non-inferiority trial. The determination of non-inferiority usually requires that the experimental arm retains more than some prespecified fraction of the effect of the active control arm. The test procedures are generally designed to maintain a desired type I error rate under the assumption that the effect of the active control in the non-inferiority trial has been unbiasedly estimated. Particular synthesis methods have been proposed in many papers.[1,13–16]

There are reasons why it is appropriate to require that a new experimental agent not only be efficacious but also retain some minimal part of the effect of the active control therapy. When it is unethical to do a placebo-controlled trial, it is probably also unethical to use a therapy that has very little effect in a clinical trial. Additionally, the Reinventing Regulation of Drugs and Medical Devices of April 1995 stated that[17]

> In evaluating the safety of a new drug or Class III device, the Agency weighs the demonstrated effectiveness of the product against its risks to determine whether the benefits outweigh the risks. This weighing process also takes into account information such as the seriousness and outcome of the disease, the presence and adequacy of existing treatments, and adverse reaction data....
>
> In certain circumstances, however, it may be important to consider whether a new product is less effective than available alternative

therapies, when less effectiveness could present a danger to the patient or to the public. For example, it is essential for public health protection that a new therapy be as effective as alternatives that are already approved for marketing when

1. The disease to be treated is life-threatening or capable of causing irreversible morbidity (e.g., stroke or heart attack)
2. The disease to be treated is a contagious illness that poses serious consequences to the health of others (e.g., sexually transmitted disease)

This provides a basis for some indications on requiring that a new therapy have efficacy greater than some minimal threshold. In the typical synthesis testing, that threshold is regarded as a prespecified fraction of the effect of the active control. Snapinn and Jiang[18] expressed concern that a requirement that the experimental therapy in a non-inferiority trial retain more than some fraction of the effect of the active control creates a higher bar for approval than was required for the active control, and that such a requirement may prevent the approval of superior treatments to the active control.

In this section we discuss definitions for the proportion of the active control effect that the experimental therapy retains, possible corresponding sets of non-inferiority hypotheses that can be tested, frequentist and Bayesian procedures, and respective issues.

## 5.3.2 Retention Fraction and Hypotheses

Holmgren[13] introduced the use of a synthesis test for showing that the experimental therapy retained more than a prespecified fraction of the historical effect of the active control. More specifically, the inference involved the experimental to active control relative risk in the non-inferiority trial, $\beta_N$, and a historical placebo to active control relative risk, $\beta_H$. As a treatment versus placebo relative risk of 1 is equivalent to no treatment effect, in Holmgren's relative risk case, $\beta_H - 1$ represents the active control effect versus placebo. When the constancy assumption holds and $\beta_H - 1 > 0$, the proportion of the active control effect that is retained by the experimental therapy, referred to as an *retention fraction*, is given by

$$\lambda = \frac{\beta_H - \beta_N}{\beta_H - 1} \tag{5.3}$$

The definition of the proportion of the active control effect that is retained by the experimental therapy in Equation 5.3 is referred to as a "arithmetic definition" in Rothmann's paper.[1] The definition of the retention fraction has been used for relative metrics—for example, a relative risk or a hazard ratio. However, for relative metrics, how two different possible values (e.g., $a$ and $b$)

for the metric compare (or statistically compare) depends on their ratio (i.e., $a/b$) not on their difference (i.e., $a − b$).

For undesirable outcomes (i.e., smaller probabilities of "success" are better) with a prespecified retention fraction of $\lambda_o$, the null and alternative hypotheses are expressed as

$$H_o: \beta_N \geq \lambda_o + (1 − \lambda_o)\beta_H \text{ vs. } H_a: \beta_N < \lambda_o + (1 − \lambda_o)\beta_H \tag{5.4}$$

where $0 \leq \lambda_o \leq 1$. When it is assumed that $\beta_H − 1 > 0$, the alternative hypothesis is that the experimental therapy retains more than $100\lambda_o\%$ of the historical effect of the active control.

The utility of a conclusion that the experimental therapy retains more than a prespecified fraction of the "historical effect" of the active control therapy (which is the formal conclusion from rejecting the null hypothesis in Expression 5.4) depends on whether that historical estimated active control effect is relevant to the non-inferiority trial. This would require the estimation of the historical active control effect to be unbiased for the active control effect in the setting of the non-inferiority trial (i.e., the constancy assumption holds) or to have very low bias.

Statistical inferences involving non-inferiority and a retention fraction are not usually based on relative metrics, but instead on "absolute" metrics (differences have meaning). For absolute metrics, how two different possible values (e.g., $a$ and $b$) for the metric compare (or statistically compare) depends on their difference (i.e., $a − b$) and not on their ratio (i.e., $a/b$). Some absolute metrics include a difference in means, a difference in proportions, a log-relative risk, or a log-hazard ratio. The same notation of $\beta_N$ and $\beta_H$ for relative metrics will be used as the parameters for absolute metrics for comparing the experimental therapy to the active control in the non-inferiority trial and for the historical comparison of the placebo to the active control, respectively.

For a metric such as a difference in means, a difference in proportions, a log-relative risk, or a log-hazard ratio, a treatment versus placebo value of zero is equivalent to no treatment effect. The historical effect of the active control versus placebo is then given by $\beta_H$. For a log-hazard ratio where the greater the time to the event the better, $\beta_H$ would represent the placebo versus active control hazard ratio and $\beta_N$ would represent the experimental to active control log-hazard ratio in the non-inferiority trial. The definitions for $\beta_H$ and $\beta_N$ would be analogous for a relative risk or difference in proportions of an undesirable event, or a difference in means where the smaller value the better. For time to event endpoints where the smaller value the better, risk ratios or proportions of desirable events, and difference in means where the larger the value the better, the parameter $\beta_H$ would be in terms of the active control to placebo and the parameter $\beta_H$ would be in terms of the active control to the experimental. When the constancy assumption holds and $\beta_H > 0$, the proportion of the active control effect that is retained by the experimental therapy or retention fraction is given by

$$\lambda = \frac{\beta_H - \beta_N}{\beta_H} \tag{5.5}$$

The definition of the retention fraction in Equation 5.5 is referred to as a "geometric" definition in Rothmann's paper.[1] This definition is also sometimes referred to as a "log-scale" definition. When the active control is effective, the arithmetic and geometric retention fractions agree at 0 and 1. When the arithmetic retention fraction is smaller than 1, the geometric retention fraction will be between zero and the arithmetic retention fraction. When the arithmetic retention fraction is greater than 1, the geometric retention fraction will be greater than the arithmetic retention fraction. Unless conditioning on the size of the active control effect is done, there is no functional relationship between the two retention fractions. Curves relating the two definitions of the retention fraction for three different active control effect sizes are provided in the paper of Hung et al.[19]

For testing that the experimental therapy maintains more than $100\lambda_0\%$ ($0 \leq \lambda_0 \leq 1$) of the active control effect based on Equation 5.5, the null and alternative hypotheses are expressed as

$$H_o: \beta_N \geq (1 - \lambda_0)\beta_H \text{ vs. } H_a: \beta_N < (1 - \lambda_0)\beta_H \tag{5.6}$$

When it is assumed that $\beta_H > 0$ the alternative hypothesis is that the experimental therapy retains more than $100\lambda_0\%$ of the historical effect of the active control.

The null and alternative hypotheses have also been expressed simply as

$$H_o: \lambda \leq \lambda_0 \text{ vs. } H_a: \lambda > \lambda_0 \tag{5.7}$$

where $0 < \lambda_0 \leq 1$. The corresponding hypotheses in Expression 5.6 and 5.7 agree when $\beta_H > 0$, but they disagree $\beta_H < 0$ and the definition of $\lambda$ in Equation 5.5 is extended to include cases where $\beta_H < 0$.

The proper formulation and testing procedure involving a retention fraction has been an issue. When the active control has a negative effect, the alternative hypothesis in Expression 5.6 contains possibilities where the placebo is better than the experimental therapy, which in turn is better than the active control (i.e., P > E > C).[20] Formulations of the hypotheses provided by Cheng et al.[20] are those in Expression 5.7 (above) through 5.9.

$$H_o: \beta_N - (1 - \lambda_0)\beta_H \geq 0 \text{ or } \beta_H \leq 0 \text{ vs. } H_a: \beta_N - (1 - \lambda_0)\beta_H < 0 \text{ and } \beta_H > 0 \tag{5.8}$$

$$H_o: \{\beta_N - (1 - \lambda_0)\beta_H \geq 0 \text{ or } \beta_H \leq 0\} \text{ and } \{\beta_N - \beta_H \geq 0 \text{ or } \beta_H \geq 0\} \text{ vs.}$$

$$H_a: \{\beta_N - (1 - \lambda_0)\beta_H < 0 \text{ and } \beta_H > 0\} \text{ or } \{\beta_N - \beta_H < 0 \text{ and } \beta_H < 0\} \tag{5.9}$$

However, the null hypothesis in Expression 5.7 contains possibilities where E > P > C, whereas the alternative hypothesis in Expression 5.7 contains possibilities where P > E > C. Likewise, the null hypothesis in Expression 5.8 contains possibilities where E > P > C. The alternative hypothesis in Expression 5.9 is the percentage retention analog for the experimental therapy to be both better than placebo and noninferior to the active control.

### 5.3.3 Synthesis Frequentist Procedures

#### 5.3.3.1 Relative Metrics

We will consider a relative metric—for example, a relative risk. The observed experimental to active control relative risk in the non-inferiority trial, $\hat{\beta}_N$, is used to estimate $\beta_N$. The historical estimator of the placebo to active control relative risk will be denoted by $\hat{\beta}_H$. The test procedure of Holmgren[13] uses a test statistic based on applying the delta method to $\log \hat{\beta}_N - \log(\lambda_o + (1 - \lambda_o)\hat{\beta}_H)$. The test statistic is

$$Z_1 = \frac{\log \hat{\beta}_N - \log(\lambda_o + (1 - \lambda_o)\hat{\beta}_H)}{\sqrt{\text{V\^ar}(\log \hat{\beta}_N) + ((1 - \lambda_o)\hat{\beta}_H/(\lambda_o + (1 - \lambda_o)\hat{\beta}_H))^2 \text{V\^ar}(\log \hat{\beta}_H)}} \qquad (5.10)$$

The test rejects the null hypothesis in Expression 5.4 and concludes non-inferiority when $Z_1 < -z_{\alpha/2}$ for some $0 < \alpha < 1$. When $\hat{\beta}_N$ and $\hat{\beta}_H$ are independent, $Z_1$ having an approximate standard normal distribution when $\lambda = \lambda_o$ depends on $\hat{\beta}_N$ having a normal distribution, and whether $\log(\lambda_o + (1 - \lambda_o)\hat{\beta}_H)$ has an approximate normal distribution with a variance reliably estimated by $(\lambda_o + (1 - \lambda_o)\hat{\beta}_H))^2 \text{V\^ar}(\log \hat{\beta}_H)$ or whether $\text{V\^ar}(\log \hat{\beta}_N)/[((1 - \lambda_o)\hat{\beta}_H/(\lambda_o + (1 - \lambda_o)\hat{\beta}_H))^2 \text{V\^ar}(\log \hat{\beta}_H)]$ is large.

The remainder of this section will focus on absolute metrics.

#### 5.3.3.2 Absolute Metrics

Hauck and Anderson[8] noted that the choice of a 90% or 95% confidence interval may be conservative for establishing a non-inferiority margin and that choosing a non-inferiority margin $\delta$ based on the results of prior studies makes $\delta$ a random variable, and thus the variability in $\delta$ should be considered. When the effect of the active control versus placebo in the setting of the non-inferiority trial also equals $\beta_H$, then $\beta_N - \beta_H$ is the effect of experimental therapy versus placebo. On the basis of this, Hauck and Anderson provided

$$\hat{\beta}_N - \hat{\beta}_H + 1.645\sqrt{s_N^2 + s_H^2} < 0 \qquad (5.11)$$

as a one-sided test that the experimental therapy is better than placebo. When the constancy assumption holds, the left-hand side in Expression 5.11 is the upper limit for the the two-sided 90% confidence interval for the difference between the experimental therapy and a placebo. This test procedure can be rewritten to compare the upper limit of a one-sided 95% confidence interval for $\beta_N$ with $\delta^* = \hat{\beta}_H - cs_H$, where $c = 1.645\sqrt{1+s_N^2/s_H^2} - 1.645s_N/s_H$. Non-inferiority is concluded when the upper limit of the two-sided 90% confidence interval for $\beta_N$ is less than $\delta^*$. A similar procedure can also be found in papers by Fisher and colleagues.[21,22]

The use of $\delta^*$ is contingent on the constancy assumption. Hauck and Anderson recommended discounting $\delta^*$ when it is believed that there may be between-trial variability in the active control effect. As there may be disagreement in the appropriate margin, Hauck and Anderson recommend reporting the 90% or the 95% two-sided confidence interval for $\beta_N$ to allow each individual to decide for themselves whether non-inferiority has been met.

When Expression 5.11 and the constancy assumption holds, the left-hand side provides the effect or minimal effect or a greater effect versus placebo that can be ruled out. For the experimental therapy to rule out the same minimal effect versus placebo as the active control has ruled out on the basis of a one-sided 95% confidence interval requires that

$$\hat{\beta}_N < -1.645 \times \left(\sqrt{s_N^2 + s_H^2} - s_H\right) < 0$$

For a one-sided $100(1 - \alpha/2)\%$ confidence interval, 1.645 is replaced with $z_{\alpha/2}$.

*The Standard Synthesis Method.* The standard synthesis method is based on the test statistic

$$Z_2 = \frac{\hat{\beta}_N - (1-\lambda_o)\hat{\beta}_H}{\sqrt{s_N^2 + (1-\lambda_o)^2 s_H^2}} \tag{5.12}$$

The test rejects the null hypothesis and concludes non-inferiority when $Z_2 < -z_{\alpha/2}$ for some $0 < \alpha < 1$. After correcting for differences in notation, we see that the test statistics $Z_1$ in Equation 5.10 and $Z_2$ in Equation 5.12 are equivalent when $\lambda_o = 0$ or 1. When the estimators $\hat{\beta}_N$ and $\hat{\beta}_H$ are independent, normally distributed, and unbiased, or approximately so, $Z_2$ will have a standard normal distribution or an approximate standard normal distribution when $\beta_N = (1 - \lambda_o)\beta_H$ with a type I error rate of approximately $\alpha/2$ for falsely concluding that an experimental therapy that retains $100\lambda_o\%$ of the active control effect retains more than $100\lambda_o\%$ of the active control effect. When $\hat{\beta}_H$ tends to overestimate (underestimate) the effect of the active control in the setting of the non-inferiority trial, the type I error rate will be inflated (deflated). If the sampling distributions for $\hat{\beta}_N - \hat{\beta}_H$ and $\hat{\beta}_N - (1-\lambda_o)\hat{\beta}_H$ are not normal distributions, these two tests can be modified to fit the appropriate sampling distributions.

A Fieller $100(1 - \alpha)\%$ confidence interval can be determined for $\lambda$.[1] The Fieller $100(1 - \alpha)\%$ confidence interval equals $\{\lambda_o: -z_{\alpha/2} < Z_2 < z_{\alpha/2}, -\infty \leq \lambda_o \leq \infty\}$.

The null hypothesis in Expression 5.6 is rejected whenever every value in the Fieller $100(1 - \alpha)\%$ confidence interval exceeds the prespecified value for $\lambda_o$.

If it is believed that the effect of the active control may have decreased, the historical estimated effect can by discounted by using $0 < \theta < 1$.[1] The test statistic $Z_2$ in Expression 5.6 would then be replaced with the test statistic $Z_2^*$ in Equation 5.13 where

$$Z_2^* = \frac{\hat{\beta}_N - (1 - \lambda_o)\theta\hat{\beta}_H}{\sqrt{s_N^2 + (1 - \lambda_o)^2\theta^2 s_H^2}} \qquad (5.13)$$

When the design of the non-inferiority trial is independent of the estimation of the active control effect and the estimator of that effect has an approximate normal distribution, then the standard synthesis test statistic will have an approximate normal distribution. If, however, the design (e.g., the sample size) of the non-inferiority trial is dependent on the estimation of the active control effect, the numerator of the synthesis test statistic is the difference of two uncorrelated but dependent quantities and the synthesis test statistic will not have an approximate standard normal distribution across the boundary of the null hypothesis.[2] The test statistic would have the form of

$$\frac{\hat{\beta}_N - (1 - \lambda_o)\hat{\beta}_H}{\sqrt{\hat{\sigma}_{\hat{\beta}_N}^2(\hat{\beta}_H) + (1 - \lambda_o)^2 s_H^2}}$$

where the true variance for the non-inferiority trial $\sigma_{\hat{\beta}_N}^2(\hat{\beta}_H)$ is a random variable that depends on the estimated active control effect, $\hat{\beta}_H$. Rothmann[2] assessed the type I error probability for testing the hypotheses in Expression 5.6 for two confidence interval methods and the standard synthesis method when the standard error from the non-inferiority trial depends on the estimated historical active control effect.

*Delta-Method Confidence Interval Approach.* When $\lambda_o = 0$, Hasselblad and Kong[15] recommended testing the hypotheses in Expression 5.6 on the basis of the test statistic $Z_2$ in Equation 5.12. However, when $0 < \lambda_o \leq 1$, Hasselblad and Kong[15] proposed a delta-method confidence interval test procedure. The estimator of $\lambda$ is given by $\hat{\lambda} = 1 - \hat{\beta}_N/\hat{\beta}_H$ and the estimated standard error is given by $S_{\hat{\lambda}} = \sqrt{(\hat{\beta}_N/\hat{\beta}_H)^2(s_N^2/\hat{\beta}_N^2 + s_H^2/\hat{\beta}_H^2)}$. The null hypothesis in Expression 5.7 is rejected, and non-inferiority is concluded when $\hat{\lambda} - z_{\alpha/2}S_{\hat{\lambda}} > \lambda_o$. The test is equivalent to rejecting the null hypothesis in Expression 5.7 when

$$Z_3 = \frac{\hat{\beta}_N - (1 - \lambda_o)\hat{\beta}_H}{\hat{\beta}_H S_{\hat{\lambda}}} < -z_{\alpha/2} \qquad (5.14)$$

As $Z_2$ in Equation 5.12 and $Z_3$ in Equation 5.14 have the same numerators, and $Z_2$ has an approximate standard normal distribution when $\lambda = \lambda_o$, how close the distribution of $Z_3$ is to a standard normal distribution may depend on the whether $R = \hat{\beta}_H S_\lambda / \sqrt{s_N^2 + (1 - \lambda_o)^2 s_H^2}$ (i.e., $Z_2/Z_3$) tends to be close to 1.

When $\lambda_o = 0$, $R$ will be greater than 1 with probability 1 and $Z_3$ would have a distribution more concentrated near zero than the distribution for $Z_2$.[23] It was noted from simulations that fairly large sample sizes may be needed for the ratio of two independent normally distributed quantities to have an approximate normal distribution.[23] In particular the ratio of the mean of $\hat{\beta}_H$ to its (estimated) standard deviation should be greater than 8 for the test based on the delta-method confidence interval to have approximately the desired type I error rate when $\hat{\beta}_H$ unbiasedly estimates the effect of the active control in the setting of the non-inferiority trial. We comment further on the distribution of $Z_3$ in Section 5.4.2 on comparing the different analysis methods.

In Example 5.3, synthesis procedures will be performed including the determination of Fieller confidence intervals for the proportion of the active control effect retained by the experimental therapy.

## Example 5.3

To illustrate the use of some of the synthesis methods just discussed, we revisit Example 4.3. The JMEI trial studied the use of pemetrexed against the active control of docetaxel at a dose of 75 mg/m² with subjects in second-line NSCLC. The endpoint of interest was overall survival. We use the result of approach 2b in Example 4.3 for the estimation of the docetaxel effect. From that approach, the estimated docetaxel versus BSC hazard ratio was 0.842, with a corresponding standard error for the log-hazard ratio estimator of 0.095 (95% confidence interval for the hazard ratio of 0.698, 1.015). From the FDA Oncologic Drugs Advisory Committee meeting,[12] the estimated pemetrexed versus docetaxel hazard ratio was 0.992 in the JMEI study with corresponding standard error for the log-hazard ratio estimator of 0.099, which is determined from the 95% confidence interval of 0.817–1.204. Then the indirect estimate of the pemetrexed versus BSC hazard ratio is given by 0.992 × 0.842 = 0.835 with a standard deviation for the corresponding log-hazard ratio estimator of $\sqrt{(0.099)^2 + (0.095)^2} = 0.137$. This leads to a 95% confidence interval for the pemetrexed versus BSC hazard ratio of 0.638–1.093.

For $\lambda_o = 0.5$, we have $Z_2 = \dfrac{\ln 0.992 - (1 - 0.5)\ln(1/0.842)}{\sqrt{(0.099)^2 + (1 - 0.5)^2(0.095)^2}} = -0.856$, which would correspond with a one-sided $p$-value of 0.195. Here, since the 95% confidence interval for the docetaxel versus BSC includes 1 (the upper limit is 1.015), and the pemetrexed versus docetaxel estimated hazard ratio is close to 1, the 95% Fieller confidence interval for $\lambda$ as defined in Equation 5.5 is $-\infty$ to $\infty$. That is that the 95% confidence interval does not rule out any possibilities for $\lambda$. A 90% Fieller confidence interval for $\lambda$ is –1.01 to 3.55.

If the estimated docetaxel effect was discounted by 20% (i.e., $\theta = 0.8$), the resulting value of the test statistic would be $Z_2^* = -0.724$ with a corresponding one-sided $p$-value of 0.23.

For the delta-method confidence interval approach with no discounting on the estimated docetaxel effect, the value for $\hat{\lambda}$ is $1.047 = 1 - \ln 0.992/\ln (1/0.842)$ and the value for $S_{\hat{\lambda}}$ is 0.576. The delta-method 95% confidence interval for $\lambda$ is then calculated as −0.083 to 2.176. Since the lower limit of the confidence interval is less than 0.5 (and is less than 0), the null hypothesis in Expression 5.7 cannot be rejected for $\lambda_o = 0.5$ (for $\lambda_o = 0$). For $\lambda_o = 0.5$ we have $Z_2 = -0.949$, which would correspond with a one-sided $p$-value of 0.171. The value for $R$ is approximately 0.902.

Discounting both the estimate and the corresponding standard deviation need not be conservative when determining whether the experimental therapy is indirectly superior to placebo.[24] This may arise when there is still an adequate amount of uncertainty on whether the active control is effective. For example, suppose the estimated placebo versus active control log-hazard ratio is $\hat{\beta}_H = 0.211$ with a corresponding standard error of 0.17. Using standard normal multipliers, this leads to a confidence interval for $\beta_H$ of 0.88–1.72. For suppose the estimated experimental versus active control log-hazard ratio is $\hat{\beta}_N = -0.163$ with corresponding standard error of 0.10. Then the indirect 95% confidence interval for the experimental versus placebo hazard ratio is 0.47, 1.01, which does not rule out that the experimental therapy is ineffective. When incorporating 50% discounting of the historical control effect, the indirect 95% confidence interval for the experimental versus placebo hazard ratio is 0.59, 0.99, which does rule out that the experimental therapy is ineffective. This seeming contradiction is due to not only having the uncertainty of the control effect more concentrated near zero when the active control is effective, but also more concentrated near zero when the active control is ineffective. As a solution to alleviate this, Odem-Davis[24] proposed discounting only the estimated active control effect without discounting the corresponding standard error. In Example 5.3, discounting the estimated active control effect by 50% without altering the standard error leads to an indirect 95% confidence interval for the experimental versus placebo hazard ratio of 0.52, 1.13.

*Non-inferiority to the Active Control and Indirect Superiority to Placebo.* The desired outcome of a non-inferiority trial is the demonstration that the experimental therapy is effective (superior to placebo) and noninferior to the active control. The hypotheses in Expression 5.9 correspond with this desired outcome and the hypotheses can be tested with simultaneous tests of (a) indirectly showing that the experimental therapy is better than placebo and (b) that the experimental therapy is noninferior to the control therapy as defined by $\beta_N - (1 - \lambda_o)\beta_H < 0$. For example, for some $0 < \alpha < 1$ and an appropriate and perhaps conservative estimation of the control effect in the non-inferiority trial, the first test may conclude that the experimental therapy is indirectly better than placebo when $\hat{\beta}_N - \hat{\beta}_H + z_{\alpha/2}\sqrt{s_N^2 + s_H^2} < 0$; a generalization of Expression 5.11 and the experimental therapy would be concluded to be noninferior to the control therapy when $Z_2 < -z_{\alpha/2}$.

*Testing Based on a Surrogate Evaluation of Constancy.* Wang and Hung[25] proposed a two-stage active control testing (TACT) method, where an assessment of constancy of the outcome distribution for the active control arm is used as a substitute for assessing the constancy of the effect of the active control. An overall assumption is made that there is no change in the underlying distribution for outcomes on a placebo arm between the historical trials and the non-inferiority trial (if the trial had a placebo arm). On the basis of this assumption, the active control effect in the non-inferiority would be the same as in the historical trial if the underlying distribution for the active control outcomes remains the same. The TACT method is presented by Wang and Hung[25] for the proportion of undesirable events. The first stage of the TACT method consists of an interim analysis that compares the event proportions for the active control arm between the non-inferiority trial and the historical trials. If there is strong evidence that the underlying event proportion for the active control is greater in the non-inferiority trial than historically, further testing of non-inferiority at the final analysis is deemed futile and a superiority comparison may be the only adequate comparison. If the trial continues, then this comparison of active control event proportions is repeated at the second/final stage. Again, if there is strong evidence that the underlying event proportion for the active control is greater in the non-inferiority trial than historically, a superiority comparison may be the only adequate comparison. If there is strong evidence that the underlying event proportion for the active control is smaller in the non-inferiority trial than historically, a synthesis method is used to test for non-inferiority. Otherwise, a 95–95 two confidence interval method is used.

An assessment of any deviation from constancy in the distribution in outcomes in the active control arm is used as a surrogate for assessing the constancy assumption. As there is no placebo arm in the non-inferiority trial, the constancy assumption cannot directly be verified. Such a surrogate comparison only checks whether the prognosis was similar between subjects on the active control arm in the non-inferiority trial and the subjects on the active control arm in the historical trials. The prognosis of subjects could be better in the non-inferiority trial without improving the effect of the active control and possibly reducing its effect. Potential differences on effect modifiers are not considered, including nonprognostic covariates that may be effect modifiers. Some indications have factors that are not prognostic but are effect modifiers. Additionally, it may not be true that patients having better prognosis are more likely to benefit from a therapy. For example, HER-2 overexpression is associated with poorer outcomes and shorter survival in breast cancer. However, the effect of trastuzumab, a HER-2 inhibitor, is greater among women whose tumors overexpress HER-2 than those whose tumors do not overexpress HER-2.

### 5.3.4 Synthesis Methods as Prediction Interval Methods

The test criterion in various synthesis methods can be expressed as a comparison of the observed difference between the experimental therapy and the active control from the non-inferiority trial to an upper or lower limit of a prediction interval for that observed difference determined under the assumption that the null hypothesis is true. For example, when the constancy assumption is true in that $E(\hat{\beta}_H)$ is the effect of the active control in the non-inferiority trial and the historical estimator of the active control effect and the estimator for the difference in effects in the non-inferiority trial are independent and normally distributed, then

$$\hat{\beta}_H \pm z_{\alpha/2}\sqrt{\text{Var}(\hat{\beta}_H) + \text{Var}(\hat{\beta}_N)}$$

is a $100(1 - \alpha)\%$ prediction interval for the observed value of $\hat{\beta}_N$ when the experimental therapy has the same efficacy as a placebo. Non-inferiority (or any efficacy in this case) is concluded when $\hat{\beta}_N < \hat{\beta}_H - z_{\alpha/2}\sqrt{\text{Var}(\hat{\beta}_H) + \text{Var}(\hat{\beta}_N)}$, or in other words when the observed value for $\hat{\beta}_N$ is less than the lower limit of the $100(1 - \alpha)\%$ prediction interval for the observed value of $\hat{\beta}_N$ (determined under the assumption that the experimental therapy has the same efficacy as a placebo).

We will discuss synthesis methods as prediction interval methods for both fixed-effects and random-effects models for the active control effect.

*Fixed-Effects Model.* Consider a fixed-effects model for the active control effect where it is assumed that the active control effect is constant across all trials, including in the setting of the non-inferiority trial. When the experimental therapy retains $100\lambda\%$ of the effect of the active control therapy, a $100(1 - \alpha)\%$ prediction interval for the observed value of $\hat{\beta}_N$ is given by $(1-\lambda)\hat{\beta}_H \pm z_{\alpha/2}\sqrt{(1-\lambda)^2 \text{Var}(\hat{\beta}_H) + \text{Var}(\hat{\beta}_N)}$. Non-inferiority (i.e., the experimental therapy retains more than $100\lambda\%$ of the effect of the active control therapy) is concluded when $\hat{\beta}_N < (1-\lambda)\hat{\beta}_H - z_{\alpha/2}\sqrt{(1-\lambda)^2 \text{Var}(\hat{\beta}_H) + \text{Var}(\hat{\beta}_N)}$, or in other words when the observed value for $\hat{\beta}_N$ is less than the lower limit of the $100(1 - \alpha)\%$ prediction interval for the observed value of $\hat{\beta}_N$ (determined under the assumption that the experimental therapy retains exactly $100\lambda\%$ of the effect of the active control therapy).

*Random-Effects Model.* Consider a random-effects model for the active control effect where it is assumed that the same random-effects model holds for all trials, including in the setting of the non-inferiority trial. For the case where a random-effects model is used for the effect of the active control, the same notation will be used as in Section 4.3.3. Thus $\gamma$ and $\gamma_{k+1}$ will be used in place of $\beta_H$ and $\beta_N$, respectively. Parameters and random variables for the non-inferiority trial will be subscripted by $k + 1$. When the experimental therapy has the same effect as a placebo and the same random-effects model that applies for the historical studies of the active control also applies in the non-inferiority trial, we have $\gamma_{k+1} = \gamma + \eta_{k+1}$ and $\hat{\gamma}_{k+1} = \gamma + \eta_{k+1} + \varepsilon_{K+1}$, where $\eta_{k+1} \sim N(0, \tau^2)$ and $\varepsilon_{k+1} \mid \eta_{k+1} \sim N(0, \sigma_{k+1}^2)$ are uncorrelated, $\mu$ is the global mean active control effect across studies, and $\sigma_{k+1}^2 = \text{Var}(\hat{\gamma}_{k+1} \mid \varepsilon_{k+1})$. If $\sigma_1, \ldots, \sigma_k$, $\sigma_{k+1}$ and $\tau^2$ are known, then $\hat{\gamma}_{k+1} - \hat{\gamma}$ has a normal distribution with mean equal to zero and variance $1 / \sum_{i=1}^{k} (\tau^2 + \sigma_i^2)^{-1} + \tau^2 + \sigma_{k+1}^2$. In practice, $\sigma_1, \ldots,$ $\sigma_k$, $\sigma_{k+1}$, and $\tau^2$ are not known and then a $100(1 - \alpha)\%$ prediction interval for the observed value of $\hat{\gamma}_{k+1}$ is given by $\hat{\gamma} \pm w_{\alpha/2} \sqrt{1 / \sum_{i=1}^{k} (\hat{\tau}^2 + \hat{\sigma}_i^2)^{-1} + \hat{\tau}^2 + \hat{\sigma}_{k+1}^2}$, where $w_{\alpha/2}$ is the $100(1 - \alpha/2)$ percentile (or an approximation thereof) of the distribution for $(\hat{\gamma}_{k+1} - \hat{\gamma}) / \sqrt{1 / \sum_{i=1}^{k} (\hat{\tau}^2 + \hat{\sigma}_i^2)^{-1} + \hat{\tau}^2 + \hat{\sigma}_{k+1}^2}$. Under certain assumptions or conditions, $(\hat{\gamma}_{k+1} - \hat{\gamma}) / \sqrt{1 / \sum_{i=1}^{k} (\hat{\tau}^2 + \hat{\sigma}_i^2)^{-1} + \hat{\tau}^2 + \hat{\sigma}_{k+1}^2}$ may have an approximate standard normal distribution or a $t$ distribution where

the approximate degrees of freedom may be determined on the basis of a Satterthwaite approximation or on the basis of simulations.

Non-inferiority (or any efficacy in this case) is concluded when the observed value for $\hat{\beta}_N$ is less than the lower limit of the $100(1 - \alpha)\%$ prediction interval for the observed value of $\hat{\beta}_N$ (determined under the assumption that the experimental therapy has the same efficacy as a placebo, i.e., $\hat{\beta}_N = \hat{\gamma}_{k+1}$).

When the experimental therapy retains $100\lambda_o\%$ of the effect of the active control therapy (i.e., $\beta_N = (1 - \lambda_o)(\gamma + \eta_{k+1})$), a $100(1 - \alpha)\%$ prediction interval for the observed value of $\beta_N$ is given by

$(1 - \lambda_o)\hat{\gamma} \pm w_{\alpha/2} \sqrt{(1 - \lambda_o)^2 / \sum_{i=1}^{k} (\hat{\tau}^2 + \hat{\sigma}_i^2)^{-1} + (1 - \lambda_o)^2 \hat{\tau}^2 + \hat{\sigma}_{k+1}^2}$. Non-inferiority (i.e., the experimental therapy retains more than $100\lambda_o\%$ of the effect of the active control therapy) is concluded when $\hat{\beta}_N < (1 - \lambda_o)\hat{\gamma} - w_{\alpha/2}$

$\sqrt{(1 - \lambda_o)^2 / \sum_{i=1}^{k} (\hat{\tau}^2 + \hat{\sigma}_i^2)^{-1} + (1 - \lambda_o)^2 \hat{\tau}^2 + \hat{\sigma}_{k+1}^2}$.

When $w_{\alpha/2} \approx z_{\alpha/2}$, this test procedure can be expressed as comparing a synthesis-like test to $-z_{\alpha/2}$. The test statistic would be given by

$$Z_4 = \frac{\hat{\beta}_N - (1 - \lambda_o)\hat{\gamma}}{\sqrt{(1 - \lambda_o)^2 / \sum_{i=1}^{k} (\hat{\tau}^2 + \hat{\sigma}_i^2)^{-1} + (1 - \lambda_o)^2 \hat{\tau}^2 + \hat{\sigma}_{k+1}^2}}$$

The appropriateness of using standard normal critical values may be influenced by the sampling distribution for the estimated active control effect and by whether the sizing of the non-inferiority trial depended on the estimation of the active control effect.

### 5.3.5 Addressing the Potential for Biocreep

When the first non-inferiority trial has established a criterion for non-inferiority that the experimental therapy retains more than $100\lambda_o\%$ of the effect of the active control in that trial, it may be reasonable that all future non-inferiority trials have the same or more stringent criterion regardless of the active control therapy used. For example, suppose that better than 50% retention of the effect of the original active control (A) was required in the first non-inferiority trial. If another therapy (B) is to be used as an active control in a non-inferiority trial and a better than 50% retention of the effect of A is still relevant, the non-inferiority criterion for B should be such that it guarantees that if the experimental therapy (C) is noninferior to B, then C is noninferior to A by retaining more than 50% of the effect of A.

Let $\hat{\beta}_{P:A}$, $\hat{\beta}_{B:A}$, and $\hat{\beta}_{C:B}$ be estimated differences between placebo and A, between B and A, and between C and B. Then when constancy holds across all trials in the effects of A and B, a synthesis test would conclude that C retains more than $100\lambda_o\%$ of the effect of A when

$$\frac{(\hat{\beta}_{C:B} - \hat{\beta}_{B:A}) - (1 - \lambda_o)\hat{\beta}_{P:A}}{\sqrt{\hat{V}ar(\hat{\beta}_{C:B}) + \hat{V}ar(\hat{\beta}_{B:A}) + (1 - \lambda_o)^2 \hat{V}ar(\hat{\beta}_{P:A})}} < -z_{\alpha/2}$$

Discounting can be incorporated when there are concerns that the effect of B (and/or A) may have decreased. If $\hat{\beta}_{B:A}$ favors or greatly favors B over A, a more stringent standard may be needed, with B as the ultimate reference instead of A.

### 5.3.6 Bayesian Synthesis Methods

A general discussion on Bayesian and frequentist methods and how they compare is given in the Appendix in Sections A.2 and A.3.

As noted by Rothmann,[2] the distribution of a synthesis test will depend on whether the sizing of the non-inferiority trial is dependent on the estimation of the active control effect. Consider the case where a random variable $Y$ has a normal distribution with mean $\mu_Y$ and variance $\sigma_Y^2$, and the conditional distribution of $X$ given $Y = y$ is a normal distribution with mean $\mu_X$ and variance $\sigma_X^2(y)$. For a two-arm non-inferiority trial, $Y$ represents the estimator of the active control effect and $X$ represents the estimated difference in effects between the experimental and active control arms in the non-inferiority trial. The density for the joint distribution of $X$ and $Y$ is then given by

$$f(x,y) = (2\pi\sigma_Y\sigma(y))^{-1}(\exp\{(-1/2)[(y - \mu_Y)^2/\sigma_Y^2 + (x - \mu_X)^2/\sigma^2(y)]\})$$

The joint density function $f(x,y)$ cannot be factored into separate functions of $x$ and $y$. However, for independent prior distributions for $\mu_Y$ and $\mu_X$ with respective densities $h_Y(\mu_Y)$ and $h_X(\mu_X)$, the joint posterior density for $\mu_Y$ and

$\mu_X$ will factor into the product of the marginal posterior densities. The joint posterior density for $\mu_Y$ and $\mu_X$ is given by

$$\frac{h_X(\mu_X)h_Y(\mu_Y)(\exp\{(-1/2)[(y-\mu_Y)^2/\sigma_Y^2+(x-\mu_X)^2/\sigma^2(y)]\})}{\int_{-\infty}^{\infty}\int_{-\infty}^{\infty}h_X(\mu_X)h_Y(\mu_Y)(\exp\{(-1/2)[(y-\mu_Y)^2/\sigma_Y^2+(x-\mu_X)^2/\sigma^2(y)]\})d\mu_Y\,d\mu_X}=$$

$$\frac{h_Y(\mu_Y)(\exp\{(-1/2)(\mu_Y-y)^2/\sigma_Y^2\})}{\int_{-\infty}^{\infty}h_Y(\mu_Y)(\exp\{(-1/2)(\mu_Y-y)^2/\sigma_Y^2\})d\mu_Y}$$

$$\times\frac{h_X(\mu_X)(\exp\{(-1/2)(\mu_X-x)^2/\sigma^2(y)\})}{\int_{-\infty}^{\infty}h_X(\mu_X)(\exp\{(-1/2)(\mu_X-x)^2/\sigma^2(y)\})d\mu_X}$$

Thus, unlike a frequentist evaluation of non-inferiority, a Bayesian evaluation of non-inferiority does not depend on whether the sizing of non-inferiority trial depended on the estimation of the active control effect.

Simon[14] discussed a Bayesian approach to non-inferiority analysis. The initial model relates a patient outcome, $y$, with treatment (experimental, control, or placebo). For continuous data the model is

$$y = \chi + \beta x + \gamma z + \varepsilon$$

where $x$ and $z$ are treatment indicators and $\varepsilon$ is the random deviation from the mean. The indicator $x = 1$ if the treatment is the control therapy; otherwise $x = 0$. The indicator $z = 1$ if the treatment is the experimental therapy; otherwise $z = 0$. Per Simon's setup, the larger values of $y$ are better outcomes. The mean outcome for the experimental therapy, control therapy, and placebo are $\chi + \gamma$, $\chi + \beta$, and $\chi$, respectively. The errors are assumed to be independent and normally distributed with mean 0 and some common variance. Let $h$ denote the joint prior distribution for $\chi$, $\beta$, and $\gamma$. Then for the sample means from the non-inferiority trial $\bar{y}_E$ and $\bar{y}_C$, the joint posterior distribution satisfies

$$g(\chi,\beta,\gamma|\bar{y}_E,\bar{y}_C) \propto f_E(\bar{y}_E|\chi,\gamma)f_C(\bar{y}_C|\chi,\beta)h(\chi,\beta,\gamma)$$

When $\chi$, $\beta$, and $\gamma$ are modeled with independent prior distributions, $h$ can be replaced with the product of the marginal prior densities.

When the sample means are modeled as having normal distributions and independent normal distributions are chosen for the prior distributions of $\alpha$, $\beta$, and $\gamma$, the joint posterior distribution for $(\chi, \beta, \gamma)$ is a multivariate normal distribution. Various posterior probabilities can be determined. These include for "positive" or desirable outcomes (e.g., response):

(a) The probability that the experimental therapy is better than placebo (i.e., $P(\gamma > 0)$).

(b) The probability that the experimental therapy is better than the control therapy (i.e., $P(\gamma > \beta)$).

(c) The probability that the experimental therapy is better than both the control therapy and placebo (i.e., $P(\gamma > \beta$ and $\gamma > 0)$).

(d) From Simon's paper,[14] the probability that the experimental therapy retains more than $100k\%$ of the control therapy's effect and the control therapy is better than placebo (i.e., $P(\gamma - k\beta > 0$ and $\beta > 0)$).

(e) The probability that the experimental therapy retains more than $100k\%$ of the control therapy's effect and the control therapy is better than placebo, or the experimental arm is better than both the control therapy and placebo (i.e., $P(\gamma - k\beta > 0$ and $\beta > 0) + P(\gamma > 0$ and $\beta < 0)$).

Note that the probability statements in (a)–(e) do not involve $\chi$. The inequalities in the probability statements would be reversed for "negative" or undesirable outcomes (e.g., adverse events, time-to-death/overall survival). In (e), the experimental therapy may have adequate efficacy when the experimental therapy retains more than some minimal fraction of the effect of the control therapy when the control therapy is effective, or when the experimental therapy is more effective than both the placebo and the control therapy when the control therapy is not effective. Additional comments on (e) are given below.

The posterior probabilities in (d) and (e) involve the experimental therapy retaining more than a minimal fraction of the control therapy's effect. The definitions for the fraction of the control therapy's effect retained by the experimental therapy, the retention fraction, require that the control therapy has a positive effect ($\beta > 0$). In such situations, the retention fraction is a measure of the relative efficacy of the experimental therapy versus the control therapy. Since the parameter space for $(\gamma, \beta)$ is $-\infty < \gamma < \infty$, $-\infty < \beta < \infty$, and includes possibilities where the effect of the active control is zero or negative, the retention fraction is not defined for some possible $(\gamma, \beta)$. In general, it can be problematic dealing with new parameters (a function of the original parameters) that do not exist everywhere over the original/underlying parameter space. This is particularly true when the estimator of the new parameter is a function of estimators of the original parameters. The variance for the original parameters and their sampling distributions would incorporate possibilities for which the new parameter is not defined. Inference on the new parameter should consider such issues. Here when $\beta \leq 0$ (the placebo is as effective or more effective than the control therapy), the desired possibilities for $(\gamma, \beta)$ may be that the experimental therapy has any efficacy (i.e., $\gamma > 0$). When $\beta > 0$ and the experimental therapy has some fixed advantage over placebo, the proportion of the control therapy's effect that is retained by the experimental therapy increases without bound as the effect of the control therapy decreases toward zero. It is thus reasonable for any fixed $\gamma > 0$ and $-\infty < a < b < \infty$ that the relative efficacy of the experimental therapy versus the control therapy is larger when $\beta = a$ than when $\beta = b$, even when $a$ (and possibly also $b$) is negative. The probability in (e) would consider any case of $(\gamma, \beta)$ where $\gamma > 0$ and $\beta \leq 0$ as providing greater relative efficacy of the experimental therapy versus the control therapy than any case of $(\gamma, \beta)$ where $\gamma > 0$ and $\beta > 0$.

For undesirable outcomes as was used in the earlier definitions of $\beta_N$ and $\beta_H$, the probability statements (a)–(e) are given by

1. $P(\beta_N - \beta_H < 0)$
2. $P(\beta_N < 0)$
3. $P(\beta_N < 0 \text{ and } \beta_N - \beta_H < 0)$
4. $P(\beta_N - (1 - k)\beta_H < 0 \text{ and } \beta_H > 0)$
5. $P(\beta_N - (1 - k)\beta_H < 0 \text{ and } \beta_H > 0) + P(\beta_N - \beta_H < 0 \text{ and } \beta_H < 0)$

In Example 5.4 various posterior probabilities of interest will be determined in a hypothetical example involving a log-hazard ratio for overall survival.

### Example 5.4

Consider the following hypothetical example for overall survival. The prior distribution for $\beta_H$, the placebo versus control therapy log-hazard ratio, is modeled as a normal distribution with mean 0.2 and standard deviation 0.1. On the basis of a noninformative prior distribution and the study results comparing the experimental and control arms in the non-inferiority trial, the posterior distribution for $\eta$, the experimental versus control log-hazard ratio $\beta_N$ is modeled modeled as a normal distribution with mean –0.10 and standard deviation 0.08. Then we have the following probabilities:

The probability that the experimental therapy is better than placebo: $P(\beta_N - \beta_H < 0) = 0.990$
The probability that the experimental therapy is better than the control therapy: $P(\beta_N < 0) = 0.894$
The probability that the experimental therapy is better than both the control therapy and placebo: $P(\beta_N < 0 \text{ and } \beta_N - \beta_H < 0) = 0.891$
The probability that the experimental therapy retains more than 50% of the control therapy's effect and the control therapy is better than placebo: $P(\beta_N - \beta_H/2 < 0 \text{ and } \beta_H > 0) = 0.964$
The probability that the experimental therapy retains more than 50% of the control therapy's effect and the control therapy is better than placebo, or the experimental arm is better than both the control therapy and placebo: $P(\beta_N - \beta_H/2 < 0 \text{ and } \beta_H > 0) + P(\beta_N - \beta_H < 0 \text{ and } \beta_H < 0) = 0.981$

Additionally, there is a 0.975 probability that the experimental therapy retains more than 61.4% of the control therapy's effect and the control therapy is better than placebo, or the experimental arm is better than both the control therapy and placebo. There is also a 0.025 probability that the experimental therapy retains more than 990% of the control therapy's effect and the control therapy is better than placebo. Therefore for $\lambda = 1 - \beta_N/\beta_H$, $0.95 = P(0.614 < \lambda < 9.90, \beta_H > 0)$. Thus 0.614–9.90 is a 95% credible interval for $\lambda$, the proportion of the control therapy's effect that is retained by the experimental therapy, when additionally requiring that the control therapy has an effect.

There are six possible orderings for the effects of the experimental therapy, active control, and placebo. Table 5.2 provides the posterior probability for each possible ordering on overall survival of the experimental therapy, control therapy, and placebo. There is only a 0.005 posterior probability that the placebo is better than both the active control and the experimental therapy. The bulk of the probability, 0.973, corresponds with the orderings of E > C > P and C > E > P.

*Effect Retention Likelihood Plot.* As a graphical tool for assessing the relative efficacy of an experimental therapy to the active control therapy, Carroll[26] proposed the use of an effect retention likelihood plot, which plots the posterior probability that the experimental therapy retains more than a given retention fraction against that given retention fraction between 0 (i.e., indirect superiority to placebo) and 1 (i.e., superiority to the active control). According to Carroll, the use of an effect retention likelihood plot would be part of a stepwise approach where first the non-inferiority trial would be sized to indirectly demonstrate that the experimental therapy is better than placebo; when the data are analyzed, the posterior probability that the experimental therapy is superior to placebo is determined, and if sufficiently high, then the relative efficacy of the experimental therapy to the control therapy is assessed using the effect retention likelihood plot.

Analogous plots to the effect retention likelihood plot can also be constructed of the posterior probability that the difference in effects of the experimental therapy and the active control therapy (or the indirect effect of the experimental therapy versus placebo) is greater than any prespecified value. Additionally, when noninformative prior distributions are used for the effects, the posterior probabilities will equal or approximately equal 1 minus the corresponding one-sided *p*-value. Therefore the one-sided *p*-values can be substituted for the corresponding posterior probabilities in such plots. Example 5.5 gives a modified version of Carroll's effect retention likelihood plot for approach 2b in the Examples 4.3, 5.2, and 5.3.

**TABLE 5.2**

Posterior Probability for Each Possible Ordering

| Order[a] | Probability Statement | Posterior Probability |
|---|---|---|
| E > C > P | $P(\beta_N < 0, \beta_H > 0)$ | 0.874 |
| E > P > C | $P(\beta_N - \beta_H < 0, \beta_H < 0)$ | 0.017 |
| C > E > P | $P(\beta_N > 0, \beta_N - \beta_H < 0)$ | 0.099 |
| C > P > E | $P(\beta_H > 0, \beta_N - \beta_H > 0)$ | 0.004 |
| P > E > C | $P(\beta_N - \beta_H > 0, \beta_N < 0)$ | 0.003 |
| P > C > E | $P(\beta_H < 0, \beta_N > 0)$ | 0.002 |

a The ">" sign represents "better than" or "superior to."

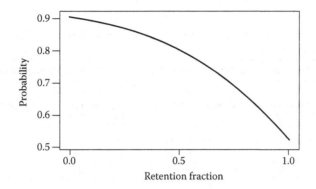

**FIGURE 5.1**
Probability that the true effect retention exceeds a given value between 0 and 1.

## Example 5.5

We will revisit the previous example involving pemetrexed, docetaxel, and BSC in NSCLC based on approach 2b. Consider noninformative prior distributions on the pemetrexed versus docetaxel log hazard ratio, $\beta_N$, and the BSC versus docetaxel log hazard ratio, $\beta_H$. Then $\beta_N$ has a normal posterior distribution with mean −0.008 and standard deviation 0.099, and $\beta_H$ has an independent normal posterior distribution with mean 0.172 and standard deviation 0.095. Let $\Lambda = 1 - \beta_N/\beta_H$. Figure 5.1 provides a plot of $P(\Lambda > \lambda, \beta_H > 0) + P(\beta_N - \beta_H, \beta_H < 0)$ versus $\lambda$, which is a modified version of Carroll's effect retention likelihood plot. For $\lambda = 0, 0.25, 0.5, 0.75$, and 1, the respective probability is 0.905, 0.868, 0.803, 0.690, and 0.526.

### 5.3.7 Application

Example 5.6 revisits the non-inferiority comparison of pemetrexed versus docetaxel discussed in Example 5.3 using all approaches discussed in Example 4.3 in Section 4.2.4. This allows for a comparison of the results from each approach in estimating the docetaxel effect. Additionally, for approaches 4a and 4b in estimating the docetaxel effect, which involves a nonnormal sampling distribution for the estimated docetaxel effect, a Bayesian analysis will be done.

### Example 5.6

For each approach discussed in Section 4.2.4, Table 5.3 provides the estimates and 95% confidence interval for the indirect pemetrexed versus BSC hazard ratio along with the one-sided $p$-value for indirectly testing that pemetrexed is superior to BSC and the one-sided $p$-value for testing that pemetrexed is noninferior to docetaxel at 75 mg/m$^2$ (D75) by retaining more than 50% of the effect of D75 versus BSC. These calculations are based on the "constancy assumption" that the effects are constant across trials. The indirect estimate of the pemetrexed versus BSC

**TABLE 5.3**

Estimates, Confidence Intervals, and $p$-values of Pemetrexed versus BSC Hazard Ratio by Approach

| | Pemetrexed vs. BSC Hazard Ratio | | One-Sided $p$-value | |
|---|---|---|---|---|
| Approach | Estimate | 95% Confidence/ Credible Interval | Pemetrexed Better than BSC | 50% Retention |
| 1 | 0.555 | (0.337, 0.916) | 0.011 | 0.026 |
| 2a | 0.737 | (0.510, 1.067) | 0.052 | 0.110 |
| 2b | 0.835 | (0.638, 1.093) | 0.095 | 0.195 |
| 3 | 0.650 | (0.438, 0.963) | 0.016 | 0.048 |
| 4a | 0.670 | (0.485, 0.974) | 0.019 | 0.053 |
| 4b | 0.699 | (0.499, 1.027) | 0.033 | 0.077 |

log-hazard ratio equals the pemetrexed versus D75 estimate from the JMEI study plus the estimated D75 versus BSC log-hazard ratio from the particular approach (hazard ratios provided in Table 4.3). For approaches 1, 2a, 2b, and 3, the standard error for the indirect log-hazard ratio estimator is the square root of the sum of the variances. The corresponding $p$-values for approaches 1, 2a, 2b, and 3 are based on synthesis test statistics. Results based on approach 2b are provided in Example 5.3.

For approaches 4a and 4b, simulations were performed to determine the indirect estimate and 95% credible interval for the pemetrexed versus BSC hazard ratio and to determine the posterior probabilities for the respective one-sided null hypotheses in testing that pemetrexed superior to BSC and that pemetrexed is noninferior to D75.

As in Example 4.3 in Section 4.2.4, let $\beta$ denote the common true D75/D100 versus BSC/V/I log-hazard ratio ($\beta = -\beta_H$). Also, define $\hat{\beta}_1$, $\hat{\beta}_2$, $\hat{\beta}_3$, and $\hat{\beta}_4$, and ($Z_1$, $Z_2$), $Z_3$, $Z_4$ as in Example 4.3. Let $\hat{\beta} = \min\{\hat{\beta}_1, \hat{\beta}_2, \hat{\beta}_3, \hat{\beta}_4\}$ denote the minimum observed D75/D100 versus BSC/V/I log-hazard ratio (maximum observed effect) and let $W = \min\{0.138Z_1, 0.138Z_2, 0.221Z_3, 0.235Z_4\}$. Then $\hat{\beta} = \beta + W$. Thus, $\beta = \hat{\beta} - W$, and because the distribution of $W$ does not depend on the value of $\beta$, it makes sense that the posterior distribution of $\beta$ is the distribution of $y - W$ given that $\hat{\beta} = y$. This will be true when a flat, improper prior distribution is selected for $\beta$.

Let $f_W(\cdot)$ denote the density for $W$ and $f(\cdot|\beta)$ denote the density for $\hat{\beta}$ given the true value $\beta$. Then for $-\infty < y < \infty$, $f(y|\beta) = f_W(y - \beta)$. For a flat, improper prior distribution for $\beta$ (i.e., the "density" equals a positive constant over the parameter space), the posterior density for $\beta$ is given by $g(\beta|y) = f_W(y - \beta)$ for $-\infty < \beta < \infty$. Thus, given the value of $y$ for $\hat{\beta}$, the posterior distribution of $\beta$ is simulated through random values of $W$, where for each replication $\beta = \hat{\beta} - W$ is calculated. On the basis of the results of the JMEI study, the posterior distribution for $\beta_N$, the pemetrexed versus D75 log-hazard ratio is modeled as having a normal distribution with mean ln 0.992 and standard deviation 0.099. The pemetrexed versus BSC log-hazard ratio is equal to $\beta + \beta_N$ (i.e., $\beta_N - \beta_H$). On the basis of 100,000 simulations, the posterior distribution of $\beta + \beta_N$ has mean $-0.400$ ($0.670 = \exp(-0.400)$) with 2.5th and 97.5th percentiles of $-0.724$ ($0.485 = \exp(-0.724)$) and $-0.026$

(0.974 = exp (−0.026)), respectively. Zero was the 98.1st percentile of the posterior distribution of $\beta + \beta_N$, which leads to the "*p*-value" (i.e., the posterior probability that pemetrexed is inferior to BSC) of 0.019 = 1 − 0.981. The retention fraction is given by $\lambda = 1 + \beta_N/\beta$ when $\beta < 0$. Among the 100,000 replications, 94.7% had $\lambda > 0.5$ and $\beta < 0$, or $\beta + \beta_N < 0$ and $\beta > 0$. The posterior probability of the complement event of 5.3% is used in Table 5.3 as a one-sided *p*-value for testing for more than 50% retention of the docetaxel effect.

For approach 4b, 31,899 of the 100,000 replications had $\hat{\beta}_4 = \min\{\hat{\beta}_1, \hat{\beta}_2, \hat{\beta}_3, \hat{\beta}_4\}$. Conditioning on $\hat{\beta}_4 = \min\{\hat{\beta}_1, \hat{\beta}_2, \hat{\beta}_3, \hat{\beta}_4\}$, the posterior distribution of $\beta + \beta_N$ has a mean of −0.358 (0.699 = exp(−0.358)) with 2.5th and 97.5th percentiles of −0.695 (0.499 = exp(−0.695)) and 0.026 (1.027= exp (0.026)), respectively. Zero was the 96.7th percentile of the posterior distribution of $\beta + \beta_N$, which leads to a one-sided *p*-value of 0.033 = 1 − 0.967. Among the 31,899 replications, 92.3% had $\lambda > 0.5$ and $\beta < 0$, or $\beta + \beta_N < 0$ and $\beta > 0$. The posterior probability of the complement event of 7.7% is used in Table 5.3 as a one-sided *p*-value for testing for more than 50% retention of the docetaxel effect.

In all, three of the six approaches provided one-study evidence that pemetrexed is more effective than BSC by having one-sided *p*-values less than 0.025. The one-sided *p*-values for each approach was greater than 0.025 for testing that pemetrexed retained more than 50% of the docetaxel effect.

### 5.3.8 Sample Size Determination

A sample size formula is provided by Simon,[14] which conditions on the estimated active control effect and its corresponding estimated standard error. This sample size formula also applies to the standard synthesis method when the estimation of the active control effect is known before the non-inferiority trial is sized. We will illustrate determining the sample size or event size and the caveats. We will use the term "power" in this context, although we are truly referring to conditional power as the estimated active control effect and its corresponding estimated standard error have already been determined.

Let $\sigma_N$ denote the standard deviation for $\hat{\beta}_N$, and let $\tilde{\beta}_H$ denote the estimate for the active control effect. For some $0 \le \lambda_o \le 1$ and desired power of $1 - \beta$ at the difference of effects between the experimental and active control arms of $\beta_{N,a}$, $\sigma_N$ must satisfy

$$z_\beta \sigma_N = [(1 - \lambda_o)\tilde{\beta}_H - \beta_{N,a}] - z_{\alpha/2}\sqrt{\sigma_N^2 + (1 - \lambda_o)^2 s_H^2} \quad (5.15)$$

We are only interested in positive solutions of $\sigma_N$. When the solution for $\sigma_N$ is zero, then the power of $1 - \beta$ at $\beta_{N,a}$ is only achieved asymptotically. When the only solutions for $\sigma_N$ are negative or complex numbers, a power of $1 - \beta$ at $\beta_{N,a}$ can never be achieved regardless of the sample size. We will elaborate further on this last point at the end of Example 5.8.

Suppose $\sigma_N > 0$ satisfies Equation 5.15. For a one-to-one randomization involving a continuous variable where the population variances within each arm is $\sigma^2$, the required overall sample size (*n*) is given by

$$n = 4\sigma^2/\sigma_N^2 \qquad (5.16)$$

For a $k$ to 1 randomization (experimental to control), the required overall sample size is given by $n = ((k+1)^2/k)(\sigma^2/\sigma_N^2)$.

For a one-to-one randomization involving a time-to-event variable where a log-hazard ratio-based method is used to compare arms, the required number of events ($r$) is given by

$$r = 4/\sigma_N^2 \qquad (5.17)$$

For a $k$ to 1 randomization (experimental to control), the required overall sample size is given by $r = ((k+1)^2/k)/\sigma_N^2$.

Sample size and event size calculations are provided in Examples 5.7 and 5.8 for cases involving continuous and time-to-event endpoints.

### Example 5.7

Consider a continuous variable where smaller values are more desirable and the estimated mean difference between placebo and the active control is 4.5 with a corresponding standard error of 0.6. An experimental therapy is required to demonstrate better than 60% retention of the active control effect. For the standard synthesis method or a Bayesian synthesis method, Table 5.4 provides the solutions for $\sigma_N$ in Equation 5.15 and the overall sample size for a one-to-one randomization based on Equation 5.16 when 90% power is desired for $\beta_{N,a} = -0.5$, 0, or 0.5, possibilities where the experimental therapy is slightly more effective than the active control, has the same effect as the active control, or is slightly less effective than the active control, respectively; $\alpha = 0.05$ and it is assumed that the population variance in each arm is 100. For $\beta_{N,a} = -0.5$, Equation 5.15 becomes

$$(1.282)\sigma_N = [(1-0.6)(4.5)+0.5] - 1.96\sqrt{\sigma_N^2 + (1-0.6)^2(0.6)^2}$$

The solution to the equation is $\sigma_N = 0.685$. Then applying Equation 5.16 gives

$$n = 4 \times 100/(0.685)^2 = 853 \text{ as the overall sample size.}$$

### TABLE 5.4

Sample Sizes by Assumed Mean Differences for Experimental and Active Control Arms for 90% Power

| Assumed Mean Difference in Non-Inferiority Trial (Exper.–Control) | Standard Error in Non-Inferiority Trial | Sample Size |
|---|---|---|
| −0.5 | 0.685 | 853 |
| 0 | 0.524 | 1459 |
| 0.5 | 0.357 | 3142 |

**TABLE 5.5**

Event Sizes by Assumed Experimental versus Active Control Hazard
Ratios for 80% Power

| Assumed Hazard Ratio (Exper./Control) | Standard Error in Non-Inferiority Trial | Event Size |
|---|---|---|
| 0.9 | 0.0885 | 511 |
| 1 | 0.0443 | 2037 |
| 1.1 | No positive solution | No event size can provide 80% power |

## Example 5.8

Consider a time-to-event variable where longer values are more desirable. The placebo versus active control hazard ratio is 1.40 with a corresponding standard error for the log-hazard ratio of 0.1. The experimental therapy is required to demonstrate better than 50% retention of the active control effect. For the standard synthesis method or a Bayesian synthesis method, Table 5.5 provides the solutions for $\sigma_N$ in Equation 5.15 and the overall required number of events for a one-to-one randomization based on Equation 5.17 when 80% power is desired for $\beta_{N,a} = \ln 0.9$, 0, or $\ln 1.1$, possibilities where the experimental therapy has a 10% lower instantaneous risk of an event than the active control, has the same instantaneous risk as the active control, or has a 10% greater instantaneous risk of an event than the active-control, respectively (where $\alpha = 0.05$). For $\beta_{N,a} = \ln 0.95$, Equation 5.15 becomes

$$(0.842)\sigma_N = [(1 - 0.5)(\ln 1.4) - \ln 0.9] - 1.96\sqrt{\sigma_N^2 + (1 - 0.5)^2 (0.1)^2}$$

The solution to the equation is $\sigma_N = 0.0885$. Then applying Equation 5.16 gives

$$n = 4/(0.0885)^2 = 511 \text{ as the overall number of events.}$$

In Example 5.8, note that 80% power cannot be achieved at $\beta_{N,a} = \ln 1.1$, regardless of the sample size. This is attributable to the knowing of the estimated active control effect along with the fixed, nonzero standard error beforehand and that powering a comparison involving a difference of two parameters is usually based on an assumed difference for those two parameters where the standard error for the estimated difference can be chosen as any positive value. Here the known standard error for $(1 - \lambda_o)\hat{\beta}_H$ establishes a positive lower bound for the standard error of $\hat{\beta}_N - (1 - \lambda_o)\hat{\beta}_H$.

The power at $\beta_{N,a} = \ln 1.1$ is maximized at approximately 9.5% for 441 events. When conditioned on the estimated active control effect and its corresponding standard error, the power for a given of $\beta_{N,a}$ need not be monotone in the sample/event size.

For a two $100(1 - \alpha)\%$ confidence interval approach, sample/event size formulas can be obtained from Equation 5.15 by replacing $\sqrt{\sigma_N^2 + (1 - \lambda_o)^2 s_H^2}$ with $\sigma_N + (1 - \lambda_o)s_H$ and reducing. This leads to Equation 5.18

$$\sigma_N = \frac{[(1 - \lambda_o)(\tilde{\beta}_H - z_{\alpha/2} s_H)] - \beta_{N,a}}{z_\beta + z_{\alpha/2}} \tag{5.18}$$

For a one-to-one randomization when $\beta_{N,a} < [(1 - \lambda_o)(\tilde{\beta}_H - z_{\alpha/2} s_H)]$, Equations 5.16 and 5.17 will apply for determining the sample size for a continuous variable and the event size for a time-to-event variable, respectively. For $\beta_{N,a} < [(1 - \lambda_o)(\tilde{\beta}_H - z_{\alpha/2} s_H)]$, the sample size or event size can be determined that provides a given power greater than $\alpha/2$. For $\beta_{N,a} > [(1 - \lambda_o)(\tilde{\beta}_H - z_{\alpha/2} s_H)]$, the power will always be less than $\alpha/2$. Again, as previously stated, the term "power" in this context is truly "conditional power." As $[(1 - \lambda_o)(\tilde{\beta}_H - z_{\alpha/2} s_H)]$ is an already observed value, $\beta_N < [(1 - \lambda_o)(\tilde{\beta}_H - z_{\alpha/2} s_H)]$ does not reflect an alternative hypothesis that specifies when the experimental therapy has acceptable efficacy. Likewise, for the standard synthesis method, when the conditional power is greater than or less than $\alpha/2$ does not necessarily correspond to exactly when the experimental therapy has acceptable efficacy.

## 5.4 Comparing Analysis Methods and Type I Error Rates

### 5.4.1 Introduction

There have been criticisms between factions involving two–confidence interval methods and synthesis methods. One criticism about the synthesis method is that a margin is not specified at the time of the design of the non-inferiority trial. Also, synthesis-like thresholds as in Equation 5.28 are decreasing as the sample size increases (standard error decreases) for the non-inferiority trial.[5] Conversely, for a synthesis test statistic, a two–confidence interval approach does not prespecify a critical value for testing. In fact, the associated critical value changes with the standard error in the non-inferiority trial in a nonmonotone fashion. The absolute value of the critical value increases as the standard error for $\hat{\beta}_N$ decreases, peaks when the standard errors for $\hat{\beta}_N$ and $(1 - \lambda_o)\hat{\beta}_H$ are equal, and then decreases as the standard error for $\hat{\beta}_N$ decreases toward zero.

Another criticism on the synthesis method is that it relies on the constancy assumption to maintain the desired type I error rate. However, adjustments can be made to a synthesis method to account for concerns involving deviations from constancy.[1,16,27] Additionally, although a 95–95 method is conservative under the constancy assumption, the amount of adjustment made by using the lower limit of a two-sided 95% confidence interval for $(1-\lambda_o)\beta_H$ as the true value is independent of the concerns involving deviations from constancy and depends on the sample size (standard error) in the non-inferiority trial, which has nothing to do with deviations from constancy. The smaller the standard error for $\hat{\beta}_N$, the smaller the adjustment. By consistently applying a 95–95 method without any additional adjustment, there will be instances when the 95–95 method makes too great an adjustment and instances when the 95–95 method makes too little an adjustment.

We will also consider the type I error rate of these procedures. A significant limitation to many of these type I error calculations is that the models ignore that the selection of the active control therapy was not done at random, but was based on the selected therapy having the best or one of the best estimated effects. Some type I error rate calculations are also based on the assumption that the sizing of the non-inferiority trial did not depend on the estimation of the active control effect. Although this is sometimes true, usually the active control effect is evaluated on the basis of previous trials and the results influence the sizing of the non-inferiority trial.

### 5.4.2 Comparison of Methods

In this subsection, we compare the two–confidence interval and synthesis test procedures. Comparisons include a ratio of respective test statistics and the difference in thresholds that are compared with a confidence interval from the non-inferiority trial.

*Ratio of Respective Test Statistics.* In examining the potential conservatism of two–confidence interval procedures relative to a synthesis method, the two–confidence interval procedure of Equation 5.2 can be expressed in the form of a test statistic.[27] The test based on the synthesis method statistic is designed to maintain a desired type I error rate when certain assumptions hold for testing hypotheses involving the effects of the active control versus placebo and the active control versus the experimental therapy. When $\alpha = \gamma$, the test procedure in Equation 5.2 would conclude non-inferiority when

$$Z_4 = \frac{\hat{\beta}_N - (1-\lambda_o)\hat{\beta}_H}{s_N + (1-\lambda_o)s_H} = Z_2/Q < -z_{\alpha/2} \qquad (5.19)$$

where $Q = \dfrac{s_N + (1-\lambda_o)s_H}{\sqrt{s_N^2 + (1-\lambda_o)^2 s_H^2}}$ and $Z_2$ is the standard synthesis method test

statistic given in Equation 5.12. The value for $Q$ is necessarily greater than or equal to 1, which follows from the triangle inequality applied to a right triangle having for the lengths of its legs $s_N$ and $(1-\lambda_o)s_H$, and thus having $\sqrt{s_N^2 + (1-\lambda_o)^2 s_H^2}$ for the length of the hypotenuse (see Figure 5.2). As the sum of the length of the legs is greater than the length of the hypotenuse, we have

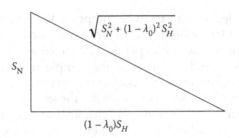

**FIGURE 5.2**
Right triangle representation of standard errors.

that $Q \geq 1$ (equality holding only when a given leg has length zero). The largest possible value for $Q$ is $\sqrt{2} \approx 1.414$, which occurs when $s_N = (1 - \lambda_o)s_H$. Wiens[28] expressed the ratio $Q$ as a function of $\lambda_o$ when $s_N = s_H$ that is equivalent to

$$t(\lambda_o) = \frac{2 - \lambda_o}{\sqrt{1 + (1 - \lambda_o)^2}}$$

The function $t$ is decreasing in $\lambda_o$ over $0 \leq \lambda_o \leq 1$ with $t(0) = \sqrt{2}$ and $t(1) = 1$. The standard synthesis method and the 95–95 approach are equivalent when $\lambda_o = 1$. When $s_N = s_H$ and $\lambda_o = 0$, the standard synthesis method test statistic equals $\sqrt{2}$ multiplied by test statistic for the 95–95 approach.

The *point estimate method*, a two–confidence interval method where $\gamma = 0$, concludes non-inferiority when

$$Z_5 = \frac{\hat{\beta}_N - (1 - \lambda_o)\hat{\beta}_H}{s_N} = R_{PE} \times Z_2 < -z_{\alpha/2} \tag{5.20}$$

where $R_{PE} = \sqrt{1 + (1 - \lambda_o)^2 s_H^2 / s_N^2}$. The factor $R_{PE}$ increases from 1 to $\infty$, as $s_H^2 / s_N^2$ increases from 0 to $\infty$.

Recall that the Hassalblad and Kong method concludes non-inferiority when

$$Z_3 = \frac{\hat{\beta}_N - (1 - \lambda_o)\hat{\beta}_H}{\sqrt{s_N^2 + (\hat{\beta}_N / \hat{\beta}_H)^2 s_H^2}} = R_{HK} \times Z_2 < -z_{\alpha/2} \tag{5.21}$$

where $R_{HK} = \dfrac{\sqrt{s_N^2 + (1 - \lambda_o)^2 s_H^2}}{\sqrt{s_N^2 + (\hat{\beta}_N / \hat{\beta}_H)^2 s_H^2}}$. We see from Equation 5.21 that when $\beta_N = (1 - \delta_o)\beta_H$ and $\mathrm{Var}(\hat{\beta}_N / \hat{\beta}_H)$ is small, $R_{HK} \approx 1$ in distribution and $Z_3$ will have an approximate standard normal distribution when $Z_2$ has a standard normal distribution.

*Differences in the Thresholds.* The standard synthesis method can also be compared with the two–confidence interval method by the threshold that is compared with the confidence interval from the non-inferiority trial. The standard synthesis method for when the null hypothesis is rejected can be expressed as

$$\hat{\beta}_N + z_{\alpha/2} s_N < (1 - \lambda_o)\hat{\beta}_H - z_{\alpha/2} \times (\sqrt{s_N^2 + (1 - \lambda_o)^2 s_H^2} - s_N) \qquad (5.22)$$

The two–confidence interval method where $\alpha = \gamma$ rejects the null hypothesis when

$$\hat{\beta}_N + z_{\alpha/2} s_N < (1 - \lambda_o)(\hat{\beta}_H - z_{\alpha/2} s_H) \qquad (5.23)$$

The right-hand side of Equation 5.22 is larger than the right-hand side of Equation 5.23 by

$$z_{\alpha/2} \times (s_N + (1 - \lambda_o)s_H - \sqrt{s_N^2 + (1 - \lambda_o)^2 s_H^2}) \qquad (5.24)$$

Expression 5.24, which can also be found in the Appendix of Fleming,[29] can be thought of as the adjustment made by the two–confidence interval method compared with the synthesis method. Such an adjustment may be in hopes of addressing any bias in the estimation of the active control effect and any deviation from constancy. In terms of $S = (1 - \lambda_o)s_H$, Expression 5.24 will be between 0 and $z_{\alpha/2}S$, which occur at $s_N = 0$ and as $s_N \to \infty$, respectively.

*Unified Test Statistic.* For indirectly testing that the experimental therapy is effective (i.e., more effective than placebo), Snapinn and Jiang[16] compared the power of synthesis and two–confidence interval methods through a unified approach where the test statistic is given by

$$U(w, v) = \frac{\hat{\beta}_N - (1 - w)\hat{\beta}_H}{\sqrt{\text{Var}(\hat{\beta}_N) + (1 - w)^2 \text{Var}(\hat{\beta}_H) + 2v(1 - w)\sqrt{\text{Var}(\hat{\beta}_N)\text{Var}(\hat{\beta}_H)}}} \qquad (5.25)$$

where $v > 0$ is a variance inflation factor and $0 < w < 1$ is a discounting factor. For a given $(w, v)$, non-inferiority is concluded when $U(w, v) < -1.96$. We see from Equation 5.25 that the standard synthesis statistic given in Equation 5.12 equals $U(\lambda_o, 0)$, the two–confidence interval equivalent test statistic in Equation 5.19 equals $U(\lambda_o, 1)$, and $U(1, v)$ provides the test statistic for a superiority test of the experimental therapy to the active control, regardless of the value of $v$. Snapinn and Jiang noted that failing to account for violations of assay sensitivity and constancy can lead to an inflated type I error rate, which increases the risk of claiming an ineffective therapy as effective.

Departures from assay sensitivity are given by the amount $a = E(\hat{\beta}_N) - \beta_{N,\text{ideal}}$ and departures from constancy by the amount $c = E(\hat{\beta}_H) - \beta_{C,P,N}$, where $\beta_{N,\text{ideal}}$ is the true treatment difference between the active control and the experimental arms under ideal trial situations and $\beta_{C,P,N}$ is the actual effect of the active control in the non-inferiority trial. For a given pair of values for the departures from assay sensitivity and constancy $(a, c)$ and $\beta = \beta_H = \beta_N$ and fixed values for the variances of $\hat{\beta}_N$ and $\hat{\beta}_H$, Snapinn and Jiang determined the values $w_S$ and $w_F$ so that the calibrated synthesis test statistic $U(w_S, 0)$ and the calibrated fixed margin approach test statistic $U(w_F, 1)$ maintain a 0.025 type I error rate. For various cases studied involving departures from assay sensitivity and constancy, Snapinn and Jiang found that the calibrated synthesis method based on the statistic $U(w_S, 0)$ had greater power than the calibrated fixed-margin approach based on $U(w_F, 1)$. The difference in power (or in the determined sample sizes) became more profound as $\text{Var}(\hat{\beta}_H)/\text{Var}(\hat{\beta}_N)$ increased.

### 5.4.3 Asymptotic Results

Hettmansperger[30] studied the properties of confidence intervals for the difference in medians where the confidence limits are differences of the limits of individual confidence intervals for the medians. The results are summarized in greater detail in Chapter 12 on means and medians and also apply to other comparisons (e.g., a difference in means and a log-hazard ratio). We apply the results here to two–confidence interval non-inferiority approaches. The $100(1 - \alpha)\%$ two-sided confidence interval for $\beta_N - (1 - \lambda_o)\beta_H$ of $(L, U) = \hat{\beta}_N - (1 - \lambda_o)\hat{\beta}_H \pm z_{\alpha/2}\sqrt{s_N^2 + (1 - \lambda_o)^2 s_H^2}$ can be expressed by $(L, U) = (L_N - U_H, U_N - L_H)$, where $(L_N, U_N) = \hat{\beta}_N \pm z_{\alpha_N/2}s_N$ is a $100(1 - \alpha_N)\%$ confidence interval for $\beta_N$ and $(L_H, U_H)$ is a $100(1 - \alpha_H)\%$ confidence interval for $(1 - \lambda_o)\beta_H$ of $(L_H, U_H) = (1 - \lambda_o)\hat{\beta}_H \pm z_{\alpha_H/2}(1 - \lambda_o)s_H$. Relating the notation here to the notation of the two–confidence interval approach in Equation 5.2, we have $\alpha_N = \alpha$ and $\alpha_H = \gamma$.

We will assume that $\hat{\beta}_N$ and $\hat{\beta}_H$ have normal distributions. Let $k = \dfrac{s_N^2}{s_N^2 + (1 - \lambda_o)^2 s_H^2}$ $\left(\text{or } k = \dfrac{\text{Var}(\hat{\beta}_N)}{\text{Var}(\hat{\beta}_N) + (1 - \lambda_o)^2 \text{Var}(\hat{\beta}_H)}\right)$. For $0 < \alpha < 1$, it is easy to show that choosing

$$z_{\alpha_N/2} = z_{\alpha/2}\sqrt{k} \text{ and } z_{\alpha_H/2} = z_{\alpha/2}\sqrt{1 - k} \tag{5.26}$$

leads to a synthesis based $100(1 - \alpha)\%$ two-sided confidence interval for $\beta_N - (1 - \lambda_o)\beta_H$ of $(L, U) = (L_N - U_H, U_N - L_H) = \hat{\beta}_N - (1 - \lambda_o)\hat{\beta}_H \pm z_{\alpha/2}\sqrt{s_N^2 + (1 - \lambda_o)^2 s_H^2}$.

**TABLE 5.6**

$z_{\alpha_N/2}$ and $z_{\alpha_H/2}$ Based on Equation 5.26

| k | $\alpha/2 = 0.0005$ | | $\alpha/2 = 0.005$ | | $\alpha/2 = 0.025$ | |
|---|---|---|---|---|---|---|
| | $z_{\alpha_N/2}$ | $z_{\alpha_H/2}$ | $z_{\alpha_N/2}$ | $z_{\alpha_H/2}$ | $z_{\alpha_N/2}$ | $z_{\alpha_H/2}$ |
| 1/2 | 2.327 | 2.327 | 1.822 | 1.822 | 1.386 | 1.386 |
| 2/3 | 2.687 | 1.900 | 2.103 | 1.487 | 1.600 | 1.132 |
| 3/4 | 2.850 | 1.645 | 2.231 | 1.288 | 1.697 | 0.980 |
| 1 | 3.291 | 0 | 2.576 | 0 | 1.960 | 0 |

The two–confidence interval test procedure based on $\alpha_N$ and $\alpha_H$ in Equation 5.26 is equivalent to the standard synthesis test with critical value $-z_{\alpha/2}$.

For $k = 1/2, 2/3, 3/4$, and 1, and $\alpha/2 = 0.0005, 0.005$, and 0.025, Table 5.6 gives the multipliers $z_{\alpha_N/2}$ and $z_{\alpha_H/2}$ based on Equation 5.26 for the two–confidence interval procedure that is equivalent to the synthesis test. For $k = 1/2$ and $\alpha/2 = 0.0005$, the standard synthesis test would be equivalent to a two–confidence interval procedure based on two 98% confidence intervals ($z_{\alpha_N/2} = z_{\alpha_H/2} = 2.327$). For $k = 0, 1/4$, and $1/3$, the values for $z_{\alpha_N/2}$ and $z_{\alpha_H/2}$ are the values for $z_{\alpha_H/2}$ and $z_{\alpha_N/2}$ in Table 5.6 for $k = 1, 3/4$, and $2/3$, respectively.

The equations in Equation 5.26 give one possibility for $z_{\alpha_N/2}$ and $z_{\alpha_H/2}$ that leads to $(L, U) = (L_N - U_H, U_N - L_H)$ being a $100(1 - \alpha)\%$ two-sided confidence interval for $\beta_N - (1 - \lambda_o)\beta_H$. It can be shown that $(L, U) = (L_N - U_H, U_N - L_H)$ will be a $100(1 - \alpha)\%$ two-sided confidence interval for $\beta_N - (1 - \lambda_o)\beta_H$ provided

$$z_{\alpha/2} = z_{\alpha_N/2}\sqrt{k} + z_{\alpha_H/2}\sqrt{1-k} \tag{5.27}$$

For $k = 0, 1/4, 1/4$, and 1, Table 5.7 gives the values for $z_{\alpha/2}$ used in the standard synthesis test and one-sided type I error rates ($\alpha/2$) under the constancy assumption for the 95–95 approach and a 95–80 approach based on Equation 5.27. When $k = 1/2$, a two 95% confidence interval approach is equivalent to a standard synthesis method that targets a one-sided type I error rate of 0.0028. For the 95–80 method, $z_{\alpha/2}$ has an umbrella shape in $k$ with its maximum at $k = 0.7004$ of $z_{0.0096} = 2.342$. For $0.1606 < k < 1$, we have $z_{\alpha/2} > 1.96$ and for $k < 0.1605$, we have $z_{\alpha/2} < 1.96$. For $k = 0, 1/4$, and $1/3$, the

**TABLE 5.7**

$z_{\alpha/2}$ Values Used in Standard Synthesis Test and One-Sided Type I Error Rates

| k | $\alpha_N/2 = \alpha_H/2 = 0.025$ | | $\alpha_N/2 = 0.025, \alpha_H/2 = 0.10$ | |
|---|---|---|---|---|
| | $z_{\alpha/2}$ | $\alpha/2$ | $z_{\alpha/2}$ | $\alpha/2$ |
| 1/2 | 2.772 | 0.0028 | 2.292 | 0.0109 |
| 2/3 | 2.732 | 0.0031 | 2.341 | 0.0096 |
| 3/4 | 2.678 | 0.0037 | 2.338 | 0.0097 |
| 1 | 1.960 | 0.0250 | 1.960 | 0.0250 |

values for $z_{\alpha/2}$ and $\alpha/2$ are the same as those in Table 5.7 for $k = 1, 3/4$, and 2/3, respectively. Rothmann et al.[1] provided a graph of the type I error for the 95–95 approach by the ratio of the standard deviations (i.e., $(1 - \lambda_0)s_H/s_N$).

For a one-to-one randomization and a time-to-event endpoint, Equation 10 of Rothmann et al.[1] gives the non-inferiority threshold for a $100(1 - \alpha_N)\%$ confidence interval of $\beta_N$ that is consistent with the standard synthesis method when $\alpha_N/2 = \alpha/2$. Non-inferiority is concluded at a targeted level of $\alpha/2$ for the standard synthesis method when the upper limit of the two-sided $100(1 - \alpha)\%$ confidence interval for $\beta_N$ is less than

$$z_{\alpha/2}s_N + (1 - \lambda_0)\tilde{\beta}_H - z_{\alpha/2} \times \sqrt{s_N^2 + (1 - \lambda_0)^2 s_H^2} \qquad (5.28)$$

where $\tilde{\beta}_H$ is the placebo versus active control estimate of the active control effect. For a two–confidence interval approach, the use of the threshold in Equation 5.28 is equivalent to choosing $z_{\alpha_H/2}$ ($\alpha_H/2$) by

$$z_{\alpha_H/2} = z_{\alpha/2}(1 - \sqrt{k}) / \sqrt{1-k} = z_{\alpha/2}\sqrt{1-k} / (1 + \sqrt{k}) \qquad (5.29)$$

For $k = 0, 1/4, 1/3, 1/2, 2/3, 3/4$, and 1, Table 5.8 gives the values for $z_{\alpha_H/2}$ and $\alpha_H/2$ for a two–confidence interval procedure where $\alpha_N/2 = \alpha/2$, which is equivalent to the standard synthesis method with critical value $-z_{\alpha/2}$, which targets a one-sided type I error rate of $\alpha/2$ under the constancy assumption. When $k = 1/2$ and $\alpha/2 = 0.025$, an approach that compares a two-sided 95% confidence interval for $\beta_N$ to a two-sided 58.3% confidence interval for $(1 - \lambda_0)\beta_H$ is equivalent to the standard synthesis test with critical value $-1.96$ ($\alpha/2 = 0.025$).

When $z_{\alpha_H/2} = 0$, the two–confidence interval method is referred to as the "point estimate method." The point estimate method would compare a con-

TABLE 5.8

$z_{\alpha_H/2}$ and $\alpha_H/2$ Based on Equation 5.29

| k | $\alpha/2 = 0.0005$ | | $\alpha/2 = 0.005$ | | $\alpha/2 = 0.025$ | |
|---|---|---|---|---|---|---|
| | $z_{\alpha_H/2}$ | $\alpha_H/2$ | $z_{\alpha_H/2}$ | $\alpha_H/2$ | $z_{\alpha_H/2}$ | $\alpha_H/2$ |
| 0 | 3.291 | 0.0005 | 2.576 | 0.0050 | 1.960 | 0.0250 |
| 1/4 | 1.900 | 0.0287 | 1.487 | 0.0685 | 1.132 | 0.1289 |
| 1/3 | 1.703 | 0.0443 | 1.333 | 0.0912 | 1.015 | 0.1552 |
| 1/2 | 1.363 | 0.0864 | 1.067 | 0.1430 | 0.812 | 0.2084 |
| 2/3 | 1.046 | 0.1478 | 0.819 | 0.2065 | 0.623 | 0.2667 |
| 3/4 | 0.882 | 0.1890 | 0.690 | 0.2450 | 0.525 | 0.2997 |
| 1 | 0 | 0.5 | 0 | 0.5 | 0 | 0.5 |

fidence interval for $\beta_N$ from the non-inferiority trial with $(1-\lambda_o)\tilde{\beta}_H$. For the point estimate method by Equation 5.27,

$$z_{\alpha/2} = z_{\alpha_N/2}\sqrt{k} \qquad (5.30)$$

The point estimate method is equivalent to the standard synthesis method using a critical value of $-z_{\alpha_N/2}\sqrt{k}$ that targets a significance level of $\Phi(-z_{\alpha_N/2}\sqrt{k})$. For fixed $\alpha_N/2$, $z_{\alpha/2}$ in Equation 5.30 is increasing in $k$. Under the constancy assumption, the associated type I error rate $\alpha/2$, as a function of $k$, is decreasing in $k$, ranging from 0.5 at $k = 0$ down to $\alpha_N/2$ at $k = 1$.

If it is desired to have $\alpha_N = \alpha_H$, then from Equation 5.27 set

$$z_{\alpha_N/2} = z_{\alpha_H/2} = z_{\alpha/2}/(\sqrt{k}+\sqrt{1-k}) \qquad (5.31)$$

For $k = 1/2$, $2/3$, $3/4$, and 1, and $\alpha/2 = 0.0005$, $0.005$, and $0.025$, Table 5.9 gives the common values for $z_{\alpha_N/2} = z_{\alpha_H/2}$ and $\alpha_N = \alpha_H$ based on Equation 5.31 that yield two–confidence interval procedures equivalent to the standard synthesis test with critical value $-z_{\alpha/2}$. For $k = 1/2$ and $\alpha/2 = 0.025$, $z_{\alpha_N/2} = z_{\alpha_H/2} = 1.386$ and $\alpha_N/2 = \alpha_H/2 = 0.0829$, corresponding to a two 83.4% confidence interval approach being equivalent to the standard synthesis method that targets a one-sided type I error rate of 0.025. As the right-hand side of Equation 5.31 is symmetric in $k$, the values for $z_{\alpha/2}$ and $\alpha/2$ for $k = 0$, $1/4$, and $1/3$ are the same as the values in Table 5.9 for $k = 1$, $3/4$, and $2/3$, respectively.

If it is desired to have equal-length confidence intervals, then for $0 < k < 1$,

$$z_{\alpha_N/2} = z_{\alpha/2}/[2\sqrt{k}] \text{ and } z_{\alpha_H/2} = z_{\alpha/2}/[2\sqrt{1-k}] \qquad (5.32)$$

For $k = 1/2$, $2/3$, $3/4$, and $k \to 1$, and $\alpha/2 = 0.0005$, $0.005$, and $0.025$, Table 5.10 gives the values for $z_{\alpha_N/2}$ and $z_{\alpha_H/2}$ ($\alpha_N$ and $\alpha_H$) based on Equation 5.32 that provide an equal-length two confidence interval procedure equivalent to the standard synthesis test with critical value $-z_{\alpha/2}$. For $k = 3/4$ and $\alpha/2 = 0.025$, a 74.2–95 two–confidence interval procedure is based on equal-length

**TABLE 5.9**

Values for $z_{\alpha_N/2} = z_{\alpha_H/2}$ and $\alpha_N = \alpha_H$ Based on Equation 5.31

| k | $\alpha/2 = 0.0005$ $z_{\alpha_N/2} = z_{\alpha_H/2}$ $(\alpha_N/2 = \alpha_H/2)$ | $\alpha/2 = 0.005$ $z_{\alpha_N/2} = z_{\alpha_H/2}$ $(\alpha_N/2 = \alpha_H/2)$ | $\alpha/2 = 0.025$ $z_{\alpha_N/2} = z_{\alpha_H/2}$ $(\alpha_N/2 = \alpha_H/2)$ |
|---|---|---|---|
| 1/2 | 2.327 (0.0100) | 1.822 (0.0343) | 1.386 (0.0829) |
| 2/3 | 2.361 (0.0091) | 1.848 (0.0323) | 1.406 (0.0798) |
| 3/4 | 2.409 (0.0080) | 1.886 (0.0297) | 1.435 (0.0757) |
| 1 | 3.291 (0.0005) | 2.576 (0.0050) | 1.960 (0.0250) |

**TABLE 5.10**

Values for $z_{\alpha_N/2}$, $z_{\alpha_H/2}$, $\alpha_N/2$ and $\alpha_H/2$ Based on Equation 5.32

| | $\alpha/2 = 0.0005$ | | $\alpha/2 = 0.005$ | | $\alpha/2 = 0.025$ | |
|---|---|---|---|---|---|---|
| $k$ | $z_{\alpha_N/2}$ ($\alpha_N/2$) | $z_{\alpha_H/2}$ ($\alpha_H/2$) | $z_{\alpha_N/2}$ ($\alpha_N/2$) | $z_{\alpha_H/2}$ ($\alpha_H/2$) | $z_{\alpha_N/2}$ ($\alpha_N/2$) | $z_{\alpha_H/2}$ ($\alpha_H/2$) |
| 1/2 | 2.327 | 2.327 | 1.822 | 1.822 | 1.386 | 1.386 |
| | (0.0100) | (0.0100) | (0.0343) | (0.0343) | (0.0829) | (0.0829) |
| 2/3 | 2.015 | 2.850 | 1.577 | 2.231 | 1.200 | 1.697 |
| | (0.0220) | (0.0022) | (0.0573) | (0.0128) | (0.1150) | (0.0448) |
| 3/4 | 1.900 | 3.291 | 1.487 | 2.576 | 1.132 | 1.960 |
| | (0.0287) | (0.0005) | (0.0685) | (0.0050) | (0.1289) | (0.0250) |
| →1 | 1.645 | → ∞ | 1.288 | → ∞ | 0.980 | → ∞ |
| | (0.0500) | (→ 0) | (0.0989) | (→ 0) | (0.1635) | (→ 0) |

confidence intervals and is equivalent to the standard synthesis method that targets a one-sided type I error rate of 0.025. For $k \to 0$ and $k = 1/4$ and $1/3$, the values for $z_{\alpha_N/2}$ and $z_{\alpha_H/2}$ (for $\alpha_N/2$ and $\alpha_H/2$) are, respectively, the values for $z_{\alpha_H/2}$ and $z_{\alpha_N/2}$ (for $\alpha_H/2$ and $\alpha_N/2$) in Table 5.10 for $k \to 1$, and $k = 3/4$ and $2/3$.

### 5.4.4 More on Type I Error Rates

When the margin is truly fixed, not based on an estimation of the active control effect, the use of a two-sided confidence interval from the non-inferiority trial in testing the hypotheses will have a one-sided type I error probability of 0.025, provided no bias is introduced by the conduct of the trial. When the "margin" is a function of the estimated active control effect, statistical hypotheses based on such a margin are surrogate hypotheses. It is hoped that the rejection of the surrogate null hypothesis will imply that there is substantial evidence that the experimental therapy is effective and not unacceptably worse than the active control therapy. For testing the surrogate hypotheses, the one-sided type I error rate (referred to by Hung et al.[5] as a conditional or within-trial type I error rate) is 0.025 for using a two-sided 95% confidence interval from the non-inferiority (provided no bias is introduced by the conduct of the trial). If the "margin" is based on the lower limit of the two-sided 95% confidence interval for $(1 - \lambda_o)\beta_H$ and the constancy assumption holds, then the type I error rate for testing the surrogate hypotheses will be greater than 0.025 for the standard synthesis method based on retaining better than $100(1 - \lambda_o)\%$ of the active control effect with critical value −1.96. The standard synthesis method is not designed to test such surrogate hypotheses. It should also be noted that the "conditional type I error rate" is based on a random possibility (or possibilities) in the parameter space that depends on the estimated active control effect.

An across-trials or unconditional type I error rate for testing the hypotheses in Expression 5.6 is also considered in the study by Hung et al.[5] When the sizing of the non-inferiority trial is independent of the estimation of the active control effect and no bias is introduced by the conduct of the trial, the 95–95 approach will have a type I error rate for testing the hypotheses in Expression 5.6 that is less than 0.025, while type I error rate for the standard synthesis method will equal 0.025. For $U = \hat{\beta}_N + z_{0.025}s_N$ and $\theta = \beta_N - (1 - \lambda_o)\beta_H$, the across-trials type I error rate for a test comparing $U$ with a random threshold $l(\hat{\beta}_H)$ is defined by Hung et al.[5] as

$$P(U < l(\hat{\beta}_H) | H_o : \theta = 0) = \int P(U < l(\tilde{\beta}_H) | H_o : \theta = 0; \tilde{\beta}_H) v(\tilde{\beta}_H) d\tilde{\beta}_H$$

Odem-Davis[24] compared across-trials type I error rates for the standard synthesis method, the standard synthesis method with various discounting, the 95–95 two–confidence interval method, and various bias-only adjusted synthesis methods for cases where the effect of the active control in the non-inferiority trial is smaller than the historical effect by some known fraction. The 95–95 two–confidence interval method has a type I error rate less than the analogous synthesis method. Odem-Davis observed that the type I error rate for the 95–95 two–confidence interval method was more sensitive to changes in the historical variance. Additionally, when the historical variance is small and the estimator of the active control effect in the setting of the non-inferiority trial has a small bias favoring overestimating the true effect, the 95–95 two–confidence interval method may have a type I error rate greater than that of a synthesis method that accounts for the bias. A synthesis method that does not closely account for the bias would be even more likely to lead to a false positive than the 95–95 two–confidence interval method.

Wang, Hung, and Tsong[3] examined the impact of deviations from constancy on a range of procedures. Included are: (a) a synthesis method indirectly testing superiority of the experimental therapy to placebo, (b) a method that uses a non-inferiority margin or a random threshold based on 50% of the estimated control effect, (c) a random threshold based on 50% of the lower limit or the 95% confidence interval for the control effect, (d) use of a non-inferiority margin or a random threshold based on 20% of the estimated control effect, (e) a random threshold based on 20% of the lower limit of the 95% confidence interval for the control effect. For the methods based on random thresholds, non-inferiority would be concluded if the 95% confidence interval for the difference in the effects of the experimental therapy and the active control lies entirely below the random threshold. The method in (a) had the largest type I error rates, then (b), (c), (d), and (e) in that order.

When there is a deviation in the effect of the active control in the non-inferiority trial relative to the historical effect of the active control by $a$ and the estimator of the historical effect of the active control is normally distributed, the approximate type I error rate is given by (from Equation 5.12, 5.19, and 5.20, respectively) $\Phi\left(-z_{\alpha/2}+\dfrac{(1-\lambda_{o})a}{\sqrt{\mathrm{Var}(\hat{\beta}_{N})+(1-\lambda_{o})^{2}\,\mathrm{Var}(\hat{\beta}_{H})}}\right)$ for the standard synthe-

sis method, $\Phi\left(-z_{\alpha/2}/Q+\dfrac{(1-\lambda_{o})a}{\sqrt{\mathrm{Var}(\hat{\beta}_{N})+(1-\lambda_{o})^{2}\,\mathrm{Var}(\hat{\beta}_{H})}}\right)$ for a two $100(1-\alpha)\%$

confidence interval method, and $\Phi\left(-z_{\alpha/2}/R_{\mathrm{PE}}+\dfrac{(1-\lambda_{o})a}{\sqrt{\mathrm{Var}(\hat{\beta}_{N})+(1-\lambda_{o})^{2}\,\mathrm{Var}(\hat{\beta}_{H})}}\right)$

for the point estimate method.

Hung et al.[31] simulated the across-trials trial type I error rate for the point estimate, 95–95, and standard synthesis methods based on proportions of undesirable outcomes. For the simulations, the true active control rate is 14% or 15%, which is less than the placebo rate by 4%; the sample size is 7500 or 10,000 per arm. The sample sizes for the active control and the placebo in the historical comparison are assumed to be roughly equal to the sample sizes in the non-inferiority trial. In each case, the 95–95 and the standard synthesis methods had type I error rates of approximately 0.003 and 0.025, respectively. The unconditional type I error rate for the point estimate method ranged from 0.050 to 0.058.

For the 95–80, 95–85, 95–90, and 95–95 methods, Figure 1 of Hung et al.[5] gives across-trials type I error rate curves as the ratio of the variance of the historical estimate of the active control effect to the variance in the non-inferiority trial ranges from 0.5 to 10.

Additionally, for both fixed-effects and random-effects meta-analyses, Wang, Hung, and Tsong[3] compared the type I error rates for 50% and 80% retention based on log-relative risks under the constancy assumption of three methods: one method using the lower limit of the 95% confidence interval for the control effect as the true effect of the control therapy, a standard synthesis method, and the Hasselblad and Kong procedure. For the random-effects model, the true effect for the active control in the non-inferiority trial is assumed to equal the true global mean effect across studies. The 95% lower confidence interval method maintained type I error rates below the target of 0.025 in all studied cases. The standard synthesis method maintains a proper type I error rate except for cases where the within-trial variability was much smaller than the between-trial variability. The Hasselblad and Kong method consistently had a slightly higher type I error rate than the synthesis method in the random-effects cases. This increase above the desired type I error rate may be largely due to using a normal sampling distribution for the estimated active control effect when an appropriate *t*-distribution would be a better choice.[32,33]

For the fixed-effects meta-analysis, the standard synthesis method maintained approximately the desired type I error rate. When basing the non-inferiority margin on the lower limit of the 95% confidence interval for the active control effect, the simulated type I error rates ranged from 0.0015 to 0.0115 with the type I error rates increasing as the percentage of retention increased. The Hasselblad and Kong method had simulated type I error rates ranging from 0.0009 to 0.0386 with the type I error rates decreasing as the percentage of retention increases.

### 5.4.4.1 Non-Inferiority Trial Size Depends on Estimation of Active Control Effect

As previously stated, the distribution of these synthesis test statistics and the corresponding across-trials type I error rates depend on whether the sizing of the non-inferiority trial was influenced by the estimation of the active control effect.

Let $X$ represent the estimated difference in the effects of the active control and experimental arms in the non-inferiority trial and let $Y$ represent the historical estimated active control effect. Suppose $Y$ is normally distributed with mean $\mu_Y$ and standard deviation $\sigma_Y$ and $X$ is normally distributed with mean $\mu_X$ and standard deviation $\sigma_X = \sigma_X(Y)$. That is, the standard deviation may be chosen based on the value for $Y$. The hypotheses are

$$H_o: \mu_X = (1 - \lambda_o)\mu_Y \text{ vs. } H_a: \mu_X < (1 - \lambda_o)\mu_Y$$

For $0 < \alpha < 1$ and $0 < \gamma < 1$, $H_o$ is rejected by a two–confidence interval procedure whenever $X + z_{\alpha/2}\sigma_X < (1 - \lambda_o)(Y - z_{\gamma/2}\sigma_Y)$. For a given strategy $\sigma_X(\bullet)$, the probability of rejecting the null hypothesis at $(\mu_X, \mu_Y)$ is

$$\int_{-\infty}^{\infty} \Phi\left(-z_{\alpha/2} + \frac{(1-\lambda_o)(y - z_{\gamma/2}\sigma_Y) - \mu_X}{\sigma_X(y)}\right) \cdot \varphi\left(\frac{y - \mu_Y}{\sigma_Y}\right) dy / \sigma_Y \qquad (5.33)$$

Based on Expression 5.33, Rothmann[2] showed that over all possible strategies $\sigma_X(\bullet)$, the supremum one-sided type I error probability is $(\alpha + \gamma)/2 - \alpha\gamma/4 > \alpha/2$. This supremum occurs as $\sigma_X(y) \to 0$ for $y > \mu_Y$ and as $\sigma_X(y) \to \infty$ for $y < \mu_Y$. Conversely, the infimum one-sided type I error probability across all possible strategies is $\alpha\gamma/4$. This infimum occurs as $\sigma_X(y) \to 0$ for $y < \mu_Y$ and as $\sigma_X(y) \to \infty$ for $y > \mu_Y$. However, practical strategies have type I error probabilities between those two extremes. For $\alpha = \gamma = 0.05$, the infimum and supremum type I error probabilities are 0.0006 and 0.0494, respectively. This is a wider range than the range for the type I error rate when the design of the non-inferiority trial is independent of the estimation of the active control effect of 0.0028–0.025.

A common strategy sizes the non-inferiority trial to have a desired power (e.g., 80% or 90% power) when there is no difference in the effects of the experimental and active control therapies. In an example, Rothmann determined the type I error probabilities for the 95–95 method when $\sigma_X(\bullet)$ is based on the 80% and 90% conditional power at $\mu_X = 0$. When $\alpha = 0.05$, Rothmann also determined for 80% and 90% conditional power at $\mu_X = 0$ the values for $\gamma$ that leads to a maximum type I error probability of 0.025 over a likely range of values for $(1 - \lambda_0)\mu_Y$ or that leads to a type I error rate of 0.025 based on a distribution for the possible values of $(1 - \lambda_0)\mu_Y$.

For a synthesis test, $H_0$ is rejected whenever $\dfrac{X - (1 - \lambda_0)Y}{\sqrt{\sigma_X^2(Y) + (1 - \lambda_0)^2 \sigma_Y^2}} < -z_{\alpha/2}$ for

some $0 < \alpha < 1$. For a given strategy $\sigma_X(\bullet)$, the probability of rejecting the null hy-

pothesis at $(\mu_X, \mu_Y)$ is $\displaystyle\int_{-\infty}^{\infty} \Phi\left( \dfrac{(1 - \lambda_0)y - \mu_X - z_{\alpha/2}\sqrt{\sigma_X^2(y) + \sigma_Y^2}}{\sigma_X(y)} \right) \cdot \varphi\left( \dfrac{y - \mu_Y}{\sigma_Y} \right) dy / \sigma_Y.$

As with the two–confidence interval approach, Rothmann determines the infimum and supremum type I error probabilities over all possible strategies for $\sigma_X(\bullet)$ for the synthesis approach. When $\alpha = 0.05$, infimum and supremum type I error probabilities are 0.0006 and 0.049, respectively. This is the same range for the type I error probabilities as with the two–confidence interval approach when $\alpha = \gamma = 0.05$. A critical value for the test can be determined that gives a maximum type I error probability of 0.025 over a likely range of values for $(1 - \lambda_0)\mu_Y$ or that leads to a type I error rate of 0.025 based on a distribution for the possible values of $(1 - \lambda_0)\mu_Y$.

In the examined settings, Rothmann noted that strategies based on having adequate power when there is no difference in effects between the experimental and active control therapies have smaller type I error rates than the type I error rate in an analogous case when the sizing of the non-inferiority trial is independent of the estimation of the active control effect. This will be likewise seen in Example 5.9 in Section 5.4.4.2, which examines the type I error rate under a model that introduces regression-to-the mean bias.

As noted in Section 5.3.6, for the Bayesian setting, the analysis does not depend on whether the sizing of the non-inferiority is independent or dependent on estimation of the active control effect. In both cases, the joint posterior distribution of the differences in effects of the experimental therapy versus the active control therapy and the active control therapy versus placebo factors into the product of the marginal posterior distributions when independent prior distributions are used. In the frequentist setting, the joint density function does not factor into the marginal density functions unless the sizing of the non-inferiority trial does not depend on the estimation of the active control effect.

### 5.4.4.2 Incorporating Regression to Mean Bias

Example 5.9 illustrates the impact that regression-to-the-mean bias can have on the type I error rates for a 95–95 method, the standard synthesis method, and a bias-corrected synthesis method. Also considered in simulating the type I error rates is whether the sizing of the non-inferiority trial is independent or dependent on the estimation of the active control effect.

### Example 5.9

Consider that there are five previous studied therapies for an indication. For $i = 1$, ..., 5, let $X_i$ denote the observed placebo versus active control log-hazard ratio for the $i$-th therapy from either one clinical trial or a meta-analysis of clinical trials. Suppose $X_1,...,X_5$ is a random sample from a normal distribution having mean $\ln(4/3)$ and standard deviation of 0.1. A new experimental therapy is to be compared in a clinical trial to that therapy among the five therapies that has the largest observed effect. We will first assume that the sizing of the non-inferiority trial is independent of the estimation of the active control effect. We are interested in testing the hypotheses in Expression 5.6 when $\lambda_o = 0$ and when the true placebo versus active control log-hazard ratio in the non-inferiority trial is $\ln(4/3)$. Table 5.11 provides simulated type I error rates for both the 95–95 and standard synthesis methods when the true placebo versus experimental hazard ratio is 1 (i.e., the experimental therapy is ineffective) for various standard errors for the experimental versus active control log-hazard ratio, as the standard error goes to 0 and as the standard error goes to infinity. Note that the regression-to-the-mean bias and the type I error rates depend on the common historical standard deviation and the standard error in the non-inferiority trial and do not depend on the value for the common effect of the five previously studied therapies.

As the standard error for the estimated difference in the non-inferiority trial decreases, the simulated type I error rate for the standard synthesis method increases from 0.025 toward about 0.12. The type I error rate for the 95–95 method is "U-shaped" in the standard error from the non-inferiority trial. That is, as the standard error in the non-inferiority trial decreases, the type I error rate decreases to some minimum value at a standard error between 0.1 and 0.2, and increases

**TABLE 5.11**

Simulated Type I Error Rates—Independent Design

| Non-Inferiority Trial Standard Error | Simulated Type I Error Rate[a] | |
|---|---|---|
| | 95–95 Method | Standard Synthesis Method |
| 0 | 0.1194 | 0.1194 |
| 0.03 | 0.0359 | 0.1161 |
| 0.04 | 0.0260 | 0.1145 |
| 0.05 | 0.0200 | 0.1114 |
| 0.0707 | 0.0157 | 0.1066 |
| 0.1 | 0.0119 | 0.0920 |
| 0.2 | 0.0130 | 0.0638 |
| $\to \infty$[b] | 0.025 | 0.025 |

[a] Each type I error rate is based on 100,000 simulations.
[b] Type I error rate for the "$\to \infty$" case determined mathematically. The simulated type I error rate for this case was 0.0245.

toward a value of about 0.12 as the standard error decreases toward zero. The 95–95 method maintains a type I error rate below 0.025 unless the standard error in the trial is very small.

We next consider the case where the sizing of the non-inferiority trial depends on the estimated active control effect. For both analysis methods, Table 5.12 provides simulated type I error rates when the true placebo versus experimental hazard ratio is 1, where the non-inferiority trial is sized based on a conditional power ranging from 3% to 100% from using the 95–95 method under the assumptions that the experimental and active control therapies have the same effect and the true placebo versus active control log-hazard ratio in the non-inferiority trial is ln(4/3).

Comparing Tables 5.11 and 5.12, when the sizing of the non-inferiority trial depends on the estimated active control effect, the type I error rates tend to be slightly smaller than when the sizing of the non-inferiority trial is independent of the estimated active control effect. The 95–95 method achieved type I error rates smaller than 0.025 for every practical choice for power. The simulated type I error rates for the standard synthesis method exceeded 0.025 and were increasing in the conditional power.

In general, the amount of regression-to-the-mean bias (and the increase to the type I error rates of the standard synthesis method and the 95–95 method) increases as there is an increase in the number of previously studied investigational agents that potentially could have produced results so as to be selected as the active control and/or as the standard deviations for the estimated effects of the previously studied investigational agents increases. When the standard errors for the previously studied investigational agents are equal, the regression-to-the-mean bias will be at its greatest when the effects of those previously studied investigational agents are equal. Reproducibility across studies in the effect size of the selected active control can provide assurance that the size of the regression-to-

**TABLE 5.12**

Simulated Type I Error Rates—Dependent Design

| | Simulated Type I Error Rate[a] | |
|---|---|---|
| **Conditional Power (%)** | **95–95 Method** | **Standard Synthesis Method** |
| 3 | 0.0225 | 0.0269 |
| 10 | 0.0138 | 0.0428 |
| 20 | 0.0113 | 0.0528 |
| 30 | 0.0101 | 0.0589 |
| 40 | 0.0095 | 0.0631 |
| 50 | 0.0094 | 0.0668 |
| 60 | 0.0092 | 0.0702 |
| 70 | 0.0090 | 0.0739 |
| 80 | 0.0091 | 0.0776 |
| 90 | 0.0095 | 0.0824 |
| 99.99 | 0.0143 | 0.0988 |
| Just below 100[b] | 0.1177 | 0.1193 |

[a] Each type I error rate is based on 100,000 simulations.
[b] Power of $1 - \beta$ where $z_\beta = 1000$.

the-mean bias is small or negligible. However, in the absence or nonexistence of a reproduced estimated effect size across studies—for example, when there is only one historical study that estimates the active control effect—regression-to-the-mean bias in the estimated active control effect should be accounted for in the non-inferiority analysis.

Under the assumption that the previously studied therapies had the same treatment effect, the bias in using the maximum estimated effect can be determined and does not depend on the size of the common effect. In this example, the bias in using the maximum estimated effect is approximately 0.116 from Table 4.1, which from Expression 5.24 is approximately the adjustment made by the 95–95 method (compared with the synthesis method) of 0.115. The type I error rate will now be examined for a synthesis method that accounts for this bias by subtracting 0.116 from the maximum estimated effect in the numerator of the test statistic. For the case where the sizing of the non-inferiority trial is independent of the estimation of the active control effect, Table 5.13 provides simulated type I error rates for a bias-corrected synthesis method when the true placebo versus experimental hazard ratio is 1 for various standard errors for the experimental versus active control log-hazard ratio, as the standard error goes to 0, and as the standard error goes to infinity.

We note that the type I error rates in Table 5.13 for the bias-corrected synthesis method are all smaller than 0.025 and are much smaller than those type I error rates given in Table 5.11 for the standard synthesis method without a bias correction. The skewed, nonnormal distribution for the maximum of a normal random sample, which the test does not account for, is the reason why the type I error rates for the bias-corrected synthesis method are not equal to 0.025 but smaller than 0.025. Additionally, the type I error rates in Table 5.13 for the bias-corrected synthesis method are increasing in the studied standard errors for the non-inferiority trial, not decreasing in the studied standard errors for the non-inferiority trial as in Table 5.11 for the standard synthesis method without a bias correction. Also, in this example, for most (not all) of the studied standard errors for the non-inferiority trial, the bias-corrected synthesis method had a smaller simulated type I error rate than the 95–95 method without bias correction. For the case where the standard

**TABLE 5.13**

Simulated Type I Error Rates for a Bias-Corrected Synthesis Test—Independent Design

| Non-Inferiority Trial Standard Error | Simulated Type I Error Rate[a] |
|---|---|
| 0 | 0.0047 |
| 0.03 | 0.0053 |
| 0.04 | 0.0057 |
| 0.05 | 0.0066 |
| 0.0707 | 0.0078 |
| 0.1 | 0.0116 |
| 0.2 | 0.0193 |
| $\to \infty$[b] | 0.025 |

[a] Each type I error rate is based on 100,000 simulations.
[b] Type I error rate for the "$\to \infty$" case determined mathematically. The simulated type I error rate for this case was 0.0245.

error for the non-inferiority trial equaled 0.1, the common standard error for the previous studies, the type I error rates for the bias-corrected synthesis and 95–95 methods are similar (0.0116 and 0.0119, respectively).

Table 5.14 provides simulated type I error rates when the true placebo versus experimental hazard ratio is 1 for a bias-corrected synthesis method where the non-inferiority trial is sized on the basis of a conditional power ranging from 3% to 100% from using the 95–95 method under the assumptions that the experimental and active control therapies have the same effect and the true placebo versus active control log-hazard ratio in the non-inferiority trial is ln(4/3).

As in the independent design case, we note that the type I error rates in Table 5.14 for the bias-corrected synthesis method are all smaller than 0.025 and are much smaller than those type I error rates given in Table 5.12 for the standard synthesis method without a bias correction. Additionally, the type I error rates in Table 5.14 for the bias-corrected synthesis method are decreasing in the conditional power, not increasing in the conditional power as in Table 5.12 for the standard synthesis method without a bias correction. Also, in this example for 90% or greater conditional power, the bias-corrected synthesis method had a smaller simulated type I error rate than the 95–95 method without bias correction. For 80% or smaller conditional power, the bias-corrected synthesis method had a greater simulated type I error rate than the 95–95 method without bias correction.

Suppose it is desired that the experimental therapy retain more than $100\lambda_o\%$ of the historical effect of the active control (for some $0 \leq \lambda_o \leq 1$) and there are five previously studied therapies with the same treatment effect that are estimated with a common standard error. Extending from Example 5.9, the regression-to-the mean bias by selecting the therapy with the greatest observed effect as the active control is approximately equal to the adjustment

**TABLE 5.14**

Simulated Type I Error Rates for a Bias Corrected Synthesis Test—Dependent Design

| Conditional Power (%) | Simulated Type I Error Rate[a] |
|---|---|
| 3 | 0.0240 |
| 10 | 0.0194 |
| 20 | 0.0166 |
| 30 | 0.0145 |
| 40 | 0.0134 |
| 50 | 0.0123 |
| 60 | 0.0113 |
| 70 | 0.0104 |
| 80 | 0.0095 |
| 90 | 0.0085 |
| 99.99 | 0.0065 |
| Just below 100[b] | 0.0048 |

[a] Each type I error rate is based on 100,000 simulations.
[b] Power of $1 - \beta$, where $z_\beta = 1000$.

made by the 95–95 method in Expression 5.24 when the standard error in the non-inferiority trial equals the $(1 - \lambda_o)$ multiplied by the common standard error for the five previously studied therapies.

Example 5.10 evaluates the type I error rate of false conclusions of efficacy of the experimental therapy under the previous model in Chapter 3, where the likelihood that the active control is truly effective depends on the probability a random agent for that indication is truly effective and the power for concluding effectiveness when the agent is effective.

## Example 5.10

Often, when a therapy has been concluded as effective for an indication based on one or more clinical trials (generally, at least two clinical trials), it may become unethical to conduct further placebo-controlled trials for that indication. The therapy concluded as effective would be given as an "active" control in future trials. The active control may or may not be effective. As we have seen earlier, the likelihood that the active control is truly effective depends on the statistical significance of the results, the probability a random agent for that indication is truly effective, and the power for concluding effectiveness when the agent is effective. Equation 3.1 gives the probability that the active control is ineffective. Under this paradigm, we will evaluate the type I error rate of falsely concluding that the experimental therapy has any efficacy when it has zero efficacy and the type I error rate of falsely concluding that the experimental therapy has efficacy more than half the efficacy of the active control when the experimental therapy has efficacy exactly half that of the active control. For ease, it is assumed that the sizing of the non-inferiority trial is independent of the estimation of the active control effect.

The active control for the non-inferiority trial will be evaluated based on achieving a favorable one-sided $p$-value less than 0.025 from a single trial. It is understood that in practice, usually a more stringent criterion than a one-sided $p$-value less than 0.025 would be used for having a therapy as an active control in future clinical trials. The evaluation of the type I error rate in a non-inferiority trial can also be done for a more stringent significance level than 0.025. At the beginning of the historical trial, the estimated active control effect is assumed to have a normal distribution with standard deviation $s_H$. When the active control is truly effective, the actual power is assumed to be 90%, which makes the true effect of the active control approximately $3.24s_H$. Given that the $p$-value is less than 0.025, the conditional mean, median, and standard deviation for the estimated effect are $3.44s_H$, $3.37s_H$, and $0.84s_H$, respectively. When the "active" control is truly ineffective (has zero effect), the conditional mean, median, and standard deviation for the estimated effect given a $p$-value of less than 0.025 are $2.33s_H$, $2.24s_H$, and $0.34s_H$, respectively.

For the cases where the active control has its assumed effect and when the active control has zero effect, the type I error rate for a conclusion of any efficacy from a non-inferiority trial was evaluated for the standard synthesis method, the standard synthesis method with 50% discounting, the 95–95 method, and the 95–95 method with 50% discounting under the assumption that the active control effect has not changed. For these methods, Tables 5.15 and 5.16 provide the simulated type I error rates for a false conclusion of any efficacy when the experimental

**TABLE 5.15**

Simulated Type I Error Rates Conditioned on Active Control Having Its Assumed Effect with Historical Power of 90% and Significance Level of 0.025

| Ratio of Variances | Synthesis Method 0% Retention | Synthesis Method 0% Retention 50% Discounting | 95–95 Method 0% Retention | 95–95 Method 0% Retention 50% Discounting |
|---|---|---|---|---|
| $\to 0^+$ | 0.025 | 0.025 | 0.025 | 0.025 |
| 0.1 | 0.0270 | 0.0075 | 0.0078 | 0.0033 |
| 0.25 | 0.0270 | 0.0031 | 0.0044 | 0.0008 |
| 0.5 | 0.0279 | 0.0011 | 0.0034 | 0.0002 |
| 1 | 0.0269 | 0.0003 | 0.0027 | 0.00001 |
| 2 | 0.0286 | 0.00003 | 0.0032 | 0 |
| 4 | 0.0271 | 0.00001 | 0.0045 | 0 |
| 10 | 0.0282 | 0 | 0.0076 | 0 |
| $\to \infty$ | 0.0278 | 0.0000001 | 0.0278 | 0.0000001 |

therapy has zero efficacy for various ratios of the variance of the estimated difference in effects from the non-inferiority trial to the variance of the historically based estimate of the active control effect. For each case, 100,000 simulations were used on the underlying distribution for the estimated effect. Thus, when the active control is effective (is ineffective), there were typically 90,000 cases (2,500 cases) with a *p*-value of less than 0.025 that were further used to simulate the type I error rate of falsely concluding that the experimental therapy has efficacy when the experimental therapy has zero effect in the non-inferiority trial.

The simulated type I error rates have associated margins of error with some simulated rates greater than the true rate. Some reversals appear to occur in the order of the simulated type I error rates. For example, when the active control is truly ineffective and a 95–95 method is used with 50% discounting, the simulated type I error rates when the ratio of variances is 0.5 and 1 are 0.0452 and 0.0404,

**TABLE 5.16**

Simulated Type I Error Rates Conditioned on Active Control Being Ineffective

| Ratio of Variances | Synthesis Method 0% Retention | Synthesis Method 0% Retention 50% Discounting | 95–95 Method 0% Retention | 95–95 Method 0% Retention 50% Discounting |
|---|---|---|---|---|
| $\to 0^+$ | 0.025 | 0.025 | 0.025 | 0.025 |
| 0.1 | 0.1033 | 0.0572 | 0.0347 | 0.0300 |
| 0.25 | 0.1619 | 0.0844 | 0.0457 | 0.0337 |
| 0.5 | 0.2524 | 0.1274 | 0.0611 | 0.0452 |
| 1 | 0.3353 | 0.1494 | 0.0646 | 0.0404 |
| 2 | 0.4651 | 0.2323 | 0.1051 | 0.0570 |
| 4 | 0.5796 | 0.3217 | 0.1466 | 0.0666 |
| 10 | 0.7221 | 0.5199 | 0.2913 | 0.1233 |
| $\to \infty$ | 1 | 1 | 1 | 1 |

respectively. Based on the general pattern for the simulated type I error rates for this procedure, it is likely that the true type I error rates in these two cases are ordered in the reverse direction. Additionally, it appears that when the active control is effective, the synthesis method has a type I error rate between 0.025 and 0.0278 and that the deviations out of this range by the simulated type I error rates are due to random error.

When the active control is truly effective, the methods based on 50% discounting appear to have type I error rates for concluding any efficacy that decreases as the ratio of the variances increases. The type I error rates for the 95–95 method without discounting was U-shaped in the ratio of the variances, with the type I error rate exceeding 0.025 only when the ratio of variances is quite large. When the active control is truly ineffective for all methods, the limiting type I error rate is 1 as the ratio of variances goes to infinity. In all cases, the limiting type I error rate is 0.025 as the ratio of variances goes to zero.

In the cases studied when the active control was truly effective, the standard synthesis method with 50% discounting had a smaller type I error rate for a false conclusion of any efficacy when the experimental therapy has zero efficacy than the 95–95 method without discounting. However, when the active control is truly ineffective, the order was reversed with the 95–95 method without discounting having the smaller type I error rate. For these two methods, when the likelihood that the active control is truly effective is considered, the standard synthesis method with 50% discounting will have the smaller type I error rate when it is highly likely that the active control is effective, and the larger type I error rate when it is highly unlikely that the active control is effective.

For the four analysis methods, Tables 5.17 and 5.18 provide the type I error rates for a false conclusion of any efficacy when the experimental therapy has zero efficacy based on various probabilities that the active control is ineffective when the ratio of the variances is 1 and 4. The probability that the active control is ineffective is based on Equation 3.1 and the selected probability that a random therapy is effective. The simulation-based type I error rate is then a convex combination of the corresponding conditional type I error rates in Tables 5.15 and 5.16. When the probability a random therapy is effective equals 0.25, the probability is 0.0769 that a therapy that achieves favorable statistical significance at a one-sided 0.025 level is ineffective. Then for the standard synthesis method when the ratio of variances is 1, the simulation-based type I error rate for a false conclusion of any efficacy

**TABLE 5.17**

Simulation-Based Type I Error Rate for a False Conclusion of Any Efficacy When Ratio of Variances = 1

| Probability Random Therapy Is Effective | Probability Active Control Is Ineffective | Standard Synthesis Method | 95–95 Method | Standard Synthesis Method with 50% Discounting | 95–95 Method with 50% Discounting |
|---|---|---|---|---|---|
| 0.1 | 0.2000 | 0.0886 | 0.0151 | 0.0301 | 0.0081 |
| 0.25 | 0.0769 | 0.0506 | 0.0075 | 0.0118 | 0.0031 |
| 0.5 | 0.0270 | 0.0352 | 0.0044 | 0.0043 | 0.0011 |
| 0.75 | 0.0092 | 0.0297 | 0.0033 | 0.0017 | 0.0004 |
| 0.9 | 0.0031 | 0.0278 | 0.0029 | 0.0008 | 0.0001 |

**TABLE 5.18**

Simulation-Based Type I Error Rate for a False Conclusion of Any Efficacy When Ratio of Variances = 4

| Probability Random Therapy Is Effective | Probability Active Control Is Ineffective | Standard Synthesis Method | 95–95 Method | Standard Synthesis Method with 50% Discounting | 95–95 Method with 50% Discounting |
|---|---|---|---|---|---|
| 0.1 | 0.2000 | 0.1376 | 0.0329 | 0.0643 | 0.0133 |
| 0.25 | 0.0769 | 0.0696 | 0.0154 | 0.0248 | 0.0051 |
| 0.5 | 0.0270 | 0.0420 | 0.0083 | 0.0087 | 0.0018 |
| 0.75 | 0.0092 | 0.0322 | 0.0058 | 0.0030 | 0.0006 |
| 0.9 | 0.0031 | 0.0288 | 0.0049 | 0.0010 | 0.0002 |

when the experimental therapy has zero efficacy is 0.9231 × 0.0269 + 0.0769 × 0.3351 = 0.0506. When the probability that a random therapy is effective is only 10% and the ratio of variances is 4, the simulation-based type I error rate for the 95–95 method without discounting is 0.0329, larger than a desired one-sided level of 0.025. When the probability a random therapy is effective is small, the type I error rate for a false conclusion of any efficacy for the 95–95 method without discounting will be larger than an intended level of 0.025. It is thus important in settings where "success" is rare to consider using a more stringent criterion than a margin based on the lower limit of the 95% confidence interval of the active control effect.

For cases where the active control has its assumed effect and when the active control has zero effect, the simulated type I error rates for a conclusion of efficacy of more than half the efficacy of the active control when the experimental therapy has efficacy exactly half that of the active control are provided in Table 5.19 for

**TABLE 5.19**

Simulated Conditional Type I Error Rates for Testing for Better 50% Retention of Active Control Effect

| Ratio of Variances | Active | | Inactive | |
|---|---|---|---|---|
| | Synthesis Method 50% Retention | 95–95 Method 50% Retention | Synthesis Method 50% Retention | 95–95 Method 50% Retention |
| →0⁺ | 0.025 | 0.025 | 0.025 | 0.025 |
| 0.1 | 0.0269 | 0.0137 | 0.0572 | 0.0300 |
| 0.25 | 0.0273 | 0.0091 | 0.0844 | 0.0337 |
| 0.5 | 0.0279 | 0.0067 | 0.1274 | 0.0452 |
| 1 | 0.0272 | 0.0043 | 0.1494 | 0.0404 |
| 2 | 0.0282 | 0.0034 | 0.2323 | 0.0570 |
| 4 | 0.0268 | 0.0030 | 0.3217 | 0.0666 |
| 10 | 0.0280 | 0.0036 | 0.5199 | 0.1233 |
| →∞ | 0.0278 | 0.0278 | 1 | 1 |

the standard synthesis and 95–95 methods based on 50% retention. When the active control has zero effect, half the active control effect equals zero effect. As with Tables 5.15 and 5.16, for each case, 100,000 simulations were used in the underlying distribution for the estimated effect. Simulations are based on no change in the active control effect. For the standard synthesis method and the 95–95 method, the limiting conditional type I error rate is $0.0278 = 0.025/0.9$ as the ratio of variances goes to infinity when the active control is truly effective. For both methods, the limiting conditional type I error rate is 0.025 as the ratio of variances goes to zero, and when the active control is truly ineffective the limiting type I error rate is 1 as the ratio of variances goes to infinity. When the probability that a random therapy is effective equals 0.25 (leading to a probability of 0.0769 that the active control is ineffective from Table 5.17) and the ratio of variances is 4, the standard synthesis method for 50% retention has a simulation-based type I error rate of $0.9231 \times 0.0268 + 0.0769 \times 0.3217 = 0.0495$ and the 95–95 method has a simulation-based type I error rate of $0.9231 \times 0.0030 + 0.0769 \times 0.0666 = 0.0079$.

## 5.5 A Case in Oncology

For the arithmetic and geometric definitions of the retention fraction, the 95–95 approach and the synthesis method will be applied to an example in two trials comparing capecitabine (the experimental therapy) with the Mayo clinic regimen of 5-fluorouracil (5-FU) with leucovorin in previously untreated metastatic colorectal. The non-inferiority analyses on overall survival were not prospectively planned.[34] A retrospective evaluation of the power for the studies will also be discussed.

The use of capecitabine for first-line metastatic colorectal cancer was approved by the FDA in April 2001. The basic information provided for this example can be found from several sources.[1,34,35] Capecitabine is an oral fluoropyrimidine—a prodrug of 5′-deoxy-5-fluorouridine, which is converted to 5-FU. Capecitabine was compared with the Mayo clinic regimen of 5-FU with leucovorin (5-FU + LV) in two clinical trials, each involving about 600 subjects having untreated metastatic colorectal cancer. The Mayo clinic regimen of 5-FU + LV consists of 20 mg/m$^2$ leucovorin by intravenous bolus followed by 425 mg/m$^2$ 5-FU by intravenous bolus on days 1–5 every 28-day cycle. Capecitabine was given 1250 mg/m$^2$ twice daily for 14 days followed by 1-week rest (3-week cycle) for each trial.

Each study was designed to demonstrate noninferior response rates for capecitabine versus 5-FU + LV and resulted in demonstrating superior response rates for capecitabine versus 5-FU + LV. The non-inferiority analysis on overall survival was retrospectively done for each study. Additionally, exploratory analyses will be performed on the basis of a fixed-effects meta-analysis of the two studies.

**TABLE 5.20**

Summary of Results on a Meta-Analysis on Overall Survival Comparing
5-FU + LV with 5-FU

| Log Hazard Ratio[a] | Standard Error | 95% Confidence Interval for Hazard Ratio[a] |
|---|---|---|
| 0.2341 | 0.0750 | (1.091–1.464) |

[a] Hazard ratios are 5-FU + LV/5-FU.

Bolus 5-FU by itself has not demonstrated an effect on overall survival for first-line metastatic colorectal cancer, whereas the addition of leucovorin to bolus 5-FU appears to improve overall survival (see Table 5.20). There is an assumption in the trials that the use of the fluoropyrimidine capecitabine does not require the additional use of leucovorin to improve its effect. A systematic review was done to find those clinical trials that compared similar regimens of 5-FU + LV as the Mayo clinic regimen of 5-FU + LV to the same regimen minus leucovorin. For each capecitabine trial, capecitabine would be regarded as noninferior to 5-FU + LV on overall survival if capecitabine retains greater than 50% of the historical survival effect of 5-FU + LV relative to 5-FU alone.

### 5.5.1 Applying the Arithmetic Definition of Retention Fraction

We will apply the arithmetic definition of the retention fraction given in Equation 5.3 to the determination of a non-inferiority margin and in the application of the synthesis method. Here, the effect of 5-FU + LV on overall survival is defined by the 5-FU versus 5-FU + LV hazard ratio minus 1.

The results of a random-effects meta-analysis on the overall survival log-hazard ratio of 5-FU + LV versus 5-FU is provided in Table 5.20. A summary of the overall survival results from the capecitabine clinical trials is provided in Table 5.21.

The results of non-inferiority analyses on overall survival based on the results of the random-effects meta-analysis of the effect of adding leucovorin to 5-FU are provided in the FDA Medical-Statistical review for Xeloda.[34]

**TABLE 5.21**

Summary of Overall Survival Results for Two Capecitabine Clinical Trials

| Study | Total Number of Deaths | Log Hazard Ratio[a] | Standard Error | 95% Confidence Interval for the Hazard Ratio[a] |
|---|---|---|---|---|
| Study 1 | 533 | −0.0036 | 0.0868 | (0.841, 1.181) |
| Study 2 | 533 | −0.0844 | 0.0867 | (0.775, 1.089) |
| Combined[b] | 1066 | −0.0440 | 0.0613 | (0.849, 1.079) |

[a] Hazard ratios are capecitabine/5-FU + LV.
[b] A fixed-effects meta-analysis of studies 1 and 2.

Both a 95–95 approach and a synthesis method with a test statistic similar to Equation 5.10 were used.

On the basis of the 95–95 approach, the non-inferiority efficacy threshold for the capecitabine versus 5-FU + LV hazard ratio is 1.091. The upper limits of the 95% confidence interval for the capecitabine versus 5-FU + LV hazard ratio are 1.181 and 1.089 for studies 1 and 2, respectively. Thus, study 1 fails to meet this criterion for determining efficacy, whereas study 2 barely succeeds. For the fixed-effects meta-analysis of studies 1 and 2, the upper limit of the 95% confidence interval for the capecitabine versus 5-FU + LV hazard ratio is 1.079, satisfying the non-inferiority efficacy threshold as determined by the 95–95 approach. The non-inferiority threshold for the capecitabine versus 5-FU + LV hazard ratio is 1.045 ($=1 + (1.09 - 1)/2$) based on the 95–95 approach and a retention fraction of 50%. Studies 1 and 2 and their combined analysis all fail to satisfy this threshold with the upper limits of their respective 95% confidence intervals for the capecitabine versus 5-FU + LV hazard ratio above 1.045.

When targeting a one-sided type I error rate of 0.025 and assuming that $Z_1$ in Equation 5.10 has an approximate standard normal distribution at the boundary of the null hypothesis in Expression 5.4, the critical value for the synthesis method is –1.96. For studies 1 and 2, the values for the test statistic $Z_1$ in Equation 5.10 for a retention fraction of 50% are –1.32 and –2.16, respectively. From this synthesis method, study 1 fails to demonstrate that capecitabine retains more than 50% of the historical effect of 5-FU + LV relative to 5-FU (–1.32 > –1.96), whereas study 2 demonstrates that capecitabine retains more than 50% of the historical effect of 5-FU + LV relative to 5-FU (–2.16 < –1.96). For the combined analysis, $Z_1 = -2.26$.

A Fieller lower confidence limit for the retention fraction can be determined by setting $Z_1 = -1.96$ and solving for the unspecified retention fraction, $\lambda_0$. From this, we see that study 1 (study 2) demonstrated that capecitabine retains at least 10% (61%) of the historical effect of 5-FU + LV on overall survival. The combined analysis demonstrated that capecitabine retains at least 64% of the historical effect of 5-FU + LV on overall survival.

### 5.5.2 Applying the Geometric Definition of Retention Fraction

We will apply the geometric definition of the retention fraction given in Equation 5.5 to the determination of a non-inferiority margin and in the application of the synthesis method using the test statistic given in Equation 5.12. Here, the effect of 5-FU + LV on overall survival is defined by the 5-FU versus 5-FU + LV log-hazard ratio.

As the two definitions agree at 0% retention, they agree on the efficacy threshold for the capecitabine versus 5-FU + LV hazard ratio of 1.091. The non-inferiority threshold for the capecitabine versus 5-FU + LV hazard ratio is 1.044 ($= \exp(\ln(1.091)/2)$) based on the 95–95 approach and a retention fraction of 50%. The conclusions are the same as those drawn earlier with the 95–95 approach using the arithmetic definition of the retention fraction.

For this application, it was argued by Rothmann[2] that $Z_2$ has an approximate standard normal distribution at the boundary of the null hypothesis in Expression 5.6 when the constancy assumption holds. Thus, when targeting a one-sided type I error rate of 0.025, the critical value for the standard synthesis method is −1.96.

For studies 1 and 2, the values for the test statistic $Z_2$ in Equation 5.12 for a retention fraction of 50% are −1.28 and −2.13, respectively. From the standard synthesis method, study 1 fails to demonstrate that capecitabine retains more than 50% of the historical effect of 5-FU + LV relative to 5-FU (−1.28 > −1.96), whereas study 2 demonstrates that capecitabine retains more than 50% of the historical effect of 5-FU + LV relative to 5-FU (−2.13 < −1.96). For the combined analysis, $Z_1 = -2.24$.

A Fieller lower confidence limit for the retention fraction can be determined by setting $Z_1 = -1.96$ and solving for the unspecified retention fraction, $\lambda_0$. From this, we see that study 1 (study 2) demonstrated that capecitabine retains at least 9% (59%) of the historical effect of 5-FU + LV on overall survival. The combined analysis demonstrated that capecitabine retains at least 62% of the historical effect of 5-FU + LV on overall survival. For these percentages to apply for the effect of 5-FU + LV on overall survival for the individual capecitabine trials, the statistical model that is used to estimate the historical effect of 5-FU + LV on overall survival must also be the appropriate model to unbiasedly estimate the effect of 5-FU + LV on overall survival for that capecitabine trial. For study 2 (the combined analysis) using a normalized test statistic, noninferior overall survival would be demonstrated by capecitabine versus 5-FU + LV if the estimator of the historical survival effect were reduced by 18% (24%).

When testing for better than 50% retention by capecitabine of the 5-FU + LV effect on overall survival (relative to 5-FU), the adjustment that the 95–95 approach makes on the threshold for the log-hazard ratio of capecitabine versus 5-FU + LV relative to the synthesis method is 0.058 as per Expression 5.24. The "threshold" for the hazard ratio of capecitabine versus 5-FU + LV is 1.107 greater than the threshold of 1.044 from the 95–95 approach by 0.063. For indirectly testing that capecitabine is superior to 5-FU on overall survival, the adjustment that the 95–95 approach makes on threshold for the log-hazard ratio of capecitabine versus 5-FU + LV relative to the synthesis method is 0.092. The "threshold" for the hazard ratio of capecitabine versus 5-FU + LV is 1.196 greater than the threshold of 1.091 from the 95–95 approach by 0.105.

### 5.5.3 Power of Such Procedures

Conditioning on the results from the estimation of the historical effect size of 5-FU + LV on overall survival, Table 5.22 provides the power for the individual studies and the combined analyses at various capecitabine versus 5-FU + LV hazard ratios for these two methods of analysis. For each case in

**TABLE 5.22**

Power Is Provided for Individual Studies and Combined Analyses at Various True Hazard Ratios of Capecitabine versus 5-FU + LV

| | Method[a] | | | |
|---|---|---|---|---|
| | Non-Inferiority Threshold of 1.044 | | Standard Synthesis Method with 50% Retention of the Control Effect | |
| True Hazard Ratio[b] | 533 Deaths (%) | 1066 Deaths (%) | 533 Deaths (%) | 1066 Deaths (%) |
| 1.05 | 2 | 2 | 9 | 12 |
| 1.00 | 7 | 10 | 22 | 35 |
| 0.95 | 19 | 34 | 42 | 67 |
| 0.90 | 40 | 68 | 67 | 91 |
| 0.85 | 66 | 92 | 86 | 99 |

[a] Based on the geometric definition of the proportion of effect retained.
[b] For overall survival of capecitabine versus 5-FU + LV.

Table 5.22, as the alternative becomes a slightly more and more advantageous effect for capecitabine, the power increases greatly. Also, for each alternative and fixed number of deaths, the power is slightly higher for the standard synthesis method than for the 95–95 approach with threshold of 1.044.

Conditioning on the results from the estimation of the historical effect size of 5-FU + LV on overall survival, Table 5.23 provides, for a one-to-one randomization, the number of events to have 90% power at various capecitabine versus 5-FU + LV hazard ratios for these two methods of analysis. For each case in Table 5.23, as the alternative becomes a slightly more and more advantageous effect for capecitabine, the number of events sharply decreases. This helps illustrate the importance of a proper choice of the alternative to size the non-inferiority trial. Also, for each hazard ratio, the number of events is smaller for the test procedure that uses a normalized test statistic than for the test procedure that is based on a two 95% confidence interval approach.

**TABLE 5.23**

For Each Method, Number of Events Is Provided to Have 90% Power for Various True Hazard Ratios of Capecitabine versus 5-FU + LV (1:1 Randomization)

| | Method | |
|---|---|---|
| True Hazard Ratio[a] | Non-Inferiority Cutoff of 1.044 | Standard Synthesis Method with 50% Retention of the Control Effect |
| 1.00 | 22,669 | 7,192 |
| 0.95 | 4,721 | 2,113 |
| 0.90 | 1,908 | 1,030 |
| 0.85 | 995 | 606 |

[a] For overall survival of capecitabine versus 5-FU + LV.

# References

1. Rothmann, M. et al., Design and analysis of non-inferiority mortality trials in oncology, *Stat. Med.*, 22, 239–264, 2003.
2. Rothmann, M., Type I error probabilities based on design-stage strategies with applications to non-inferiority trials, *J. Biopharm. Stat.*, 15, 109–127, 2005.
3. Wang, S.-J., Hung, H.M.J., and Tsong, Y., Utility and pitfalls of some statistical methods in active controlled clinical trials, *Control Clin. Trials*, 23, 15–28, 2002.
4. Temple, R. and Ellenberg, S.S., Placebo-controlled trials and active-controlled trials in the evaluation of new treatments: Part 1. Ethical and scientific issues, *Ann. Intern. Med.*, 133, 455–463, 2000.
5. Hung, H.M.J., Wang, S.-J., and O'Neill, R., Issues with statistical risks for testing methods in non-inferiority trial without a placebo arm, *J. Biopharm. Stat.*, 17, 201–213, 2007.
6. U.S. Food and Drug Administration, Guidance for industry: Non-inferiority clinical trials (draft guidance), March 2010.
7. Sankoh, A.J., A note on the conservativeness of the confidence interval approach for the selection of non-inferiority margin in the two-arm active-control trial, *Stat. Med.*, 27, 3732–3742, 2008.
8. Hauck, W.W. and Anderson, S., Some issues in the design and analysis of equivalence trials, *Drug Inf. J.*, 33, 109–118, 1999.
9. Lawrence, J., Some remarks about the analysis of active control studies, *Biom. J.*, 47, 616–622, 2005.
10. Gupta, G. et al., Statistical review experiences in equivalence testing at FDA/CBER, *Proc. Biopharm. Sec.*, American Statistical Association Alexandria, VA, 1999, 220–223.
11. The ASSENT-2 Investigators, Single bolus tenecteplase compared with front-loaded alteplase in acute myocardial infarction: ASSENT-2 double-blind randomised trial, *Lancet*, 354, 716–22 1999.
12. U.S. Food and Drug Administration Oncologic Drugs Advisory Committee meeting July 27, 2004, transcript at http://www.fda.gov/ohrms/dockets/ac/04/transcripts/2004-4060T1.pdf.
13. Holmgren, E.B., Establishing equivalence by showing that a specified percentage of the effect of the active control over placebo is maintained, *J. Biopharm. Stat.*, 9, 651–659, 1999.
14. Simon, R., Bayesian design and analysis of active control clinical trials, *Biometrics*, 55, 484–487, 1999.
15. Hasselblad, V. and Kong, D.F., Statistical methods for comparison to placebo in active-control trials, *Drug Inf. J.*, 35, 435–449, 2001.
16. Snapinn, S. and Jiang, Q., Controlling the type 1 error rate in non-inferiority trials, *Stat. Med.*, 27, 371–381, 2008.
17. Clinton, B. and Gore, A., Reinventing regulation of drugs and medical devices, *National Performance Review*, April 1995.
18. Snapinn, S. and Jiang, Q., Preservation of effect and the regulatory approval of new treatments on the basis of non-inferiority trials, *Stat. Med.*, 27, 382–391, 2008.

19. Hung, H.M.J., Wang, S-J,. and O'Neill, R.T., A regulatory perspective on choice of margin and statistical inference issue in non-inferiority trials, *Biom. J.*, 47, 28–36, 2005.

20. Chen, G., Wang, Y-C., and Chi, Y.H.G., Hypotheses and type I error in active-control non-inferiority trials, *J. Biopharm. Stat.* 14, 301–313, 2004.

21. Fisher, L.D., Active control trials: What about a placebo? A method illustrated with clopidogrel, aspirin and placebo, *J. Am. Coll. Cardiol*, 31, 49A, 1998.

22. Fisher, L.D., Gent, M., and Büller, H.R., Active-control trials: How would a new agent compare with placebo? A method illustrated with clopidogrel, aspirin, and placebo, *Am. Heart J.*, 141, 26–32, 2001.

23. Rothmann, M.D. and Tsou, H., On non-inferiority analysis based on delta-method confidence intervals, *J. Biopharm. Stat.*, 13, 565–583, 2003.

24. Odem-Davis, K.S., Current issues in non-inferiority trials, dissertation, University of Washington, Department of Biostatistics, 2010.

25. Wang, S.-J. and Hung, H.M.J., TACT method for non-inferiority testing in active controlled trials, *Stat. Med.*, 22, 227–238, 2003.

26. Carroll, K.J., Active-controlled non-inferiority trials in oncology: Arbitrary limits, infeasible sample sizes and uninformative data analysis. Is there another way? *Pharm. Stat.*, 5, 283–293, 2006.

27. Snapinn, S.M., Alternatives for discounting in the analysis of non-inferiority trials, *J. Biopharm. Stat.*, 14, 263–273, 2004.

28. Wiens, B., Choosing an equivalence limit for non-inferiority or equivalence studies, *Control Clin. Trials*, 1–14, 2002.

29. Fleming, T.R., Current issues in non-inferiority trials, *Stat. Med.*, 27, 317–332, 2008.

30. Hettmansperger, T.P., Two-sample inference based on one-sample sign statistics, *Appl. Stat.*, 33, 45–51, 1984.

31. Hung, H.M.J. et al., Some fundamental issues with non-inferiority testing in active controlled clinical trials, *Stat. Med.*, 22, 213–225, 2003.

32. Follmann, D.A. and Proschan, M.A., Valid inferences in random effects meta-analysis, *Biometrics*, 55, 732–737, 1999.

33. Larholt, K., Tsiatis, A.A, and Gelber, R.D., Variability of coverage probabilities when applying a random effects methodology for meta-analysis, Harvard School Public Health Department of Biostatistics, unpublished, 1990.

34. FDA Medical-Statistical review for Xeloda (NDA 20-896), dated April, 23, 2001.

35. FDA/CDER New and Generic Drug Approvals: Xeloda product labeling, at http://www.fda.gov/cder/foi/label/2003/20896slr012_xeloda_lb1.pdf.

# 6

## Three-Arm Non-Inferiority Trials

## 6.1 Introduction

When designing a study to show that an experimental therapy is effective, it is sometimes possible to include a third arm in the study to obtain data on both a concurrent placebo control and an active control. Earlier in this book, we considered two-arm non-inferiority trials having only an active control when the use of a placebo control is unethical or problematic—for example, if an effective treatment is available for a disease with obvious discomfort or irreversible morbidity, it may be difficult to obtain permission from an ethical review board to include a placebo control and most likely impossible to obtain informed consent from potential study subjects. Alternatively, when a placebo control is ethical, the comparison of an experimental treatment to a placebo control is the gold standard and inclusion of an active control is generally not required. However, there are situations in which inclusion of both a placebo control and an active control are ethically and scientifically defensible.

A three-arm trial involving concurrent active and placebo controls may evolve in one of two ways—an active control may be added to a placebo-controlled trial where the objective is the demonstration of superior efficacy of the experimental therapy relative to placebo or a placebo control is added to a two-arm non-inferiority trial.

When the standard therapy has a large, important effect, use of placebo-controlled trials without a concurrent standard therapy arm may allow claims of effectiveness for drugs that are substantially less effective than standard therapy. Also, failure to demonstrate superiority of an experimental therapy to placebo can either be due to the experimental therapy being ineffective or due to the trial lacking assay sensitivity. Additional use of an active control arm (i.e., a three-arm trial) can assist in determining whether the study has assay sensitivity. If the active control is demonstrated to be superior to placebo, then the trial has assay sensitivity. If neither the active control nor the experimental therapy is demonstrated to be superior to placebo, the trial may have lacked assay sensitivity.

Reasons for including a placebo in a non-inferiority trial include uncertainty that the active control is effective in the setting of the non-inferiority trial, great heterogeneity in the active control effect in previous trials, and lack of complete medical understanding of the indication being studied.[1] When the active control has not consistently shown an effect over placebo, it may be necessary to additionally include a placebo arm (i.e., a three-arm trial) to assess assay sensitivity. Unless the experimental arm is demonstrated to be superior to the active control, omission of a placebo arm may prevent the determination of whether the clinical trial had assay sensitivity and provide results that are difficult to interpret.

As expressed by Temple and Ellenberg,[2] a three-arm non-inferiority trial is optimal because: (1) assay sensitivity can be assessed within the trial, (2) the effect of the experimental therapy can be directly evaluated, and (3) the effects of the experimental and active control arms can be compared. Additionally, since the effect of the active control therapy versus placebo can be estimated within the three-arm non-inferiority trial, there are no issues on applying an estimate of the active control effect from one study (or many studies) to another study. The three-arm non-inferiority trial can also be sized so as to have an estimate of the active control effect versus placebo that has the desired precision. When previous studies are used to estimate the active control effect, the precision of the estimate of the active control effect is what it is—you are stuck with it.

Pigeot et al.[3] proposed that in addition to an experimental and active control arm, a third arm of subjects receiving placebo should be used whenever possible in a non-inferiority study. The use of a three-arm trial may be considered when:

1. The use of a placebo control is ethical
2. The use of a placebo control will not preclude potential study subjects from giving informed consent to participate in the study
3. An active control is available that will provide additional information about the benefit and/or risk of the experimental medication compared with other available therapy

It may be rare that all these criteria are met. Potential situations include a disease that does not result in discomfort, mortality, or irreversible morbidity (perhaps for short-term studies of chronic indications in which there are limited acute symptoms, or for acute diseases with relatively mild symptoms). Another potential situation is when the active control has not been studied in a clinical trial in the disease under investigation but is hypothesized to work. A third potential situation is a disease area in which the active control is believed to confer benefit, but for whatever reason does not always show an advantage in direct comparisons to placebo (i.e., the studies lack assay sensitivity). An example of the first situation is a study of mild infections that are usually self-limiting without irreversible morbidity. An example of the second situation is a study of hypertension in a special population—for example, organ transplant recipients—when the only available active comparators are drugs that have been shown active as first-line treatment for mild to moderate hypertension but never studied comprehensively in the special population. An example of the third situation is in studies of dementia, in which assay sensitivity is commonly low. In each situation, it is conceivable that an experimental therapy could be studied in a three-arm trial, with comparisons to both placebo and an active control. In the second situation, a secondary advantage is that definitive information can be obtained on the comparator treatment compared with placebo, thereby advancing medical knowledge.

When a placebo-controlled trial is ethical, three-arm non-inferiority trials have also been designed to assess a new schedule of an approved drug to the approved schedule. The comparison of new schedule against placebo will address whether the new schedule is efficacious, whereas including the approved schedule will provide important information on how the two schedules compare in efficacy and safety. If the new schedule is efficacious but is clearly inferior to the approved schedule, use of the new schedule may not be appropriate.

It is important that the analysis objectives for all three arms be well understood and documented before the design is considered. If the purpose of the active control and placebo arms is to establish assay sensitivity, this will require a different analysis strategy than if the purpose is to establish the retention of at least some portion of the efficacy of the active control. Analyses and issues with each of these situations are discussed in the next section.

As in all non-inferiority trials, a secondary objective of superiority of the experimental treatment compared with the active control can be considered. This can be done by evaluating the difference between the experimental and active control therapies without considering the placebo arm, or by using the results of all three arms in estimating the percentage of the active control effect that is retained by the experimental therapy. Establishing assay sensitivity may require the extra step of comparing the active control to placebo. There is no agreement on whether this step requires an adjustment for multiple comparisons.

In each of these situations, a parallel group design is generally used and is assumed in the following sections. Crossover designs with two or three periods can be considered if treatment effects are not irreversible and if an appropriate washout period can be incorporated. Allocation of subjects to the various treatment groups need not be balanced: especially if the active control and experimental treatment are both thought to provide substantial efficacy compared with placebo, but the non-inferiority margin is relatively small compared with the effect versus placebo, the number of subjects allocated to receive placebo can be smaller than the number allocated to receive either of the other two treatments. Unless the active control and the experimental treatment are hypothesized to have substantially different efficacy or safety profiles, planned sample sizes in these two groups will generally be balanced.

## 6.2 Comparisons to Concurrent Controls

Several authors have proposed analyses to interpret three-arm non-inferiority trials. Before we introduce the details of these analyses, we will discuss the general aspects of such analyses. Because of the multiple comparisons problem that will occur with three arms, it is important to prospectively describe analysis methods and control of the type I error rate. With three arms in the trial, there are three potential pairwise treatment comparisons that can be considered: active versus placebo, active versus experimental, and experimental versus placebo. A risk–benefit advantage in the last comparison is necessary to deem an experimental product to be worthy of use, but may not be sufficient if the experimental therapy does not have at least a comparable risk–benefit profile as the active control. Analyses can also be considered that include all three treatments in a single hypothesis test, as will be discussed in analyzing the percentage of effect retained.

### 6.2.1 Superiority over Placebo

When the primary objective of the three-arm trial is to establish superiority of the experimental therapy over placebo in a direct comparison, the active control arm will be used to assess assay sensitivity by comparing the active control to placebo. If neither the experimental treatment nor the active control demonstrate efficacy beyond that provided by the placebo, the study might not have been able to identify any benefit provided by the experimental treatment because of design problems, poor study conduct, and/or random variability. Rather than a conclusion of lack of efficacy, a conclusion of lack of assay sensitivity might be appropriate. Alternatively, if the active control demonstrates benefit over placebo but the experimental treatment does not, then a conclusion of lack of sufficient benefit from the experimental treatment would be appropriate. Such data are useful when the active and experimental treatment yield similar responses and it is necessary to determine whether this is due to equal efficacy or equally ineffective treatment in a given study. This design and analysis strategy is common when treatments known to be active often do not demonstrate benefit in individual trials, such as in certain neurological indications.

If the active control is included to demonstrate assay sensitivity by showing that the active control is superior to placebo, some thought must be given to the desired control of the type I error rate. If comparison of both the active control to placebo and the experimental treatment to placebo are made at the full $\alpha$ level, there is an inflated chance that at least one of these two treatments will falsely be considered superior to placebo, if neither is. However, this might not be sufficient to disregard such an analysis strategy. The hypotheses can be structured such that the only hypothesis used for making a conclusion is the comparison of the experimental treatment to placebo, and the comparison of active control to placebo is presented for descriptive purposes and only interpreted in the event that the primary comparison is not positive. Testing the experimental treatment versus placebo at the full $\alpha$ level will control the probability that an ineffective treatment is considered effective, even though there is an inflated chance of rejecting at least one true null hypothesis. Note that a non-inferiority comparison in this case is of relatively low interest, and a non-inferiority margin $\delta$ might not even be proposed a priori. If a margin is proposed, a fixed sequence test would be a useful approach to multiplicity, with the first comparison being the experimental therapy to placebo and the second comparison being the non-inferiority comparison of the experimental therapy to the active control. Again, the comparison of the active control to placebo would be used to establish assay sensitivity, but will not be included in the $\alpha$-preserving multiple comparison procedure.

Concluding superiority of the experimental treatment compared with the active control can be a secondary objective of a trial in which the primary comparison is the experimental treatment to placebo. The comparison of the experimental treatment to the active control will generally follow the fixed-sequence approach, so other hypotheses are first tested and, conditional on demonstrating sufficient benefit, the experimental treatment is compared with the active control using the full two-sided $\alpha$ level. As an example, the first comparison could be a direct comparison of the experimental treatment versus placebo at the full $\alpha$ level. If this comparison shows significance favoring the experimental treatment, the second comparison would be the non-inferiority comparisons, also at the full $\alpha$ (or one-sided $\alpha/2$) level. If this again is positive, the third comparison might be to demonstrate that experimental treatment is superior to the active control, again at the full $\alpha$ level. This third comparison would be two sided; thus, it is possible with this strategy to demonstrate that the experimental treatment is worse or better than the active control.

Another use of an active control as a third arm in a non-inferiority trial might be to establish favorable risk–benefit of the experimental treatment compared with the active control. When the active control consistently demonstrates efficacy compared with placebo but is also associated with considerable toxicity, a less efficacious but better tolerated experimental treatment might be preferable and therefore should be considered so future patients can choose among treatments with various levels of efficacy and tolerability. Statistical methodology for combining measures of efficacy and toxicity are not commonly used, although they have been proposed.[4–6] Therefore, such comparisons of risk–benefit will generally be made more informally. This analysis strategy will use an $\alpha$-preserving strategy for assessment of efficacy, and conclusions pertaining to risk and risk–benefit will be made without formal hypothesis testing.

## 6.2.2 Non-Inferior to Active Control

*Inferences on Means.* When planning a three-arm non-inferiority study, one might be concerned about the relative efficacy of the experimental treatment versus placebo compared with the efficacy of the active treatment versus the control. Simply demonstrating that some small amount of efficacy over placebo is obtained can be insufficient if the magnitude of efficacy is small compared with the efficacy of the active control (when the active control is beneficial). In such a situation, the ratio

$$\lambda = \frac{\mu_E - \mu_P}{\mu_C - \mu_P} \tag{6.1}$$

is commonly used to measure the proportion of efficacy retained by the experimental treatment. The ratio $\lambda$ is often called the preservation ratio, the retention fraction, or one of many other similar phrases. It is the ratio of the effect of the experimental therapy to the effect of the control therapy and is presented in Equation 6.1 for the means of continuous data. This ratio will be of limited value unless both the active control and experimental treatment are shown superior to placebo in the three-arm trial, since a negative value will be difficult to interpret, and a positive value will be even more difficult to interpret when both the numerator and denominator are negative. A value of $\lambda$ that is positive (assuming that the denominator is positive) will imply that the experimental treatment is superior to placebo; however, a value of $\lambda$ that is close to 0 will imply that the experimental treatment provides relatively little benefit compared with the active control. A situation with positive but small $\lambda$ might lead to the conclusion that the active control should be the standard and the experimental treatment should not be widely adopted although it provides some benefit over placebo. When $\lambda > 1$, the experimental treatment provides more benefit than the active control.

An analysis strategy for a three-arm trial is to directly base inference on $\lambda$. The hypotheses are given as

$$H_o: \lambda \leq \lambda_o \text{ vs. } H_a: \lambda \geq \lambda_o \tag{6.2}$$

where $0 \leq \lambda_o \leq 1$ is prespecified. For continuous data and an overall assumption that $\mu_C - \mu_P > 0$, Pigeot et al.[3] rewrote the hypotheses in Equation 6.2 based on Equation 6.1 as

$$H_o: \mu_E - \lambda_o\mu_C - (1 - \lambda_o)\mu_P \leq 0 \text{ vs. } H_a: \mu_E - \lambda_o\mu_C - (1 - \lambda_o)\mu_P > 0 \tag{6.3}$$

The best linear unbiased estimator for the unknown quantity in the hypotheses of Expression 6.3 is found by substituting the sample means for the population means: $\psi(\hat{\mu}) = \bar{X}_E - \lambda_o \bar{X}_C - (1 - \lambda_o)\bar{X}_P$, where $\hat{\mu}$ is the vector of sample means. Under the null hypothesis, with the assumption of normality and homogenous variances, the test can use the statistic

$$T_{\text{BLUE}} = \frac{\bar{X}_E - \lambda_o \bar{X}_C - (1 - \lambda_o)\bar{X}_P}{\hat{\sigma}\sqrt{\dfrac{1}{n_E} + \dfrac{\lambda_o^2}{n_C} + \dfrac{(1 - \lambda_o)^2}{n_P}}} \tag{6.4}$$

where $\hat{\sigma}$ is the pooled estimate of variance and $n_P$, $n_C$, and $n_E$ are the corresponding sample sizes. Under $H_o$, $T_{\text{BLUE}}$ in Equation 6.4 follows a $t$ distribution with $n_E + n_{Ct} + n_P - 3$ degrees of freedom.

Pigeot et al.[3] discussed values of $\lambda_o$ and suggested $\lambda_o = 0.8$ could be appropriate. This implies that the experimental treatment retains 80% or more of the efficacy of the active control. Values of $\lambda_o = 0.5$, or even less, may be appropriate depending on the indication and the amount of efficacy and toxicity of the active control.

Hasler, Vonk, and Hothorn[7] considered the continuous case under the assumption of unequal, unknown variances. The test statistic becomes

$$\frac{\bar{X}_E - \lambda_o \bar{X}_C - (1 - \lambda_o)\bar{X}_P}{\sqrt{\dfrac{S_E^2}{n_E} + \dfrac{\lambda_o^2 S_C^2}{n_C} + \dfrac{(1 - \lambda_o)^2 S_P^2}{n_P}}}$$

and the critical value is based on a $t$ distribution with degrees of freedom, $\nu$, equal to the greatest integer less than or equal to

$$\frac{\left(\dfrac{s_E^2}{n_E} + \dfrac{\lambda_o^2 s_C^2}{n_C} + \dfrac{(1 - \lambda_o)^2 s_P^2}{n_P}\right)^2}{\dfrac{s_E^4}{n_E^2(n_E - 1)} + \dfrac{\lambda_o^4 s_C^4}{n_C^2(n_C - 1)} + \dfrac{(1 - \lambda_o)^4 s_P^4}{n_P^2(n_P - 1)}}$$

Koch and Tangen[8] provided a sample size formula for three-arm trials. Pigeot et al.[3] also discussed optimal allocation ratios of active control to experimental to placebo. In general, the active control and experimental treatment should have the same sample size. With equal sample sizes for the first two treatments, the optimal allocation ratio becomes $1:1:k_P$. Pigeot et al. showed that

$$k_P = \frac{(1 - \lambda_o)\sqrt{2 + 2\lambda_o}}{1 + \lambda_o^2} \tag{6.5}$$

was optimal for $\lambda_o < 1$. In other words, the optimal ratio depends only on the value of $\lambda_o$. For $\lambda_o = 0.8$, the ratio is 1:1:0.23, or approximately 9:9:2. For $\lambda_o = 0.5$, the optimal ratio is 1:1:0.69, or approximately 3:3:2. For $\lambda_o = 0.3213$, the ratio is approximately 1:1:1; whereas for even smaller values of $\lambda_o$, more subjects are required in the placebo group than in the other two groups. We have from Equation 6.5 that $k_P$ is decreasing in $\lambda_o$ for $0 < \lambda_o < 1$ and as $\lambda_o$ increases toward 1, $k_P$ decreases toward 0. For $\lambda_o < 0.5$, we recommend equal sample sizes in all three groups, as the power loss will not be large compared to the complexity of unequal ratios and the potential ethical concern of enrolling more subjects in the placebo group than the other groups. In the general case (the experimental and active control arms need not have equal sample sizes), as shown by Pigeot et al.,[3] the optimal allocation across the experimental, active control, and placebo arms is $1:\lambda_o:1 - \lambda_o$. Optimal allocation ratios when the variances are unequal among the three arms lead to larger allocation for the study arms with larger variances.[7] The procedure described by Hasler, Vonk, and Hothorn's paper[7] required that the ratio of variances among treatments be known, something that in practice is not known precisely. A two-stage sample size recalculation procedure for the three-arm testing problem is described by Schwartz and Denne.[9] The optimal allocation ratios for the two-stage procedure are the same as those for the fixed sample size case.

*Inferences on Proportions.* Similar methodologies have been proposed for demonstrating that a sufficient proportion of the effect of the active control was maintained when considering binomial data rather than continuous data.[10,11] Letting $\hat{p}_P$, $\hat{p}_C$, and $\hat{p}_E$ be the observed success proportions of desired outcomes for the placebo, active control, and experimental treatment groups, respectively, inference is made on the quantity $\lambda$, modified from Equation 6.1 for binary data $\left( \lambda = \dfrac{p_E - p_P}{p_C - p_P} \right)$. The hypotheses are expressed as

$$H_o: p_E - \lambda_o p_C - (1 - \lambda_o)p_P \leq 0 \text{ vs. } H_a: p_E - \lambda_o p_C - (1 - \lambda_o)p_P > 0 \qquad (6.6)$$

where $0 \leq \lambda_o \leq 1$ is prespecified. A Wald-type test statistic is $\psi/\sigma(\psi)$, where $\psi = \hat{p}_E - \lambda_o \hat{p}_C - (1 - \lambda_o)\hat{p}_P$ is again an alternative (linear) way of expressing the relative treatment effect (with $\psi = 0$ corresponding to $\lambda = \lambda_o$) and $\hat{\sigma}^2(\psi) = \hat{p}_E(1 - \hat{p}_E)/n_E + \lambda_o^2 \hat{p}_C(1 - \hat{p}_C)/n_C + (1 - \lambda_o)^2 \hat{p}_P(1 - \hat{p}_P)/n_P$ is the estimated standard error of $\psi$. Asymptotically, $\psi/\sigma(\psi)$ will be distributed as a standard normal when $\lambda_o$ is the correct retention fraction.

Tang and Tang[10] compared the type I error rate and power for the Wald-type test and the similar test using the null restricted maximum likelihood estimates of the true proportions when estimating the standard error, in analogous fashion to that of Farrington and Manning[12] for the two-sample inference on a risk ratio. For a desired one-sided test with $\alpha = 0.05$, simulations were performed to assess the type I error rate when $\lambda_o = 0.6$ and 0.8; $(p_P, p_C) = (0.1, 0.8), (0.2, 0.7), (0.3, 0.6),$ and $(0.4, 0.5)$; and $p_E = \lambda_o p_C + (1 - \lambda_o)p_P$ (which appears in Tang and Tang[10] as $p_E = \lambda_o p_C - (1 - \lambda_o)p_P$). The overall sample sizes were 30, 60, 90, and 120 with allocations to the experimental, active control, and placebo groups of 1:1:1, 2:2:1, and 3:2:1, respectively. According to their results, the simulated one-sided type I error probabilities ranged overall from 0.043 to 0.060 with the use of the restricted maximum likelihood estimates maintaining the desired one-sided type I error rate of 0.05 better than the Wald's test. The power when using the restricted maximum likelihood estimates to estimate the standard error was consistently slightly less than the power when using the sample proportions.

Kieser and Friede[11] also investigated the type I error rate for a three-arm non-inferiority test on proportions when using the Wald-type test and the analogous test based on the null restricted maximum likelihood estimates of the true proportions when estimating the standard error. Their calculations were based on the actual probabilities from the corresponding binomial distributions, not simulations, and differed from those of Tang and Tang.[10] All cases in Tang and Tang's study[10] were considered along with additional cases. Desired one-sided levels of $\alpha = 0.025$ and 0.05 were considered with $\lambda_o = 0.6$ and 0.8; $p_P = 0.05, 0.10, \ldots, 0.50$; $p_C - p_P = 0.05, 0.10, \ldots, 0.95$ (only those cases where $p_C \leq 1$); and $p_E = \lambda_o p_C + (1 - \lambda_o)p_P$. The overall sample sizes were 30, 60, 90, 120, 180, 240, and 300 with allocations to the experimental, active control, and placebo groups of 1:1:1, 2:2:1, and 3:2:1. Both procedures tended to have actual type I error rates above the desired rates of 0.025 and 0.05. The inflation was more profound for the Wald-type test. Interestingly, cases that had the greatest actual type I error rate for the Wald-type test (as high as 0.212) had the actual type I error rate maintained under the desired level of 0.025 or 0.05 when using the restricted maximum likelihood estimates to estimate the standard error.

Kieser and Friede further proposed sample size calculations to achieve a given power. Because power estimates depend on the variances, which differ under the null and alternative hypotheses, Kieser and Friede proposed several sample size formulae. The one with the best properties has for the overall sample size $N = (1 + k_E + k_C)(z_\alpha \tau_0 + z_\beta \tau_1)^2 \psi_1^{-2}$, where the allocation of subjects to the experimental, active control, and placebo groups is $k_E{:}k_C{:}1$, $\tau_i^2 = (1 - \lambda_o)^2 p_{P,i}(1 - p_{P,i}) + (\lambda_o^2/k_C)p_{C,i}(1 - p_{C,i}) + (1/k_E)p_{E,i}(1 - p_{E,i})$, where for $i = 0$ the proportions are under the null hypothesis and for $i = 1$ the proportions are in the alternative hypothesis. However, even this formula can be incorrect, so simulations are advised to confirm the power before conducting the study. In addition, the ratio $k_E{:}k_C{:}1$ does not in general have a unique point that maximizes power, so investigation of various values (with $k_C > k_E > 1$ often holding) is advised.

Additionally, because the hypotheses in Expression 6.2 assume that the active control is superior to placebo, Kieser and Friede[11] recommended testing first that the active control is superior to placebo at the full $\alpha$, and if superiority is concluded, proceed to testing for non-inferiority at the full $\alpha$. They further discussed how to size the trial under this testing sequence to achieve the desired power of concluding non-inferiority.

The inconsistent calculations reported by Tang and Tang[10] and Kieser and Friede[11] suggest additional caution against planning based on direct application of published results. Simulations or calculations before a new study should always be done to confirm the performance of testing plans, especially for newer research. Koch and Tangen[8] illustrated nonparametric analysis of covariance with three-arm non-inferiority trials. Nonparametric analysis has the advantage of robustness when the distributions of the data are not as expected. Using analysis of covariance further allows for variance reduction due to inclusion of factors affecting the variability of results.

*Inferences on a Log-Hazard Ratio.* For time-to-event endpoints, we will assume that the three underlying time-to-event distributions have proportional hazards. Let $\beta_{P/E}$ denote the placebo versus experimental log-hazard ratio and let $\beta_{P/C}$ denote the placebo versus active control log-hazard ratio. Then, for $\beta_{P/C} > 0$, the retention fraction is given by

$$\lambda = \beta_{P/E}/\beta_{P/C}$$

The hypotheses given in Expression 6.2 are to be tested where $0 \le \lambda_o \le 1$ is prespecified. The hypotheses can be reexpressed as

$$H_o: \beta_{P/E} - \lambda_o\beta_{P/C} \le 0 \text{ vs. } H_a: \beta_{P/E} - \lambda_o\beta_{P/C} > 0 \tag{6.7}$$

Alternatively, the hypotheses can be expressed around $\beta_{C/E} + (1 - \lambda_o)\beta_{P/C} = \beta_{P/E} - \lambda_o\beta_{P/C}$, where $\beta_{C/E}$ is the active control versus the experimental log-hazard ratio. All three treatment arms can be modeled using a Cox proportional hazards model. Let $\hat{\beta}_{P/E}$ and $\hat{\beta}_{P/C}$ denote the Wald estimators of the $\beta_{P/E}$ and $\beta_{P/C}$ from a Cox model. The test statistic is given by

$$\frac{\hat{\beta}_{P/E} - \lambda_o\hat{\beta}_{P/C}}{\sqrt{\text{Var}(\hat{\beta}_{P/E}) + \lambda_o^2\text{Var}(\hat{\beta}_{P/C}) - 2\lambda_o\text{Cov}(\hat{\beta}_{P/E}, \hat{\beta}_{P/C})}} \tag{6.8}$$

Software packages provide estimates of the variances and covariance under the assumption that $\beta_{P/E} = \beta_{P/C} = 0$. In practice, when the true log-hazard ratio is not far from zero, a good estimate of the denominator in Expression 6.8 is given by

$$\sqrt{1/r_E + \lambda_o^2/r_C + (1 - \lambda_o)^2/r_P} \tag{6.9}$$

where $r_E, r_C,$ and $r_P$ denote the number of events in the experimental, active control, and placebo arms, respectively. From Expressions 6.8 and 6.9, we have the test statistic

$$Z = \frac{\hat{\beta}_{P/E} - \lambda_o\hat{\beta}_{P/C}}{\sqrt{1/r_E + \lambda_o^2/r_C + (1 - \lambda_o)^2/r_P}} \tag{6.10}$$

The test rejects the null hypothesis in Expression 6.7 and concludes non-inferiority when $Z > z_{\alpha/2}$.

A similar test statistic to Equation 6.10 was used by Mielke, Munk, and Schacht[13] under the assumption of underlying exponential distributions. There, the estimator $\hat{\beta}_{P/E}$ ($\hat{\beta}_{P/C}$) is equal to the difference in the natural logarithms of the maximum likelihood estimators of the means of the experimental (active control) and placebo arms.

For all of these three-arm non-inferiority cases, a Fieller $100(1 - \alpha)$ confidence interval for $\lambda$ can be found by treating $\lambda_0$ as unknown and setting the test statistic equal to $\pm z_{\alpha/2}$ (or the analogous values from the appropriate $t$ distribution) and solving for $\lambda_0$.[3,8] If all the values in the confidence interval are greater than zero, superiority of the experimental arm to the placebo arm is concluded. If all the values in the confidence interval are greater than $\lambda_0$, non-inferiority of the experimental arm to the active control arm is concluded. If all the values in the confidence interval are greater than 1, superiority of the experimental arm to the active control arm is concluded.

*Capturing All Possibilities of Efficacy.* In determining whether the experimental therapy is efficacious or has adequate efficacy, the possibility that $\mu_C \leq \mu_P < \mu_E$ should be included, but is not included, in the non-inferiority inference. For a two-arm non-inferiority trial, superiority of the experimental therapy to the active control therapy is intended to imply non-inferiority and that the experimental therapy is effective (i.e., due to the assumption that the control therapy is "active"). Although the possibility that $\mu_P < \mu_C < \mu_E$ is included in the alternative hypothesis in Expression 6.3 by having the overall assumption that $\mu_P < \mu_C$ and that $\mu_E - \lambda_0\mu_C - (1 - \lambda_0)\mu_P > 0$ for some pre-specified $0 \leq \lambda_0 \leq 1$, the possibility that $\mu_C \leq \mu_P < \mu_E$ is excluded. The possibility of $\mu_C \leq \mu_P < \mu_E$ accounts for one-sixth of the overall, unrestricted parameter space for $(\mu_P, \mu_C, \mu_E)$, and the estimation of $\mu_E - \lambda_0\mu_C - (1 - \lambda_0)\mu_P$ that is done, including the modeling of the uncertainty in that estimation, does not preclude $\mu_C \leq \mu_P < \mu_E$. Order restricted or constrained inference is not done. The aforementioned test procedures do not estimate or model the estimation of $\mu_E - \lambda_0\mu_C - (1 - \lambda_0)\mu_P$ under the restriction that $\mu_P < \mu_C$.

Having as the alternative hypothesis

$$H_a: \{(\mu_P, \mu_C, \mu_E) : \mu_E - \lambda_0\mu_C - (1-\lambda_0)\mu_P > 0, \mu_P < \mu_C\} \cup$$
$$\{(\mu_P, \mu_C, \mu_E) : \mu_C \leq \mu_P < \mu_E\} \tag{6.11}$$

seems appropriate for three-arm inferences. The null hypothesis would be the complement within the unrestricted parameter space. Koch and Röhmel[14] also argued that the goal of testing in the three-arm setting should be to demonstrate that the experimental therapy is both efficacious (better than placebo) and noninferior to the active control therapy. They proposed to achieve this by performing separate tests for superiority to placebo and non-inferiority to the active control. The overall alternative hypothesis of Koch and Röhmel[14] is

$$H_a: \{(\mu_P, \mu_C, \mu_E): \mu_E > \max\{\mu_P, \mu_C - \delta\}\} \tag{6.12}$$

where $\delta > 0$ is the prespecified non-inferiority margin. To provide a some-what analogous form as the alternative hypothesis in Expression 6.11, the alternative hypothesis in Expression 6.12 can be expressed as $H_a: \{(\mu_P, \mu_C, \mu_E): \mu_E > \mu_C - \gamma \geq \mu_P, \text{ for any } 0 < \gamma \leq \delta\} \cup \{(\mu_P, \mu_C, \mu_E): \mu_C \leq \mu_P < \mu_E\}$.

After demonstrating that the experimental therapy is efficacious and non-inferior to the active control therapy, Koch and Röhmel suggested additionally testing whether (1) the active control is superior to placebo and (2) whether the experimental therapy is superior to the active control.

## 6.3 Bayesian Analyses

In this section, we will consider non-inferiority analysis based on three-arm studies using Bayesian approaches. For proportions and means, the param-eters are modeled based on independent random samples and independent prior distributions (for a joint prior distribution for the mean and variance). In addition to those posterior probabilities provided in Section 5.3.6, the pos-terior probability that the active control is superior to placebo can be deter-mined. A list of interested posterior probabilities include:

(a) The posterior probability that the experimental therapy is superior to placebo.

(b) The posterior probability that the active control is superior to placebo.

(c) The posterior probability that the experimental therapy is superior to the active control therapy.

(d) The posterior probability that the experimental therapy is superior to both the active control therapy and placebo.

(e) The posterior probability that the experimental therapy retains more than $100\lambda_0\%$ of the active control therapy's effect and the active con-trol therapy is superior to placebo.

(f) The posterior probability that the experimental therapy retains more than $100\lambda_0\%$ of the active control therapy's effect and the active control therapy is superior to placebo, or the experimental arm is superior to both the active control therapy and placebo (i.e., the experimental therapy is superior to placebo and noninferior to the active control therapy).

The determination that the experimental therapy is effective requires that the posterior probability in (a) is close to 1 (e.g., greater than 0.975). The posterior probabilities given in (e) and (f) are two possibilities for determining whether the experimental therapy is not unacceptably worse than the active control. We will discuss testing for means, proportions, and log-hazard ratios based on the calculation of the posterior probabilities in (e) and (f).

*Inferences on Means When the Variances Are Unknown.* In each arm, we will consider a joint Jeffreys prior density, $h$, where $h(\theta) \propto \theta^{-3/2}$ for $-\infty < \mu < \infty$, and $\theta > 0$ where $\theta = \sigma^2$ is the variance. For a given treatment arm, $X_1, X_2, \ldots, X_n$ is a random sample from a normal distribution with mean $\mu$ and variance $\theta$. Then the joint posterior density is given by

$$g(\mu, \theta \mid x_1, x_2, \ldots, x_n) \propto \theta^{-1/2} \exp\left\{-\frac{(\mu - \bar{x})^2}{2\theta/n}\right\} \times \theta^{-n/2-1} \exp\left\{-\frac{1}{2}\sum_{i=1}^{n}(x_i - \bar{x})^2/\theta\right\}$$

(6.13)

We see from Expression 6.13 that the joint density factors into the product of an inverse gamma marginal distribution for $\theta$ and a normal conditional distribution for $\mu$ given $\theta$. The inverse gamma distribution has shape and scale parameters equal to $n/2$ and $\sum_{i=1}^{n}(x_i - \bar{x})^2/2$, respectively, with a mean equal to $\sum_{i=1}^{n}(x_i - \bar{x})^2/(n-2)$ and a variance equal to $2\left(\sum_{i=1}^{n}(x_i - \bar{x})^2\right)^2 / [(n-2)^2(n-4)]$. Note that $\theta$ has an inverse gamma distribution with parameters $n/2$ and $\sum_{i=1}^{n}(x_i - \bar{x})^2/2$ if and only if $1/\theta$ has a gamma distribution with parameters $n/2$ and $2/\sum_{i=1}^{n}(x_i - \bar{x})^2$ with mean equal to $n/\sum_{i=1}^{n}(x_i - \bar{x})^2$.

Given $\theta$, $\mu$ has a normal distribution with mean equal to $\bar{x}$ and variance equal to $\theta/n$. For simulating probabilities involving $\mu$, a random value for $1/\theta$ can be taken from the gamma distribution with parameters $n/2$ and $2/\sum_{i=1}^{n}(x_i - \bar{x})^2$, and then a random value for $\mu$ can be taken from a normal distribution having mean $\bar{x}$ and variance $\theta/n$. The above procedure is quite similar to a procedure provided by Ghosh et al.[15]

The probability in (f) (see preceding list) is the posterior probability of the alternative hypothesis given in Expression 6.11. If $P(\mu_E - \lambda_o\mu_C - (1 - \lambda_o)\mu_P > 0, \mu_P < \mu_C$ or $\mu_C \leq \mu_P < \mu_E)$ exceeds some threshold (e.g., 0.975), the experimental therapy is concluded to be noninferior to the active control therapy and to be efficacious. The probability in (e) is the posterior probability of the alternative hypothesis given in Expression 6.3. The analogs of the frequentist tests would be based on $P(\mu_E - \lambda_o\mu_C - (1 - \lambda_o)\mu_P > 0)$, or $P(\mu_E - \lambda_o\mu_C - (1 - \lambda_o)\mu_P > 0, \mu_P < \mu_C)$, or both of $P(\mu_P < \mu_C)$ and $P(\mu_E - \lambda_o\mu_C - (1 - \lambda_o)\mu_P > 0)$. If the posterior probability exceeds the preselected threshold (or in the last case both probabilities exceed the threshold), non-inferiority of the experimental therapy would be concluded.

An alternative way of calculating posterior probabilities in this case is provided by Gamalo et al.[16] They discussed the use of a generalized $p$-value (i.e., the posterior probability of a one-sided null hypothesis) and a generalized confidence interval (i.e., a credible interval) for $\mu_E - \lambda_0 \mu_C - (1 - \lambda_0)\mu_P$ when the variances are unknown and are not assumed equal. For arm $i$, $i$ = C, E, P, let $n_i$ denote the number of subjects on that arm, $\bar{x}_i$ denote the observed sample mean, and $s_i$ denote the observed standard deviation (i.e., $s_i^2$ has the form $\sum_{j=1}^{n_i}(x_{j,i} - \bar{x})^2/(n_i - 1)$). The means $\mu_C$, $\mu_E$, and $\mu_P$ are independent where for $i$ = C, E, P, $\mu_i$ has a posterior distribution equal to the distribution of

$$\bar{x}_i + T_i s_i / \sqrt{n_i}$$

where $T_i$ has a $t$ distribution with $n_i - 1$ degrees of freedom.

Similar results are reported from applying the procedure described by Gamalo et al.[16] as with applying the procedure of Hasler, Vonk, and Hothorn[7] in testing the hypotheses in Expression 6.3. The advantage of the Bayesian procedure is that the uncertainty of the active control is superior to placebo (i.e., $\mu_P < \mu_C$) and the possibility that $\mu_C \le \mu_P < \mu_E$ can be included directly into the testing procedure. That is, posterior probabilities like those in (e) and (f) can be calculated. Gamalo et al.[16] validated the type I error rate of their procedure in testing the hypotheses in Expression 6.3 with simulations based on a model that includes modeling the variances.

The hypotheses given in Expression 6.12 can also be tested either based on the posterior probability of $\mu_E > \max\{\mu_P, \mu_C - \delta\}$ or based on $\min\{P(\mu_E > \mu_P),$ $P(\mu_E > \mu_C - \delta)\}$. Comparing $\min\{P(\mu_E > \mu_P), P(\mu_E > \mu_C - \delta)\}$ to a threshold of 0.975 would be analogous to two separate one-sided tests at level 0.025 that the experimental therapy is superior to placebo (i.e., $\mu_E > \mu_P$) and that the experimental therapy is noninferior to the active control therapy (i.e., $\mu_E > \mu_C - \delta$).

The two-arm versions of these Bayesian approaches are discussed in Section 12.2.4, along with examples that calculate posterior probabilities and credible intervals for the difference in means.

*Inferences on Proportions.* For each arm we will consider a beta prior distribution for the probability of a success. For a random sample of $n$ binary observations where $x$ are successes, a beta prior distribution with parameters $\alpha$ and $\beta$ for $p$, the probability of success, leads to a posterior distribution for $p$ with parameters $\alpha + x$ and $\beta + n - x$. A Jeffreys prior distribution has $\alpha = \beta = 0.5$.

For proportions where a "success" is desirable, the probability in (f) is the posterior probability that $p_E - \lambda_0 p_C - (1 - \lambda_0)p_P > 0$, $p_P < p_C$ or $p_C \le p_P < p_E$. If that probability exceeds some threshold (e.g., 0.975), then the experimental therapy is concluded to be noninferior to the active control therapy and efficacious. The probability in (e) is the posterior probability of the alternative hypothesis given in Expression 6.6. As with means, the analogs of the frequentist tests would be based on $P(p_E - \lambda_o p_C - (1 - \lambda_o)p_P > 0)$, or $P(p_E - \lambda_o p_C - (1 - \lambda_o)p_P > 0, p_P < p_C)$, or both of $P(p_P < p_C)$ and $P(p_E - \lambda_o p_C - (1 - \lambda_o)p_P > 0)$. If the posterior probability exceeds the preselected threshold (or in the last case both probabilities exceed the threshold), non-inferiority of the experimental therapy would be concluded.

*Inferences Based on Log-Hazard Ratios.* Once again we will assume that the three underlying time-to-event distributions have proportional hazards. Asymptotically, $(\hat{\theta}_{P/E}, \hat{\theta}_{P/C})$ can be modeled as having a bivariate normal distribution with mean $(\theta_{P/E}, \theta_{P/C})$ and variances denoted by $\sigma^2_{P/E}$ and $\sigma^2_{C/E}$ and correlation denoted by $\rho$. Using this bivariate normal distribution with $\sigma^2_{P/E}$, $\sigma^2_{C/E}$, and $\rho$ treated as known values and a constant, improper prior density for $(\theta_{P/E}, \theta_{P/C})$ leads to a joint posterior distribution for $(\theta_{P/E}, \theta_{P/C})$ with mean $(\hat{\theta}_{P/E}, \hat{\theta}_{P/C})$, variances $\sigma^2_{P/E}$ and $\sigma^2_{C/E}$, and correlation $\rho$. Unlike the estimator of the standard error of a sample mean from a normal random sample, the estimators of standard error for a log-hazard ratio are quite stable in the vast majority of applications. Likewise, the estimator of $\rho$ is fairly stable. Therefore, for the Bayesian model, with slight crudeness, we use estimates of $\sigma^2_{P/E}$ and $\sigma^2_{C/E}$ and correlation $\rho$ as the true values. A parallel to using the test statistic in Equation 6.10 would use $1/r_P + 1/r_E$, $1/r_P + 1/r_C$, and $1/\sqrt{(1 + r_P/r_C)(1 + r_P/r_E)}$ as the values for $\sigma^2_{P/E}$ and $\sigma^2_{C/E}$ and $\rho$. Then the posterior probability of $\theta_{P/E} - \lambda_o \theta_{P/C} \leq 0$ would equal exactly the one-sided $p$-value from using the test statistic in Equation 6.10.

For time-to-event endpoints where the event is undesirable (i.e., longer times are more desirable), the probability in (f) is the posterior probability that $\theta_{P/E} - \lambda_o \theta_{P/C} > 0$, $\theta_{P/C} > 0$, or $\theta_{P/C} \leq 0 < \theta_{P/E}$. If that probability exceeds some threshold (e.g., 0.975), then the experimental therapy is concluded to be noninferior to the active control therapy and to be efficacious. The probability in (e) is the posterior probability of the alternative hypothesis given in Expression 6.7. As with means and proportions, the analogs of the frequentist tests would be based on $P(\theta_{P/E} - \lambda_o \theta_{P/C} > 0)$, or $P(\theta_{P/E} - \lambda_o \theta_{P/C} > 0, \theta_{P/C} > 0)$, or both $P(\theta_{P/C} > 0)$ and $P(\theta_{P/E} - \lambda_o \theta_{P/C} > 0)$. If the posterior probability exceeds the preselected threshold (or in the last case both probabilities exceed the threshold), non-inferiority of the experimental therapy would be concluded.

Example 6.1 illustrates the use of a frequentist and a Bayesian method in a three-arm testing involving proportions.

## Example 6.1

To demonstrate the calculation of the various posterior probabilities of (a)–(f) and some of the issues, we consider the example of the reporting rates of adverse events in Tang and Tang's study of patients with functional dyspepsia.[10] The motivation for doing non-inferiority testing is not clear and we do not recommend doing non-inferiority testing in this fashion. For this example, we will assume that the greater the reporting rate of adverse events the better. Adverse events were reported in 7 of 61 patients randomized to placebo, 10 of 59 patients randomized to cisapride (the active control), and 12 of 58 patients randomized to simethicone (the experimental therapy). For testing the hypotheses in Expression 6.6 with $\lambda_o = 0.8$, the one-sided $p$-value = 0.234 for the Wald-type test as given by Tang and Tang.[10] It should be noted that the 95% confidence intervals for the difference in rates between the simethicone and placebo arms, and between the cisapride and placebo arm are −0.039 to 0.224 and −0.070 to 0.179, respectively. Thus, neither the active control nor experimental therapy demonstrated a higher underlying reporting rate of adverse events. Also, for every $-\infty < \lambda_o < \infty$, the value for Wald-type test statistic is between −1.96 and 1.96. Therefore, ignoring that the active control was not demonstrated to be "superior" to placebo, the Fieller 95% confidence interval for the retention fraction is $-\infty$ to $\infty$.

Jeffreys' prior distributions were used for each of $p_E$, $p_C$, and $p_P$. Posterior probabilities and credible intervals were approximated from 100,000 simulations. The simulated 95% credible intervals for the difference in rates between the simethicone and placebo arms, and between the cisapride and placebo arm were similar to the Wald's 95% confidence intervals and are (−0.038 to 0.224) and (−0.070 to 0.180), respectively. The simulated 95% credible interval for the difference in rates between the simethicone and cisapride arms is (−0.104 to 0.178). Simulated posterior probabilities of interest are given in Table 6.1.

Note that, in (f), $p_E > p_P \geq p_C$ implies $p_E - 0.8p_C - 0.2p_P > 0$. From Table 6.1, the simulated posterior probability of $p_E - 0.8p_C - 0.2p_P > 0$ and $p_C > p_P$, or $p_E > p_P \geq p_C$ equals 0.738, which is far smaller than 0.975. The experimental arm has not demonstrated the combination of non-inferiority and efficacy (i.e., adverse events reporting rates greater than placebo). The direct analog of the one-sided $p$-value of .234 of the Wald-type test in testing the hypotheses in Expression 6.6 is the simulated posterior probability for $p_E - 0.8p_C - 0.2p_P > 0$ of 0.763 (i.e., the simulated posterior probability of the null hypothesis is 0.237 = 1 − 0.763). However, in 2.5% of the simulations, $p_E - 0.8p_C - 0.2p_P > 0$ and $p_P > p_E > p_C$. The uncertainty that $p_P > p_E > p_C$ is not accounted for in the Wald's-type test of the hypotheses in Expression 6.6.

**TABLE 6.1**

Simulated Posterior Probabilities of Interest

| Event | Simulated Posterior Probability |
| --- | --- |
| (a) $p_E > p_P$ | 0.916 |
| (b) $p_C > p_P$ | 0.806 |
| (c) $p_E > p_C$ | 0.696 |
| (d) $p_E > \max\{p_C, p_P\}$ | 0.667 |
| (e) $p_E - 0.8p_C - 0.2p_P > 0$ and $p_C > p_P$ | 0.589 |
| (f) $p_E - 0.8p_C - 0.2p_P > 0$ and $p_C > p_P$, or $p_E > p_P \geq p_C$ | 0.738 |
| $p_E - 0.8p_C - 0.2p_P > 0$ and $p_P > p_E > p_C$ | 0.025 |
| $p_E - 0.8p_C - 0.2p_P > 0$ | 0.763 |

# References

1. Koch, G.G., Comments on 'current issues in non-inferiority trials' by Thomas R. Fleming, *Stat. Med.*, 27, 333–342, 2008.
2. Temple, R. and Ellenberg S.S., Placebo-controlled trials and active-controlled trials in the evaluation of new treatments: Part 1. Ethical and scientific issues, *Ann. Intern. Med.*, 133, 455–463, 2000.
3. Pigeot, I. et al., Assessing non-inferiority of a new treatment in a three-arm clinical trial including a placebo, *Stat. Med.*, 22, 883–899, 2003.
4. Letierce, A. et al., Two-treatment comparison based on joint toxicity and efficacy ordered alternatives in cancer trials, *Stat. Med.*, 22, 859–868, 2003.
5. Jennison, C. and Turnbull, B.W., Group sequential tests for bivariate response: Interim analyses of clinical trials with both efficacy and safety endpoints, *Biometrics*, 49, 741–752, 1993.
6. Thall, P.F. and Cheng, S.-C., Treatment comparisons based on two-dimensional safety and efficacy alternatives in oncology trials, *Biometrics*, 55, 746–753, 1999.
7. Hasler, M., Vonk, R., and Hothorn, L.A., Assessing non-inferiority of a new treatment in a three-arm trial in the presence of heteroscedasticity, *Stat. Med.*, 27, 490–503, 2008.
8. Koch, G.G. and Tangen, C.M., Nonparametric analysis of covariance and its role in non-inferiority clinical trials, *Drug Inf. J.*, 33, 1145–1159, 1999.
9. Schwartz, T.A. and Denne, J.S., A two-stage sample size recalculation procedure for placebo- and active-controlled non-inferiority trials, *Stat. Med.* 45, 3396–3406, 2006.
10. Tang, M.-L. and Tang, N.-S., Tests of non-inferiority via rate difference for three-arm clinical trials with placebo, *J. Biopharm. Stat.*, 14, 337–347, 2004.
11. Kieser, M. and Friede, T., Planning and analysis of three-arm non-inferiority trials with binary endpoints, *Stat. Med.*, 26, 253–273, 2007.
12. Farrington, C.P. and Manning, G., Test statistics and sample size formulae for comparative binomial trials with null hypothesis of non-zero risk difference or non-unity relative risk, *Stat. Med.*, 9, 1447–1454, 1990.
13. Mielke, M., Munk, A., and Schacht, A., The assessment of non-inferiority in a gold standard design with censored, exponentially distributed endpoints, *Stat. Med.*, 27, 5093–5110, 2008.
14. Koch, A. and Röhmel, J., Hypothesis testing in the 'gold standard' design for proving the efficacy of an experimental treatment relative to placebo and a reference, *J. Biopharm. Stat.*, 14, 315–325, 2004.
15. Ghosh, P. et al., Assessing non-inferiority in a three-arm trial using the Bayesian approach, Technical report, Memorial Sloan-Kettering Cancer Center, 2010.
16. Gamalo, M. et al., A Generalized *p*-value approach for assessing non-inferiority in a three-arm trial, *Stat. Methods Med. Res.* Published online February 7, 2011.

# 7

## Multiple Comparisons

## 7.1 Introduction

Multiple comparisons pose a problem in any clinical trial. There are many aspects to this, including exploring multiple treatment arms, multiple efficacy measurement endpoints, and multiple timepoints, but all lead to the same problem: the chance of falsely concluding efficacy is inflated without proper recognition of multiplicity.

In non-inferiority testing, the roles of the null and alternative hypotheses are in some ways reversed, which can cause confusion at first glance. In non-inferiority testing, the type I error is the probability of concluding non-inferiority when the active control is markedly superior to the experimental treatment; in superiority testing, the type I error is the probability of concluding superiority of one treatment when the effects of the treatments are identical. This may lead to some confusion about the interpretation of type I and type II errors. However, when the type I error is properly recognized as the probability of rejecting a null hypothesis that is true, and the type II error is the probability of not rejecting a null hypothesis that is false, the confusion dissipates. This is the same in non-inferiority or superiority testing.

Control of the type I error rate can be defined in different ways. The most common for clinical trials is control of the familywise error (FWE) rate—the probability of rejecting at least one true null hypothesis. In the case of testing multiple endpoints for non-inferiority, this means concluding non-inferiority at least one time when non-inferiority is not true. Control in the strong sense requires that the FWE is controlled at the claimed $\alpha$ level or less for every possibility in the parameter space (i.e., no matter which null hypotheses are true and which are false). This is in contrast to control in the weak sense, which requires that the FWE is controlled at or below the claimed level only when all null hypotheses are true. Control of the FWE in the strong sense is most commonly used in a regulatory setting. Thus, in the rest of this chapter, we do not continually state "in the strong sense" when we refer to control of the FWE, although this is implied. Other definitions of type I error rate can also be considered, including comparisonwise error rate (CWE: controlling nominal comparisons without adjustment for multiplicity), experimentwise error rate (EWE: controlling the expected number of true null hypotheses that are rejected), and false discovery rate (FDR: controlling the number of rejected null hypotheses that are true). These other definitions are not commonly used for efficacy endpoints in clinical trials.

In this chapter, we outline some of the common multiple comparison issues in non-inferiority testing along with discussing some solutions. The solutions are meant to be representative rather than comprehensive and allow illustration of the issues. For notation, we will consider comparing $I$ experimental treatments to a single control, and use the additional subscript $i$ to denote the experimental treatment being considered, as in the mean of the $i$th treatment, $\mu_{E,i}$.

## 7.2 Comparing Multiple Groups to an Active Control

Multiple experimental treatments can be compared to one active control with known efficacy. This might occur when multiple doses of an experimental treatment are considered, when multiple schedules or routes of administration are examined, when multiple experimental treatments are compared to an active control in a single study for efficiency, or when alternative regimens of the active control therapy are compared with the standard regimen. In each case the family of hypotheses for which the type I error rate is to be controlled must be considered, and an appropriate adjustment must be made for control of the FWE in the strong sense. Two general structures are possible: attempting to find a subset of treatments that are noninferior to the control or attempting to show that all treatments are noninferior to the control. In the first case, the hypotheses are written as $H_o$: $\mu_C - \mu_{E,i} > \delta_i$ for all $i$ versus $H_a$: $\mu_C - \mu_{E,i} \leq \delta_i$ for at least one $i$. In the second case, the hypotheses are written as $H_o$: $\mu_C - \mu_{E,i} > \delta_i$ for at least one $i$ versus $H_a$: $\mu_C - \mu_{E,i} \leq \delta_i$ for all $i$. In the first case, an intersection–union test will result and the standard multiple comparisons procedures will be considered.[1] In the second case, a union–intersection test will result and often no adjustment to the nominal type I error rate will be needed.[2] Note that each experimental treatment might have the same non-inferiority margin to the control, particularly in the indirect determination of efficacy relative to a placebo, but we have used notation to allow for the possibility that the non-inferiority margin may differ across experimental treatments. In a dose-ranging or frequency-ranging study of an approved agent, the non-inferiority margin for evaluating whether a particular regimen has efficacy that is not unacceptably worse than the standard regimen may depend on the reduction in dose or frequency.

### 7.2.1 Unordered Treatments: Subset Selection

When comparing multiple experimental treatments that have no a priori ordering of importance in an intersection–union test, many options are available for controlling the FWE. One option is to treat the various comparisons of experimental treatments to the active control as unrelated comparisons and make no adjustment. This would be acceptable if each experimental treatment was compared to the active control in an independent clinical trial, the difference being that the comparisons will be correlated when the same active control group is used for all comparisons. This is generally considered inappropriate when the same control is being used for comparison to multiple experimental treatments in a confirmatory trial, as it controls the CWE only and ignores the larger family of comparisons as required for control of the FWE. In a proof-of-concept study in which any conclusion must be confirmed in a subsequent trial, control of only the CWE may be appropriate.

If adjustment is employed, then a simple method will be the Bonferroni adjustment, in which each of the $I$ experimental treatments is compared to the active control in a test at the $\alpha/I$ level. If confidence intervals are used for the decision process, a common two-sided confidence interval of $100 \times (1 - \alpha/I)\%$ is appropriate. Non-inferiority will be concluded for a given comparison if the upper bound of the confidence interval is less than $\delta_i$. This will control the type I error rate and has the advantage of producing valid simultaneous confidence intervals; however, more powerful procedures are generally available.

The Holm procedure is related to the Bonferroni procedure.[3] The Holm procedure will control the FWE and, in addition, will be uniformly more powerful than the Bonferroni procedure. The Holm procedure is usually described in terms of $p$-values. The $p$-values for the various comparisons are ordered resulting in $p_{(1)}, \dots p_{(I)}$. The smallest $p$-value, $p_{(1)}$, is compared to $\alpha/I$. If $p_{(1)} < \alpha/I$, the hypothesis associated with $p_{(1)}$ is rejected and $p_{(2)}$ is compared to $\alpha/(I - 1)$. Again, if $p_{(2)} < \alpha/(I - 1)$, the hypothesis associated with $p_{(2)}$ is rejected and $p_{(3)}$ is compared to $\alpha/(I - 2)$. Testing continues in this fashion as long as hypotheses are rejected. When a hypothesis is not rejected because $p_{(i)} > \alpha/(I - i + 1)$, the procedure stops without conclusions on untested hypotheses. For non-inferiority testing, which typically does not explicitly rely on $p$-values, an analogous procedure will produce confidence intervals with decreasing nominal confidence. That is, based on maintaining a familywise one-sided type I error rate of $\alpha/2$, if non-inferiority is concluded on the basis of at least one of the $100 \times (1 - \alpha/I)\%$ two-sided confidence intervals, the procedure moves to the next step. At the second step, if non-inferiority is concluded for at least two of the $I$ comparisons based on $100 \times (1 - \alpha/(I - 1))\%$ two-sided confidence intervals, the procedure moves to the next step (and so on). When the procedure stops owing to not concluding non-inferiority for at least $j$ of the $I$ endpoints, the $j - 1$ comparisons from the previous step are the ones on which non-inferiority is concluded. Because of the complexities of this procedure, it is probably easier to do the testing with the $p$-values and calculate confidence intervals after testing has determined the experimental treatments that are noninferior to the control. These confidence intervals will be at the level $(1 - \alpha/(I - j)) \times 100\%$ for the endpoint associated with $p$-value $p_{[j+1]}, j = 0, \ldots, I - 1$, if the null hypothesis associated with that endpoint was rejected. For comparisons not concluded noninferior, no associated confidence interval will be determined.

A common multiple comparison procedure when testing more than one experimental group to a single control is Dunnett's test.[4] Dunnett's test calculates a confidence interval according to the form $\bar{x}_C - \bar{x}_C \pm d_\alpha(I - 1, f) \times se$, where $f$, the degrees of freedom, is the total number of observations minus the number of treatment groups, and $I$ is the number of treatment groups including the control. Tables for the critical values $d$ are available in many textbooks or from statistical packages including the PROBMC function of SAS®. The advantage of using Dunnett's test is that it is more powerful because it considers the correlation among various comparisons.

### 7.2.2 Ordered Treatments: Subset Selection

When the various comparisons have an a priori ordering, more structure can be added to gain efficiency. Multiple doses of an experimental treatment are often considered to have a monotone efficacy effect: larger doses have better efficacy. When comparing various doses of an experimental treatment to a single active control to determine which, if any, doses are noninferior to the active control, a fixed sequence approach can be considered. That is, the highest dose (most likely to be efficacious) is compared to the active control; if non-inferiority is concluded, the next highest dose is compared, and so on. As long as non-inferiority is concluded on all previous comparisons, the tests can be made at the full $\alpha$ level. The tests can be conducted by calculating two-sided $100 \times (1 - \alpha)\%$ confidence intervals for the comparison of each dose to the active control or by calculating another test statistic. Once one dose is not shown to be noninferior to the active control, no further testing can be considered. In addition, confidence intervals for further comparisons are difficult to interpret because of the lack of a procedure to produce valid simultaneous confidence intervals corresponding to the fixed sequence testing procedure. The fixed sequence procedure can also be used when comparing multiple experimental treatments other than doses, when the order of importance or likelihood of a positive conclusion can be prespecified. A modification to the fixed sequence procedure is the fallback procedure[5,6]; some of the type I error rate is reserved for later comparisons in the event that an early comparison does not result in rejecting the null hypothesis. In dose comparison designs, this will be especially appropriate when higher doses may lead to tolerability issues, which may negate the intended or on-target efficacy advantage.

### 7.2.3 All-or-Nothing Testing

The preceding discussion involves intersection–union testing. For union–intersection testing, different strategies are required. Since these designs are less common than those involving intersection–union tests, we present a motivating example (Example 7.1).

#### Example 7.1

Consider a study of three consistency lots. The objective is to demonstrate that the product can be consistently produced without substantial impact on time immunogenicity of the product. Rather than the usual equivalence criteria, for this example we will consider that if each lot is concluded to be noninferior to a control, manufacture consistency will be concluded. That is, rather than showing that each lot induces similar protection to every other lot, just showing that each lot induces adequate protection in comparison to the control will be sufficient. Then the hypotheses can be written as $H_o$: $\max_i(\mu_C - \mu_{E,i}) \geq \delta$ versus $H_a$: $\max_i(\mu_C - \mu_{E,i}) < \delta$. Consider testing these hypotheses by testing separately for $i = 1, 2, 3$, $H_{o,i}$: $\mu_C - \mu_{E,i} \geq \delta$ versus $H_{a,i}$: $\mu_C - \mu_{E,i} < \delta$ each at the nominal level $\alpha$. If each null hypothesis for an individual lot is rejected at the nominal level $\alpha$, then global noninferiority (manufacture consistency) will be concluded. If exactly one of the lots is not noninferior to the control, the FWE will be less than or equal to the nominal level $\alpha$. If more than one lot is not noninferior to the control, the FWE will be less than the nominal level $\alpha$, since more than one error must be made to incorrectly conclude consistency. Thus, making each comparison at the nominal level $\alpha$ may be conservative.

## 7.3 Non-Inferiority on Multiple Endpoints

Multiple comparisons involving testing of several endpoints frequently arise in non-inferiority clinical trials of a single experimental treatment and a single active control. To achieve efficiency, many endpoints will typically be collected in a single clinical trial, and many of these endpoints could be used for inference. Maintaining the FWE is critical to support the conclusions of the study, and that is the primary topic of this subsection. There are, however, other issues to consider when performing non-inferiority testing on multiple endpoints with the same control therapy. Using the most effective agent relative to an endpoint helps guard against biocreep. However, for a given patient population or disease, the most effective agent for one endpoint may not be the most effective agent for another endpoint. Additionally, determining a margin for the most important clinical endpoint is a difficult enough task. This most important clinical endpoint may often have the best-quality data on the effect of the active control, with less-quality data available on other efficacy endpoints. The active control may not consistently demonstrate an effect on the additional endpoints, or may not demonstrate an effect of consistent magnitude, even if the active control has a real effect.

It is common to begin addressing the multiple comparisons issue by assigning priority among the comparisons: one comparison is (or very few comparisons are) called primary, a small number are called secondary, and maybe a few are called tertiary, with other endpoints not being part of a formal inferential or exploratory strategy. Testing begins on the primary comparison(s) with secondary endpoints being considered only if the comparison of the primary endpoint(s) results in significance.[7] This general structure is common in clinical trials, both those designed to demonstrate non-inferiority and those designed to demonstrate superiority. However, it is only a start: the set of primary hypotheses can contain more than one test, and a decision on how to move from primary to secondary comparisons (and eventually to tertiary comparisons) needs to be addressed to control the FWE.

In non-inferiority clinical trials, the purpose of the secondary endpoints must be considered. An obvious strategy is to consider a formal assessment of non-inferiority for all endpoints (primary, secondary, tertiary, etc.), which requires non-inferiority margins to be prespecified for all endpoints. This may be required for regulatory approvals for endpoints beyond the primary one(s). Alternatively, secondary and tertiary endpoints might be reported with point estimates and confidence intervals, but not formally tested for non-inferiority (or superiority). This strategy would be considered if providing information to patients and physicians is sufficient without explicit non-inferiority conclusions, and especially if it is not possible to select appropriate margins for non-inferiority assessments on these endpoints.

In the following sections, we first consider multiple endpoints in a single family, such as a family of primary comparisons that will be the sole basis of conclusions of the study. We then consider multiple families of comparisons, including how to evaluate secondary tests after first confirming efficacy in a primary test.

### 7.3.1 Multiple Endpoints in a Single Family

We start by considering multiple comparisons in a single family of comparisons or endpoints. These procedures will be appropriate when inference is desired only on prespecified primary endpoints and form the basis of discussions of multiple comparisons in more than one family. Some methods will be similar as in the prior discussion of comparing multiple experimental treatments to a control. However, other procedures will be available, as the comparisons of multiple endpoints have different inherent structures and often more comparisons than was considered in the prior discussion. Because a family, especially a primary family, will generally not have many comparisons, this discussion will assume that very few comparisons are of interest within a single family.

When there is a priori ordering of the multiple endpoints, the fixed sequence procedure can be considered, as discussed in the previous section.

With such an approach, one primary endpoint becomes the most important primary endpoint, and other endpoints are not even considered unless non-inferiority is demonstrated for the first endpoint. If non-inferiority is demonstrated on an endpoint, the next endpoint (again from the prespecified ordering) will be tested. It may seem illogical that a less important endpoint that is called "primary" might not be tested at all, depending on the results of the previous comparisons. If this is a concern, an alternative is to only call the first primary endpoint "primary" and label other endpoints "secondary," by placing them in a separate family of endpoints. This change in labels has no impact on the testing process, the power, or the interpretation of results.

An obvious alternative to the fixed sequence strategy, to avoid some of the problems mentioned above, is to save some of the $\alpha$ for subsequent testing, as in the fallback test described earlier. With the fallback, all comparisons can be considered even if one or more endpoints do not result in a conclusion of non-inferiority.

The Holm procedure, described earlier, can also be used for a single family of comparisons, and will control the FWE.

Hochberg[8] proposed a procedure based on the Holm procedure, but using a step-up rather than a step-down approach. That is, the null hypothesis is rejected and non-inferiority is concluded for the endpoint associated with the largest $p$-value for which $p_{(j)} \leq \alpha/(I - j + 1)$, and for all endpoints with smaller $p$-values. By definition, this will include all endpoints for which the Holm procedure concludes non-inferiority, and maybe more; thus, the Hochberg procedure is uniformly more powerful than the Holm procedure. Again, caution must be used as the Hochberg procedure does not always control the FWE in the strong sense.

More generally, a multiple comparison procedure that is a closed testing procedure will control the FWE in the strong sense.[9] A closed testing procedure considers all possible nonempty subsets of hypotheses. With $J$ endpoints, there will be $\sum_{k=1}^{J} \binom{J}{k} = 2^J - 1$ subsets to consider. Within each subset, a global null hypothesis is considered: the null hypothesis is that non-inferiority exists in none of the endpoints of that subset, and the alternative hypothesis is that non-inferiority exists in at least one endpoint in that subset. The global hypothesis in each subset is tested, but is a very flexible procedure: any testing method that controls the type I error rate for that subset of tests can be used if it is prespecified. When all such subsets have been considered, individual endpoints are considered by determining whether the global null hypothesis has been rejected in all subsets containing that endpoint. If so, non-inferiority is concluded for that endpoint. Many tests for a single family of hypotheses discussed in this section can be written as closed tests, including Bonferroni, fixed sequence, fallback, and Holm.

In practice it is uncommon to consider all $2^J - 1$ subsets of hypotheses in a closed testing procedure, as shortcut methods reduce the number of required comparisons while producing the same conclusions.[10] The Bonferroni, Holm, fixed sequence, and fallback tests are examples of shortcut procedures that result in identical conclusion to a closed test.

When a closed testing procedure is used, it is generally desired that the tests be $\alpha$-exhaustive—that is, the type I error rate should be equal to and not less than $\alpha$ when testing the global null hypothesis in each subset. The Bonferroni procedure, for example, does not accomplish this, as the type I error rate for a given subset will be bounded above by $\sum_{k=1}^{J} \alpha_k I(k)$, where $I(k)$ is an indicator function that equals 1 if endpoint $k$ is in the subset and 0 otherwise. This will be strictly less than the nominal $\alpha$ level except for the subset containing all endpoints. For this reason, it is always possible to find a procedure that is more powerful than the Bonferroni procedure for a single family of hypotheses but still controls the FWE in the strong sense.

### 7.3.2 Multiple Endpoints in Multiple Families

When multiple endpoints are grouped into multiple families in decreasing order of importance, the multiple comparison procedures to control the FWE become more complex. Multiple comparison procedures typically can be described as gatekeeping strategies. We consider two gatekeeping strategies: serial gatekeeping and parallel gatekeeping.[11]

Serial gatekeeping strategies require all endpoints in a given family to be found significant before endpoints in the next family can be considered. For non-inferiority clinical trials, it is required that non-inferiority be demonstrated on every endpoint in a family before any endpoint in the next family can be considered. This is similar in philosophy to the fixed sequence approach, which requires non-inferiority to be demonstrated on each endpoint before the next endpoint is considered, but with multiple endpoints in a given step. Testing within a family of endpoints can be conducted in any way that controls the FWE in the strong sense at the $\alpha$ level in that family: Bonferroni, fixed sequence, fallback, Holm, etc. As long as non-inferiority is demonstrated for each endpoint in a family, the next family of endpoints is tested in any way that controls the FWE in the strong sense at the $\alpha$ level in that family. It is advisable that the methods in the various families are consistent, but this is not required.

Parallel gatekeeping strategies differ from serial gatekeeping strategies in that they require non-inferiority to be demonstrated on at least one endpoint, rather than all endpoints, in a family before the subsequent family of endpoints is considered. However, if non-inferiority is demonstrated for some, but not all, endpoints in a given family, subsequent families of endpoints will be tested at a smaller type I error level to account for the lack of a non-inferiority conclusion in that prior family. In addition, testing within a given family of endpoints may have lower power than in a series gatekeeping strategy, as more powerful $\alpha$-exhaustive procedures for early families of endpoints can be used only in a serial gatekeeping procedure.

Consider a non-inferiority trial with two primary, one secondary, and two tertiary endpoints, each with a non-inferiority margin defined. A serial gatekeeping strategy can be defined that uses for the family of primary endpoints an $\alpha$-exhaustive procedure. (This includes the Holm, fixed sequence, Bonferroni, and other possible approaches.) A parallel gatekeeping strategy cannot use an $\alpha$-exhaustive procedure for this first family, but is limited to Bonferroni and other procedures with generally lower power for the family of primary endpoints. The tradeoff is that the parallel gatekeeping strategy can consider secondary endpoints if at least one primary endpoint shows non-inferiority; the serial gatekeeping strategy can consider secondary endpoints only if both primary endpoints show non-inferiority. Note that an $\alpha$-exhaustive procedure can be used for the last family of endpoints, as no further testing is involved after consideration of these endpoints.

### 7.3.3 Further Considerations

When choosing a multiple comparison procedure, the importance of a conclusion for each endpoint and the corresponding power should be considered.

For multiple endpoints in a single family, if there is a priori ordering of importance, then a method that exploits that ordering, such as the fixed sequence or fallback, will be preferred. If all endpoints have similar importance, then a procedure that does not require a priori ordering will be most appropriate—for example, the Holm procedure or the Hochberg procedure. A weighted version of the Holm procedure can be used if there is some a priori preference among endpoints, but not a complete ordering.

For multiple families of hypotheses, the gatekeeping structure can be used very flexibly with a number of associated procedures within individual families. Choosing between a serial and parallel gatekeeping procedure should be based on the relative importance of showing non-inferiority for all endpoints in a given family versus the improved ability to consider endpoints in a less important family.

In either case, the order of importance of endpoints is critical. It is not possible to give advice that will be flexible enough to apply to all situations while rigorous enough to be helpful, but there are some general thoughts. Primary endpoints should be those on which regulatory approval (or widespread acceptance, if regulatory approval is not a goal of the trial) will be based. Secondary endpoints should be those on which further conclusions will be based. Tertiary endpoints are correspondingly less important.

## 7.4 Testing for Both Superiority and Non-Inferiority

It is possible that some endpoints will be tested for non-inferiority and, conditional on demonstrating this, be tested for superiority. Such a strategy is an example of multiple comparisons that is unique to non-inferiority studies.

Testing of superiority and non-inferiority in the same study has been discussed extensively in the literature. There are two ways to testing for both non-inferiority and superiority: if non-inferiority is concluded, it is logical to take the next step and ask whether the experimental treatment is not just noninferior to but also superior to the active control; alternatively, when superiority is the goal of the active-controlled trial but is not concluded, it is natural to try to examine non-inferiority as an alternative to further understand how the two treatments arms compare.

### 7.4.1 Testing Superiority after Achieving Non-Inferiority

It is generally, but not universally, agreed that testing for superiority after concluding non-inferiority can be acceptable if the testing plan is prespecified. This is often justified as a fixed sequence approach to multiple testing in which it is possible to test hypotheses in a prespecified order at the full type I error level as long as all prior hypotheses were rejected[12] or that the same confidence interval is used for each test. However, once a hypothesis is not rejected, all testing stops and claims cannot be made on untested null hypotheses. The addition of multiple endpoints will complicate the testing scheme. The easiest scenario to discuss is two endpoints, a primary and a secondary, in which both are to be tested for non-inferiority and then superiority. A fixed sequence test can be set up in two ways, as provided in Table 7.1.

The difference in the two approaches is whether to test more than one endpoint for non-inferiority before testing any endpoint for superiority, or to test one endpoint both for non-inferiority and superiority before testing the next endpoint for non-inferiority. The solution will be based on the

**TABLE 7.1**

Fixed Sequence Approaches

| Testing Order | Alternative A Hypothesis | Alternative B Hypothesis |
|---|---|---|
| 1 | Test primary endpoint for non-inferiority | Test primary endpoint for non-inferiority |
| 2 | Test secondary endpoint for non-inferiority | Test primary endpoint for superiority |
| 3 | Test primary endpoint for superiority | Test secondary endpoint for non-inferiority |
| 4 | Test secondary endpoint for superiority | Test secondary endpoint for superiority |

relative advantages of superiority on an important endpoint versus the consequences of not concluding (by not testing) non-inferiority on a less important endpoint.

It may be unusual for labeling to explicitly state the superiority of one product over another. Demonstrating superiority over a truly active control in a single trial is much stronger evidence of any efficacy (relative to placebo) than superior efficacy to the active control. In such cases, it makes more sense to test non-inferiority on all endpoints that would be the basis for conclusions before considering any superiority claims. However, superiority testing can be used outside of the regulatory channels, in publications in the medical literature. In addition, if there is debate over the proper non-inferiority margin to use, a superiority conclusion will effectively end the debate and lead to a much quicker determination that the experimental treatment is also superior to a putative placebo.

Incorporating superiority determinations, if desired, can be done using the framework of gatekeeping procedures. The superiority comparison can be inserted into a serial gatekeeping approach, either in the same family with the corresponding non-inferiority comparison or in a subsequent family of comparisons. If the superiority comparison is in the same family as the corresponding non-inferiority comparison, superiority must be demonstrated on all endpoints in a given family before non-inferiority can be considered for subsequent families of comparisons. If superiority comparisons are in subsequent families relative to the corresponding non-inferiority comparisons, non-inferiority must be demonstrated on all endpoints for a given family before the subsequent corresponding superiority comparisons are considered. With either process, the simple testing of superiority, conditional on demonstrating non-inferiority, becomes quite complicated. Non-inferiority and superiority comparisons can also be considered for parallel gatekeeping strategies, as discussed in the following example.

Dmitrienko et al.[13] proposed a potential alternative that would allow investigation of both superiority of the primary endpoint and non-inferiority of the secondary endpoint, based on a finding of non-inferiority for the primary endpoint in a manner more complex than the usual gatekeeping approach. The proposed procedure allows branching or consideration of hypotheses in multiple directions, based on whether a single hypothesis or family of hypotheses is rejected. Consider the diagram in Figure 7.1. A clinical trial that has one primary and one secondary endpoint is considered, and a non-inferiority margin is prespecified for each of these two endpoints. Based on a finding of non-inferiority for the primary endpoint, two tests are considered: non-inferiority for the secondary endpoint and superiority for the primary endpoint. Based on then concluding non-inferiority for the secondary endpoint, superiority can be considered for the primary endpoint. An adjustment will need to be made for the two comparisons. In this case, a Bonferroni adjustment is easy to implement: if non-inferiority is concluded for the primary endpoint at a one-sided level of $\alpha$, the superiority for the

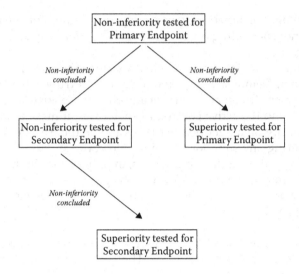

**FIGURE 7.1**
A testing sequence involving two endpoints.

primary endpoint and non-inferiority for the secondary endpoint are each tested at a one-sided level of $\alpha/2$ (or otherwise such that the levels sum to $\alpha$). If both are rejected, then superiority for the secondary endpoint can be tested at the one-sided level of $\alpha$; if only non-inferiority for the secondary endpoint is concluded, then superiority for the secondary endpoint can be tested at the one-sided level of $\alpha/2$; otherwise, superiority for the secondary endpoint cannot be tested. However, further testing would cease if non-inferiority for the primary endpoint was not concluded, or if either superiority for the primary endpoint or non-inferiority for the secondary endpoint was not concluded.

Proof of the control of FWE is shown by considering the closure of all potential hypotheses. In the example, the closure contains 15 nonempty subsets, but the shortcut illustrated in Figure 7.1 provides an equivalent test.

This method can be expanded to a larger number of endpoints, including co-primary and co-secondary endpoints.

## 7.4.2 Testing Non-Inferiority after Failing Superiority

Multiple comparison problems also occur when a study designed to demonstrate superiority of an experimental treatment over an active control is transformed post hoc into a non-inferiority comparison because the superiority conclusion was not demonstrated. Attempting to salvage a failed superiority comparison with a non-inferiority conclusion is difficult. For example, prespecification of the non-inferiority margin is required for any non-inferiority analysis. In attempting to salvage a failed superiority trial, it is unlikely that a margin can be chosen objectively, as the confidence interval for the difference in effects between the experimental and active control therapies is already known. Unless a small, negligible difference is ruled out (as would be the case when a superiority test just fails), a post hoc non-inferiority analysis will be difficult to defend.

An additional issue with using a non-inferiority test to salvage a failed superiority test is the choice of analysis set including imputation methods. Several authors have proposed that with a prespecified margin, it is trivial to test in either order: non-inferiority then superiority, or superiority then non-inferiority.[14,15] This is typically justified by noting that a finding of superiority automatically implies a finding of non-inferiority, and that any conclusion involving superiority and/or non-inferiority can be simultaneously based on the same confidence interval. However, this ignores the fact that superiority testing and non-inferiority testing often differ in the details.[16] For example, non-inferiority testing often relies more heavily on a per-protocol analysis set than superiority testing, and missing data may need to be imputed under the null hypothesis being considered. When some other aspect of the testing changes, such as imputation of missing data or choice of analysis set, the closure argument no longer allows simultaneous testing. We acknowledge that this is possible only when one can conclude superiority but not non-inferiority on the same data. Although this may seem illogical on its face, there is nothing to prevent this outcome when different analysis sets are used—especially when many subjects are removed from the per-protocol analysis set, the margin is very small or imputation strongly affects the conclusions. Thus, we recommend testing non-inferiority first, then superiority, and that the results be robust with respect to appropriate analysis sets.

# References

1. Berger, R.L. and Hsu, J.C., Bioequivalence trials, intersection–unions tests and equivalence confidence sets, *Stat. Sci.*, 11, 283–319, 1996.
2. Berger, R.L., Multiparameter hypothesis testing and acceptance sampling, *Technometrics*, 24, 295–300, 1982.
3. Holm, S., A simple sequentially rejective multiple test procedure, *Scand. J. Stat.*, 6, 65–70, 1979.
4. Dunnett, C., New tables for multiple comparisons with a control, *Biometrics*, 20, 482–491, 1964.
5. Wiens, B.L., A fixed sequence Bonferroni procedure for testing multiple endpoints, *Pharm. Stat.*, 2, 211–215, 2003.
6. Wiens, B.L. and Dmitrienko, A., The fallback procedure for evaluating a single family of hypotheses, *J. Biopharm. Stat.*, 15, 929–942, 2005.
7. O'Neill, R.T., Secondary endpoints cannot be validly analyzed if the primary endpoint does not demonstrate clear statistical significance, *Control. Clin. Trials*, 18, 550–556, 1997.

8. Hochberg, Y., A sharper Bonferroni procedure for multiple tests of significance, *Biometrika*, 75, 800–802, 1988.
9. Marcus, R., Peritz, E., and Gabriel, K.R., On closed testing procedures with special reference to ordered analysis of variance, *Biometrika*, 63, 655–660, 1976.
10. Grechanovsky, E. and Hochberg, Y., Closed testing procedures are better and often admit a shortcut, *J. Stat. Plan. Infer.*, 76, 79–91, 1999.
11. Dmitrienko, A., Offen, W.W., and Westfall, P.H., Gatekeeping strategies for clinical trials that do not require all primary endpoints to be significant, *Stat. Med.*, 22, 2387–2400, 2003.
12. Westfall, P.H. and Krishen, A., Optimally weighted, fixed sequence and gatekeeper multiple testing procedures, *J. Stat. Plan. Infer.*, 99, 25–40, 2001.
13. Dmitrienko, A. et al., Tree-structured gatekeeping procedures in clinical trials with hierarchically ordered multiple objectives, *Stat. Med*, 26, 2465–2478, 2007.
14. Dunnett, C.W. and Gent, M., An alternative to the use of two-sided tests in clinical trials, *Stat. Med*, 15, 1729–1738, 1996.
15. Morikawa T. and Yoshida M., A useful testing strategy in phase III trials: Combined test of superiority and test of equivalence, *J. Biopharm. Stat.*, 5, 297–306, 1995.
16. Wiens, B.L., Something for nothing in non-inferiority/superiority testing: A caution, *Drug Inf. J.*, 35, 241–245, 2001.

# 8

## Missing Data and Analysis Sets

## 8.1 Introduction

Issues involving missing data are often linked with issues involving the choice of the proper analysis set. However, it is also true that missing data issues are often confused with issues involving the proper choice of analysis set. Consider a randomized, double-blind, two-arm clinical trial in which some subjects drop out at randomization before undergoing study therapy or any other therapy, and have no follow-up for the study endpoint. Because these subjects should be fairly distributed between arms, they may be excluded from the analysis without compromising the integrity of the randomization. Whether to include such subjects in the analysis is an analysis set issue. If these subjects were included in the analysis [i.e., as in an intent-to-treat (ITT) analysis], the imputation or representation of their values for the endpoint should not depend on treatment arms (since such subjects should have been fairly distributed between arms) and should consider the actual adherence or nonadherence to therapy. A variation of this "imputation under the null" can be used for non-inferiority trials and will be discussed later.

According to the ITT principle, all subjects should be followed to the endpoint or the end of study with the comparisons based on the "as-randomized" treatment groups (i.e., based on the ITT population). This allows for an unbiased analysis. Missing data violate the ITT principle and can undermine both the integrity of the randomization and confidence in the results. Additionally, selective follow-up of subjects can weaken the quality of the data from the high quality expected from a randomized clinical trial to that obtained from an observational study.

The purpose of accounting for missing data is not to retrospectively change the design or objective of the clinical trial or the adherence to therapy or the protocol of any subject. Rather, the purpose is to account for all subjects with respect to the ITT principle. For a subject with a missing outcome, the objective is to adequately represent the missing outcome based on what would have been the expected outcome had the outcome been measured.

## 8.2 Missing Data

Missing data are an important feature of clinical trials, and at least some missing data must be anticipated in non-inferiority trials. However, because missing data may introduce or add uncertainty in the results, every effort should be made to limit the amount of missing data. In any clinical trial, a subject has the right to withdraw from the study, at any time and for any reason, without prejudice. In many clinical trials, at least a few subjects will miss assessments for various reasons as well, while remaining in the study and having assessments at later time points. The fact that data are missing can be very indicative of the outcome, so ignoring missing data can lead to incorrect inference on and biased estimates of treatment effects. Many methods for addressing missing data are specifically developed for studies with multiple assessments over time (called "longitudinal" studies), whereas other methods are applicable when only a single assessment is planned.

Per EMA guidelines,[1] the amount of missing data may be affected by: (1) the nature of the endpoint, (2) the length of follow-up (the greater the length of follow-up, the greater the proportion of missing data), (3) the therapeutic indication, and (4) treatment modalities, and thus the design of the study should consider these factors. As an example, a mortality endpoint is a hard (objective) endpoint that does not require physical visits, and should be less susceptible to missing data.

Philosophically, missing data can be separated into two types: data that exist but are unknown, and data that do not exist. Consider evaluation of a subject's serum hemoglobin levels. If a serum sample is obtained but lost in transit to the laboratory, the hemoglobin level at the time of the blood draw exists but is unknown. Alternatively, if a subject dies before the scheduled assessment, the hemoglobin level does not exist (or is zero) at the time of the scheduled assessment. Many methods for handling missing data were developed originally for sample surveys, in which answers to specific questions were left blank by respondents. Such data exist but are unknown. These data are easier to handle than data that do not exist and with better developed methods and a longer history of their application to clinical trials.

### 8.2.1 Potential Impact of Missing Data

The Code of Federal Regulations (21 CFR 314.126)[2] provides seven characteristics of an adequate and well-controlled trial, which includes that "adequate measures are taken to minimize bias on the part of the subjects, observers, and analysts of the data" and that "There is an analysis of the results of the study adequate to assess the effects of the drug." The first characteristic listed above is usually associated with the need for a double-blind design in a clinical trial. However, missing data are generally outcome-related and may also be treatment-related and thus, introduce bias. Per the second characteristic of an adequate and well-controlled trial listed above, the analysis of the results should address the impact of missing data. Therefore, how to minimize the amount of missing data and how to handle missing data in the analysis are critical in the design and conduct of a clinical trial.[1] Additionally, as all analysis methods have unverifiable assumptions, even when there is no missing data, the existence of missing data increases the number of unverifiable assumptions that need to be made and, thus, affect the interpretation of the results.

Missing data introduce a bias in the estimation of a treatment effect or difference in treatment effects that affects the comparability of study groups and the relation of the study results to the external target population.[1] Statistically, missing data that are not correctly addressed alters the type I error rate of statistical tests, the precision and reliability of the estimates, and the confidence level of confidence intervals.

Clinical trials should be designed and conducted to minimize the amount of missing data. The statistical analysis should address missing data whenever possible and not ignore the missingness. In some instances, the existence of missing data may make the results more similar between arms and in other instances the missingness may introduce a bias favoring one of the study arms. Carroll[3] describes a situation in oncology in which the discontinuation of tumor assessments may bias the results in favor of the more toxic, less effective treatment arm in a comparison of progression-free survival. In general, the assay sensitivity of the clinical trial may be impaired because of missing data.

The ability to measure and collect data on an endpoint may affect the role of the endpoint. If one endpoint is difficult to measure or collect, whereas a second endpoint is as meaningful and easier to measure and collect data on, the second endpoint would likely be a better choice for the primary endpoint. Thus, it may be necessary to choose as the primary endpoint that clinically meaningful outcome that is less susceptible to missing data. Additionally, the appropriate amount of follow-up should be considered. Reducing the length of follow-up for an endpoint should also decrease the amount of missing data. For a non-inferiority trial there should be reliable data on the effect of the active control on whatever primary endpoint is selected.

It is important to follow-up subjects and to collect relevant data. Off study treatment is not the same as off study. Outcomes should still be measured and collected from subjects after they stop study treatment. Going off study treatment is part of being on a treatment arm and is often treatment-related. Failure to collect information from subjects after they have stopped study therapy leads to a conditional analysis of subjects conditioned on continuing therapy.[4] While this may be relevant to patients after they have been on one of the study therapies for a while, it does not provide relevant information at the time when an initial treatment decision needs to be made. Data should continue to be collected from subjects after they have discontinued study therapy.

Additionally, ignoring that a subject has died prior to the evaluation of the endpoint by not accounting for that in the determination of the endpoint creates a hypothetical endpoint – an endpoint for which some subjects do not have any value.[4] Complete information or follow-up has been done on the subject. In the analysis a subject who died prior to the evaluation of the endpoint is represented by subjects on the same arm that have evaluations for the endpoint, or at best the analysis is conditioned on subjects not dying. This would particularly be problematic if the experimental therapy increases the risk of death.

The International Conference on Harmonization (ICH) E-9 guidance[5] states: "Because bias can occur in subtle or unknown ways and its effect is not measurable directly, it is important to evaluate the robustness of the results and primary conclusions of the trial. Robustness is a concept that refers to the sensitivity of the overall conclusions to various limitations of the data, assumptions, and analytic approaches to data analysis." Sensitivity analyses should study the limitations of the data and the potential impact of missing data. When a subject has an outcome of interest that is missing, but related or correlated outcomes are collected, a reasonable imputation of the missing outcome may be possible when using the collected information and prior measured outcomes on that subject. For subjects for whom there is no reasonable imputation for the missing outcome, a variety of imputation techniques should be considered to examine the sensitivity of the results, including imputation under the null hypothesis and imputation techniques that would assume that the missingness if unaddressed introduce a bias favoring the experimental arm.

Missing data in the historical studies of the active control therapy can obscure the active control effect. Differing behavior of missing data in those historical studies can also contribute to variability of the active control effect across the historical studies. Missing data in the non-inferiority trial not only obscure the comparison of the study arms but also undermine the confidence in the believed or assumed effect of the active control therapy and in the selected non-inferiority margin.

The best approach to missing data is to prevent it. Missing data should be kept to a minimum with the missing data addressed in the analysis, not be ignored.

### 8.2.2 Preventing Missing Data

The best approach to missing data is prevention. An ounce of prevention is better than a pound of imputation. Fleming[6] identifies eight factors that should be recognized and addressed in improving the prevention of missing data: (1) proper distinction in the study protocol between nonadherence and nonretention, (2) the misuse of the term "withdrawal of consent," (3) informing subjects of the impact of missing data in the Informed Consent process, (4) lack of integrity that the ITT principle preserves the integrity of the randomization, (5) sample size adjustments for dropouts only address the variability and not the bias induced by missing data, (6) lack of specification in the protocols of the performance standards and targeted levels of data capture needed to achieve high quality trial conduct, (7) implementation of procedures during enrollment and follow-up to enhance the likelihood of achieving high levels of retention and (8) monitoring/overseeing of the quality of both trial conduct and data capture.

As mentioned earlier, discontinuing treatment is not the same as discontinuing study. Going off study treatment is an outcome in itself. Failure to include information from subjects while off study therapy leads to a conditional analysis of subjects conditioned on continuing therapy. Additionally, the only valid reasons to identify a subject as "off study" are withdrawal of consent, or achievement of all required efficacy and safety information.[6] The term "withdrawal of consent" is often used to justify or explain why a subject is no longer being followed for outcomes. However, "withdrawal of consent" means that the subject no longer wishes to participate in any way in the trial (including directly or indirectly ascertaining that subject's outcomes).

If a subject understands how missing data reduces the value of the information obtained from the subject and the potential bias in the estimated treatment effect that can result, then the subject may be more willing to have their data measured and collected. Additionally, investigators need to understand the necessity to collect data from subjects even after the subjects have discontinued their randomized therapy. Protocols also provide misleading information about correcting for dropouts. The convention to increase the sample size in a clinical trial based on an assumed fraction of "dropouts" does not address the bias that is produced by missing data, and simply produces a more precise biased estimate of the treatment effect. The mention of targeted levels of retention will both encourage investigators to collect information, and also help them evaluate their success or lack of success in doing so. The absence of such information places the quality of follow-up on the arbitrary willingness of the investigator.

There have been successful approaches to achieving high levels of retention in settings where low retention levels were anticipated. A study in Kampala, Uganda on mother-to-child transmission (MCT) of HIV achieved high levels of retention for every determination of the rate of MCT of HIV including 95.5% at 18 months, even though many mother-infant pairs did not have addresses and lived far from the hospital in Kampala, Uganda.[6,7] The high retention levels were achieved by assigning to each study participant a Health Visitor who would create a rapport with the subject, answer questions, periodically visit the subject and their family, and encouraged the subject's participation in the clinical trial.

### 8.2.3 Missing Data Mechanisms

Despite best efforts, most studies have some missing data. Analyses of missing data is well developed in the statistical literature.[8,9] Most of the methods have been implicitly developed with the intention of application to superiority trials rather than to non-inferiority trials and the application to non-inferiority trials is not always straightforward.

Missing data might be missing due to study discontinuation or due to a one-off situation in which some data are missing but other subsequent or simultaneous data are available. The difference is that in one case, missing data are intermittent, and in the other, missing data are terminal. It is common to have both in any given study.

Missing data are often categorized into three distinct classes: missing completely at random (MCAR), missing at random (MAR) and missing not at random (MNAR). The last is non-ignorable missingness, whereas the first two are considered ignorable missingness because methods have been developed to obtain valid inference and unbiased point estimates when data are ignorable. The various categories are assigned based on whether the unobserved response data is related to other aspects of the study: baseline characteristics and demographics, observed response data or the unobserved response data. The factors that cause a data point to be missing are referred to as the missing data mechanism.

*MCAR.* With data that are MCAR, the missing data mechanism is unrelated to the outcome of interest, any observed or unobserved covariate, and other measurable quantity. Specifically, data are missing not because of any relationship to the unobserved data or even to the observed outcome data. Data that are MCAR are the easiest to handle in the analysis, since they can essentially be ignored for proper inference. However, MCAR is the most restrictive assumption, and one that is often not applicable, so there is a risk in assuming that data are MCAR.

*MAR.* Data that are MAR have some relationship between the observed outcome data and the missing data mechanism. Conditioned on all observed measurements (pre- and post-baseline measurements), missingness is random. That is, values of observed data give insight into the chance that subsequent data are missing. However, there is no such relationship between the unobserved data and the chance that data are missing. MAR is a slightly weaker assumption about the missing data mechanism than MCAR and as such requires more careful analysis and interpretation, but with care can provide valid inference.

*MNAR.* Data that are MNAR have some relationship between the unobserved outcome value and the missing data mechanism. Missing data are likely to have a different distribution than observed data, and the distribution is not predictable based only on observed outcome values or covariates. MNAR is the weakest assumption, but the assumption that is hardest to incorporate into an analysis. However, since the value of the data is unobserved, it is not possible to accurately predict which values will be missing MNAR, or what the unobserved values will be, or the relationship between the unobserved values and the missing data mechanism.

At any scheduled assessment, it is possible that some or all measurements are not obtained. This might be due to human error—forgetting to take a measurement—or due to lack of supplies—not having a sufficient supply of tubes to collect serum. Such missing data are commonly MCAR, but if a measurement is missing because of a subject's health state—not feeling well enough at a particular visit to fill out a form for quality of life data, for example—the data might easily be MNAR. The reason for such missingness must be collected prospectively to accurately assess the missing data mechanism. A discussion of missingness in uncomplicated bacterial infections and long-term cardiovascular endpoint studies are provided in Example 8.1.

### Example 8.1

*Uncomplicated bacterial infections*: A subject who discontinues a study of an uncomplicated infection may discontinue for one of many reasons, some of which are suggestive of positive outcomes and some of which are not. Subjects who are cured of such an infection might decide not to return for evaluations because they no longer have symptoms, and a clinic visit is an inconvenience. A subject who is not cured might be more likely to return for a clinic visit to request further treatment, but also might choose a different health care provider if the first treatment was unsuccessful. In short-term studies (such as a week or so of oral therapy), discontinuation due to a major life event such as moving to another city would be rare. Therefore a subject who discontinues the study might discontinue due to the outcome, in which case the data would be MNAR. If subjects who are cured and subjects who fail discontinue at the same rate, the data could be MAR or MCAR.

*Long-term cardiovascular endpoint study*: A study in which subjects are followed up for a long period of time (years in many cases) will inevitably have subjects who discontinue for administrative reasons, as well as for reasons of mortality, adverse reactions, etc. It is tempting to assume many of the dropouts are MCAR (or at least ignorable) but because any dropout might be related to unobserved outcome, this is again a risky assumption. In such a study, understanding the missing data mechanism is vital to understanding the treatment effect.

### 8.2.4  Assessing Missing Data Mechanisms

Distinguishing among missing data mechanisms is vital to using the appropriate analysis and supporting analyses. Distinguishing MCAR from MAR is generally easier than distinguishing ignorable from non-ignorable missing data.

In the event that data are thought to be MCAR (or it is thought a likely scenario), an assessment of whether the data might be MAR should be undertaken. Since the difference between MCAR and MAR is in the relationship between the missingness and the observed responses, efforts should focus on investigating this relationship.

Informal visual approaches can often provide a preliminary assessment of the relationship between the missingness and the observed responses. A graphical approach to assessing this relationship with longitudinal data is to graph over time the outcomes for those who have complete data and those who discontinue at various time points (including only data up to the point at which the discontinuation occurred, of course). Subsetting the subjects into a small number of groups (three to five groups often provide informative results) based on meaningful study benchmarks will allow a quick visual assessment of potential relationships. Such graphs can provide assessment of whether the trajectory of response or baseline values differed between those with complete data and those with incomplete data.

Inferential assessments of the relationship between missingness and the observed data are also possible. Under the null hypothesis that data are MCAR, various test statistics can be developed. Simplistically, testing for differences in early response between subjects with complete data and subjects with incomplete data can provide for a test – albeit, perhaps not an optimal test. Incorporating covariates into this assessment can improve the test. Correlation assessments of presence of outcome measurements to baseline characteristics and, after accounting for this relationship, of presence of outcome to the previous measurement will provide insight. Logistic regression can be used for the multivariate investigations. This will give information on whether subjects with higher or lower outcomes, or a better or worse intermediate response, are more likely to have missing data at the subsequent assessment, something allowed under MAR but not under MCAR. Other tests specifically developed for missing data assessments are available as well.[10] The ability of such tests to differentiate the type of missingness is unclear due to lack of power, but they can be used to demonstrate the lack of MCAR.

Distinguishing between MAR and MNAR is more difficult. By definition, MNAR data are missing in part because of the unobserved value—but without observing the value, it can be difficult to establish this association. Heuristic arguments can be used, more convincingly if they argue for MNAR than if they argue against MNAR. Other markers of disease progression or regression can be used if collected after discontinuation. In the situation that subjects who discontinue (or otherwise have missing data) tend to experience events at a higher or lower rate after discontinuation than subjects who continue, one can conclude MNAR. Subjects who discontinue because of adverse events might also be considered to have data that are MNAR, especially if the specific adverse events are related to or affect the outcome measurement (such as when the outcome is quality of life). This can be handled more easily in the analysis than MNAR in general, but if discontinuation due to reported adverse events is common, it might also be common that discontinuation occurs because of adverse events that are not reported, which is harder to handle.

## 8.2.5 Analysis of Data when Some Data Are Missing

Analysis in the presence of missing data typically focuses on two objectives: unbiased estimation of treatment effect, and appropriate inferential analysis relative to the null hypothesis. Both objectives require care in imputing the missing data to avoid a bias in estimation of means, but the second objective also requires care in accurately determining the reliability of the estimated treatment effect or treatment difference, to avoid a situation in which the point estimates are unbiased but the hypothesis test results in an inflated chance to reject a true null hypothesis.

When a subject has a missing measurement that is ignored in the analysis, that subject is represented in the analysis by subjects on the same treatment arm that have a measurement. If their prognosis is not similar to those other subjects on the same arm that have measurements, their missingness may be informative. For some imputation methods (including inverse probability weighting methods), a subject with a missing measurement is represented by a subset of subjects on the same treatment arm that are believed to have a similar prognosis to the subject with the missing measurement. It should be understood that the strategy used to handle missing data instead of alleviating bias may introduce more bias in the same direction or may change the direction of the bias. The reason for discontinuation and any relevant post dropout information on the subject may be useful in imputing or representing a subject's missing outcome.

According to EMA guidelines,[1] the particular way of handling missing data should be primarily based on providing "an appropriately conservative estimate for the comparison of primary regulatory interest" and not on the "properties of the method under particular assumptions."

### 8.2.5.1 Handling Ignorable Missing Data in Non-Inferiority Analyses

When data are ignorably missing, valid analysis is possible (and probably fairly easy) with a few precautions.

Complete case analysis is analysis of only subjects who have no missing data. A completers analysis is equivalent to applying a particular imputation method to the missing data that has treatment arm as the sole factor. When discontinuations and other missing data are MCAR and also independent of covariates, then complete case analysis will be appropriate for inference and estimation. When discontinuations and other missing data are MCAR but not independent of covariates, then complete case analysis will be appropriate for inference and estimation only if the relevant covariates are included appropriately in the analysis model. When discontinuations and other missing data are not MCAR, then complete case analysis will not be appropriate for inference and estimation, as type I error rates will be higher or lower than advertised and treatment effects will be under- or overestimated. Since MCAR is a very strong assumption and the testing methodologies are imprecise at best, it is generally wise to avoid heavy reliance on complete case analysis. However, such an analysis can be reported as a sensitivity analysis.

Use of covariates in the analysis can be somewhat controversial, especially when the covariates are not pre-specified. It is standard to pre-specify stratification factors in the randomization process as covariates in the analysis, but post hoc inclusion of other covariates in the analysis is known to have the potential to inflate the type I error. When covariates are known to be associated with the likelihood of missing data, they should be included in the model, even if the association was not known or predicted before study start. To comply with this inconsistent pair of assertions, reporting multiple analyses will be required. The pre-specified analysis will generally serve as the primary analysis, with the model adding covariates labeled as a supportive or sensitivity analysis. Too much missing data or inconsistent results from the supportive and sensitivity analyses may invalidate the assumptions of the primary analysis and make the corresponding results impossible to interpret. An alternative that might be acceptable in some cases is to investigate the relationship between baseline covariates and missingness in the blinded data set, decide which covariates should be included in the primary analysis, finalize the analysis plan, and then unblind. Imputation using baseline covariates and/or post-baseline measurements may also be used.

A complete case analysis violates the intent-to-treat principle and is thus subject to bias. This approach is not recommended as the primary analysis for a confirmatory trial.[1] When data are MAR but not MCAR, complete case analysis will generally be inappropriate and an analysis recognizing the missing data mechanism must be used. Generally this involves imputation of missing data. The process of imputation involves substituting a likely value for a missing value and a subsequent analysis of the resulting data, so it is important to choose the imputed value appropriately.

A simple method that can always be used for longitudinal data, but is rarely optimal, is to substitute using the last observation carried forward (LOCF) method. When a value is missing, the last observed value for that subject substitutes for the missing value. The resulting data set is analyzed as if every subject has complete data available. LOCF is not valid under MCAR or MAR, and is only valid when the subject's outcome does not change after dropout. When the therapeutic effect is reached rather quickly, and long-term observation follows, LOCF may be more appropriate for accurately estimating treatment effects.

LOCF is particularly prone to criticism when a disease state is progressive (progressively worsening disease, or progressively improving disease after treatment). When the subject's condition or prognosis tends to deteriorate over time an LOCF analysis likely gives optimistic imputations that may lead to understating the subject-level standard deviation. Also, if the discontinuations occur earlier on the experimental arm, a bias may potentially be induced favoring the experimental arm (particularly when the experimental

therapy has inferior or equal efficacy to the active control). Even when estimation is unbiased, LOCF may understate the standard error surrounding
the point estimate, leading to test procedures that have an inflated type I
error rate.

Koch[11] recommended incorporating imputation under the null hypothesis
for missing data in a non-inferiority trial. Therefore, a penalized imputation might be considered. When a subject discontinues, subsequent missing
observations can be imputed as the last observation plus or minus a small
amount to penalize the analysis for the missing data. For a non-inferiority
margin of $\delta > 0$ with larger values being more preferred, possible ways of
performing imputation under the non-inferiority null hypothesis include
first performing an imputation under the assumption of no treatment difference and then either (1) subtract $\delta$ from each imputed outcome in the experimental group, or (2) add $\delta$ to each imputed outcome in the control group, or
(3) subtract $\delta/2$ from each imputed outcome in the experimental group and
add $\delta/2$ to each imputed outcome in the control group (or some variation
thereof). Per Koch[11] for binary data (i.e., "0" for a failure and "1" for a success)
with a fixed margin on the difference in proportions, the imputed outcomes
should be inclusively between 0 and 1. Imputation under the non-inferiority
null hypothesis is particularly important when the evaluation of the active
control effect and the selection of the non-inferiority margin did not consider
the possibility of missing data.

Incorporating an imputation under the non-inferiority null hypothesis can
help alleviate some problems inherent in disregarding missing data when
establishing the non-inferiority margin. Consider a continuous outcome
for an endpoint of interest where the conditional effect of the active control
among subjects for which the endpoint is measured is 50 and the conditional
effect of the active control is zero among subjects for which the endpoint is
not measured. If the endpoint is not measured in 20% of subjects in both
arms, then the true effect of the active control is 40. Suppose that the evaluation of the active control effect in previous clinical trials treated missing
data as ignorable and M1, the efficacy margin, was set at 50. Suppose the
non-inferiority trial had 20% of subjects on each arm not measured for
the primary endpoint. The true effect of the active control in the setting of
the non-inferiority trial for the ITT population is 40. The efficacy margin
can be adjusted retrospectively to account for the missing data. However,
there may be a preference to pre-specify the margin without later changing it. Employing an imputation under the non-inferiority null hypothesis (based on a true mean difference of 50) without altering the margin can
achieve the same or similar results as altering the margin and ignoring the
missing data.

Another simple imputation approach, with similar advantages and disadvantages to LOCF, is to use the mean value of all observed values (or other
estimate of central tendency) to replace missing values. This might be the
mean across an individual's observed assessments, or the median across all

subjects' observed assessments at the time point in question, or some other quantity. If such a procedure is used, it is very likely that the estimated variability of outcomes will be reduced, resulting in an inflation of type I error. In addition, this method is applicable to MCAR, not MAR, if the mean is calculated from other subjects than the subject with missing data.

Rather than using the mean value, a regression-based value can be substituted for missing data. When outcomes are expected to be monotonically changing over time, an individual's assessments can be used to model disease progression (or improvement) over time. Baseline covariates and post-baseline measurements from all subjects can be used to develop more complex regression models and obtain more precision in this modeling and imputation process, but again using data from other subjects assumes a missing data mechanism closer to MCAR or MAR than MNAR.

Rank-based imputation is often useful for superiority studies but less so for non-inferiority studies. Subjects who have died (i.e., without data due to death) might have all their ranks assigned (as a tie) to the worst ranks among all subjects; subjects who discontinue for toxicity might have all their ranks assigned (as a tie) to the next worst ranks, etc. For a rank-based nonparametric analysis, this provides an appropriate data set for analysis. However, nonparametric analyses are uncommon for non-inferiority due to concerns about permutation tests and also due to lack of an appropriate rank-based margin, so this method is less useful.[12]

Mixed model repeated measures (MMRM) analysis is attracting increased interest in analysis of superiority studies. MMRM analyses use two key features that are distinct from ANOVA models: MMRM models both fixed effects and random effects, and MMRM models within-patient correlations of various data.[13] The fixed effect in MMRM is typically the efficacy or other primary outcome data of interest, whereas the random effect is the patient-specific data, often consisting primarily of the correlation of outcome data at various time points or the correlation of various outcome variables. By modeling the within-patient data as random effects, the missing data mechanism can be modeled to obtain valid inference under an MAR scenario. MMRM is now well accepted for analysis of superiority trials, but is not as common for non-inferiority trials. Although the mathematical properties suggest that MMRM will perform well in non-inferiority trials, an MMRM analysis in a non-inferiority trial should also include sensitivity analyses and assessments of the impact of different missing data mechanisms on the conclusions.

The most complex procedure for imputing missing data is multiple imputation, which is a two-step process. First, missing data are modeled using known information about the missing data mechanism. Second, this missing data model is used to impute data that are used for analysis. The process is repeated a few times (e.g., 3 to 20 times) to obtain a sample of full data sets and associated analyses. The variability among the various analyses is incorporated into the interpretation of analyses, resulting in appropriate point estimates, confidence intervals and inference. Multiple imputation is most helpful when the process used to impute data is different from the process used to analyze data, such as by using different covariates to predict missingness than are used in the analysis model, and most useful for large studies. This could happen when information used to explain missingness occurs after discontinuation, such as death predicting (backwards) study discontinuation within a short period before the death. Such information may not be included in the analysis model, but can easily be incorporated into the imputation procedure. If multiple imputation is planned, collection of auxiliary data (such as other measures of efficacy or follow-up after discontinuation, subject to ethical constraints) should be planned. Treating subjects with missing data the same way as subjects with complete data may bias the conclusion toward no difference. For a non-inferiority analysis, a penalized imputation approach can also be considered with multiple imputation. However, little is known about how this approach will affect an analysis in the presence of non-ignorable missingness.

A trite but true piece of advice is that missing data are best handled by prevention. Missing data are most easily handled when almost all data are known. Most sensitivity analyses, even when missing data are handled or imputed quite differently, will give very similar conclusions if the amount of missing data is small. This requires planning before the study starts to allow collection of auxiliary data, motivate study subjects and investigators to comply with the protocol, and perhaps gain approval from ethics committees to allow collection of data even after a subject "withdraws" from some study procedures. Efforts should be directed toward obtaining data on all subjects, even those who discontinue treatment, to best assess treatment effects when subjects are noncompliant with certain aspects of the study plan.

In summary, there is not one clear method for handling missing data that are MAR but not MCAR. Each method has promise, but each must be used with caution. Simple methods that focus on unbiased estimation may inflate the type I error rate by underestimating the variance. For example, in many settings, dropouts are likely to have poorer outcomes than subjects who remain under observation for outcomes. Then, for continuous outcomes, treating the missing outcome as ignorable or using LOCF (and other single imputation methods) treats their unknown outcomes as being more central-valued, and thus will then lead to an underestimated subject-level standard deviation. More complex methods can address the estimation of the variance of the estimated treatment difference. Various methods should be considered for most studies, with one simple method pre-specified as primary and the other methods pre-specified as sensitivity or exploratory analyses.

### 8.2.5.2 Handling Non-Ignorable Missing Data

When missing data are non-ignorable, the solutions are even less appealing. Although it might be clear that the missing data mechanism depends on the value of the missing data, it is much less clear how to use available data to recreate or impute the unobserved data. Therefore, the analysis involves some level of speculation.

Analysis of data that are MNAR typically involves modeling the missing data mechanism and the value of the missing data. These models are difficult in practical situations, and even more difficult to pre-specify. When not pre-specified, the analysis may be regarded as a sensitivity analysis or as an exploratory analysis.

For non-inferiority trials, the problems may be more severe. Most studies on missing data implicitly assume that the hypothesis test of interest is a test of equality of treatment effects. The performance of such methods in non-inferiority analyses has yet to be comprehensively explored in the statistical literature.

One model that may be of interest (perhaps as a sensitivity analysis rather than a primary analysis) is a pattern-mixture model. Such a model assumes that the possibly unobserved outcome is dependent on the missingness of the data. Mathematically, the model is often written as $f_\theta(Y, M) = f_\psi(Y \mid M) \times f_\pi(M)$, where $Y$ is the vector of outcome data and $M$ is the corresponding vector of indicators of missingness. Such a model assumes that the outcome is dependent on the missingness, and the average treatment effect over all patterns of missingness is of interest. Covariates related to the missingness can be included in the imputation model as well as a factor for the randomized treatment group.

One proposed application of the pattern-mixture model that may be applicable to non-inferiority analyses is a composite approach.[14] In this composite approach, the pattern-mixture model is factored so $f_\psi(Y \mid M)$ and $f_\pi(M)$ are considered separately. That is, the probability of missingness (with emphasis on missingness for unfavorable reasons, since missingness unrelated to outcome is MCAR) is considered separately from the treatment effect dependent on the successful completion of the study. Again, this was proposed for a superiority analysis, so a non-inferiority analysis must adapt.

Example 8.2 illustrates the use of LOCF in a non-inferiority analysis.

### Example 8.2

Consider a clinical trial of an antihypertensive agent. The primary endpoint is change from baseline to the end of the study (8 weeks after baseline) in sitting systolic blood pressure (SBP), with low numbers (changes that are large negatively) being preferred. The non-inferiority margin was prospectively set at 4 mmHg, so the lower limit of the two-sided 95% confidence interval for the difference in mean change in SBP (active control – experimental) must exceed –4.0 mmHg to conclude non-inferiority. Dropouts were included with LOCF. Results are given in Table 8.1, and show that based on the overall study population a conclusion of non-inferiority is obtained.

Consider, though, the following discontinuation pattern. Subjects with data available at week 8 showed a similar response to subjects included via LOCF imputation in the active control group, but such subjects showed less antihypertensive effect in the experimental group. Using only the subset of subjects with complete data, a conclusion of non-inferiority would not be obtained.

Among subjects who did not complete the study, a nearly equal number were for administrative reasons such as inconvenient study visits. Many more subjects discontinued the experimental group because of adverse events or lack of efficacy (14.0% vs. 1.3%). It is quite possible that subjects who discontinued because of adverse events had a fairly large antihypertensive effect, enough to affect the group mean by 1 mmHg. So, the discontinuation is related to outcome, and implies that the missing data are not MCAR. In addition, subjects who do not complete the study do not receive benefit at week 8, so the inclusion of such data in the primary analysis via LOCF is questionable (for superiority or non-inferiority).

A pattern-mixture model composite approach such as suggested by Shih and Quan,[14] does not fully confirm the conclusions of the primary analysis, due both to the lack of conclusion of non-inferiority on antihypertensive effect and also the increase in discontinuations for adverse reasons. Therefore the conclusion of the primary analysis is questioned by the sensitivity analysis because of the impact of missing data.

The impact of missing data in a study in which the primary endpoint is binary is easier to assess than when the primary endpoint is continuous due to the finite number of possible imputations. For $n_E$ and $n_C$ subjects in the experimental and active control groups, respectively, with $m_E$ and $m_C$ of those subjects with missing data, there are $2^{m_C+m_E}$ possible imputation patterns but only $(m_E + 1) \times (m_C + 1)$ potential unique results. A "tipping point" graphical sensitivity analysis, similar to that proposed for general study of missing data, can be used to assess the proportion of the $(m_E + 1) \times (m_C + 1)$ potential unique results that will result in a conclusion of non-inferiority. A grid is constructed with horizontal length $m_E + 1$ and vertical height $m_C + 1$. Grid box $(a, b)$ corresponds to having $a$ of the $m_E + 1$ missing subjects in

**TABLE 8.1**

Summary of Trial Results and Reasons for Discontinuation

| Treatment Group | Active Control | Experimental |
|---|---|---|
| *Overall Study Population* | | |
| N | 150 | 150 |
| Mean | −8.0 | −8.2 |
| Standard deviation | 14.2 | 13.8 |
| Confidence interval | (−3.0, 3.4) | |
| *Completers* | | |
| N | 140 | 120 |
| Mean | −8.0 | −7.2 |
| Standard deviation | 14.2 | 13.8 |
| Confidence interval | (−4.2, 2.6) | |
| *Discontinuation* | *Active Control* | *Experimental* |
| Administrative reasons | 8 | 9 |
| Lack of efficacy | 0 | 2 |
| Adverse events | 2 | 19 |

the experimental group imputed as a success and $b$ of the $m_C + 1$ missing subjects in the active control group imputed as a success, $0 \le a \le m_E$ and $0 \le b \le m_C$. Each of the $(m_E + 1) \times (m_C + 1)$ resulting squares in the grid is shaded, say shaded black if the conclusion would not be of non-inferiority, gray if the conclusion would be of non-inferiority and white (i.e., no shading) if the conclusion would be of superiority of the experimental treatment compared to the control.[15] Such a figure will easily represent not only how many of the $(m_E + 1) \times (m_C + 1)$ potential unique results would lead to each conclusion, but also how different the pattern of missing successes would have to be to change the conclusion of the primary analysis.

A worst case analysis assigns the worst possible outcome for the missing outcome. This usually alters the definition of the endpoint to include dropout or discontinuation as the worst outcome. This can mix tolerability and safety with efficacy and usually does not address the bias of missing data for the original endpoint. When outcomes tend to deteriorate over time, a worst case analysis may favor the treatment arm where discontinuation occurs earlier. For time to event endpoints (where worst case means assigning an event at discontinuation of follow-up), a worst case analysis may favor the treatment arm where the discontinuation occurs later.

A worst comparison analysis assigns the best possible outcome to missing values in the control group while assigning the worst possible outcome to missing values in the experimental group. A worst comparison analysis provides a bound on the true internal value of the estimated difference. Such an analysis is conservative, and the conservatism increases as the amount of missing data increases.

Missing data must be considered in any clinical trial. Even though most applications of missing data analysis have inherently assumed a superiority hypothesis, missing data must also be considered in non-inferiority analyses.

Simple methods are encouraged, due to both ease of interpretation and lack of extensive experience applying more sophisticated methods to non-inferiority analyses. Methods should emphasize the identification of a missing data mechanism (MCAR, MAR or MNAR). If MNAR can be convincingly ruled out (which may be unlikely), then the simple methods are sufficient. Otherwise, interpretation will require creative use of more sophisticated methods along with justification that such methods are applicable. Imputation under the null hypothesis and/or even less favorable imputation for the experimental arm should be considered.

## 8.3 Analysis Sets

The analysis set (the set of subjects who contribute to the analysis result) used for analysis of non-inferiority clinical trials has been debated, and is an evolving issue. In this section we will define several analysis sets used in clinical trials, and discuss their advantages and disadvantages.

### 8.3.1 Different Analysis Populations

Several analysis sets (or analysis populations) have been defined for analyzing the results of clinical trials. It is common to pre-specify two or three analysis sets in the analysis plan, with one of those sets identified for primary inference. Analysis based on the other pre-specified analysis sets are also presented to provide support for the primary endpoint by checking the robustness of the results: if results are similar for all analysis sets, it provides some assurance that the results are not strongly dependent on the choice of the analysis set. Additionally, a different analysis set is often used for investigation of safety and tolerability. Although names and definitions differ, the following give some outline to various analysis sets that are commonly proposed:

*ITT Set.* The phrase "intention to treat" was originated by Sir Austin Bradford Hill,[16] and the term "intent-to-treat" is commonly used. The ITT set comprised all subjects according to their randomly assigned treatment groups. The advantage of using the ITT analysis is that inclusion in the analysis is based on randomization rather than a post-randomization (and post-treatment) event.

Modifications to the ITT set may be allowed under certain circumstances. Subjects who were retrospectively identified as not having the disease under study may be removed from the analysis set if the information was collected before randomization but not interpreted until after randomization. This can occur when it requires time to complete certain laboratory procedures, but it is necessary for the subject not to have their treatment delayed until the results are known. Such subjects would be randomized, but are removed from the analysis if the laboratory results do not confirm the diagnosis. However, any modifications to the ITT set move the inference further away from the randomization on which it is based. Modifications based on data obtained after the initiation of treatment have the potential to introduce bias in the inference.

*Per-Protocol Set.* This analysis set includes all subjects who in retrospect met the inclusion and exclusion criteria of the clinical trial, received at least a minimal amount of randomized study drug (or, in some cases, subjects might be required to have nearly perfect adherence to the medication plan to be included), and had assessments to allow evaluation of the primary endpoint. In other words, the per-protocol (PP) analysis set used only the subjects whose behavior adhered to protocol. Also, a per protocol analysis alters the outcomes from protocol violators to missing and treats the missingness as ignorable.

*Completers.* This analysis set includes subjects who had assessments to allow evaluation of the primary endpoint of the study. This might be a super-set of the per protocol subset, or might be equal to the per protocol subset, depending on the definition. In addition, a completers analysis set might be defined for a particular secondary endpoint that is independent of the primary endpoint. This analysis set can also be described as an "Evaluable Subjects Set". For a time-to-event endpoint, a completer can be considered as a subject who is followed to the event or to the end of study.

*Safety Set.* This analysis set is intended for the summaries of safety and toler-ability data. Unlike the ITT analysis set, subjects who receive a treatment that is inconsistent with the randomized treatment are often summarized according to treatment actually received, as an important safety event in a subject assigned to one treatment but receiving the other could change the safety conclusions. In addition, it is generally required that a subject receive at least one dose of study medication, and sometimes required that a subject had a follow-up assessment of safety or tolerability, to be included in this analysis set.

Other possible analysis sets/populations and definitions have been pro-posed: all subjects screened, all subjects randomized, all subjects who took at least one dose of study medication, and all subjects with any follow-up. Each has its advantages and disadvantages, but the key is that exclusion from the statistical analysis of some patients who were randomized to treatment may induce bias. To the extent that an analysis set avoids bias from such exclu-sions (i.e., still maintains the fairness of randomization), it can be consid-ered as adhering to the ITT principle. Additionally, it is mandatory that the analysis set be clearly defined before study start and that this definition is consistently applied.

For the remainder of this discussion we will use terminology from the ICH guidance document E-9[5] and call the primary analysis set/population the "full analysis set."

### 8.3.2 Influence of Analysis Population on Conclusions

In choosing the appropriate strategy for building the full analysis set, the influence on the results and the potential to bias results toward a conclusion of non-inferiority will be of interest. There is also the possibility that investi-gators' knowledge that all subjects are receiving an "active" therapy may bias outcome evaluations toward success at a subject-level, which may reduce the sensitivity to detecting a difference between treatments.[17] Too much missing data, which may introduce a large bias, and/or bias in the study conduct may invalidate the analysis regardless of selected analysis population even when the results are similar across analysis populations.

For analysis of clinical trials aimed at showing the superiority of one treat-ment over another, it is nearly universal to base the full analysis set on the ITT set. For a non-inferiority analysis, the full analysis set has often been based on the PP set, rather than the ITT set. More recently, international guid-ance documentation[18] discussed the use of PP analysis sets in non-inferiority analyses and suggested providing separate, co-primary analyses using the PP and the ITT sets.

Depending on the method of analysis, use of the ITT set can provide unbiased estimation and comparisons of the endpoints in the setting of the clinical trial. If the clinical trial setting represents medical practice, subjects in the trial are randomly selected from subjects in medical practice, and full follow-up is obtained, then use of the ITT set will provide unbiased estimates and comparisons on the use of the treatments in practice. Limitations to this include selecting subjects who are not representative of medical practice and accounting for missing data – especially from subjects who discontinue the trial but would have continued to be observed by a physician in medical practice.

Part of the ITT principle is that subjects be followed until an event or end of study. A subject is said to be lost to follow-up if the individual is not followed to the endpoint or to the end of study. Loss to follow-up can rarely be assumed to be random – that is subjects lost to follow-up are different from subjects not lost to follow-up, particularly with respect to the distribution of the primary outcome. Even if the numbers that are lost to follow-up are similar between arms, the unobserved outcomes may not be similar between arms. It is much more important to keep loss-to-follow-up rates to a minimum than to have similar numbers that are lost to follow-up. The greater the amount of loss to follow-up, the greater the potential of substantial bias, even if rates are similar between groups.

One rationale for using the PP analysis set is that it may more closely follow the scientific hypothesis that a subject with the disease of interest, who receives a particular treatment, will exhibit improvement compared with a subject not receiving that treatment. If a subject does not have the disease under study, or does not receive the treatment, the subject is not part of the target population for examining the scientific hypothesis.

Another reason for the recommendation to use the PP set for non-inferiority analyses is that deviations from the protocol in randomization, conduct or evaluation might make the outcomes for the treatment groups more similar. In other words, sloppiness in trial conduct or other deviations from the planned procedures may bias the results toward no difference between the arms. In an extreme case, all subjects on both treatment arms could discontinue treatment immediately upon randomization, resulting in all subjects receiving the same treatment. Producing outcomes that are more similar between the groups has the effect of making a superiority analysis conservative, since no difference between treatments is in the null hypothesis. However, producing outcomes that are more similar between the groups might have the effect of making a non-inferiority analysis anticonservative since no difference between treatments is in the null hypothesis.

Consider the following situations to illustrate the relative conservativeness of the ITT and PP analysis sets:

*Subjects are treated with study treatment that is not their randomized study treatment.* This can be caused by several kinds of errors, including a problem with the randomization process, the site personnel delivering medication inconsistent with the randomized study medication and packaging errors. In each case, analyzing the subject according to randomized treatment is not appropriate for a non-inferiority study as subject outcomes will tend to be more similar across the "as randomized" study arms resulting in an analysis biased toward a conclusion of non-inferiority.

*Subjects do not comply with their randomized treatment assignment.* If this problem is equally distributed between treatment groups, then again there is a fear that the resulting outcomes are more similar to each other than randomization would dictate. If this problem is confined to one treatment group, then it might be an indication of a problem with the tolerability of the treatment or the existence of investigator bias if the blinding is not maintained. In either case, analyzing according to randomized treatment group will be conservative for a superiority analysis but not necessarily for a non-inferiority analysis.

*Subjects who discontinue the study early.* One problem unique to active-controlled trials is that the active control and perhaps other effective therapies are often available commercially, so subjects who discontinue the experimental therapy are then able to take a treatment that has known efficacy (i.e., a "rescue" therapy). This can encourage subjects to discontinue early more often than in the setting of a placebo-controlled study for an indication with no known effective therapy. Whether to continue to follow subjects in non-inferiority trials after discontinuing study drug or to use the subsequent efficacy data for primary inference has been an issue. The outcomes from the two treatment groups of subjects may be more similar than if rescue therapy were not available. Often, during historical placebo-controlled studies (data on which the margin is based), the active control was not available as rescue treatment. This would lead to efficacy and non-inferiority margins that are too large relative to the design and conduct of the non-inferiority trial. In such a case, the non-inferiority analysis may be invalidated by excessive use of rescue therapy or need to appropriately consider the effect of the active control therapy relative to a placebo followed by rescue therapy. Failure to collect data upon using rescue therapy limits the ability to examine the robustness of conclusions and the value of switching treatment based on lack of early efficacy. Additionally, if other treatments are available and were available during the historical comparison of the active control to placebo, including data from subjects who discontinue study medication is consistent with the design of these studies and on how the non-inferiority margin is based. We therefore strongly advocate collecting data on all subjects even after discontinuation of study treatment and using the data appropriately in the analysis.

*Subjects are not evaluated for the primary endpoint.* The problem of missing data must be considered for all clinical trials. Most methodology for missing data has been developed implicitly assuming that the test of interest has a null hypothesis of no difference. Thus, new methodology is needed to appropriately test the null hypothesis of a nonzero difference. Ignoring subjects who do not have the primary endpoint evaluation should be recognized as inappropriate: if the missingness is associated with treatment, then valuable information on the missing data is not used. In that sense, the PP analysis set is not appropriate because of elimination of subjects who are not evaluated for the primary endpoint. However, utilizing such subjects is also problematic: if all subjects in both treatment arms are simply counted as treatment failures for a binary endpoint, this will bias the conclusions toward non-inferiority. Thus, a way to account for missing data without biasing the conclusions is required but not necessarily readily available.

Under the belief that the PP analysis set addresses these problems, the PP analysis set has been commonly used instead of or in addition to the ITT analysis set. A concern is whether the PP analysis set is always the appropriate way to address these and other issues. Some authors have questioned the wisdom of using the per protocol concept for analyzing non-inferiority clinical trials. A discussion of several antibiotic non-inferiority clinical trials concluded that use of the ITT analysis set does not systematically lead to smaller estimates of treatment effect in these trials (see Section 2.5 for more details).[19] A hybrid ITT/PP analysis set that excludes noncompliant subjects as with the PP set while addressing the impact of missing data (based on maximum likelihood) due only to lack of efficacy using an ITT approach was also proposed as a compromise.[20] More aggressively campaigning against the use of the PP analysis sets, Hauck and Anderson[21] noted that standards required the use of PP analysis set for the null hypothesis $H_o$: $\mu_C - \mu_E > \delta$ versus $H_a$: $\mu_C - \mu_E \leq \delta$ for any value of $\delta > 0$, making the null hypothesis of equality the only point at which the ITT analysis set is favored. Such a discontinuity is difficult to justify for only one point, making the assumption faulty. Wiens and Zhao[22] expanded this idea and concluded that the arguments for using the ITT set for superiority analyses apply equally well to non-inferiority analyses, and therefore the ITT analysis set should be considered. Furthermore, the PP approach is not the universally best choice for a sensitivity analysis and therefore should not be a standard adjunct analysis, much less the standard co-primary analysis.

Other authors have proposed basing the analysis on the treatment actually received, regardless of whether it was the randomized treatment. These proposals include analyzing subsets of the ITT set, incorporating some ideas of the PP analysis set[23]; and structural models that account for non-compliance and consider treatment received as a time-varying covariate, and thus have some of the advantages of the PP analysis set and the ITT analysis set.[24] However, other authors such as Lee et al.[25] determined that analysis by treatment actually received can produce counterintuitive results that obscure rather than enlighten.

Our conclusion is that the agreement on an appropriate full analysis set for analysis of a non-inferiority clinical trial is not currently widespread in the clinical trials community. Momentum seems to be moving toward an approach that favors something closer to the ITT analysis set and moves away from the PP approach. In the absence of unanimity, the statistician preparing a non-inferiority analysis must rely on sound statistical principles to develop a comprehensive strategy. Pre-specification of the analysis strategy is essential. This requires understanding potential issues before the study begins and preparation for all eventualities in the analysis plan. Even though most trials will have surprises that cannot be foreseen, the planning for foreseeable issues will be vital. Wiens and Zhao[22] outlined four areas that will be problematic: missing data, ineligible subjects, poor study conduct, and sensitivity analysis, with missing data posing the biggest concern. We outline some ideas on ineligible subjects below; the other three topics are covered elsewhere in this book.

Ineligible study subjects should be considered differently in the analysis of non-inferiority clinical trials than in superiority clinical trials. Subjects who cannot benefit from the treatment (i.e., they do not have the disease under study) may tend to make the treatments look more similar to each other, if such subjects are equally randomized to the treatments under study. If such subjects will receive treatment once the treatment is approved for marketing, excluding these subjects provides information inconsistent with real-world application. It has been advocated that subjects enrolled in a clinical trial be offered the study drug as long as it does not pose a safety concern, and should be aggressively followed even after discontinuation of study drug to avoid bias.[26] Although this might seem less important in a non-inferiority clinical trial, and even misguided as noted in the above example of subjects receiving the active control after discontinuation of study drug, it is better to obtain the information and not use it than to need the information and not have it. Example 8.3 discusses whether to include in the non-inferiority analysis subjects from whom cultures were not obtained in a study of uncomplicated bacterial infections.

### Example 8.3

Cultures from subjects enrolled in a study of uncomplicated bacterial infections must be obtained at enrollment to confirm the pathogen causing the symptoms. Similar symptoms can be associated with fungal or viral infections, and such infections are not meant to be treated by antibacterial drugs. The determination of the pathogen takes 48 hours but treatment must be commenced immediately for ethical reasons. If the pathogen is determined to be one that is not susceptible to the study treatments based on preclinical results, or if no pathogen is found, it is common to discontinue the subject from the study and ignore the subject in any efficacy analyses. However, this might not be the best course of action. When the new treatment is approved for marketing, it will be prescribed based on empirical symptoms rather than on cultures. Thus, it may be of interest for the subject to be offered the chance to stay in the study and even continue to receive study medication until symptoms resolve. It is likely that the informed subject will not choose to remain on study medication if told that it will likely not impact the symptoms. The subject who chooses to discontinue study medication and start a different course of treatment should continue to receive follow-up evaluations in accordance with the protocol. The next question becomes how to use the subject in the analysis. For the primary analysis, it may be possible to remove the subject since the exclusion was in place before randomization—even though it was not known until after randomization.[27] It may be of benefit to report the success rate in the primary analysis and also among subjects treated empirically, to give the physician information on success rates under clinical trial situations and under practical situations.

### 8.3.3 Further Considerations

The preference of analysis set in non-inferiority is not universal. Using a PP analysis set and an ITT analysis set is common. However, it is possible that bias will persist in the analyses for each analysis set and that neither analysis set is appropriate.

The method of Sanchez and Chen[20] is one of the more promising approaches. However, it requires knowledge of the reason for missing data. Often in clinical trials the decision of a subject to discontinue is based on a series of complex and intermingled reasons, which can depend on whether the subject and the investigator of the therapy are truly blinded of the therapy the subject is receiving at the time of discontinuation. Anything less than complete and immediate efficacy can cause a subject to discontinue if the logistics of study participation is difficult, and a desire to please can cause a study subject to be less than honest with the study staff. When the reason for discontinuation is known with certainty, this method will work, but when the reasons are not known with certainty the method will have flaws.

Sensitivity analyses will be important in addressing the analysis set for non-inferiority clinical trials, particularly the impact of missing data on the analysis set of choice. In fact, by suggesting that both a PP and an ITT analysis set be considered, guidance implicitly includes a sensitivity analysis. Until resolution is found in the clinical trials community, a primary approach based on ITT with variations to account for anticipated deviations from the protocol, along with sensitivity analyses, is probably the best approach. Sensitivity analyses should be pre-specified to the extent possible. Although sensitivity analyses are recommended, the use of sensitivity analyses is not a substitute for poor trial conduct or poor adherence to the protocol, and does not salvage a poorly conducted clinical trial.

Although both the ITT and PP approaches have been criticized, we recommend that non-inferiority analyses should be performed for both the ITT and the PP analyses. In most instances, the results should be quite similar. Any notable difference in the results should be investigated and may be indicative of poor study conduct or other reasons that must be thoroughly investigated and explained. Similarity in the results of the ITT and PP analyses, while reassuring, does not imply confidence in the results. A poorly conducted trial likely introduces bias, which can be of a nearly equal size for both the ITT and PP analyses.

## References

1. European Medicines Evaluation Agency, *Guideline on Missing Data in Confirmatory Clinical Trials*, Committee for Medical Products for Human Use, 2009, at http://www.ema.europa.eu/human/ewp/177699endraft.pdf.
2. Code of Federal Regulations 21 CFR 314.126.
3. Carroll, K.J., Analysis of progression-free survival in oncology trials: Some common statistical issues. *Pharm. Stat.*, 6, 99–113, 2007.
4. Fleming, T.R., Rothmann, M.D., and Lu, H.L., Issues in using progression-free survival when evaluating oncology products. *J. Clin. Oncol.*, 27, 2874–2880, 2009.
5. International Conference on Harmonization of Technical Requirements for Registration of Pharmaceuticals for Human Use (ICH), E9: Statistical principles for clinical trials. 1998, at http://www.ich.org/cache/compo/475-272-1.html#E4.
6. Fleming, T.R., Addressing missing data in clinical trials. *Ann. Intern. Med.*, 154, 113–117 2011.
7. Jackson, J.B. et al., Intrapartum and neonatal single-dose neviparine compared with ziodovudine for prevention of mother-to-child transmission of HIV-1 in Kampala, Uganda: 18 months follow-up of the HIVNET 012 randomised trial. *Lancet* 362, 859–868, 2003.
8. Little, R.J.A. and Rubin, D.B., *Statistical Analysis with Missing Data*, John Wiley, New York, NY, 1987.
9. Little, R.J.A., Regression with missing X's: a review, *J. Am. Stat. Assoc.*, 87, 1227–1237, 1992.
10. Little, R.J.A., A test for missing completely at random for multivariate data with missing values, *J. Am. Stat. Assoc.*, 83, 1198–1202, 1988.
11. Koch, G.G., Comments on 'current issues in non-inferiority trials' by Thomas R. Fleming, *Stat. Med.*, 27, 333–342, 2008.
12. Wiens, B.L., Randomization as a basis for inference in noninferiority trials, *Pharm. Stat.*, 5, 265–271, 2006.
13. Mallinckrodt, C.H. et al., Recommendations for the primary analysis of continuous endpoints in longitudinal clinical trials, *Drug Inf. J.*, 42, 303–319, 2008.

14. Shih, W.J. and Quan, H., Testing for treatment differences with dropouts present in clinical trials—A composite approach, *Stat. Med.*, 16, 1225–1239, 1997.
15. Hollis, S., A graphical sensitivity analysis for clinical trials with non-ignorable missing binary outcome, *Stat. Med.*, 21, 3755–3911, 2002.
16. Hill, A.B., *Principles of Medical Statistics*, 7th ed., Lancet, London, 1961.
17. Snapinn, S.M., Noninferiority trials. *Curr. Control Trials Cardiovasc. Med.* 1, 19–21, 2000.
18. International Conference on Harmonization of Technical Requirements for Registration of Pharmaceuticals for Human Use (ICH) E-10: Guidance on choice of control group in clinical trials, 2000, at http://www.ich.org/cache/compo/475-272-1.html#E4.
19. Brittain, E. and Lin, D., A comparison of intent-to-treat and per-protocol results in antibiotic non-inferiority trials, *Stat. Med.*, 24, 1–10, 2005.
20. Sanchez, M.M. and Chen, X., Choosing the analysis population in non-inferiority studies: Per protocol or intent-to-treat, *Stat. Med.*, 25, 1169–1181, 2006.
21. Hauck, W.W. and Anderson, S., Some issues in the design and analysis of equivalence trials, *Drug Inf. J.*, 33, 177–224, 1999.
22. Wiens, B.L. and Zhao, W., The role of intention to treat in analysis of noninferiority studies, *Clin. Trials*, 4, 286–291, 2007.
23. Stewart, W.H., Basing intention-to-treat on cause and effect criteria, *Drug Inf. J.*, 38, 361–369, 2004.
24. Robins, J.M., Correction for non-compliance in equivalence trials, *Stat. Med.*, 17, 269–302, 1998.
25. Lee, Y.J., Ellenberg, J.H., Hirtz, D.G. and Nelson, K.B., Analysis of clinical trial data by treatment actually received:Is it really an option? *Stat. Med.*, 10, 1595–1605, 2002.
26. Peto, R., Pike, M.C., Armitage, P., Breslow, N.E., Cox, D.R., Howard, S.V., Mantel, N., McPherson, K., Peto, J., Smith, P.G., Design and analysis of randomized clinical trials requiring prolonged observation of each patient. I. Introduction and design, *Brit. J. Cancer*, 34, 585–612, 1976.
27. Gillings, D. and Koch, G., The application of the principle of intention-to-treat to the analysis of clinical trials, *Drug Inf. J.*, 1991.

# 9

## Safety Studies

### 9.1 Introduction

Statistical hypotheses and testing to rule out a prespecified increased risk of an adverse event is statistically similar to that in the determination of non-inferior efficacy. Examples include establishing the safety of a test treatment compared to placebo or establishing the safety of a test compound compared to an active control, both with the objective of ruling out an important increase in the rates of adverse events. Less common, but possible, is the comparison of a test compound to an active control with inference desired on the event rate of the test compound compared to a putative placebo. Because the design and analysis are dependent on the objectives, and the objectives can vary, it is vital to prespecify and define the study objectives.

There may be uncertainty about the safety of a drug at the time of approval. Some adverse events are infrequent or are long-term adverse outcomes that may not be discovered during the clinical trials that led to approval. Additionally, the risk–benefit profile can change based on changes in supportive care, the nature of the disease, the standards or in the understanding of the risks. A change to an unfavorable risk-benefit assessment may alter or remove the indication or intended use. New evidence on safety may be sufficient to provide caution in the use of the product but not sufficient to lead to an unfavorable or uncertain risk–benefit profile. Some changes in risks can be addressed through introduced changes in medical practice. Subjects who are more at risk of a particular known adverse event can either not be recommended or provided the drug or may be more thoroughly monitored on their risk of experiencing the adverse event while receiving the drug.

The U.S. Food and Drug Administration (FDA) Amendments Act of 2007,[1] which expanded the authority of the FDA during postmarketing, provides situations in which a postapproval study on safety may be required. A postapproval study on the safety of a drug may be required to assess a known serious risk, or a signal of a serious risk, or to identify an unexpected serious risk when data indicate the potential for a serious risk. The source of a safety signal may be clinical trials, adverse event reports, postapproval studies, peer-reviewed biomedical literature, postmarket data, or other scientific data.

Evaluating whether data suggest a safety signal that was not prespeci-fied for the investigation may be associated with substantial error and bias. Although the efficacy of a drug is based on the intended effects of the exper-imental agent or regimen, its safety profile usually involves unintended, harmful effects. If the rate of these unintended, harmful effects is too great, the risk–benefit profile may be unfavorable. However, unlike efficacy analy-ses that prespecify the endpoints to be tested and with the overall type I error rate maintained at a desired level, standard safety analyses usually involve multiple tests, sometimes on nonprespecified adverse events, without any multiplicity adjustment. Thus there is an exploratory nature to the standard safety analyses that are conducted in a clinical trial. Additionally, owing to the multiple testing, the most impressive differences between arms in an adverse event will tend to be randomly high, and more likely than not will have a smaller observed difference in a subsequent identically designed and conducted clinical trial. Therefore, when the safety signal is evaluated on the basis of ongoing or previous trials, any meta-analysis used to formally test whether an unacceptable increased risk can be ruled out should not include the results from the clinical trial that identified the potential safety risk as the analysis that identified a potential safety signal is conditionally biased and potentially represents a random high.

Retrospective meta-analysis may be used to identify safety signals. If random-effects meta-analyses are done, the results should be viewed with care. Increasing the variability and altering the weighting of the studies can obscure the determination of a safety signal or in what subgroup a safety signal may be present.

There are three criteria or questions to be considered when assessing the reliability of an exploratory safety analysis[2]:

(1) Is it unlikely that such events can be explained by chance?

(2) Is the safety risk biologically plausible?

(3) Can independent, prospectively obtained data be identified to con-firm the finding?

These criteria are illustrated to clinical trials in aortic stenosis in Fleming's paper.[2]

Safety clinical trials prospectively performed to "confirm" a safety sig-nal do not generally have the primary hypotheses of no increased risk (as the null hypotheses) and an increased risk (as the alternative hypoth-esis). Although such hypotheses and their testing are of interest, the pri-mary hypotheses generally involve whether there is an increased risk of a clinically unacceptable size. This threshold of increased risk is the non-inferiority or safety margin. The margin should be selected so that ruling out an increased risk as large as the safety margin rules out every unac-ceptable increased risk. For such a safety trial to provide meaningful, interpretable results regarding the potential safety risk(s), (1) enrollment of subjects with the target population of interest (e.g., subjects for whom excess risk by the experimental agent or regimen is a concern or uncertain) should be of sufficient size so that the required number of events will be achieved in a timely fashion; (2) adherence and use of the experimental regimen should be consistent with the real world; (3) the number of subjects on the control arm who cross over to the experimental agent or regimen should be minimized; and (4) long-term retention/follow-up of nearly all randomized subjects should be done.[2]

## 9.2 Considerations for Safety Study

Safety studies have special considerations for extrapolation to the larger population. If an active- or placebo-controlled study does not show an important safety differential, it is not immediately apparent that the safety signal does not exist. The study may lack assay sensitivity and not be able to distinguish a difference in the safety endpoint between the two treatments. It is vital to plan and conduct the study appropriately so that it will have assay sensitivity.

Safety studies should minimize the risk to trial subjects. Patients for whom there is clear evidence of harm or a negative risk–benefit profile should not be included as subjects in the trial. The need for a safety study suggests that either the experimental drug has an increased safety risk, no meaningful increase in risk, or no increased risk. There is generally no conveyance of a potential improvement (decrease) in risks. There should be careful monitoring of such safety studies to make sure that subjects are not being harmed. Internal or external adverse safety data may arise during a study that will require termination.

If there is an expectation that a drug induces an increased risk of an adverse event, subjects may be monitored more thoroughly for the event and for symptoms, or monitored more thoroughly for intermediate outcomes that are associated with an increased risk of the adverse event. It is likely (and natural) that the study drug would be discontinued for a subject who has an intermediate outcome associated with an increased risk of the adverse event due to a concern, a belief, or an assumption that the drug may increase the risk of the adverse event. This may prevent some subjects from experiencing the adverse event and may make the observed risks more similar between study arms. Not addressing such patient monitoring in the evaluation of the experimental drug will increase the likelihood of declaring an unsafe drug as safe (relative to that adverse event). Such a conclusion could lead to medical practice that does not discontinue the drug when an intermediate outcome associated with an increased risk of the adverse event occurs.

The study population must be chosen to accurately reflect patients who will receive the treatment when it is widely available, or that subpopulation in which the potential safety concern arises. Intentionally enrolling subjects not susceptible to the event of interest will rarely provide convincing evidence of lack of a real effect. Alternatively, enrolling only those subjects within the target population with the greatest risk of the event of interest may also be problematic. As subjects will not be representative of the target population, the safety margin may not be relevant to the estimated difference in risk and the corresponding confidence interval.

In this section, we will discuss several aspects of planning and conducting the non-inferiority safety study, including choice of study population, selection of endpoint, and study conduct issues.

### 9.2.1 Safety Endpoint Considerations

Safety is commonly assessed through observation of discrete events such as deaths or events of morbidity such as stroke or myocardial infarction, resulting in a comparison of proportions or hazard ratios. Less common is assessment of continuous values such as serum glucose levels or change in body weight. In this section, we assume that the discrete events are of interest; however, the ideas may also apply to assessment of continuous data.

The endpoint used to assess comparative safety must be well defined and accepted as an appropriate endpoint to assess relative safety. This requires clear definition and perhaps (in the case of a surrogate endpoint or a biomarker) requires confirmation that the endpoint is of clinical consequence. An independent, blinded adjudication committee should be considered to evaluate every suspected event and determine, consistently and clearly, whether the event met the definition of the safety signal. The general outline of the process is not importantly different from the adjudication process used commonly for superiority studies.

Composite endpoints are attractive when the concern is about a group of related events but can pose problems. A vague concern about cardiovascular issues might lead a sponsor to consider incidence of stroke, myocardial infarction, heart failure, or mortality, and the incidence of any event (or the time to the first such event) might be a candidate for the safety endpoint. It must be recognized that in a safety context, this can be problematic. First, an advantage in one component can counter a negative effect in another, leading one to conclude that the safety is identical when in fact it is not. An advantage on one component and a disadvantage on another is not necessarily a negative for the test treatment, but must be known to fully evaluate the comparative safety. Second, increasing the incidence rate by adding components on which there is no difference may obscure a real difference in other components. When inference is based on a simple difference in event rates, the variance of the estimate is based on the event rate, with maximum variance occurring when the incidence rate is 50%. Increasing a low incidence rate by adding components for which there is no difference may therefore increase the variance without changing the difference, resulting in less power to draw a conclusion.

As in any analysis of a composite endpoint, sensitivity analyses are required to understand the contribution of the various components. With a non-inferiority hypothesis, an increase in any single component may cause concern, even if the other components show no effect or even a beneficial effect. Setting a non-inferiority margin for each component of the composite is not practical and defeats the purpose of the composite endpoint, which is to increase power by increasing the event rate. These issues with using a composite endpoint must be understood, and planning must account for them, for a successful safety study with a composite endpoint.

Other decisions about safety endpoints may include hard endpoints (such as mortality or irreversible morbidity) or less important endpoints (such as symptoms or laboratory values that may imply a hard endpoint is more likely in the near future). The more definitive endpoints are preferred, but when rare will require a large sample size to rule out a difference of interest.

Trial conduct will be critical to ensure that all safety events are identified, which requires careful monitoring by the investigator and appropriate training. Again, this is required to ensure that events are observed and that they are consistently assessed and evaluated. Of importance is the decision on whether to prompt study subjects for evidence of a transient symptom for a serious safety event. Doing so can increase the number of events considered, but this may increase the noise without uncovering any new events of concern.

The analysis set choice for a safety endpoint with a non-inferiority analysis may be different than for an efficacy endpoint. The intention-to-treat approach is generally considered inappropriate for safety analyses—including subjects according to randomized treatment when some subjects receive a treatment to which they were not randomized can mask a safety signal. For this reason, intention to treat is rarely used for safety analyses. However, the per-protocol analysis may also be inappropriate as excluding subjects (especially those who experience the safety event) for any reason can also mask a safety signal. The most appropriate analysis will consider and may include all subjects treated, according to treatment actually received. If the safety event can be a residual effect occurring well after treatment is discontinued, then including follow-up for a lengthy period after discontinuation may also be necessary.

### 9.2.2 Design Considerations

A discussion of non-inferiority analyses for safety endpoints needs to consider several potential comparisons. An investigational treatment may be compared to placebo or to an active control, in each case to assess whether there is a difference in safety risks. A comparison to an active control may be to rule out an increase compared to the active control, or estimate the difference versus placebo via an indirect comparison. In this subsection, we discuss various designs. In any given situation, the question of interest will generally be known; thus, the solution will be a straightforward choice of the appropriate design to answer the question of interest.

The metric used in the comparison (e.g., a difference in proportions or a relative risk) should provide the necessary assessment of the safety risk. For example, for a non-contagious event, a 1% increase in the incidence means that 1 additional person out of every 100 will get the event regardless of whether the increase is from 0% to 1%, from 5% to 6%, or from 20% to 21%. The impact at a health-care level (absolute increase in financial cost or morbidity) is the same regardless of the background rate.

### 9.2.3 Possible Comparisons

#### 9.2.3.1 *Ruling Out a Meaningful Risk Increase Compared to Placebo*

When developing a new treatment, there may be concerns about certain adverse events caused by the new drug treatment. These concerns may be based on effects in animal models, on effects of other treatments that are chemically similar, on data observed in early studies, or on hypothetical or logical concerns. Regardless of the reason, these concerns must be addressed to fully understand and characterize the safety of the product.

The inferential framework is straightforward: letting the event rate be denoted by $p_k$ for $k$ = E or P, the null and alternative hypotheses may be written as

$$H_o: p_E - p_P \geq \delta \text{ vs. } H_a: p_E - p_P < \delta$$

for an appropriate $\delta > 0$. In other words, the null hypothesis is that the investigational treatment increases the event rate by some difference $\delta$, and the alternative hypothesis is that the investigational treatment increases the rate by less than $\delta$ (or has no effect, or decreases the event rate). The safety margin for increased risk or harm may depend on the benefit of the product. The parallels to non-inferiority testing for efficacy are immediately obvious. A possible disadvantage to expressing the hypotheses about a risk difference includes lack of robustness to an incorrect estimate of the rate in the control group: an increase of 5 percentage points may seem inconsequential when the placebo event rate is 30%, but not when placebo event rate is 3%. This disadvantage can be exacerbated when the patient population changes over time.

Alternatively, the hypotheses can be expressed as a ratio of event rates (risk ratio):

$$H_o: p_E/p_P \geq \delta \text{ vs. } H_a: p_E/p_P < \delta$$

for an appropriate $\delta > 1$. The relative risk is often perceived as being more consistent across different patient populations with different event rates than the risk difference. However, the risk ratio is not robust to a change in event rates, particularly for fairly rare events. A 50% increase when the placebo event rate is 1% (which affects 1 out of every 200 subjects) is quite different from a 50% increase when the placebo event rate is 10% (which affects 1 out of every 20 subjects).

If time to an undesirable event is of interest, then the hypotheses can be expressed as a hazard ratio, $\theta$:

$$H_o: \theta \geq \delta \text{ vs. } H_a: \theta < \delta$$

for an appropriate $\delta > 1$. Basing non-inferiority on hazard ratios is beneficial when the time to an event is important—that is, not only whether an event that occurs is of interest but also when that event occurs. However, the interpretation of an estimated hazard ratio depends on whether the cumulative hazards are fairly proportional over time.

Although the mathematics and interpretations of the various forms of the hypotheses are quite different, many fundamentals are similar. In each case, the null hypothesis is that the investigational treatment induces more events of concern than does placebo (or sooner), and the alternative hypothesis is that the difference, if it exists, is not a safety concern.

Because the investigational treatment is being compared to placebo, the margin of interest (whether it be based on differences, relative rates, or hazard ratios) cannot be based on a putative placebo argument. Instead, it must be based on ruling out an increase of importance. This can be difficult with an adverse event of mortality or irreversible morbidity as it is difficult to state that any increase is acceptable. However, because of the inability to definitively conclude that two rates are identical, and the lack of reason to conjecture that the test treatment will be superior to placebo, a small margin may be necessary to represent some maximum allowed risk. The conclusion of noninferior safety would be with respect to the particular safety margin.

As there is subjectivity involved with selecting a safety margin, there is bound to be disagreement in the selected margin. Any increase in the risk of a serious adverse event is undesirable. To evaluate the margin, various aspects of the comparative treatments must be considered. If the test treatment offers irreversible efficacy advantages over placebo, these may counter the increase in adverse events. However, if the test treatment offers only reversible or symptomatic advantages over placebo, there would be no willingness to accept any increase in mortality or high morbidity.

### 9.2.3.2 Ruling Out a Meaningful Risk Increase
#### Compared to an Active Control

Comparisons of adverse event rates between two active treatments are conceptually similar to comparisons for efficacy. Event rates can be compared using differences, ratios, or hazard ratios. The purpose is to rule out an important increase in events in the experimental arm compared to the control arm, so the null hypothesis is that the experimental arm has a higher event rate than the control arm by at least the amount of the margin. The alternative hypothesis is that the experimental arm has an event rate that is lower than the control arm, identical to the control arm, or greater than the control arm by less than the margin.

As always, setting the margin will be an exercise of high importance. The putative placebo effect is again irrelevant, so the margin will be based on differences of importance, as discussed in the previous section. However, at this point, the parallels with the safety comparison to placebo described in the previous section cease. Relative efficacy can be very important in determining the margin of safety between two active treatments. An investigational treatment that has superior efficacy on an endpoint of high morbidity may allow a larger safety margin than an investigational treatment that has identical efficacy or has an advantage only on symptomatic endpoints. An investigational treatment that has no advantage over the active control (or has an efficacy disadvantage) will not support a large margin and may not support any positive margin at all on safety endpoints. It is unlikely that ease of administration of the test treatment or cost will affect the margin—a relatively inexpensive oral medication may have different acceptable safety margins compared with an expensive parenteral medication, but it is difficult to allow an increase in important adverse events due solely to cost.

Figure 9.1 demonstrates how confidence intervals can be interpreted to rule out an important increase in adverse events. Three situations are illustrated, with parentheses noting the bounds of the confidence intervals. The solid vertical line represents the point of equality (risk ratio = 1 or hazard ratio = 1). The area to the right of the solid line represents an increase in events in the investigational group compared to the reference group. The dashed vertical line represents the point of concern, $\delta$, below which an important difference would be excluded.

In the first (top) situation, the confidence interval extends from the area of decreased risk to a point above the area of concern. In this case, an important difference cannot be excluded, even though a decreased risk cannot be excluded either. A wide confidence interval suggests that the event rate was not adequate to establish the precision of the risk with sufficient accuracy.

In the second (middle) situation, the confidence interval extends from the area of decreased risk to the area of elevated risk, but does not extend above the point of concern. Such a confidence interval is sufficient to rule out an important elevation of risk.

In the third (bottom) situation, the confidence interval lies entirely in the area of elevated risk above the point of concern. This situation clearly demonstrates an important safety signal. Depending on the nature of the elevated risk and the benefits conferred by the investigational treatment, it may be unlikely that such a treatment would be adopted for general use.

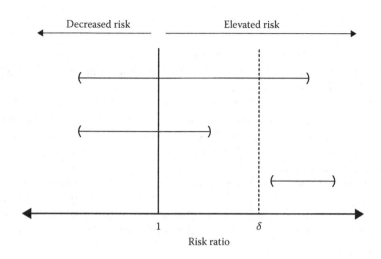

**FIGURE 9.1**
Some possible outcomes for a safety evaluation.

### 9.2.3.3 Indirect Comparison to Placebo

Another option for non-inferiority safety studies is to compare safety of a test and control treatment with the intention of indirectly comparing the test treatment to placebo. The purpose is to determine whether the test treatment has a safety signal compared to placebo. This assumes that the active control has been compared to placebo, directly or indirectly, so that the relative safety is known. As in all putative placebo efficacy analyses, the implication is that the comparison of the active control to placebo will consistently produce an effect of known magnitude, enabling the indirect comparison of the experimental treatment to placebo.

If the active control has a safety concern compared to placebo, but better efficacy than the experimental therapy, the purpose of the safety analysis may be to show that the experimental therapy is more safe than the active control, preferably with a magnitude approaching that of the magnitude of the difference between the active control and placebo. It will never be possible to definitively rule out the lack of a safety signal of the experimental therapy versus placebo, but the magnitude of the difference between the experimental therapy treatment and placebo can be shown to be small enough to be of little interest. The margin, therefore, may be based on the "putative placebo" difference. However, the form of the hypotheses differs: the null hypothesis will be that the experimental and control therapies have similar risks or that the experimental therapy has a much greater risk than the control therapy, and the alternative hypothesis will be that the experimental therapy has a lower risk of an event and the magnitude is large enough to be important. With risk differences, the hypotheses can be written as

$$H_o: p_E - p_C \geq -\delta \text{ vs. } H_a: p_E - p_C < -\delta$$

where $\delta > 0$. Setting $\delta$ to be large enough results in a conclusion that the safety event incidence rate of the active control approximates that of placebo, if the null hypothesis is rejected. The hypotheses can be tested on the basis of an appropriate-level confidence interval for the difference in risks. Failing to reject the null hypothesis does not imply a safety concern for the experimental therapy—it may be safer than, or as safe as, the control therapy, but it has a safety signal compared to the putative placebo. This testing paradigm differs from other testing in this book in one important aspect: equality, be it a difference of zero or a ratio of 1, is part of the null space rather than the alternative space. However, the study can only be successful if the rate in the experimental group is lower than in the active control group; thus, basing the test statistic on the assumption that the rates differ by an amount of $\delta$ is still required.

If the active control does not have a safety signal compared to placebo, then the non-inferiority safety study comparing the experimental therapy to the active control therapy will be conducted to rule out an increase in the safety risk of the experimental treatment compared to placebo, or an estimate of the magnitude of the increase. The hypotheses to be tested start with the assumption that the test treatment has an important effect compared to the active control, hoping to reject in favor of the alternative that the test treatment does not have an important increase. For risk differences, the hypotheses are

$$H_o: p_E - p_C \geq \delta \text{ vs. } H_a: p_E - p_C < \delta$$

for an appropriate $\delta > 0$.

## 9.3 Cardiovascular Risk in Antidiabetic Therapy

A recent guidance from the FDA on evaluating cardiovascular risk in anti-diabetic therapies[3] uses some of the principles discussed earlier and is summarized here to illustrate the concepts.

Patients with diabetes are at increased risk for mortality and various morbidities. Of importance, diabetic patients are at increased risk for cardiovascular morbidity, including stroke, myocardial infarction, acute coronary syndrome, revascularization, and other events, whether fatal or not. Such events can lead to mortality, hospitalization, or irreversible physical limitations. Although improving the glycemic control (measured by HbA1c, a generally accepted primary endpoint for efficacy) is logically expected to improve long-term outcomes, it may not always accurately predict effects on mortality or irreversible morbidity. Thus, the desire to study such long-term outcomes is justified.

The guidance proposes two non-inferiority boundaries for evaluation of cardiovascular risk, both based on the upper bound of a two-sided 95% confidence interval on the risk ratio. If the upper bound for increased risk is less than 1.3, the new treatment can be reasonably assumed to not induce an important increase in cardiovascular events. If the upper bound for increased risk is greater than 1.8, the new treatment cannot be assumed safe without further study. If the upper bound is between 1.3 and 1.8, conclusions are more difficult, and the FDA guidance suggests that further study is necessary, perhaps after the investigational product is approved.

Put into hypothesis testing terminology, the guidance suggests that two sets of hypotheses are tested:

$$H_{o1} : \theta \geq 1.8 \text{ vs. } H_{a1} : \theta < 1.8$$

and

$$H_{o2} : \theta \geq 1.3 \text{ vs. } H_{a2} : \theta < 1.3$$

where $\theta$ is the risk ratio of the experimental therapy (in the numerator) to the control therapy (either a placebo or an active comparator, if appropriate). If the first null hypothesis, $H_{o1}$, is not rejected, the possibility of an important safety signal cannot be ruled out, and approval is unlikely. If both null hypotheses are rejected, an important safety signal is unlikely and approval is possible, given that efficacy and other safety data support such approval. If only the first null hypothesis ($H_{o1}$) is rejected and the second ($H_{o2}$) is not, then further study is required. In this situation, other safety and efficacy data may allow approval of the product, but the sponsor will be obligated to conduct further study to rule out an important increase of 30% in cardiovascular risk.

With multiple hypotheses being tested, and possibly multiple attempts at testing, it is necessary to consider the type I error rate. We consider testing on the basis of two-sided 95% confidence intervals. When $\theta \geq 1.8$, the probability is at most 0.025 that $\theta \geq 1.8$ is rejected on the basis of the data used for the consideration of the approval. The probability of concluding that $\theta < 1.3$ is much less than 0.025. In the event that it is falsely concluded that $\theta < 1.8$ but $\theta < 1.3$ is not concluded, it is quite unlikely that a later safety study, if properly designed and conducted, would conclude $\theta < 1.3$ (the probability being much less than 0.025).

When $1.3 < \theta < 1.8$, concluding that $\theta < 1.8$ is not an error and is also not automatic. Any given test of $\theta \geq 1.3$ versus $\theta < 1.3$ would maintain the desired type I error rate or a smaller rate. Because there would likely be two opportunities (the pre- and post-approval analyses) to conclude that $\theta < 1.3$, the overall the type I error rate would be a little less than 0.05 (when $\theta$ is slightly larger than 1.3) or smaller.

In calculating confidence intervals on risk ratios, the number of events (particularly for rare events) must be adequate to result in a sufficiently narrow confidence interval. Observing few events will result in a confidence interval that is wide, which may result in not being able to rule out an important increase in events even if the observed rates are similar. To obtain an adequate number of events, the guidance document recommends enrolling patients at increased risk of the event. This serves a second purpose, which is to study patients who are at higher risk, since some such patients will inevitably receive the drug once it is approved for marketing. However, enrolling patients at increased risk of the event can also make the results less generalizable: enrolling patients with different risk levels than the general population may provide results that do not easily extrapolate to the general population, whether those rates are higher or lower.

We direct the reader to Chapter 13 in this book for determining confidence intervals for non-inferiority testing of time-to-event endpoints and to Chapter 7 on multiple testing.

## References

1. Food and Drug Administration Amendments Act of 2007 http://frwebgate .access.gpo.gov/cgi-bin/getdoc.cgi?dbname=110_cong_public_laws&docid=f: publ085.110.
2. Fleming, T.R., Identifying and addressing safety signals in clinical trials, *New Engl. J. Med.*, 359, 1400–1402, 2008.
3. Guidance for industry: Diabetes mellitus—evaluating cardiovascular risk in new antidiabetic therapies to treat type 2 diabetes, United States Food and Drug Administration, Silver Spring, MD, 2008.

# 10

## Additional Topics

## 10.1 Introduction

In this chapter, we discuss additional topics that may be involved in the design, analysis, and interpretation of the results of a non-inferiority trial. Many of these topics are well developed for superiority trials but less understood for non-inferiority trials. We discuss issues involving the consistency of non-inferiority across subgroups in Section 10.2. The relationship between non-inferiority inferences on a surrogate endpoint and the corresponding clinical benefit endpoint are discussed in Section 10.3. Adaptive designs (mostly involving trial monitoring) and group sequential trials are discussed in Section 10.4. Section 10.5 provides a brief discussion on equivalence comparisons.

The effects of therapies (e.g., the active control and experimental therapies in a non-inferiority trial) may vary across meaningful subgroups. A non-inferiority inference involves concluding that the effectiveness of the experimental therapy is both superior to placebo and not unacceptably worse than the active control. To formally make such an inference, or to check for consistency in those inferences, across subgroups would require an understanding of the effect of the active control relative to placebo in the investigated subgroups along with the estimated difference in effects between the active control and experimental therapies from the non-inferiority trial(s). There are various scenarios in which the effects (relative to placebo) of the active control therapy and/or the experimental therapy, as well as the differences in their effects, may vary across subgroups. It may also be the case that different subgroups may have different "non-inferiority margins" due to varying effects of the active control.

A surrogate endpoint is an endpoint used as a substitute for a clinical benefit endpoint. The objective of using a surrogate endpoint is that specific inferences on the surrogate endpoint imply specific inferences on the clinical benefit endpoint. It is therefore important that treatment effects on the surrogate endpoint are related to treatment effects on the clinical benefit endpoint. In a superiority trial with a rather good surrogate endpoint that represents the sole pathway toward clinical benefit, superiority on the surrogate endpoint would imply superiority on the clinical benefit endpoint. For a non-inferiority trial on a surrogate endpoint, if the experimental therapy is noninferior to the active control on the surrogate endpoint (relative to an appropriate non-inferiority margin), then the experimental therapy will have an effect on the clinical benefit endpoint that is superior to placebo and not unacceptably worse than the active control therapy. To efficiently determine the non-inferiority margin on the surrogate endpoint and the size of the non-inferiority trial not only requires understanding of the effect of the active control on both the surrogate and clinical benefit endpoints, but also requires understanding on the relationship between the size of a treatment effect on the surrogate endpoint and the size of a treatment effect on the clinical benefit endpoint.

Adaptive clinical trial designs allow for the modification of some aspect of the design based on an interim look of the data. Possible modifications include altering the sample size or the timing of an analysis, changing endpoints, changing inclusion/exclusion criteria, or dropping or modifying treatment arms. We will focus mostly on designs in which the timing of the definitive analysis is random.

Two therapies would be concluded "equivalent" if the difference in their effects lies within the lower and upper boundaries. The experimental therapy effect on the endpoint of interest would either be slightly better than, the same as, or slightly worse than the active control. Mathematically, equivalence can be considered a two-sided non-inferiority inference—that is, the experimental therapy is noninferior to the active control therapy and the active control therapy is noninferior to the experimental therapy.

---

## 10.2 Interaction Tests

Interaction effects with treatment are a concern in any clinical trial. By "interaction effects with treatment," we mean that the size of the relative treatment effects varies across relevant disjoint subgroups of subjects. The impact of a baseline or demographic characteristic or time-varying covariate on the outcome can have a major influence on the interpretation of the clinical trial results. Like other areas discussed in this book, interaction tests must be approached differently for non-inferiority clinical trial analyses than for superiority clinical trial analyses.

In their simplest form, interaction tests look for differences in the effect on outcomes in two or more distinct subgroups of study subjects. The presence of interaction itself does not automatically negate the findings of a clinical trial, but the form of interaction will impact the interpretation of results and for which subjects benefit can be expected from the experimental therapy. Two types of interaction are generally considered: qualitative and quantitative. In this book we will define these terms as follows.

An *interaction* is any heterogeneity of treatment effect across subgroups of subjects. A *qualitative interaction* is an interaction where there is heterogeneity in the state of efficacy (or relative efficacy) across subgroups of subjects. States of efficacy or relative efficacy include the existence or nonexistence of efficacy, the order between arms of the extent of efficacy, and the type of comparisons between arms on the extent of efficacy (inferior, noninferior, or superior). As a relevant example, if the experimental therapy is noninferior to the control therapy with respect to primary endpoint for some subgroups, but not for other subgroups, this is a qualitative interaction. A *quantitative interaction* is an interaction that is not qualitative.

In a superiority analysis for an active-controlled trial, a qualitative interaction generally means that one treatment is superior in one subgroup and the other treatment is superior in a mutually exclusive subgroup. Thus, the conclusion of the study depends on the subgroups considered. For non-inferiority analyses, a qualitative interaction means that non-inferiority exists in one subgroup and does not exist in a mutually exclusive subgroup.[1]

Obviously, qualitative interaction is a concern. If qualitative interaction exists, a finding of non-inferiority among the entire study population is not sufficient to demonstrate that the experimental treatment is adequate for general use.

An interaction can be examined graphically as shown in Figure 10.1. For each subgroup, the estimated difference (active control minus experimental)

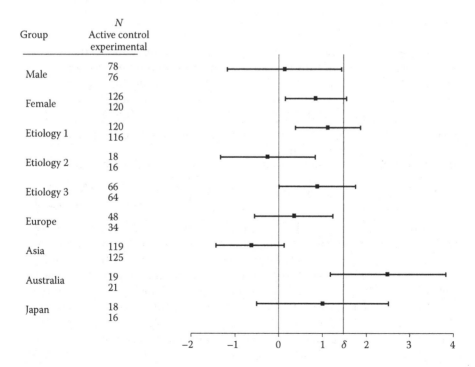

**FIGURE 10.1**
Exploratory plot to informally look for qualitative or quantitative interaction.

is shown with a line representing the two-sided 95% confidence interval (95% CI). The axis at the bottom shows the scale with the margin (1.5 units in the example) also marked. In this example, geographic region is quickly identified as a potential source of heterogeneity, with non-inferiority being demonstrated in some subgroups (bound of the 95% CI less than $\delta$), observed in most (point estimate less than $\delta$), but not observed in Australia (point estimate larger than $\delta$). This is an informal assessment of interaction effects, and there may be various interpretations.

It is also important to note that there may be interaction effects between the experimental and control therapies, but not between the experimental therapy and a placebo, and vice versa. This distinction is important if the most important purpose of the trial is to indirectly demonstrate that the experimental therapy is efficacious (better than placebo). Historical information on the relative efficacy of the active control and placebo in various subgroups must be known to fully interpret the interaction between the experimental therapy and placebo—information that may not be well known. If the putative placebo comparison is of less interest and the non-inferiority comparison is based on clinically relevant differences only, the interpretations of interaction conclusions is more straightforward—a conclusion of a qualitative interaction implies that there is no uniform conclusion on non-inferiority across subgroups.

### 10.2.1 Test Procedures

Interaction tests are common, and the test for any interaction is identical for superiority and non-inferiority studies. For continuous data, letting $D_i$ be the estimated difference in effects in the $i$th stratum and $\delta_i$ be the true difference, $i = 1, \ldots, I$, the test of any interaction is the test of the null hypothesis $H_o$: $\delta_1 = \delta_2 \ldots = \delta_I$ versus the alternative of at least one inequality. An appropriate test statistic is $W = \sum_{i=1}^{I} \frac{(D_i - \bar{D})^2}{\sigma_i^2}$, where $\sigma_i$ is the standard deviation within the $i$th stratum and $\bar{D}$ is the mean of the $D_i$ values weighted by the inverses of the subgroup variances (standard errors squared): $\sum_{i=1}^{I} \frac{D_i}{\sigma_i^2} \left( \sum_{i=1}^{I} \frac{1}{\sigma_i^2} \right)^{-1}$.

Under the null hypothesis, $W$ has a central $\chi^2$ distribution with $I - 1$ degrees of freedom.[2] Rejection of $H_o$ implies that an understanding of the interaction is required to fully interpret the full results of the clinical trial. However, failure to reject $H_o$ does not imply anything about the magnitude or meaningfulness of the differences in effects across subgroups. Tests for interaction effects tend to have little power at meaningful alternatives, and thus the interaction effects may be rather large even if $H_o$ is not rejected. These limitations will also be true for tests of a qualitative interaction when comparing the experimental and active control therapies. The test procedure for an interaction is basically the same as the test procedure for heterogeneous effects across studies provided in Section 4.3.1.

Wiens and Heyse proposed several likelihood ratio–type tests for a qualitative interaction, where a common non-inferiority margin across a partition of subgroups is used, on the basis of quantities $Q_{LR}$, $Q_{LR}^-$, and $Q_{LR}^+$ defined as follows:

$$Q_{LR}^+ = \sum_{i=1}^{I} \frac{(D_i - \delta)^2 I(D_i > \delta)}{\sigma_i^2}$$

$$Q_{LR}^- = \sum_{i=1}^{I} \frac{(D_i - \delta)^2 I(D_i < \delta)}{\sigma_i^2}$$

$$Q_{LR} = \min\left(Q_{LR}^+, Q_{LR}^-\right)$$

In these calculations, $I(\bullet)$ is an indicator function that equals 1 if the argument is true and 0 otherwise.[1] The test statistic $Q_{LR}^+$ tests the null hypothesis that non-inferiority exists in all strata versus the alternative hypothesis that non-inferiority does not exist in at least one stratum. Such hypotheses can be written as

$$H_{oQ+}: \delta_i < \delta \text{ for all } i \tag{10.1}$$

$$H_{aQ+}: \delta_i > \delta \text{ for at least one } i$$

The test statistic $Q_{LR}$ tests the null hypothesis that either non-inferiority exists in all strata or the experimental treatment is markedly inferior to the active control in all strata; the alternative is that this is true in some strata but not in others. Such hypotheses can be written as

$$H_{oQ}: \delta_i > \delta \text{ for all } i \text{ or } \delta_i < \delta \text{ for all } i \tag{10.2}$$

$$H_{aQ}: \delta_i > \delta \text{ for at least one } i \text{ and } \delta_i < \delta \text{ for at least one } i$$

Thus, both $Q_{LR}$ and $Q_{LR}^+$ are testing for the existence of qualitative interaction. For non-inferiority, the test statistic $Q_{LR}$ does not seem to make much sense because the additional area in the null region is an area in which it will not be possible to conclude non-inferiority, but in equivalence analyses $Q_{LR}$ might be appropriate. Critical values for both tests are given in Table 1 of Gail and Simon,[2] who discussed this test without the "$-\delta$" in the formulae for either $Q_{LR}^+$ or $Q_{LR}^-$ for superiority trials.

When the point estimate of treatment effect in every subgroup has the same directional relationship to the non-inferiority margin, either $Q_{LR}^+$ or $Q_{LR}^-$ will be zero (and therefore $Q_{LR}$ will be zero). Using the data from Figure 10.1, the test statistic $Q_{LR}^{+}$ will be zero when testing interaction between treatment and gender and between treatment and etiology, but neither $Q_{LR}^{+}$ nor $Q_{LR}^{-}$ will be zero when testing interaction between treatment and geographic region. Therefore, $Q_{LR}$ will be zero in the interaction test for treatment and gender and treatment and etiology, but not for the interaction test of treatment and geographic region.

Note that in both cases, the tests start with the null hypothesis of no qualitative interaction and conclude qualitative interaction only if there is strong evidence that it exists. In both cases, observed qualitative interaction (i.e., at least one stratum with $D_i > \delta$ and at least one stratum with $D_i < \delta$) is necessary to reject the null hypothesis and conclude the existence of qualitative interaction. An alternative test, based on the "min test," assumes the existence of qualitative interaction unless there is strong evidence to the contrary.[1] However, a test will require a conclusion of non-inferiority in each stratum when based on the min test.[3] Hence, this test will have little power for the typical non-inferiority trial (and thus great uncertainty) and is not recommended.

An alternative to the likelihood ratio test is the standardized range test. For an appropriate critical value $C$, the standardized range test considers the hypotheses in Expression 10.1—analogous to $Q_{LR}$—with test statistics $\max((D_i - \delta)/\sigma_i)$ and $\min(-(D_i - \delta)/\sigma_i)$. This can be written as $Q_{SR} = \min[\max((D_i - \delta)/\sigma_i), -\min(-(D_i - \delta)/\sigma_i)]$. $H_{oQ}$ is rejected if $Q_{SR} > C$. Alternatively, the range test considers the hypotheses in Expression 10.2—analogous to $Q_{LR}^{+}$—with the test statistic $Q_{LR}^{+} = \min(-(D_i - \delta)/\sigma_i)$. $H_{oQ+}$ is rejected if $Q_{LR}^{+} < C'$. Critical values for $C$ and $C'$ are given in Table 1 of Piantadosi and Gail.[4] Furthermore, the range test is more powerful when the effect is reversed in very few subgroups, whereas the likelihood ratio test is more powerful when a few subgroups have an effect in one direction and a few in the other. For non-inferiority purposes, it is unlikely to achieve a conclusion of non-inferiority if there are many subgroups for which the true effect is against a conclusion of non-inferiority, which argues for use of the standardized range test. However, this test has not been studied extensively in the non-inferiority literature and therefore should be approached with caution. In addition, with few strata, the difference in the performance of the tests is minor, so the likelihood ratio test should perform well.[4]

### 10.2.2 Internal Consistency

Tests of interaction are important to check for internal consistency of treatment effect and to identify subgroups for which one therapy may be preferred over another. In the non-inferiority setting, historical comparisons of the active control to placebo in a subgroup are needed to completely understand the effect–effect size of the experimental therapy in that subgroup. If there are some subgroups in which the active control is more efficacious compared to placebo than others, this information must be known to completely understand the indirect comparison of the experimental treatment to placebo. Unfortunately, the detailed subgroup analyses comparing the active control to placebo are not always known, so it is often assumed that there are no interaction effects between the active control and placebo.

### 10.2.3 Conclusions and Recommendations

When the purpose of an active control study is to indirectly demonstrate the superiority of the experimental therapy compared to placebo, the interaction between the active control and the experimental therapies must be considered in the context of the putative placebo effect.

- The easiest outcome to interpret is an outcome that suggests no interaction effect between the active control and experimental therapies. The conclusion in this case is that any interaction effect between the experimental therapy and placebo is identical to the interaction effect between the active control and placebo.

- A conclusion of quantitative interaction between the active control and experimental therapies may lead to a conclusion of either qualitative or quantitative interaction between the experimental therapy and placebo or a conclusion of no interaction between the experimental therapy and placebo. The conclusion depends on the relationship between the active control and placebo from historical comparisons. Apart from the putative placebo inference, a conclusion of a quantitative interaction does not invalidate the effectiveness of the experimental therapy and may, in fact, lead to a conclusion of qualitative interaction when considering only the active control and experimental therapies.

- A conclusion of qualitative interaction between the active control and experimental therapies must be interpreted in the context of an interaction effect involving the active control and placebo. In the most obvious situation of uniform efficacy of the active control over placebo in all disjoint subgroups, qualitative interaction between the active control and the experimental therapies will lead to a conclusion of efficacy of the experimental therapy compared to placebo in some subgroups but not in other subgroups, resulting in a recommendation to use the experimental therapy only in a subset of studied patients. However, if the active control has demonstrated an interaction effect with placebo, the conclusion becomes more complicated. In the absence of a putative placebo comparison, a qualitative interaction between the active control and experimental therapies will lead to a recommendation on treatment option on the basis of the subgroup in which the individual patient belongs.

Questions about varying treatment effects across meaningful subgroups should be studied in any registration trial. Graphs are a good starting point. The $Q^+$ test is most applicable for testing for qualitative interaction but the power may be low for most non-inferiority trials to detect a qualitative interaction.

## 10.3 Surrogate Endpoints

The efficacy of an experimental therapy is evaluated on the basis of a clinical benefit endpoint, which is related to how a subject feels, functions, or survives. A surrogate endpoint is a different endpoint that is evaluated to make inferences on the clinical benefit endpoint. Surrogate endpoints are of interest when they provide reliable inference on the clinical benefit endpoint more quickly than evaluating the clinical benefit endpoint itself. This can occur when there is a strong relationship between the surrogate and clinical benefit endpoints, and the surrogate endpoint is determined much sooner than the clinical benefit endpoint. How surrogate endpoints are involved in the approval of an indication depends on the type of approval—regular approval or accelerated approval.

Regular approval (also known as full approval or traditional approval) in the United States usually involves the conclusion that the experimental therapy provides clinical benefit for the studied indication. Such a conclusion is based on demonstrating a meaningful effect on either a clinical benefit endpoint or an appropriate surrogate endpoint. For serious or life-threatening illnesses, a product may be able to receive accelerated approval in the United States. For accelerated approval, the product should provide "meaningful therapeutic benefit to patients over existing treatments," where the accelerated approval may be based on a surrogate endpoint that reasonably likely predicts clinical benefit.[5] Accelerated approval requires that experimental therapy be studied further to verify and describe its clinical benefit when "there is uncertainty as to the relation of the surrogate endpoint to clinical benefit, or of the observed clinical benefit to ultimate outcome." In essence, results on the surrogate endpoint should predict with high probability that the clinical trial or trials designed to evaluate the clinical benefit endpoint will demonstrate the clinical benefit of the experimental therapy.

It should be biologically plausible that effects on the surrogate endpoint will translate to effects on the clinical benefit endpoint. An endpoint that is correlated with a clinical benefit endpoint need not be an appropriate surrogate endpoint. Positive changes in symptoms of a disease (e.g., normalization of body temperature for subjects having community-acquired pneumonia) within a subject have been used as "signs" that a therapy may be working. However, because the symptoms are often results of the disease and not on the pathway of treating the disease, a symptom-related endpoint is unlikely to be an appropriate surrogate endpoint.[6] Improvement in symptoms can occur by treating the symptoms without treating the disease.

In another example, Bruzzi et al.[7] studied the results on objective tumor response rate and overall survival from 10 randomized trials comparing standard therapy with intensified epirubicin-containing regimens in metastatic breast cancer. There was a clear, demonstrated advantage in response rate (odds ratio, 0.60; 95% CI, 0.51, 0.72) favoring the intensified epirubicin-containing regimens that was accompanied with an unimpressive overall survival advantage that failed to reach statistical significance (hazard ratio, 0.94; 95% CI, 0.86, 1.04). Thus, for that setting, objective tumor response does not appear to be an appropriate surrogate endpoint for overall survival. It is possible that a chemotherapeutic agent may successfully destroy nonaggressive cancer cells, but have no effect on aggressive cancer cells. In so doing, the objective tumor responses may identify subjects who had less aggressive disease at baseline and thus those who will have longer overall survival. Alternatively, off-target detrimental effects can cancel the on-target effects of the chemotherapy. For example, the chemotherapeutic agent may provide little or no benefit to subjects particularly when weighing the agent's toxicities. Thus, objective tumor response would be associated with overall survival (a correlate) at the subject level, but effects on tumor response would not be associated with an effect on overall survival. If this is the case, then the objective response rate would not be an appropriate surrogate endpoint for overall survival.

Many studies have discussed when an endpoint, which is correlated with a clinical benefit endpoint, would or would not be an appropriate surrogate endpoint.[8–10] The disease may casually affect both the biomarker (i.e., the potential surrogate) and the clinical outcome, thus correlating the biomarker with clinical outcome. However, if the biomarker does not represent a mechanism through which clinical benefit is induced, then affecting the biomarker may not affect the clinical benefit endpoint. If instead there are multiple pathways in which clinical benefit can be induced and an experimental agent only affects one pathway, use of a surrogate may overpredict or underpredict the effect on the clinical benefit endpoint depending on whether the surrogate is an endpoint relevant to the actual pathway in which the experimental agent induces clinical benefit. Related to this, it should be noted that validating a surrogate endpoint for one pharmacological class of agents may not apply to another pharmacological class of agents.

It is more difficult to prove surrogacy involving non-inferiority comparisons than that involving superiority comparisons.[11,12] The use of surrogate endpoints in active-controlled trials would involve using a non-inferiority comparison on a surrogate endpoint to imply a non-inferiority comparison on the clinical benefit endpoint. Rather precise information on the relationship between effects on the surrogate endpoint and effects on the clinical benefit endpoint are needed to establish a margin for the non-inferiority comparison on the surrogate endpoint. The non-inferiority criterion on the surrogate endpoint needs to consider the uncertainty in establishing the effect of the control therapy on the clinical benefit endpoint, as well as the uncertainty in establishing the effect on the surrogate endpoint.

*Regular Approval.* Consistent with regular approval in demonstrating superiority to a placebo, Prentice[13] defined a surrogate endpoint as a "a response variable for which a test of the null hypothesis of no relationship to the treatment groups under comparison is also a valid test of the corresponding null hypothesis based on the true endpoint." This means that the surrogate endpoint must be correlated with the clinical benefit endpoint and fully capture the net effect of treatment on the clinical benefit.[8] Prentice noted that this is a rather strong, restrictive relationship between the surrogate endpoint and the clinical benefit endpoint.[14] It would be sufficient to require for surrogate (S) and clinical benefit (T) time-to-event endpoints

1. The hazard rate for T depends on S.
2. Given S, the hazard rate for T does not depend on the treatment arm (i.e., given S, the treatment-related hazard ratio equals 1).

Similar criteria can be established for other types of endpoints (e.g., continuous or binary endpoints). The second criterion requires that the effect of a treatment on the clinical benefit endpoint is completely mediated through the surrogate endpoint. Various researchers recommend verifying the second criterion through meta-analysis of relevant trials studying the (potential) surrogate and clinical benefit endpoints.[9,13,15]

For regular approval, the type I error rate for drawing conclusions on the clinical benefit endpoint from testing on the surrogate endpoint should be maintained at the desired level that would be used in a test directly on the clinical benefit endpoint. In a superiority trial, if testing at a one-sided type I error rate of 0.025 on the clinical benefit endpoint is desired, then the surrogate endpoint must be such that if the experimental therapy has zero effect on the clinical benefit endpoint, the probability is 0.025 of demonstrating superiority on the surrogate endpoint.

For an active-controlled trial, if the aim of the trial is to demonstrate any efficacy, the non-inferiority margin for the surrogate endpoint should assure that when the experimental therapy has zero effect on the clinical benefit endpoint, the probability that the experimental arm demonstrates non-inferiority to the control arm on the surrogate endpoint is at most 0.025 or whatever level is prespecified. In this setting, it is concluded that the experimental therapy has a positive effect on the clinical benefit endpoint whenever it is concluded that the experimental therapy has a noninferior effect on the surrogate endpoint. When the surrogate endpoint is acceptable for regular approval, it is sufficient to choose a non-inferiority margin that additionally guarantees that the experimental therapy has an effect on the surrogate endpoint (i.e., the non-inferiority margin is less than or equal to the effect on the surrogate endpoint that the active control can be assumed to have in the non-inferiority trial). When the aim of the trial is to demonstrate adequate efficacy (e.g., the experimental therapy retains at least some minimal amount or fraction of the active control effect), more precise information is needed on the relationship between effect sizes on the surrogate endpoint and effect sizes on the clinical benefit endpoint. Such precise information on the relationship of the effect sizes may not be known. Uncertainty on the precise relationship may lead to a more conservatively selected margin for the surrogate endpoint or invalidate the use of the surrogate endpoint in a non-inferiority trial setting.

For fixed non-inferiority margins, the requirements for the surrogate and clinical benefit endpoints have a similar appearance as the mathematical requirement for a function to be uniformly continuous. Consider the cases of using means where $\mu_{E,S}$ and $\mu_{C,S}$ are the true means for the surrogate endpoint and $\mu_{E,CB}$ and $\mu_{C,CB}$ are the true means for the clinical benefit endpoint for the experimental and active control arms, respectively. The use of the surrogate endpoint for regular approval with an associated type I error rate or significance level of 2.5%, where the non-inferiority margin on the surrogate endpoint is $\delta > 0$ and the non-inferiority margin on the clinical benefit endpoint is $\varepsilon > 0$, requires 97.5% certainty that $\mu_{C,S} - \mu_{E,S} < \delta$ to imply 97.5% certainty that $\mu_{C,CB} - \mu_{E,CB} < \varepsilon$. The value for $\varepsilon$ would represent either the entire effect of the control therapy (vs. placebo) on the clinical benefit endpoint or the amount of the effect that a therapy can be worse than the active control therapy but still have adequate efficacy. It is unlikely that 97.5% certainty that $\mu_{C,S} - \mu_{E,S} < \delta$ will be equivalent to 97.5% certainty that $\mu_{C,CB} - \mu_{E,CB} < \varepsilon$. A surrogate endpoint would still be useful and conservative when 97.5% certainty that $\mu_{C,S} - \mu_{E,S} < \delta$ was equivalent to a greater than 97.5% certainty that $\mu_{C,CB} - \mu_{E,CB} < \varepsilon$. Example 10.1 describes the approval of peg-filgrastim based on non-inferiority comparisons from two clinical trials on a surrogate endpoint.

## Example 10.1

The registration clinical trials comparing peg-filgrastim with filgrastim are examples of non-inferiority trials on the surrogate endpoint of the duration of severe neutropenia, which led to the regular approval of peg-filgrastim. Filgrastim was approved on the basis of a demonstrated improvement (reduction) in the clinical benefit endpoint of the incidence in febrile neutropenia.[16] Filgrastim also demonstrated an improvement in the duration of severe neutropenia during the first cycle of chemotherapy.[16] The duration of severe neutropenia in the first cycle is correlated with the chance of getting febrile neutropenia. It is also biologically plausible that reducing the duration of severe neutropenia decreases the likelihood of experiencing febrile neutropenia.

In each of the two non-inferiority registration trials comparing peg-filgrastim with filgrastim, the non-inferiority margin for the mean duration of severe neutropenia during the first cycle of chemotherapy was 1 day.[17] Study 1, which randomized 157 subjects, used fixed-dose peg-filgrastim, whereas study 2, which randomized 310 subjects, used a weight-adjusted dose of peg-filgrastim. The

95% CIs for the difference in the mean duration of severe neutropenia between peg-filgrastim and filgrastim were (–0.2 to 0.6) and (–0.2 to 0.4) for studies 1 and 2, respectively. Both studies succeeded in demonstrating that peg-filgrastim was noninferior to filgrastim in the mean duration of severe neutropenia during the first cycle of chemotherapy, leading to the conclusion that peg-filgrastim is effective (relative to placebo) in reducing the incidence of febrile neutropenia.

*Reasonably Likely Predicting Clinical Benefit.* As an accelerated approval requires that the experimental therapy provides meaningful benefit over available therapy, it is not likely that a non-inferiority analysis would be used for accelerated approval. The concepts involving non-inferiority and reasonably likely predicting clinical benefit can apply in making a go–no go decision based on the results of a randomized, phase 2 trial or in monitoring a long-term phase 3 trial.

When the aim is whether the experimental therapy has any efficacy on the clinical benefit endpoint, the non-inferiority margin on the surrogate endpoint should be such that demonstrating non-inferiority on the surrogate endpoint reasonably likely predicts that the plan for evaluating the efficacy of the experimental therapy will lead to the demonstration of efficacy on the clinical benefit endpoint. The predictability of the results on the surrogate endpoint will depend on the observed difference on the surrogate, its corresponding standard error, and the correlation of the comparison on the surrogate endpoint with the comparison on the clinical benefit endpoint and the precision in the later comparison. Therefore, there may not be a universal margin that will work in evaluating the surrogate endpoint independent of the sample size, the timing of the analyses, and other design features.

When comparing both endpoints using fixed margins, for the non-inferiority margin on the clinical benefit endpoint of $\varepsilon > 0$, there needs to exist a $\delta > 0$ such that demonstrating that $\mu_{C,S} - \mu_{E,S} < \delta$ will reasonably likely predict with a high enough probability that $\mu_{C,CB} - \mu_{E,CB} < \varepsilon$ will be demonstrated in the non-inferiority comparison on the clinical benefit endpoint (which may occur in a separate trial).

---

## 10.4 Adaptive Designs

Adaptive clinical trial designs are designs that allow modification of some aspect of the design during the study on the basis of accumulating data. Modifications may include changing the sample size (increasing or decreasing, including stopping study enrollment completely), dropping or adding a treatment arm, changing the test statistic, changing inclusion criteria to study only a subset of subjects, and even changing the primary endpoint. Because these modifications are made on the basis of accumulating data, the procedures must be carefully implemented to maintain the integrity of the final analysis. Proponents of adaptive designs believe that they can be more efficient than standard designs in either reducing the expected trial size for a given power or in increasing the study power for a given trial size.

Adaptive designs should not be a means to alleviate the burden of rigorous planning. Changes in the design of ongoing trials are not recommended.[18] When substantial changes are made to the design of the trial, the primary analysis may need to stratify by whether subjects were randomized before or after the change.[18] There may not be a way to correct an analysis for adaptations that affect subjects already in the trial.

When an adaptation involves external information, that external information tends to be available to the study subjects. However, this is not true when adaptations are made based on internal information. As such, there may be ethical concerns when adaptations are made based on internal data. If a sponsor deems the results as important enough to make design modifications during the trial, then information learned from the study should be important for subjects to learn.[19] However, subjects and investigators may prejudge the results if provided information on the relative treatment effect.

Properties of adaptive designs are not well understood for many potential adaptations in non-inferiority clinical trials. Properties of non-inferiority group sequential trials have been studied in greater detail, so our discussion focuses on such designs. In these designs, the study can be terminated early at prespecified time points on the basis of accumulating evidence of efficacy or of lack of efficacy. For reasons that will be discussed later, such designs are not as common in non-inferiority trials, but can easily be implemented if desired in a particular situation. Other adaptations can also be considered for non-inferiority designs, but much less experience is available with which to evaluate them. Adding or dropping a treatment arm is applicable when multiple treatment arms are tested (notably in dose-ranging trials that compare several doses of a single drug), but such designs are not commonly analyzed as non-inferiority trials. Changing the primary endpoint or primary test statistic is a difficult proposition for superiority trials, and not much is known about the effects on non-inferiority trials. Changing the sample size is possible in a non-inferiority trial, usually as a result of insufficient information being available before the start of the study to appropriately power the study.

For the rest of this section, we will focus on group sequential methods and the use of sample size reestimation based on interim results. For other issues involving adaptive designs, see the U.S. Food and Drug Administration (FDA) draft guidance.[20]

## 10.4.1 Group Sequential Designs

Interim analyses are performed to assess the efficacy and/or safety of the experimental therapy or the correctness of design assumptions in the sizing of the trial. Interim analyses should be planned on the basis of realistic objectives and conducted on complete data with no or limited data delinquency. Substantial data delinquency leads to an interim analysis based on a convenience sample. This can be particularly problematic if the clinical trial is not blinded. Group sequential designs perform interim analyses at prespecified time points to assess efficacy. Some discussions and papers on adaptive designs include group sequential analyses as a special case of adaptive designs, whereas other discussions and papers separate group sequential analyses from adaptive designs.

A blinded interim analysis, in which summaries by randomized treatment are not produced, can provide information on the appropriateness of the assumed subject-level standard deviations in each arm or in the event rate. This can lead to a change in the sample size. However, for a non-inferiority trial, the appropriate choice of a margin may also depend on the subject-level standard deviations or the event rate. Thus, the interim results may negate not only sample size assumptions but also the prespecified non-inferiority margin. For continuous data when the means differ by a margin of 5 units, the subject-level distributions between the arms would be more similar if the common subject-level standard deviation is 100 than if the common subject-level standard deviation is 1.

Group sequential designs compare treatments at a small number of prespecified intervals, and the study is stopped if efficacy has been demonstrated or if it has been determined that efficacy is unlikely to be demonstrated even if the study continues to the planned final analysis (commonly called futility). Such designs are used to obtain a final answer earlier than if the entire study was finished, thereby potentially saving financial and other resources if the study stops early and potentially marketing the drug earlier as well. Such designs also satisfy ethical concerns of limiting the number of subjects on inferior treatment if the study stops early for either a conclusion of efficacy or futility. Group sequential designs, like all adaptive designs, are most useful when prior information on the relative efficacy of the various treatments is insufficient to more efficiently design the trial in advance.

Most goals of a group sequential design in the superiority setting can also be applied to non-inferiority trials. However, in a superiority trial, ethical considerations have been used as a rationale for including interim analyses for efficacy and for harm. If the experimental therapy is highly effective compared to placebo, an interim boundary would likely be crossed and the experimental therapy can be provided broadly to patients more quickly. When the experimental therapy is harmful, an early analysis result indicative of harm would lead to a "stopping" of the trial with no further subjects provided the experimental therapy and restricting or stopping use to subjects currently receiving the experimental therapy on the trial. When an early determination of non-inferiority is not a determination of superiority, there may be no ethical reason to stop the trial and broadly provide the experimental therapy to patients unless the control therapy has great risks. Additionally, in the absence of a placebo arm, it cannot be precisely determined how much worse the experimental therapy would need to be compared to the active control to be harmful.

Group sequential tests were introduced by Pocock,[21] who proposed a method that required for an interim or final analysis a $p$-value less than a common nominal significance level to reject the null hypothesis. O'Brien and Fleming[22] proposed a method of different critical boundaries for the test statistics at each analysis that are consistent with a common boundary on the difference in cumulative sums. The nominal significance level is smaller for earlier interim analyses and increases from one analysis to the next. Often, the nominal level for the final analysis is slightly less than the overall type I error rate. The methods of Pocock[21] and O'Brien and Fleming[22] were introduced for analyses at equally spaced time points relative to the amount of information in the data, and some examples involving one-sided nominal significance levels are given below. The methodology for calculating boundaries for analyses at unequal time points while maintaining the overall type I error rate was generalized by Lan and DeMets[23] through error spending functions.

Jennison and Turnbull[24] proposed the calculation of repeated confidence intervals. With the confidence interval at interim analysis $k$, $k = 1, \ldots, K$, being of the form $(\theta_k^L, \theta_k^U)$, the repeated confidence intervals have the property that $P(\theta \in (\theta_k^L, \theta_k^U), k = 1, \ldots, K) \geq 1 - \alpha$, where $\theta$ is a parameter of interest. In theory, repeated confidence intervals can be calculated to correspond to any group sequential method, including those of Pocock, O'Brien and Fleming, and others.

The application of repeated confidence intervals to equivalence trials was discussed by Jennison and Turnbull.[25] Although Jennison and Turnbull also touched on non-inferiority analyses, the main focus of their study was equivalence. With $\theta = \mu_C - \mu_E$, the study will conclude equivalence at the $k$th interim analysis if $(\theta_k^L, \theta_k^U) \in (-\delta, \delta)$, and will conclude nonequivalence if $\theta_k^L > \delta$ or if $\theta_k^U < -\delta$ (i.e., if the confidence interval for $\theta$ lies entirely outside of the equivalence zone). By considering only the lower bound of the confidence intervals, the methods can be applied to non-inferiority analyses as well. In this type of situation, the study will conclude non-inferiority if $\theta_k^U < \delta$ at an interim analysis, and might also conclude futility if $\theta_k^L > \delta$ at the interim analysis. We point at a caveat to this process, as stopping a study for early demonstration of non-inferiority is not required for ethical reasons, but continuing the study will often be preferred: if a treatment can be shown noninferior to a control on the basis of a smaller sample size than predicted, it is quite possible that the investigational treatment can be shown superior given the full sample size. Continuing the study in spite of an early demonstration of non-inferiority may be the preferred course of action.

However, if non-inferiority is the only desired outcome, and if stopping early for non-inferiority is a desired possibility, direct application of group sequential methods designed for superiority studies will often suffice. Such designs may be important when the difference in efficacy is not expected to be large enough to be of interest, and the variances are not well known (or the population proportions are not well known for binomial data). In such a case, a group sequential design might save considerable time and enroll fewer subjects than otherwise required, while maintaining the power under less optimal parameter configurations.

When a trial continues after demonstrating non-inferiority at interim analysis to test for superiority at a later analysis, different nominal confidence intervals (supposedly on the same parameter) are used for the tests of non-inferiority and superiority. It is therefore possible that the nominal confidence interval at the final analysis does not rule out the non-inferiority margin. If this is the outcome from a single study, there is likely no persuasive evidence that the experimental therapy is noninferior to the active control. Statistically, if it is reasonable to assume that the same parameter for treatment difference is tested at both analyses, a repeated confidence interval can be used for testing superiority. However, the existence of the nominal confidence interval at the later analysis not ruling out the non-inferiority margin would introduce doubt that the same parameter for treatment difference is tested at both analyses. This could happen with a time-to-event endpoint where after short-term follow-up the two time-to-event distributions are similar, but thereafter become separate with the distribution for the control arm being noticeably better.

Consider now the details of group sequential analyses for non-inferiority conclusions. At the $i$th planned interim analysis, $i = 1, \ldots, I$, the null hypothesis is tested at the one-sided level $\alpha_i$. This corresponds to comparing the lower bound of a one-sided $(1 - \alpha_i) \times 100\%$ CI to the non-inferiority margin $\delta$. If the lower bound of the confidence interval is larger than $\delta$, a conclusion of non-inferiority is made at that interim analysis and the study concludes; otherwise, the study continues to the next planned analysis. Choosing the values of $\alpha_i$ is required to maintain the overall type I error rate, and the only difference among candidate methods is how to make this choice.

Pocock[21] proposed a boundary that stayed constant at each planned interim analysis ($\alpha_1 = \alpha_2 = \ldots = \alpha_I$). For a study with one interim analysis at the halfway point, the Pocock procedure uses $\alpha_1 = \alpha_2 = 0.0147$ for an overall one-sided significance level of 0.025. For a study with four equally spaced analyses, three interim and one final, the Pocock procedure uses $\alpha_1 = \alpha_2 = \alpha_3 = \alpha_4 = 0.0091$.

The method of O'Brien and Fleming[22] can be adapted to non-inferiority group sequential trials. O'Brien and Fleming proposed a boundary that changed with every interim analysis, specifically with $\alpha_i < \alpha_{i+1}$ for every $i$. That is, the criterion to stop early is most strict at the first interim analysis and is progressively eased until the final interim analysis. For a study with one interim analysis at the halfway point, the method would use $\alpha_1 = 0.0026$ and $\alpha_2 = 0.0240$ for an overall one-sided significance level of 0.025. For three equally spaced interim analyses plus a final analysis, the values would be $\alpha_1 = 0.00003$, $\alpha_2 = 0.0021$, $\alpha_3 = 0.0130$, and $\alpha_4 = 0.0215$.

The Pocock boundary, being constant at every analysis, yields a better chance of stopping early than the O'Brien–Fleming boundary. However, conditional on not stopping at an interim analysis, the O'Brien–Fleming boundary retains more of a chance of concluding non-inferiority at the final analysis than the Pocock boundary. Other methods can also be used, including general spending functions that allow a variety of differently shaped stopping boundaries.

For synthesis tests, when considering the historical estimation of the active control effect as random, the correlation between the synthesis test statistics at two different time points is different from the correlation between two non-inferiority test statistics based on fixed margins.[26] This would require a different error spending approach for the synthesis test than for a fixed margin procedure. The same correlation structure would be present if the results of the estimation of the historical effects were treated as fixed, nonrandom. However, this may be counterintuitive to arguments used in choosing a synthesis method due to maintaining a desired type I error rate.

## 10.4.2 Changing the Sample Size or the Primary Objective

Reestimating the sample size is an adaptive technique that has been discussed in the literature and applied successfully to clinical trials. Sample size reestimation techniques can be categorized into two distinct classes, each with specific advantages and disadvantages: blinded and unblinded. With blinded methods, comparative treatment effects are not calculated or known, and decisions are based on data (generally with focus on the variability of the responses) from the various treatment groups being combined. This can be helpful when estimating the variance for normal data or when the average response rate for binomial data is required: under the assumption of a given treatment effect, such parameters can be estimated in an interim analysis without simultaneously estimating the treatment effect. With unblinded methods, on the other hand, subjects are unblinded at the interim analysis (at least in aggregate treatment groups) and a treatment effect is estimated. This can be helpful when the treatment effect is not known in advance and a better estimate is desired to develop a sample size that maintains a desired power; however, the downside is that bias may be introduced (in an unquantifiable manner) that in turn introduces uncertainty about the final results. In general, reestimating sample size via a blinded assessment is most commonly used. When unblinded sample size reestimation methods are employed, an independent group is commonly charged with assessing the interim data and recommending changes to the sample size.[27]

Wittes and Brittain[28] introduced the concept of an internal pilot study, in which a small sample is first enrolled and results from this sample are used to refine procedures for the entire study. This is one form of a sample size reestimation clinical trial design. An internal pilot study requires one interim analysis, usually fairly early in the study. The concept of sample size reestimation has been expanded to non-inferiority trials, keeping blind to treatment effects at the interim analysis and using only blinded estimates of variability of the internal pilot study to reestimate sample size.[29] For an interim analysis, consider calculating the one-sample standard deviation at a prespecified interim analysis (i.e., the internal pilot study) according to the usual method:

$\hat{\sigma} = \sqrt{\dfrac{1}{n_1 - 1} \sum_i \left(X_i - \bar{X}\right)^2}$ . This estimate is then used with a priori estimates of

treatment effect hypothesized before the study began to obtain an updated estimate of sample size required for sufficient power. Since $\bar{X}$ in this equation is the grouped mean from all treatment groups combined, it should be clear that this estimate will be biased (an overestimate) if the treatment group means are not all identical, and unbiased otherwise. Since it is known that this bias does not appreciably affect the process for superiority studies, even when the treatment groups are different enough to be meaningful, it should be of even less concern for non-inferiority studies in which the treatment effects are identical. Simulations show an inflation of the type I error but this is negligible in situations that are clinically relevant.[29] This demonstrates that sample size reestimation can be effectively applied in the design of non-inferiority studies, but we again note that if superiority of the investigational treatment over the active control is a desired outcome, then further increases in sample size should also be considered.

Alternatively, the sample size may be reestimated based on an unblinded interim analysis that estimates the relative treatment effect or treatment difference. When using a fixed non-inferiority margin, the type I error rate is maintained at the desired type I error rate by either reducing the significance level for the final analysis, as described by Gao, Ware, and Mehta,[30] or by appropriately "weighting" the separate test statistics of the individual stages, as described by Cui, Hung, and Wang.[31] One reason why the test needs to be adjusted in the frequentist sense is that the joint distribution of the estimated effects that are respectively solely based on stage 1 data and solely based on stage 2 data does not factor into the product of the marginal distributions. For a Bayesian analysis, the likelihood function for the treatment effect does not depend on whether the sizing of stage 2 does or does not depend on the results at stage 1.

If the estimated effect is quite small at the interim analysis for a superiority trial or barely on the favorable side of the non-inferiority margin, a large sample size adjustment would be needed to obtain a high conditional power. This would make the objective of the trial the investigation for small, perhaps clinically meaningless, effects just to obtain a statistically significant result. The primary goal should not be to obtain a statistically significant result, but to obtain a statistically reliable evaluation on whether the experimental therapy is safe and provides clinically meaningful benefit.[19] The clinical relevance of a particular size of a treatment effect should be decided before initiating the clinical trial and should not be influenced by interim results.[18]

Also there is a danger in using an unreliable estimate of the treatment effect to size the remainder of the trial. Comparisons between arms of outcomes from the first stage of the trial may not represent the comparisons between arms of the same endpoint for the second stage of the trial. For a time-to-event endpoint, the duration of follow-up is different for the two analyses. Short-term effects or relative effects may not translate to long-term effects. The effects of therapy may be greater for subjects with better prognosis than those with worse prognosis. The patients with worse prognosis will have a lopsided contribution to the total number of events at the interim analysis and thus may yield an estimated effect that would tend to be smaller than that obtained at the final analysis. Additionally, sample size or event size reestimation allows for the possibility of back calculating, gaining an idea of the estimated effect at the interim analysis. Such knowledge could be used in a manner that reduces the integrity of the trial.

Alternatively, the trial can be sized for a minimally meaningful effect size and a group sequential testing procedure can be implemented. This is generally more efficient and provides a more natural relative weighting of data generated before and after any interim look.

Wang et al.[32] compared the power, type I error rate, and sample size for two group sequential approaches for testing non-inferiority and superiority with a two-stage adaptive approach. In the two-stage adaptive approach, if the trial continues after stage 1, the specific primary objective for the end of trial is chosen between non-inferiority and superiority based on the results in stage 1. Thresholds for decision making in such two-stage adaptive approaches were also considered by Shih, Quan, and Li[33] and Koyama, Sampson, and Gleser.[34]

## 10.5 Equivalence Comparisons

There are various instances where it is important to understand whether two underlying distributions are similar. The approval of generic drugs involves "equivalence" comparisons to the innovator drug. It is important to demonstrate that there is no meaningful difference in outcomes between the generic drug and the innovator drug. An equivalence margin or criterion involves how similar the underlying distributions for the control product (e.g., the innovator drug) and the experimental product (e.g., the generic drug) need to be for the experimental product to be "equivalent" (or "bioequivalent") to the control product. The criterion for equivalence may depend on the purpose of demonstrating equivalence. Having a goal that the products have similar distributions for outcomes may lead to a different criterion than that the outcomes from one treatment arm are not unacceptably worse than the outcomes from the other treatment arm.

There are a variety of approaches in evaluating the similarity of two parameters or two distributions. Four of the more common approaches in testing for equivalence or describing the similarity of two distributions involve (1) comparing a two-sided confidence interval for a comparative parameter (e.g., a difference in means or proportions, a relative risk, an odds ratio, or a hazard ratio) to equivalence limits, (2) determining the overlap of two distributions (in particular, the overlap in the density or mass functions), (3) comparing the distribution functions, and (4) comparing the variability of two random observations from different groups or distributions to the variability of two random observations from the same distribution.

We discuss various approaches to equivalence for two or more groups. The equivalence problem involving more than two groups is presented in relation to determining lot consistency. These test procedures presented for more than two groups also have simplified versions for two groups. An additional discussion on equivalence comparisons is provided in Section A.3.2 in the Appendix. A more extensive discussion of equivalence comparisons can be found in Wellek's book.[35]

### 10.5.1 Data Scales

The types of comparisons that can be performed depend on the scale of the outcome data. For data having a nominal scale (i.e., the outcomes involve unordered categories), the relevant parameters are the actual relative frequencies or probabilities for each category. Since the categories are unordered, comparisons between study arms of the distributions for such measurements involve comparing the similarity of the respective relative frequencies for each category. When considering a nominal endpoint, comparison of two groups either involves demonstrating the true or underlying relative frequencies are different between the groups or demonstrating that the true or underlying relative frequencies are similar between the groups. One measure of the similarity of two distributions of nominal categorical data is the sum over all categories of the smaller relative frequency between the two arms. This is a type of "overlap" measure.

When the data are ordered, various types of inferences (e.g., equivalence, non-inferiority, or superiority) can be made. For measurements that have an ordinal scale, additional relevant parameters would include the actual cumulative relative frequencies or cumulative probabilities for each category. For a given category, its cumulative relative frequency is the relative frequency of observations that either fall into that category or any category having less value. The cumulative relative frequencies are often used in the comparison on ordinal data between two groups.

For measurements that have an interval or ratio scale, parameters of interest include means, medians, specific percentiles, variances, or standard deviations of the distribution. For measurements having an interval scale, comparisons between study arms of the same type of parameter would involve examining differences in the respective parameters. For measurements having a ratio scale, comparisons between study arms of the same type of parameter may involve examining differences or quotients in the respective parameters.

## 10.5.2 Two One-Sided Tests Approach

Equivalence trials are also frequently conducted in phase I studies to evaluate the pharmacokinetics of drug candidates. In a typical bioequivalence trial, for example, the primary objective is to show that the experimental and control arms have equivalent pharmacokinetic profiles, often measured by the area under the plasma-concentration curve (AUC). The criterion for equivalence is based on the ratio of the two geometric mean AUCs such that the ratio is not too small and not too large. Let $\mu_E$ and $\mu_C$ be the true geometric mean AUCs for the experimental and control arms, respectively. The hypotheses of interest are

$$H_o : \mu_E/\mu_C \le \delta_1 \text{ or } \mu_E/\mu_C \ge \delta_u \text{ vs. } H_a : \delta_1 < \mu_E/\mu_C < \delta_u \tag{10.3}$$

where $\delta_1$ and $\delta_u$ are the prespecified equivalence margins. A common choice for the equivalence margins is $\delta_1 = 0.8$ and $\delta_u = 1/\delta_1 = 1.25$.

The equivalence hypothesis can be analyzed by two one-sided tests (TOST), in which the $p$-value to reject the null hypothesis of nonequivalence is the larger of the two $p$-values from the TOST.[36] Specifically, equivalence testing can be reformulated as one-sided tests of the respective sets of hypotheses

$$H_{o,1} : \mu_E/\mu_C \le \delta_1 \text{ vs. } H_{a,1} : \mu_E/\mu_C > \delta_1 \text{ and} \tag{10.4}$$

$$H_{o,2} : \mu_E/\mu_C \ge \delta_u \text{ vs. } H_{a,2} : \mu_E/\mu_C < \delta_u$$

The validity of the TOST approach has been documented by Berger,[37] and this approach is general enough to apply to continuous, discrete, or time-to-event data. The testing approach is equivalent to comparing an appropriate-level confidence interval for $\mu_E/\mu_C$ with the interval $(\delta_1, \delta_1)$. If the confidence interval lies entirely within $(\delta_1, \delta_1)$, then "equivalence" is concluded. Otherwise, equivalence is not shown. For example, performing the standard sets of tests of the respective sets of hypotheses in Expression 10.4, each at a significance level of $\alpha/2$, is equivalent to comparing a $100(1 - \alpha)\%$ two-sided confidence interval for $\mu_E/\mu_C$ with the interval $(\delta_1, \delta_1)$. As the two tests are simultaneously performed at a significance level of $\alpha/2$ and both null hypotheses needed to be rejected to conclude equivalence, the type I error rate is maintained at a level of $\alpha/2$ or less.

This TOST approach is recommended in the International Conference on Harmonization of Technical Requirements for Registration of Pharmaceutic als for Human Use E9,[27] which states that "Operationally, this (equivalence test) is equivalent to the method of using two simultaneous one-sided tests to test the (composite) null hypothesis that the treatment difference is outside the equivalence margins versus the (composite) alternative hypothesis that the treatment difference is within the margins." We will later discuss a generalized version of the confidence interval or TOST approach and its limitations when requiring similarity of more than two groups.

### 10.5.3 Distribution-Based Approaches

*Overlap Approaches.* For discrete data, the proportion of similar responses (PSR) is defined as the sum of the minimum probabilities or minimum relative frequencies across categories. For example, when there are $k$ possible outcomes, the parameter of interest is $\sum_{i=1}^{k} \min\{p_{E,i}, p_{C,i}\}$. It is easy to see that this parameter does not retain any information on any ordered relationship among the observations—that is, the possible outcomes are treated as nominal, unordered categories.

For continuous data, Rom and Hwang[38] defined the PSR to be the overlap under the density curves between the two treatments. It measures the degree of overlap (similarity) of the two distributions. More formally, the PSR is given by

$$PSR = \int_{-\infty}^{\infty} \min\{f_E(x), f_C(x)\}dx$$

where $f_E$ and $f_C$ are the underlying density functions for the experimental and control arms, respectively. A PSR close to 1 indicates similar distributions of outcomes between the two arms. In practice, when the PSR is far from 1, the means, medians, and/or variances of the two distributions will be quite different.

We will consider the special case of normal distributions having equal standard deviations. Let $\mu_E$ and $\mu_C$ denote the underlying means for the experimental and control arms, and let $\sigma$ denote the common standard deviation. Then the PSR can be expressed as a decreasing function of the absolute standardized difference in the means ($|D_S|$). That is,

$$PSR(D_S) = 2\Phi(-|D_S|/2)$$

where $D_S = (\mu_E - \mu_C)/\sigma$ and $\Phi$ is the distribution function for a standard normal distribution. Inferences on PSR then reduce to inferences on $|\mu_E - \mu_C|/\sigma$, the number of absolute standard deviation difference in the means. When $\sigma$ is known, the inference reduces to an inference on $|\mu_E - \mu_C|$ with hypotheses tested like those in Expressions 10.3 and 10.4. The analysis can then be based on a confidence interval for $\mu_E - \mu_C$ or a TOST on the difference in means. When $\sigma^2$ is unknown, the $t$ statistic can be used to make inference since its noncentrality parameter is a monotone function of the standardized difference in means.[38] Rom and Hwang[38] have also derived the PSR as a function of the means and standard deviations of two normal distributions, allowing the standard deviations to be different and unknown. They showed in this general normal case that the PSR measure provides a better tool for comparing treatments than the standard $t$ test, which only focuses on a difference in means and not a difference in the standard deviations.

In terms of equivalence margin for PSR, there is no universally agreed-upon value. Rom and Hwang[38] suggested that a PSR of at least 0.7 (70% overlap) could be used to judge whether two treatments are equivalent. Values of 0.8 or 0.9 have also been suggested. As with other designs, if one is interested in using the PSR to analyze equivalence trials, it is important to prespecify the equivalence margin and discuss its properties before the start of the study.

For two normal distributions with a common standard deviation, Table 10.1 gives for different values of $|D_S|$ the corresponding PSR. To further interpret these values of $|D_S|$ and $PSR(D_S)$, the probability that a random observation from the smaller distribution is greater than a random observation from the larger distribution and the percentile of the value of the smaller mean in the larger distribution are also provided in Table 10.1. When the two means differ by half a standard deviation (i.e., $|D_S| = 0.5$), the PSR is 0.80, and the probability that a random observation from the smaller distribution is greater than a random observation from the larger distribution is 0.36. Also, since a value of a half standard deviation below the mean is the 31st percentile of a normal distribution, the smaller mean is the 31st percentile of the larger normal distribution.

A nonparametric estimate of PSR using kernel density estimates was proposed by Heyse and Stine.[39] This nonparametric estimate avoids strong assumptions on the shape of the populations, such as normality or equal variance. Through empirical studies, they showed that nonparametric estimates of PSR are accurate for a variety of normal and nonnormal distributions. The sampling variance from the kernel-based estimate of PSR is only slightly larger than that of the normal maximum likelihood estimated variance for

**TABLE 10.1**

Comparative Characteristics of Two Normal Distributions Having a Common Standard Deviation

| Number of Standard Deviation Difference in Means | Proportion of Similar Responses | Probability that Smaller Distribution Has Greater Random Value | Percentile of Smaller Mean in the Larger Distribution |
|---|---|---|---|
| 0 | 1 | 0.50 | 50 |
| 0.25 | 0.90 | 0.43 | 40 |
| 0.5 | 0.80 | 0.36 | 31 |
| 0.75 | 0.71 | 0.30 | 23 |
| 1 | 0.62 | 0.24 | 16 |

normal data, and the kernel-based estimate may have less bias in analyzing nonnormal data.

In a pure nonparametric setting where no assumptions are made about the underlying distributions, the amount of overlap in the densities also treats the data as having a nominal scale as in the discrete case. A relationship to order in the outcomes is introduced when the densities are expressed involving a parameter for which order makes sense (e.g., the mean), as in the afore-mentioned normal case.

*Kolmogorov–Smirnov Approach.* When order makes sense (i.e., the data have an ordinal, interval, or a ratio scale), a Kolmogorov–Smirnov type of statistic is one of the possibilities for an equivalence comparison. For distribution functions of the experimental and control arms of $F_E$ and $F_C$, respectively, the hypotheses are expressed as

$$H_0 : \sup_{-\infty < x < \infty} |F_E(x) - F_C(x)| \geq \delta \text{ vs. } H_a : \sup_{-\infty < x < \infty} |F_E(x) - F_C(x)| < \delta$$

The Kolmogorov–Smirnov statistic is given by $\max_{-\infty < x < \infty} |\hat{F}_E(x) - \hat{F}_C(x)|$, where $\hat{F}_E$ and $\hat{F}_C$ are the corresponding estimated distribution functions. As equality of the distributions is in the alternative hypothesis, the common scaled version of the Kolmogorov–Smirnov test statistic would not apply for equivalence testing. Bootstrapping or simulations may be useful in studying the behavior of the Kolmogorov–Smirnov statistic in equivalence testing. Alternatively, the hypotheses in the above expression could be tested on the basis of simultaneous confidence bounds for $F_E(x) - F_C(x)$. Some rank-based tests of equivalence are provided by Wellek.[35]

## 10.5.4 Lot Consistency

Equivalence-like comparisons are also used in vaccine clinical development to evaluate the consistency of the manufacturing process. The production of vaccines or related biological products involves product lots consisting of thousands of individual doses. Licensure in the United States requires the demonstration of consistently produced lots of vaccine that produce similar biological effects. This reduces to demonstrating simultaneous "equivalence" across multiple pairs of groups or lots. In a typical lot consistency trial, there will be three arms representing three consecutive manufacturing lots of new vaccines.[40] In addition, an active control, if available, will be included to show that the new vaccine (with all lots combined) is noninferior to the active control once consistency of manufacturing is demonstrated.

A typical clinical trial evaluating lot consistency has several hundred subjects randomized to one of many (at least three) vaccine lots. Subjects are vaccinated from their respective lot and have blood samples taken when peak immune responses are expected (usually sometime between 14 and 42 days after their last dose). The blood samples are assayed, and for each lot the geometric mean concentration of some measurement (e.g., antibody levels) or equivalently the arithmetic mean of the log of the measurements is calculated.

For a continuous endpoint, such as geometric mean titers (GMT) of antibody responses, consistency of the lots has been interpreted as the range of the true means being less than a prespecified threshold or margin. For example, let $\mu_i$, $i = 1, 2, 3$ be the true GMTs of the three lots. Then the consistency hypothesis can be formulated as

$$H_o : \max_{i,j} |\mu_i - \mu_j| \geq \delta \text{ vs. } H_o : \max_{i,j} |\mu_i - \mu_j| < \delta \qquad (10.5)$$

Some approaches have assumed normally distributed data, usually with equal variances, with equivalence margins defined with respect to the differences in means. Thus, these methods fail to test for similarity with the variances or the distributions altogether. The paper by Lachenbruch, Rida, and Kou[41] provides a very good overview on lot consistency and the various procedures that can be used.

A margin must be determined that reflects that the production process will consistently produce lots that have similar biological effects. There is no universal margin or equivalence criterion; instead agreements are made between the FDA and the manufacturers on a case-by-case basis.[40] A conventional approach to demonstrating equivalence involves comparing all pairwise 90% CIs with the interval $-\delta$ to $\delta$, where $\delta$ is the equivalence margin.[36] If all 90% confidence intervals lie within $-\delta$ to $\delta$, then it is concluded that the production process will consistently produce lots that have similar biological effects. Issues with such an approach include that

(1) The pooled estimate of the within-lot variance is often used when it may not be appropriate, and by doing so the within-lot variances are assumed to be equal when it may be important to additionally demonstrate that the within-lot variances are similar to reliably conclude that the production process will consistently produce lots that have similar biological effects.

(2) Normality is assumed for the distribution for the mean and this may not be an appropriate assumption in many cases.

(3) The type I error rate may be much less than 0.05 and is dependent on the number of lots compared; the larger the number of lots compared, the smaller the type I error rate.

This confidence interval approach is equivalent to performing three TOST, one for each pair of lots and is discussed in detail by Wiens, Heyse, and Matthews.[42]

A real example of consistency lots trial is reported by Lieberman et al.[43] in evaluating the manufacturing consistency of a quadrivalent measles, mumps, rubella, and varicella vaccine (MMRV) in approximately 4000 healthy children. Since there are multiple components, the testing strategy is rather complicated, even though the design is relatively straightforward. The success of the study requires demonstration of consistency on all end-points as well as non-inferiority between the quadrivalent vaccine and the active control. Because of the high dimension of success criteria, a large sample size for consistency lots trial was required to provide sufficient power.

In this study, children 12–23 months of age were randomized to receive either one of three consistency lots of MMRV or a control of concomitant administration of MMR and varicella vaccine (MMR + V). For the assessment of consistency, both immune response rate and GMTs of measles, mumps, rubella, and varicella antibody responses at 6 weeks after vaccination were used as co-primary endpoints. The consistency margins for response rates were 5 percentage points for measles, mumps, and rubella, where expected responses are approximately 95%. However, the consistency margin was 10 percentage points for varicella because the expected response rate is approximately 90%. For GMTs, the consistency margin was 1.5-fold for all four antigens. For the consistency testing, the TSOT approach was used at the $\alpha = 0.05$ level for each pair of comparison. Manufacturing consistency was to be declared only if consistency can be concluded for both immune response rates and GMTs. Once consistent responses were established among lots of MMRV, results from the three lots of MMRV were combined and compared with the control group given MMR + V. The results presented in the paper showed that the manufacturing process of MMRV is consistent and that the antibody responses from the combined MMRV lots are noninferior to the control of MMR + V for both immune response rates and GMTs.

For location parameters $\mu_1, \ldots, \mu_k$, and equivalence margin $\delta > 0$, Giani and Finner[44] considered a test of the hypotheses in Expression 10.5 based on the range of estimators of the parameters. For means and normally distributed data, equivalence is concluded if $\bar{X}_{(k)} - \bar{X}_{(1)} < a \cdot \delta$, where $\bar{X}_{(1)} \leq \bar{X}_{(2)} \leq \cdots \leq \bar{X}_{(k)}$ are the ordered sample means of the $k$ lots and $0 < a < 1$. The value of $a$ depends on the significance level and the common within-lot standard error. For uni-modal distributions, the type I error probability under the null hypothesis is maximized when $\mu_{(1)} = -0.5\delta + c$, $\mu_{(2)} = \ldots = \mu_{(k-1)} = c$, and $\mu_{(k)} = 0.5\delta + c$ for any real number $c$.

Wiens, Heyse, and Matthews[42] proposed a test, which they referred to as the min test, which corresponds to the likelihood ratio test of Sasabuchi.[45] For each pair of lots, the test statistic for the equivalence of the two respective means is computed, and then the minimum value of these test statistics is the value of the min test statistic. The test statistic is

$$Z_{\min} = \min_{1 \leq i < j \leq k} \left[ \frac{\delta - |\bar{X}_i - \bar{X}_j|}{\sqrt{\sigma_i^2/n_i + \sigma_j^2/n_j}} \right]$$

If $Z_{min} > Z_\alpha^*$, then equivalence is concluded, where the critical value is calculated from the distribution of the range statistic for the means.[46] When the standard errors are equal (e.g., equal lot sizes and the lots are assumed to have the same within-lot variance), the min test is equivalent to the range test of Giani and Finner.[44]

When the within-lot variances or the lot sizes are not equal, the min test can be quite conservative (i.e., the min test has a type I error rate much smaller than the desired significance level). Wiens and Iglewicz[46] suggested using an adjusted critical value, $Z_\alpha^{**}$. The smallest within-lot standard error for the mean across all lots is used as the common within-lot standard error in determining the value for $Z_\alpha^{**}$. Wiens and Iglewicz[46] showed that the resulting test is still conservative, but since $Z_\alpha^{**} \le Z_\alpha^*$, the test is both less conservative and more powerful than the original min test.

Ng[47] proposed hypotheses and an equivalence test on the basis of the between-lot variability of the means for common lot sizes and common within-lot variances. Here, the null hypothesis is

$$H_o : \left( \sum_{i=1}^{k} (\mu_i - \bar{\mu})^2 \right)^{1/2} \ge \delta$$

for some margin on the variability, $\delta$. The test statistic is the standard $F$ statistic for testing for any between-lot variability. On the boundary of the null hypothesis, the test statistic has a noncentral $F$ distribution with $k - 1$ and $k(n - 1)$ degrees of freedom and noncentrality parameter $n\delta^2/\sigma^2$, where $n$ is the common lot size and $\sigma^2$ is the common within-lot variance. The critical value depends on the value of $\sigma^2$. When $\sigma^2$ must be estimated, Ng provides an iterative method for finding the critical value. The test procedure assumes that the data are normally distributed.

When the means and variances of the distributions exist and differences in values make sense, the expected difference of the average squared difference between random observations from any two distributions relative to the same expected difference when the random observations are drawn from the same distribution is an appropriate measure of the amount of difference between the two distributions. For many distributions, this would mean taking a random observation from each distribution and measuring the variability in their values. Let $X_1, \ldots, X_k$ be independent but not identically distributed random variables with $\bar{X} = \sum_{i=1}^{k} X_i/k$. Then if we randomly select two of the $k$ distributions and randomly draw an observation from each selected distribution,

$$E\left( \sum_{i<j} (X_i - X_j)^2 \middle/ \binom{k}{2} \right) = E\left( \sum_{i=1}^{k} (X_i - \bar{X})^2 /k \right)$$

represents the expected distance squared between the two observations. Let the respective means be denoted by $\mu_1, \ldots, \mu_k$ and the respective variances denoted $\sigma_1^2, \ldots, \sigma_k^2$. Then the above expected value equals

$$(2/k)\sum_{i=1}^{k} \sigma_i^2 + \sum_{i=1}^{k}(\mu_i - \bar{\mu})^2/k \tag{10.6}$$

For each $i = 1, \ldots, k$, let $W_{1i}$ and $W_{2i}$ be independent and identically distributed with variance $\sigma_i^2$. Then, if we select one of the $k$ distributions at random,

$$E\left(\sum_{i=1}^{k}(W_{1i} - W_{2i})^2/k\right) = (2/k)\sum_{i=1}^{k}\sigma_i^2 \tag{10.7}$$

represents the expected square distance between two random observations taken from that randomly selected distribution. The difference in the two expectations in Expressions 10.6 and 10.7 equals

$$\sum_{i=1}^{k}(\mu_i - \bar{\mu})^2/k$$

and thus inferences can be reduced to that statistic proposed by Ng.[47] Alternatively, instead of the difference, the inference can be based on the ratio of the expected values in Expressions 10.6 and 10.7.

Lachenbruch, Rida, and Kou[41] considered a nonparametric approach to equivalence based on a similarity coefficient for the probability densities (or probability mass functions if discrete data) provided by Cleveland and Lachenbruch.[48] For absolutely continuous distributions, the similarity coefficient is defined as

$$\gamma = \frac{1}{k}\int_{-\infty}^{\infty} \max_{1 \le j \le k}(f_i(x))dx$$

The value of $\gamma$ is between $1/k$ and 1. Smaller values of $\gamma$ correspond with greater similarity in the distributions. The null hypothesis would be $H_0: \gamma \ge \delta$, where $1/k \le \delta \le 1$ is prespecified. Preliminary simulations by Lachenbruch, Rida, and Kou[41] suggest that their test of similarity can be useful in demonstrating equivalence when distributions have much heavier tails than normal distributions or when there are differences in scale or shape of the distributions being compared.

It is very important in an evaluation of the consistency of lots to establish an appropriate definition of what it means for the lots to be consistent. There have been different approaches as to the meaning of lot consistency. For any given approach, the equivalence margins will vary from case to case, and may be dependent on the indication and the efficacy of the product.

# References

1. Wiens, B.L. and Heyse, J.F., Testing for interaction in studies of non-inferiority, *J. Biopharm. Stat.*, 13, 103–115, 2003.
2. Gail, M. and Simon, R., Testing for qualitative interactions between treatment effects and patient subsets, *Biometrics*, 41, 361–376, 1985.
3. Laska, E.M. and Meisner, M.J., Testing whether an identified treatment is best, *Biometrics*, 45, 1139–1151, 1989.
4. Piantadosi, S. and Gail, M.H., A comparison of the power of two tests for qualitative interactions, *Stat. Med.*, 12, 1239–1248, 1993.
5. U.S. Code of Federal Regulations, Title 21, Sec. 314.500-560 and Sec. 601.40–46.
6. Fleming, T.R. and Powers, J.H., Issues in non-inferiority trials: The evidence in community-acquired pneumonia, *Clin. Infect. Dis.*, 47, S108–120, 2008.
7. Bruzzi, P. et al., Objective response to chemotherapy as a potential surrogate endpoint of survival in metastatic breast cancer patients, *J. Clin. Oncol.*, 23, 5117–5125, 2005.
8. Fleming, T.R. and DeMets, D.L., Surrogate endpoints in clinical trials: Are we being misled? *Ann. Intern. Med.*, 125, 605–613, 1996.
9. Fleming, T.R., Surrogate endpoints and FDA's accelerated approval process: The challenges are greater than they seem, *Health Affair*, 24, 67–78, 2005.
10. Fleming, T.R., Objective response rate as a surrogate endpoint: A commentary, *J. Clin. Oncol.*, 23, 4845–4846, 2005.
11. Rothmann, M.D., Issues to consider when constructing a non-inferiority analysis, *ASA Biopharm. Sec. Pro.*, 1–6, 2005.
12. Fleming, T.R., Current issues in non-inferiority trials, *Stat. Med.*, 27, 317–332, 2008.
13. Prentice, R.L., Surrogate endpoints in clinical trials: Discussion, definition and operational criteria, *Stat. Med.*, 8, 431–440, 1989.
14. Prentice, R.L., Surrogate and mediating endpoints: Current status and future directions, *J. Natl. Cancer Inst.*, 101, 216–217, 2009.
15. Baker, S.G., Surrogate endpoints: Wishful thinking or reality? *J. Natl. Cancer Inst.*, 9, 502–503, 2006.
16. Neupogen product labeling available at http://www.accessdata.fda.gov/Scripts/cder/DrugsatFDA/index.cfm?fuseaction=Search.Label_ApprovalHistory.
17. Neulasta product labeling available at http://www.accessdata.fda.gov/Scripts/cder/DrugsatFDA/index.cfm?fuseaction=Search.Label_ApprovalHistory.
18. Committee for Proprietary Medicinal Products. Reflection paper on methodological issues in confirmatory clinical trials with flexible design and analysis plan, EMA, London, 2006.
19. Fleming, T.R., Standard versus adaptive monitoring procedures: A commentary, *Stat. Med.*, 25, 3305–3312, 2006.

20. Guidance for Industry: Adaptive design clinical trials for drugs and biologics (draft guidance), February 2010.
21. Pocock, S.J., Group sequential methods in the design and analysis of clinical trials, *Biometrika*, 64, 191–199, 1977.
22. O'Brien, P.C. and Fleming, T.R., A multiple testing procedure for clinical trials, *Biometrics*, 35, 549–556, 1979.
23. Lan, K.K. and DeMets, D.L., Design and analysis of group sequential tests based on the type I error spending function, *Biometrika*, 74, 149–154, 1983.
24. Jennison, C. and Turnbull, B.W., Repeated confidence intervals for group sequential clinical trials, *Control. Clin. Trials*, 5, 33–45, 1984.
25. Jennison, C. and Turnbull, B.W., Sequential equivalence testing and repeated confidence intervals, with application to normal and binary data, *Biometrics*, 49, 31–43, 1993.
26. Lawrence, J., Some remarks about the analysis of active control studies, *Biometrical J.*, 47, 616–622, 2005.
27. International Conference on Harmonization of Technical Requirements for Registration of Pharmaceuticals for Human Use (ICH), E9: Statistical principles for clinical trials, 1998, at http://www.ich.org/cache/compo/475-272-1.html#E4.
28. Wittes, J. and Brittain, E., The role of internal pilot studies in increasing the efficiency of clinical trials, *Stat. Med.*, 9, 65–72, 1990.
29. Friede, T. and Kieser, M., Blind sample size reassessment in non-inferiority and equivalence trials, *Stat. Med.*, 22, 995–1007, 2003.
30. Gao, P., Ware, J.H., and Mehta, C.R., Sample size re-estimation for adaptive sequential design in clinical trials, *J. Biopharm. Stat.*, 18, 1184–1196, 2008.
31. Cui, L., Hung, H.M.J., and Wang, S.-J, Modification of sample size in group sequential clinical trials, *Biometrics*, 55, 321–324, 1999.
32. Wang, S.J. et al., Group sequential test strategies for superiority and non-inferiority hypotheses in active controlled clinical trials, *Stat. Med.*, 20, 1903–1912, 2001.
33. Shih, W.J., Quan, H., and Li, G., Two-stage adaptive strategy for superiority and non-inferiority hypotheses in active controlled clinical trials, *Stat. Med.*, 23, 2781–2798 2004.
34. Koyama, T., Sampson, A.R., and Gleser, L.J., A framework for two-stage adaptive procedures to simultaneously test non-inferiority and superiority, *Stat. Med.*, 24, 2439–2456, 2005.
35. Wellek, S., *Testing Statistical Hypotheses of Equivalence*, Chapman & Hall/CRC Press, Boca Raton, FL, 2003.
36. Schuirmann, D., A comparison of the two one-sided tests procedure and the power for assessing the equivalence of average bioavailability, *J. Pharmacokinet. Pharm.*, 15, 657–680, 1987.
37. Berger, R.L., Multiparameter hypothesis testing and acceptance sampling, *Technometrics*, 24, 295–300, 1982.
38. Rom, D.M. and Hwang, E., Testing for individual and population equivalence based on the proportion of similar responses, *Stat. Med.*, 15, 1489–1505, 1996.
39. Heyse, J.F. and Stine, R., Use of the overlapping coefficient for measuring the similarity of treatments, *Am. Stat. Assoc. Proc. Biopharm. Sec.*, 29–32, 2000.
40. Guidance for Industry for the Evaluation of Combination Vaccines for Preventable Diseases: Production, Testing, and Clinical Studies. U.S. Department of Health and Human Services, Food and Drug Administration, Center for Biologics Evaluation and Research, April 1997.

41. Lachenbruch, P.A., Rida, W., and Kou, J., Lot consistency as an equivalence problem, *J. Biopharm. Stat.*, 14, 275–290, 2004.
42. Wiens, B.L., Heyse, J.F., and Matthews, H., Similarity of three treatments, with application to vaccine development, *Am. Stat. Assoc. Proc. Biopharm. Sec.*, 203–206, 1996.
43. Lieberman, J.M. et al., The safety and immunogenicity of a quadrivalent measles, mumps, rubella and varicella vaccine in healthy children: A study of manufacturing consistency and persistence of antibody, *Pediatr. Infect. Dis. J.*, 25, 615–622, 2006.
44. Giani, G. and Finner, H., Some general results on least favorable parameter configurations with special reference to equivalence testing and the range statistic, *J. Stat. Plan. Infer.*, 28, 33–47, 1991.
45. Sasabuchi, S., A test of multivariate normal mean with composite hypotheses determined by linear inequalities, *Biometrika*, 67, 429–439, 1980.
46. Wiens, B. and Iglewicz, B., On testing equivalence of three populations, *J. Biopharm. Stat.*, 9, 465–483, 1999.
47. Ng, T., Iterative chi-square test for equivalence of multiple treatment groups, *Am. Stat. Assoc. Proc. Biopharm. Sec.*, 2464–2469, 2002.
48. Cleveland, W. and Lachenbruch, P., A measure of divergence among several populations, *Commun. Stat.*, 33, 201–211, 1974.

# 11

## Inference on Proportions

### 11.1 Introduction

Many clinical trials use outcome variables that are binary in nature, that is, there are two possible outcomes for each subject. Without loss of generality, these two outcomes are called "success" and "failure." Other terms used are "with an event" and "without the event." Usually there are qualitative differences between these two outcomes, in that one outcome is always preferred over the other. Specifically excluded from this class of outcome variables are outcomes of a time-to-event endpoint where we are interested in the length of time until an event (as well as the occurrence of the event are of interest), or outcomes where we are interested in the magnitude in gradations between success and failure.

The simplest model of proportions is the binomial model, in which each subject has the same probability of success, $p$. When a sample of $n$ subjects is taken, the expected number of successes is $np$ and the variance is $np(1 - p)$. Often of most interest is the proportion of successes, $\hat{p} = x/n$, which then has mean $p$ and variance $p(1 - p)/n$. With large sample sizes, the normal approximation can be used to describe the distribution of $\hat{p}$ with ease of calculations and minimal loss of precision, as noted below. Since the variance estimate is not independent of the mean estimate, extension of the normal approximation to more complex situations must be used with caution.

The experimental therapy is noninferior to the control therapy if the probability of a success outcome on the experimental arm is better than or not too much worse than that of the control arm. When a "success" is a positive or desirable outcome (as the word "success" suggests), this means that the probability of a success for the experimental arm is greater than or not too much less than that for the control arm. When a "success" is a negative or undesirable outcome, this means that the probability of a success for the experimental arm is less than or not too much greater than that for the control arm. This "not too much less than" or "not too much greater than" can be expressed through a difference in the two probabilities of a success, in the ratio of the two probabilities (i.e., a relative risk), or through an odds ratio.

The choice of the appropriate scale is a very important consideration in the non-inferiority testing of proportions. For a superiority comparison, the null and alternative hypotheses do not depend on the scale or "metric" (the difference in proportions or risk difference, the relative risk, or the odds ratio) used. There is no such correspondence across respective non-inferiority hypotheses based on a risk difference, relative risk, and odds ratio. It is thus important to choose the proper metric or function when comparing proportions in a non-inferiority trial.

A non-inferiority comparison can be used to compare the efficacy of an experimental therapy with that of an active control therapy, or to compare the safety of an experimental therapy with a placebo or an active control. When using binary data, the clinical benefit and the safety risk are expressed by the difference in proportions. If an experimental therapy cures an additional 2 subjects out of every 100 subjects, then that clinical benefit of the experimental therapy is expressed by the 2% improvement in cure rates. If 10 additional subjects out of every 100 subjects incur an adverse event, then the safety risk of the experimental therapy is expressed by that 10% increase in incurring that adverse event.

The best way to describe the effect of a therapy may be different from describing the amount of benefit the therapy provides. How a drug works and the extent of the disease in the population can impact which is the appropriate metric. For example, consider a drug that prevents 60% of the subjects from getting a disease they would have otherwise acquired, independent of the background rate for the disease. That 60% prevention rate among patients at risk of the disease describes the effect of the therapy. If the background rate in acquiring the disease is 5%, then the drug prevents 3 out of every 100 treated subjects from acquiring the disease. That 3 out of 100 describes the clinical benefit and would be used as the benefit part of a risk–benefit assessment. If the background rate is 50%, then the drug prevents 30 out of every 100 treated subjects from acquiring the disease, and that 30 out of 100 would be used as the benefit part of a risk–benefit assessment.

Background disease risk rates and cure rates tend to change over time and from study to study. Also, the prevention rate or cure rate of the drug may change as the background disease rate or background disease cure rate changes from one population to another. When one population has a lower disease rate than another population, the typical extent of disease among those subjects having the disease may also be different in the two populations. Thus, the cure rate of an effective drug may be different between the two populations (lower cure rate for the population where the disease is more prevalent).

If the non-inferiority inference is not based on a risk difference, the results when positive may somehow need to be translated to a risk difference to provide an impression of the clinical benefit of the experimental therapy.

When the non-inferiority margin is developed, it will be based on the chosen scale. Ideally, the scale will be determined on the basis of the evaluation of the effect of the active control therapy compared to placebo. The choice of scale should also consider whether and how the effects or estimated effects of therapies in the studied indication vary across studies or important subgroups when based on the risk difference, relative risk, and odds ratio. If the active control therapy is routinely superior to placebo with a consistent magnitude for the treatment effect in a given scale, then that scale is appropriate for consideration for the non-inferiority analysis.

There are many other considerations that need to be made when choosing the appropriate scale. For proportions near zero, a non-inferiority efficacy margin on a risk difference of $\delta > 0$ would allow for an experimental success rate of 0 to be noninferior to any control rate less than $\delta$. This does not make sense, and a non-inferiority inference based on a risk difference should be avoided unless the minimum likely control rate is sufficiently larger than the non-inferiority margin that would be used. For negative outcomes, Dann and Koch[1] recommended using the relative risk for control rates $\leq 0.20$.

A binary outcome is modeled as having a Bernoulli distribution. The log-odds ratio is the canonical link or natural parameter for a Bernoulli distribution, and therefore better lends itself to a linear model as an adjusted analysis (e.g., better than modeling the probability of a success with a linear model). For the relative risk and the odds ratio, the true value that is being estimated by an unadjusted analysis is different from the value that is being estimated by an adjusted analysis. Therefore, the same type of analysis should be used in estimating the difference in effects in the non-inferiority trial as in evaluating the active control effect and determining non-inferiority margin.

In this chapter we will discuss non-inferiority analyses based on the risk difference (difference in proportions), the relative risk, and the odds ratios in Sections 11.2 through 11.4, respectively. Bayesian analyses will be discussed in Section 11.5. Adjusted analyses will be discussed in Section 11.6. The use of variable margins and the corresponding analyses will be discussed in Section 11.7. Non-inferiority inference based on matched-pair designs for proportions is discussed in Section 11.8.

## 11.2 Fixed Thresholds on Differences

In this section we will consider the situation in which a margin has been chosen on the basis of the difference in two proportions.

### 11.2.1 Hypotheses and Issues

For this discussion, consider a situation in which the margin has been determined on the basis of the simple difference in the probability of a success of a desirable outcome. The null and alternative hypotheses to be considered are

$$H_o: p_E - p_C \leq -\delta \text{ vs. } H_a: p_E - p_C > -\delta \tag{11.1}$$

That is, the null hypothesis is that the active control is superior to the experimental treatment by at least a quantity of $\delta \geq 0$ that is prespecified. The alternative hypothesis is that the active control is superior by a smaller amount, or the two treatments are identical, or the experimental treatment is superior. When $\delta = 0$, the hypotheses in Expression 11.1 reduce to classical one-sided hypotheses for a superiority trial. The null hypothesis in Expression 11.1 is rejected and the experimental therapy is concluded to be noninferior to the control therapy when a decrease in the proportion of success of $\delta$ or greater is statistically ruled out. If a "success" is an undesirable outcome, the roles of $p_C$ and $p_E$ in the hypotheses in Expression 11.1 would be reversed (i.e., test $H_0: p_C - p_E \leq -\delta$ vs. $H_a: p_C - p_E > -\delta$).

In the simplest application involving a desirable outcome, a confidence interval can be calculated on the difference $p_E - p_C$, and non-inferiority is concluded (the null hypothesis is rejected) if the lower bound is greater than $-\delta$.

## 11.2.2 Exact Methods

When the sample sizes are small, the performance of asymptotic methods is uncertain. In some cases, the inflation of type I error rate becomes substantial. In addition, the type I error rate may depend on the sample size in ways that are not predictable or may vary greatly across the boundary of the null hypothesis. Therefore, it is desirable to use exact methods to control the true level of the non-inferiority test.

Let us consider the setting where we are comparing an experimental treatment (E) to an active control treatment (C) using a dichotomous (success/failure) endpoint in a randomized study. Suppose there are $n_C$ and $n_E$ subjects randomized to the active control and experimental arms, respectively. Let $x$ and $y$ be the observed number of successes (responses) in the active control and experimental arms, respectively. Table 11.1 provides the notation for the breakdown of the number of responses ("successes") and nonresponses ("failures") between treatment arms.

Let $p_C$ and $p_E$ represent the true response rates of the control and experimental arms, respectively. Let $X$ and $Y$ be independent random variables representing the number of successes in the treatment arms and distributed as Binomial $(n_C, p_C)$ and Binomial $(n_E, p_E)$, respectively. To show that the experimental treatment is noninferior to the active control, we will test the hypotheses in Expression 11.1 for some fixed $\delta > 0$.

**TABLE 11.1**

Notation for Breakdown of Counts of Responses between Treatment Arms

| Treatment Arm | Response | No Response | Sample Size |
|---|---|---|---|
| Control | $x$ | $n_C - x$ | $n_C$ |
| Experimental | $y$ | $n_E - y$ | $n_E$ |
| Total | $s$ | $n_C + n_E - s$ | $n_C + n_E$ |

Given a true difference $(p_E - p_C = -\delta)$ under the null hypothesis $H_o$, the probability of observing an outcome $(x, y) = (i, j)$ is given by

$$P(X = i, Y = j \mid H_o) = \binom{n_C}{i}\binom{n_E}{j}(p+\delta)^i(1-p-\delta)^{n_C-i}p^j(1-p)^{n_E-j} \quad (11.2)$$

where $p = p_E$ is the nuisance parameter with the domain $A = [0, 1 - \delta]$. For the classical null hypothesis of no difference $(\delta = 0)$, the marginal total $(S = X + Y)$ is the sufficient statistic for the nuisance parameter $(p)$. To eliminate the effect of $p$, an exact test can be constructed conditional on this sufficient statistic, which yields the well-known Fisher's exact test.

In the case where $p_E - p_C = -\delta$ $(\delta > 0)$, there is no simple sufficient statistic for $p$ (the numbers of successes from each group are jointly minimal sufficient statistics). Therefore, the conditional argument will not simplify the problem of the nuisance parameter in testing non-inferiority hypotheses. In general, an exact test of non-inferiority can be developed on the basis of the null probability distribution given in Equation 11.2 using the unconditional sampling space consisting of all possible 2 × 2 tables given the sample sizes $(n_C, n_E)$. The exact test procedure defines the tail region (TR) of the observed table $(i, j)$ as the region of those tables that are at least as extreme as the observed table according to a predefined ordering criterion. Then the exact $p$-value is defined as

$$p\text{-value} = \max_{p \in A} P\big((X, Y) \in TR(i, j) \mid H_o, p\big) \quad (11.3)$$

The exact $p$-value calculation eliminates the nuisance parameter using the maximization principle,[2,3] which caters to the worst-case scenario. Because the maximization involves a large number of iterations in evaluating sums of binomial probabilities, the exact unconditional tests are computationally intensive, particularly with large sample sizes.

A natural ordering criterion proposed by Chan[4] used the Z-statistic based on the constrained maximum likelihood estimate (MLE) of parameters under the null hypothesis:

$$Z_1 = Z(x, y) = \frac{\hat{p}_E - \hat{p}_C + \delta}{\left\{ \tilde{p}_E(1-\tilde{p}_E)/n_E + \tilde{p}_C(1-\tilde{p}_C)/n_C \right\}^{1/2}} \quad (11.4)$$

where $\hat{p}_C$ and $\hat{p}_E$ are the observed response rates for the control and experimental treatment groups, respectively. In addition, $\tilde{p}_C$ and $\tilde{p}_E$ are the MLEs of $p_C$ and $p_E$, respectively, under the constraint $p_E - p_C = -\delta$ given in the null hypothesis. The closed-form solutions for $\tilde{p}_C$ and $\tilde{p}_E$ are given by Farrington and Manning[5] and are provided in Expression 11.6. Since large values of $Z_1$

favor the alternative hypothesis, the tail region includes those tables whose $Z_1$ statistics are larger than or equal to the $Z_1$ statistic associated with the observed table $(i, j)$, $z_{obs}$. As a result, the exact $p$-value can be obtained as

$$p\text{-value} = \max_{p \in A} P\Big((X,Y): Z_1 \geq z_{obs} \mid H_0, p\Big).$$

In summary, we can calculate the exact $p$-value in the following steps:

1. Compute the $Z_1$ statistic for all tables and order them. Let $z_{obs}$ be the calculated value of $Z_1$ for the observed table. The tail of the observed table includes those tables whose $Z_1$ statistics are larger than or equal to $z_{obs}$.
2. For a given value of the nuisance parameter $p$ in $A = [0, 1 - \delta]$, calculate the tail probability by summing up the probabilities of those tables in the tail using the probability function (Equation 11.2).
3. Repeat step 2 for every value of $p$ in its domain. Then the exact $p$-value is the maximum of the tail probability over the domain of $p$. Since the domain of $p$ is continuous, a numerical grid search (e.g., more than 1000 points) over the domain can be done to obtain the maximum tail probability. This should provide adequate accuracy for most practical uses.

For a nominal $\alpha$ level test, we reject the null hypothesis if the exact $p$-value is less than or equal to $\alpha$. To obtain the true level of the exact test, we first convert the test procedure to find the critical value given the nominal $\alpha$ level and the sample sizes $n_C$ and $n_E$. This critical value does not depend on any specific value of the nuisance parameter, and the true level is the maximum (over the domain of the nuisance parameter) null probability of those tables of which the test statistics are less than or equal to the critical value. This exact test has been implemented in commercial software.

When $\delta = 0$, $\tilde{p}_C$ and $\tilde{p}_E$ both simplify to the pooled estimate of the response rate among the two groups, and the $Z_1$ statistic in 11.4 reduces to the Z-pooled statistic for the classical null hypothesis of no difference. As a result, the exact unconditional test of non-inferiority based on Z provides a generalization of the unconditional test of the classical null hypothesis studied by Suissa and Shuster[6] and Haber.[7]

Other types of statistics may also be considered as ordering criteria. A few examples include: (1) the observed difference $D_{obs} = \hat{p}_E - \hat{p}_C$, (2) a Z-statistic with the variance (denominator of Z) estimated directly from the observed proportions, (3) a Z-statistic with the variance estimated from fixed marginal totals,[8] and (4) a likelihood ratio statistic.[9] Findings from empirical investigations show that the Z-statistic in 11.4 generally performs better than $D_{obs}$ and other Z-type statistics. Röhmel and Mansmann[10] recommended an ordering criterion, which adapts Barnard's test to the non-inferiority hypothesis. Using Barnard's ordering criterion, one creates the critical region (CR) for testing the non-inferiority hypothesis in a sequential manner starting with the most extreme outcome $(0, n_E)$, and then adds to the CR from the set of adjacent outcomes the one that increases the null probability of the new CR by the smallest amount. Barnard's test has been shown through empirical investigation to be generally the most powerful exact unconditional method for testing the classical null hypothesis of no difference between two independent proportions.[11] However, its relative performance in non-inferiority tests is generally unknown. Since Barnard's ordering criterion requires evaluating the true size of the test multiple times (once for each candidate outcome to be introduced into the CR), the computational complexity involved is substantially greater than methods using a test statistic (such as $Z$) as the ordering criterion. Additional discussions on ordering criteria were provided in previous papers.[10,12]

### 11.2.2.1 Exact Confidence Intervals

Besides a hypothesis test, a confidence interval is usually provided to assess the treatment difference. To provide a consistent confidence interval corresponding to the exact test of non-inferiority described above, a test-based confidence interval can be constructed as follows. Following the ideas of Clopper and Pearson,[13] Chan and Zhang[9] proposed to construct a two-sided $100(1 - \alpha)\%$ confidence interval for the true difference $\Delta = p_E - p_C$ by inverting the exact test procedure based on the $Z_1$ statistic for two one-sided hypotheses in 11.4, one for the lower bound and the other for the upper bound.

For example, the upper bound of the two-sided $100(1 - \alpha)\%$ confidence interval $(\Delta_U)$ is obtained by considering the one-sided hypothesis $H_o: \Delta = \Delta_o$ versus $H_a: \Delta < \Delta_o$ such that

$$\Delta_U = \sup_{\Delta_o}\left\{\Delta_o : \max_{p_C \in A} P(Z_1(X, Y; \Delta_o) \leq Z_1(x, y; \Delta_o) | \Delta_o, p_C) > \alpha/2\right\}.$$

Similarly, the lower bound of the two-sided $100(1 - \alpha)\%$ confidence interval $(\Delta_L)$ is obtained by considering the one-sided hypothesis $H_o: \Delta = \Delta_o$ versus $H_1: \Delta > \Delta_o$ such that

$$\Delta_L = \inf_{\Delta_o}\left\{\Delta_o : \max_{p_C \in A} P(Z_1(X, Y; \Delta_o) \geq Z_1(x, y; \Delta_o) | \Delta_o, p_C) > \alpha/2\right\}.$$

It was shown by Chan and Zhang[9] that the exact confidence interval based on the $Z_1$ statistic is much better than the simple tail-based confidence interval (see Santner and Snell[14]) as well as confidence intervals based on the

Z-unpooled and the likelihood ratio statistics. Also, since the exact confidence interval based on $Z_1$ is obtained by inverting two one-sided tests, it controls the error rate of each side at the $\alpha/2$ level, and hence provides consistent inference with the $p$-value from the one-sided hypothesis. In other words, if the null hypothesis in 11.1 is rejected at the one-sided $\alpha/2$ level for a specific $\delta$, then the lower bound of the two-sided $100(1 - \alpha)\%$ confidence interval for the difference $p_E - p_C$ will be greater than $-\delta$.

Exact confidence intervals have also been proposed by Agresti and Min[15] and Chen[16] by inverting the two-sided hypothesis $H_o: \Delta = \Delta_o$ versus $H_1: \Delta \neq \Delta_o$ based on the $Z_1$ statistic. The resulting confidence interval generally has a shorter width than the one obtained by inverting two one-sided tests, and therefore is very useful if the hypothesis is two-sided in nature or if estimation is of primary interest. Other methods (non-test-based) that are also useful for estimation purposes have been proposed by Coe and Tamhane[17] and Santner and Yamagami.[18] By inverting a two-sided test, the confidence interval controls the overall error rate at the $\alpha$ level but does not guarantee control of the error rate of each side at the $\alpha/2$ level. Consequently, it may potentially produce results that are inconsistent with a one-sided hypothesis test. Therefore if the criterion of showing non-inferiority is to require that the lower bound of the two-sided confidence interval for the difference $p_E - p_C$ be greater than $-\delta$, then controlling the one-sided type I error is essential, and constructing the confidence interval by inverting two one-sided tests is recommended.

Examples 11.1 and 11.2 provide $p$-values from applying some of these methods to the results of actual studies.

## Example 11.1

Fries et al.[19] conducted a challenge study to evaluate the efficacy of a recombinant protein influenza A vaccine against wild-type H1N1 virus. In this study, subjects were randomized to receive either the experimental vaccine or the placebo injection, and they were subsequently challenged intranasally with influenza virus. All subjects were closely monitored in the next 9 days for viral infection and clinical symptoms. The observed incidence rates of any clinical illness among subjects who had viral infection were $\hat{p}_P = 0.8$ (12/15) in the placebo group and $\hat{p}_E = 0.467$ (7/15) in the vaccine group. The investigators are interested in testing the classical null hypothesis of no difference—that is, $H_o: p_E = p_P$ versus $H_1: p_E < p_P$. The one-sided Fisher's exact (conditional) test yields a $p$-value of 0.064, compared with 0.0341 from the exact unconditional test based on Chan's[4] $Z$ statistic in Equation 11.4. For the same data set, Röhmel and Mansmann[10] reported a $p$-value of 0.036 from the exact unconditional test using Barnard's criterion, and a $p$-value of 0.034 when Fisher's $p$-value was used as an ordering criterion.[20] The results from these exact unconditional tests are consistent and produced smaller $p$-values than from the exact conditional test. The smaller $p$-values are due to having the probability distributed over a more extensive set of "possible outcomes $(x, y)$."

## Example 11.2

Rodary, Com-Nougue, and Tournade[21] reported a randomized clinical trial of 164 children with nephroblastoma to evaluate whether chemotherapy (new treatment) is noninferior to radiation therapy (control) based on a prespecified margin of $\delta = 0.1$ and a one-sided 5% level test. The study showed that the chemotherapy group actually had a higher success response rate ($\hat{p}_E = 0.943$ [83/88]) than the radiation group ($\hat{p}_C = 0.908$ [69/76]), and the observed difference ($\hat{p}_E - \hat{p}_C$) was 0.035. Because of the relatively few failures in both treatment groups, an approximate exact conditional method based on the odds ratio statistic (see Dunnet and Gent[8]) was used to test the non-inferiority hypothesis and yielded a $p$-value of 0.002. These data were reanalyzed by Chan[4] using the exact unconditional test based on the $Z$-statistic in Equation 11.4, and yielded a $p$-value of 0.0017. The exact unconditional test based on Barnard's criterion[10] yielded a $p$-value of 0.0011. The results from various analyses were very consistent and provide evidence that chemotherapy is noninferior to the radiation therapy at a margin of 0.1. In addition, the results suggested that the exact methods not only guarantee the level of the test, but can also be as powerful as the asymptotic method. For this example, the size of the exact unconditional test based on the $Z_1$ statistic is 4.97%, whereas the true type I error rate of the asymptotic $Z_1$ statistic could be as high as 5.78%. In addition, the two-sided 90% exact confidence interval based on the $Z_1$ statistic (see Chan and Zhang[9]) for the difference in response rate (chemotherapy–radiation therapy) was –0.035 to 0.117. Since the lower bound was much greater than –0.1, the confidence interval supports the conclusion of non-inferiority (at a one-sided 5% level).

### 11.2.3 Asymptotic Methods

When the sample size allows, use of asymptotic methods will generally be common for analysis of binary data. Historically, this has been attributable to computational needs for exact methods, as the number and complexity of computations have made asymptotic methods very attractive. In the computer age, the preference for asymptotic methods has probably outlived this explanation. Asymptotic methods can simplify/reduce a problem to its most basic components, thus making the method more understandable, particularly to nonstatisticians.

Asymptotic methods are appropriate when the sample size is large enough that the binomial distribution can be approximated by the normal distribution. In that case, the distribution of an observed proportion, $\hat{p}$, is approximately normal with mean $p$ and variance $p(1-p)/n$. A general rule of thumb is that as long as the sample size is at least 30 or 35, and the population proportion is such that at least 5 successes and at least 5 failures are expected, the normal approximation will be adequate. In Section 11.2.4, cases will be provided (including when that rule of thumb is satisfied) where the coverage probability for confidence intervals is noticeably lower than desired (i.e., the two-sided type I error rate for testing superiority is noticeably higher than desired).

The non-inferiority hypotheses can be tested using asymptotic methods analogously to the methods presented in Chapter 12 for continuous data. The test statistic of

$$Z = \frac{\hat{p}_E - \hat{p}_C + \delta}{se(\hat{p}_E - \hat{p}_C)} \tag{11.5}$$

will be distributed approximately as a standard normal variable under the null hypothesis in Expression 11.1; thus a value of $Z > z_{\alpha/2}$ will be sufficient evidence against the null hypothesis at the one-sided $\alpha/2$ level. More commonly (and equivalently), the testing procedure does not use a test statistic $Z$. Instead a two-sided $100(1 - \alpha)\%$ confidence interval for the true difference, $p_E - p_C$, is calculated. If the lower bound is larger than $-\delta$, then the null hypothesis is rejected in favor of the alternative hypothesis of non-inferiority.

There are many ways of calculating a large-sample confidence interval. The simplest expression is

$$\hat{p}_E - \hat{p}_C \pm \{z_{\alpha/2} \times se(\hat{p}_E - \hat{p}_C)\},$$

where $z_{\alpha/2}$ is the $100(1 - \alpha/2)$th percentile of the standard normal distribution and $se(\hat{p}_E - \hat{p}_C)$ is the standard error of the estimated difference in proportions. The standard error is commonly estimated by the unrestricted MLE of $\sqrt{\frac{\hat{p}_E(1 - \hat{p}_E)}{n_E} + \frac{\hat{p}_C(1 - \hat{p}_C)}{n_C}}$, which leads to a Wald's confidence interval for the true difference in proportions. However, not all possible values of $p_E$ and $p_C$ can be observed: in a study with $n_C$ and $n_E$ observations, only multiples of $1/n_E$ and $1/n_C$ can be observed. Thus, when there are large sample sizes, these $(n_E + 1)$ $(n_C + 1)$ possible outcomes are fairly dense within the unit square, the parameter space of $p_E$ and $p_C$. In cases with small sample sizes and/or probabilities of success near 0 or 1, this simple confidence interval can have suboptimal coverage probabilities and the associated test can reject the null hypothesis less often (or more often) than desired. In addition the unrestricted MLE of the variance is inconsistent with the null hypothesis, which restricts the true difference in proportions.

Hauck and Anderson[22] considered confidence intervals of the form $\hat{p}_E - \hat{p}_C \pm \{z_{\alpha/2} \times se(\hat{p}_E - \hat{p}_C) + CC\}$, where CC denotes continuity correction and $se(\hat{p}_E - \hat{p}_C)$ may or may not be based on the unrestricted MLE of the variance. Hauck and Anderson concluded that some adjustment is necessary, either in estimating the standard error or through use of a continuity correction or both, even if sample sizes are large. With minor restrictions on sample size, Hauck and Anderson recommended the unbiased estimate of standard error (i.e., using $n - 1$ in the denominators instead of $n$ as in the MLE) and also using a continuity correction of $1/\{2 \times \min(n_E, n_C)\}$. We note that this is based on two-sided coverage probabilities, not on the testing of a one-sided non-inferiority hypothesis of a nonzero difference. In addition, the estimate of standard error is not consistent with the null hypothesis, which restricts the relative values of $p_E$ and $p_C$.

Dunnett and Gent[8] discussed a test statistic with the same form as $Z$ in Equation 11.5, but with the variance estimated conditional on total number of successes. The values for $p_E$ and $p_C$ that are used to estimate the variance satisfy $p_E - p_C = -\delta$ and $n_E p_E + n_C p_C = n_E \hat{p}_E + n_C \hat{p}_C$. The solutions are $p_E = [k\hat{p}_E + \hat{p}_C - \delta]/(1 + k)$ and $p_C = [k\hat{p}_E + \hat{p}_C + k\delta]/(1 + k)$, where $k = n_E/n_C$ is the ratio of sample sizes of the two groups. Note that when $k = 1$, the solutions $p_E$ and $p_C$ satisfy $p_E + p_C = \hat{p}_E + \hat{p}_C$. Although this may be thought of as an improvement over the methods using unrestricted MLEs, they do not always produce valid estimates in the closed interval [0,1]. For example, when $\hat{p}_E = \hat{p}_C < \delta/2$ and the two groups have equal sample sizes, the solution for $p_E$ will be negative.

Testing based on the test statistic in Equation 11.5 using restricted MLEs of the standard error was discussed by Farrington and Manning.[5] That is, under the restriction that $p_E - p_C = -\delta$, the MLEs of $p_E$ and $p_C$ were found. The MLE for $p_C$ was found to be the unique solution in the open interval $(\delta, 1)$ to the third-degree maximum likelihood equation. The solution is

$$\tilde{p}_C = 2u\cos(w) - b/3a \text{ and } \tilde{p}_E = \tilde{p}_C - \delta, \qquad (11.6)$$

where
$v = b^3/(3a)^3 - bc/(6a^2) + d/(2a)$
$u = \text{sign}(v)[b^2/(3a)^2 - c/(3a)]^{\frac{1}{2}}$
$w = [\pi + \cos^{-1}(v/u^3)]/3,$

and

$$a = 1 + k,$$

$$b = -[1 + k + \hat{p}_C + k\hat{p}_E + \delta(k + 2)],$$

$$c = \delta^2 + \delta(2\hat{p}_C + k + 1) + \hat{p}_C + k\hat{p}_E,$$

$$d = -\hat{p}_C\delta(1 + \delta)$$

and $k$ is as defined above.

Although the algebra of the estimates of these MLEs appears somewhat complex, it is easily implemented in computer code. The corresponding test statistic is that given in Equation 11.4

$$Z_{FM} = Z_1 = \frac{\hat{p}_E - \hat{p}_C + \delta}{\sqrt{\dfrac{\tilde{p}_E(1-\tilde{p}_E)}{n_E} + \dfrac{\tilde{p}_C(1-\tilde{p}_C)}{n_C}}},$$

which has an approximate normal distribution for sufficient sample sizes when $p_E - p_C = -\delta$. Note that the form of the test statistic of the Farrington and Manning method is identical to that proposed by Dunnett and Gent—only the choice of $\tilde{p}_i$'s differ.

Farrington and Manning reported simulation results of three methods, which differed by how the variance is estimated. In the estimation of the variance, the methods differ only in the estimates of the proportions that are used: using the observed proportions $\tilde{p}_i$ to estimate variance (the unrestricted MLEs), conditioning on fixed marginal totals (as proposed by Dunnett and Gent), or the restricted MLEs as defined above. The simulation results showed that the method using the MLEs under the null hypothesis was preferable if an asymptotic test is desired. The method using unrestricted MLEs often gave dramatically different sample size requirements (either higher or lower), and the method of Dunnett and Gent occasionally failed to supply a meaningful number (i.e., occasionally provided estimates of proportions not between 0 and 1), which will be discussed again in Section 11.2.5.

Using the same summary data, Example 11.3 considers testing on the basis of the approaches of Hauck and Anderson, Dunnett and Gent, and Farrington and Manning.

### Example 11.3

Suppose an investigational antibiotic is being compared to a standard antibiotic in uncomplicated bacterial sinusitis. The margin was set to $\delta = 0.10$ (10 percentage points). The observed cure rates were 92/100 (92%) for the control group and 89/100 (89%) for the experimental group.

The 95% confidence interval for the difference in cure rates by Hauck and Anderson approach is (−0.117, 0.057). As −0.117 is less than −0.10, the null hypothesis would not be rejected and non-inferiority cannot be concluded.

Using the method of Dunnett and Gent, the variance will be estimated with values of $\tilde{p}_C = 0.955$ and $\tilde{p}_E = 0.855$. The resulting test statistic is 1.71 (one-sided $p$-value = 0.044), which is less than the critical value of 1.96 (for a one-sided $\alpha = 0.025$).

Using the method of Farrington and Manning, the variance will be estimated with values of $\tilde{p}_C = 0.941$ and $\tilde{p}_E = 0.841$. The resulting test statistic is 1.61 (one-sided $p$-value = 0.054), which is less than the critical value of 1.96 (for a one-sided $\alpha = 0.025$).

Regardless of the approach, non-inferiority cannot be concluded.

### 11.2.4 Comparisons of Confidence Interval Methods

For both a single proportion and a difference in proportions, we will discuss the results and conclusions of many papers that compare the probability coverage, type I error rates, and power of various methods. Most research involving a difference in proportions concerns two-sided testing. The results from two-sided testing will not directly apply to one-sided testing.

Wald's intervals for a single proportion and a difference in proportions may be the most commonly used confidence interval methods. For a single proportion, Wald's interval is given by

$$\hat{p} \pm z_{\alpha/2}\sqrt{\hat{p}(1-\hat{p})/n}.$$

For a difference in proportions, Wald's interval is given by

$$\hat{p}_E - \hat{p}_C \pm z_{\alpha/2}\sqrt{\hat{p}_E(1-\hat{p}_E)/n_E + \hat{p}_C(1-\hat{p}_C)/n_C}.$$

Many authors have discussed the poor coverage properties (and maintenance of a desired type I error probability) of Wald's confidence interval in both the single proportion and the difference in proportions settings.[23-29] We begin with some results involving a single proportion.

### 11.2.4.1 Inferences on a Single Proportion

On Wald's interval for a single proportion, Brown, Cai, and Dasgupta[29] reported that the "popular prescriptions the standard interval comes with are defective in several respects and are not to be trusted." The authors found that the 95% Wald's interval has an unpredictable, random coverage probability even for large sample sizes. More specifically, noted on the 95% Wald's interval of Brown, Cai, and Dasgupta[29] are the following:

- For a sample size or 100, the coverage probability of a 95% Wald's interval was 0.952 when $p = 0.106$ and 0.911 when $p = 0.107$.
- When $p = 0.005$, the coverage probability of a 95% Wald's interval was 0.945, 0.792, 0.945, 0.852, and 0.898 when $n = 591, 592, 953, 954$, and 1876, respectively.

The Wilson[30] interval, a popular alternative to Wald's interval, is found by inverting the standard large-sample normal test statistic (score test), which scales by the standard deviation for the sample proportion under the null hypothesis. The Wilson interval is $\{p_o : |\hat{p} - p_o|/\sqrt{p_o(1-p_o)} < z_{\alpha/2}\}$. This reduces to

$$\hat{p}\left(\frac{n}{n+z_{\alpha/2}^2}\right) + \frac{1}{2}\left(\frac{z_{\alpha/2}^2}{n+z_{\alpha/2}^2}\right) \pm z_{\alpha/2}\sqrt{\frac{1}{n+z_{\alpha/2}^2}\left[\hat{p}(1-\hat{p})\left(\frac{n}{n+z_{\alpha/2}^2}\right) + \left(\frac{1}{2}\right)\left(\frac{1}{2}\right)\left(\frac{z_{\alpha/2}^2}{n+z_{\alpha/2}^2}\right)\right]}$$

The Wilson interval has a midpoint that is a weighted average of $\hat{p}$ and 0.5 (weights $n$ and $z_{\alpha/2}^2$). In fact, the midpoint is the "proportion of successes" if $z_{\alpha/2}^2/2$ successes and $z_{\alpha/2}^2/2$ failures were added. This motivated Agresti and Coull[25] to apply the 95% Wald's confidence interval to an adjusted proportion that adds two successes and two failures to the mix ($z_{0.025}^2 \approx 4$). The resulting 95% confidence interval is

$$\tilde{p} \pm 1.96\sqrt{\tilde{p}(1-\tilde{p})/\tilde{n}}$$

where $\tilde{p} = (x+2)/\tilde{n}$ and $\tilde{n} = n+4$. Applying this idea to a $100(1-\alpha)\%$ confidence interval for arbitrary $\alpha$ yields the interval of $\tilde{p} \pm z_{\alpha/2}\sqrt{\tilde{p}(1-\tilde{p})/\tilde{n}}$, where $\tilde{p} = (x + z_{\alpha/2}^2/2)/\tilde{n}$ and $\tilde{n} = n + z_{\alpha/2}^2$. When this interval is applied to no data ($x = 0$, $n = 0$), the resulting interval is $[0, 1]$. Agresti and Coull reported substantial improvement in the coverage probability of this interval over Wald's interval for small sample sizes.

Brown, Cai, and Dasgupta[29] compared the probability coverage and interval lengths of several methods for constructing a confidence interval for a single proportion. They recommended the Wilson interval or the equal-tailed Jeffreys credible interval for small sample sizes ($n \leq 40$), and the interval of Agresti and Coull[25] for large sample sizes ($n > 40$). All of these intervals have instances where the coverage probability of the 95% interval is below 95%. For success rates fairly close to 0 and 1, the Jeffreys interval had a very small coverage probability. To improve the probability coverage in such cases, a modified version of the Jeffreys interval was proposed by Brown, Cai, and Dasgupta.[29] When $x = 0$, define the upper limit of the interval by $1 - (\alpha/2)^{1/n}$ and when $x = n$ define the lower limit of the interval by $(\alpha/2)^{1/n}$. For the details and more on the Jeffreys credible interval, see Appendices A.2.2 and A.3.

### 11.2.4.2 Inferences for a Difference in Proportions

Hauck and Anderson[22] performed simulations to evaluate the coverage probabilities of two-sided confidence interval methods based on combinations of two forms for the standard error and four forms for the continuity correction factor. The confidence intervals are expressed as

$$\hat{p}_E - \hat{p}_C \pm (z_{\alpha/2}\text{se}(\hat{p}_E - \hat{p}_C) + CC).$$

The choices for standard errors are the unrestricted MLE of the standard error and a modified version of such that replaces $n_i$ with $n_i - 1$ for $i = E, C$. The possible corrections are: (1) no correction ($CC = 0$), (2) Yates correction ($CC = 1/(2n_E) + 1/(2n_C)$), (3) a correction of Schouten et al.[31] ($CC = 1/(2 \max(n_E, n_C))$), and (4) a correction of Hauck and Anderson ($CC = 1/(2 \min(n_E, n_C))$). The cases considered had minimum expected cell counts ranging from 2 to 15 with the smallest group size ranging from 6 to 100.

As mentioned in Section 11.2.3, Hauck and Anderson[22] recommended the use of the Hauck–Anderson correction with the modified version of the standard error. When the desired confidence level was 90% or 95% and the minimum expected cell count was at least 3, that method gave coverage probabilities close to the desired level. Wald's interval with a Yates correction also performed reasonably well but was more conservative. Wald's interval without any correction did not provide adequate coverage at any sample size studied, and Hauck and Anderson recommended against its use. When the desired confidence level was 99% and the minimum expected cell count was at least 5, their recommended method and Wald's interval with a Yates correction performed equally as well. No method studied performed consistently well when the minimum expected cell count was 2. Tu[32] preferred Wald's interval with a Hauck–Anderson continuity correction for equivalence testing.

Li and Chuang-Stein[33] evaluated and compared the type I error rate in non-inferiority testing of a difference of two proportions using Wald's interval with and without a Hauck–Anderson continuity correction. Their evaluation was based on equal allocation for "sample sizes relevant to the confirmatory trials." The sample sizes were between 100 and 300. For the cases studied where all the expected cell counts (success and failures for both arms) were all greater than 15 and 2.5% is the one-sided targeted type I error rate, the estimated type I error rate for the standard Wald's interval was between 2.3% and 2.75%. Wald's interval with Hauck–Anderson continuity correction produced type I error rates consistently below 2.5%. For the cases studied where some of the expected cell counts were less than 15 and 2.5% is the one-sided targeted type I error rate, the estimated type I error rate for the standard Wald's interval could go beyond 2.75%. The inflation appeared to increase as the smallest expected cell count approached 5. In these cases, Wald's interval with Hauck–Anderson continuity correction performed fairly well and produced type I error rates below 2.75%. Li and Chuang-Stein[33] recommend using Wald's interval without a continuity correction when the expected frequency of all cell counts is at least 15. Otherwise, they recommend implementing the Hauck–Anderson correction.

Newcombe[28] compared the coverage probabilities and expected lengths of 11 methods for determining a confidence interval for the difference in proportions. A tail area profile likelihood-based method, and the methods of Mee[34] and Miettinen and Nurminen,[35] which invert test statistics that use standard errors restricted to the specified difference in proportions, all performed well but were either difficult to compute or required a computer program. Newcombe recommended a method that combined Wilson score intervals for the two proportions either with or without a continuity correction. The Newcombe–Wilson $100(1 - \alpha)\%$ confidence interval without a continuity correction is given by $(L, U)$ where

$$L = (p_E - p_C) - z_{\alpha/2}\sqrt{l_E(1 - l_E)/n_E + u_C(1 - u_C)/n_C},$$

$$U = (p_E - p_C) + z_{\alpha/2}\sqrt{u_E(1 - u_E)/n_E + l_C(1 - l_C)/n_C},$$

$l_E$ and $u_E$ are the roots of $|p_E - y/n_E| = z_{\alpha/2}\sqrt{p_E(1-p_E)/n_E}$ and $l_C$ and $u_C$ are the roots of $|p_C - x/n_C| = z_{\alpha/2}\sqrt{p_C(1-p_C)/n_C}$. For the Newcombe–Wilson interval with a continuity correction, $l_E$ and $u_E$ are the limits of the interval $\left\{p : |p - y/n_E| - 0.5/n_E \leq z_{\alpha/2}\sqrt{p(1-p)/n_E}\right\}$ and $l_C$ and $u_C$ are the limits of the interval $\left\{p : |p - x/n_C| - 0.5/n_C \leq z_{\alpha/2}\sqrt{p(1-p)/n_C}\right\}$.

Motivated by Agresti and Coull[25], Agresti and Caffo[26] proposed to use the corresponding Wald's interval after adding one success and one failure to each treatment group for the 95% confidence interval of the difference of proportions. This addition of observations performed best on the basis of their simulations using pairs of true probabilities selected randomly over the unit square (uniform distribution) and group sizes selected randomly (uniform distribution) over {10, 11, ..., 30}. The resulting 95% confidence interval is

$$\tilde{p}_E - \tilde{p}_C \pm 1.96\sqrt{\tilde{p}_E(1-\tilde{p}_E)/\tilde{n}_E + \tilde{p}_C(1-\tilde{p}_C)/\tilde{n}_C}$$

where $\tilde{p}_i = (x_i + 1)/\tilde{n}_i$ and $\tilde{n}_i = n_i + 2$ for $i = E, C$. Applying this idea to a 100(1 − α)% confidence interval for arbitrary α leads to the interval $\tilde{p}_E - \tilde{p}_C \pm z_{\alpha/2}\sqrt{\tilde{p}_E(1-\tilde{p}_E)/\tilde{n}_E + \tilde{p}_C(1-\tilde{p}_C)/\tilde{n}_C}$ where $\tilde{p}_E = (y + z_{\alpha/2}^2/4)/\tilde{n}_E$, $\tilde{p}_C = (x + z_{\alpha/2}^2/4)/\tilde{n}_C$, and $\tilde{n}_i = n_i + z_{\alpha/2}^2/2$ for $i = E, C$. When this interval is applied to no data ($x = y = 0$ and $n_i = 0$ for $i = E, C$), the resulting interval is [−1, 1]. Note that for a common sample size, the middle of the Agresti and Caffo interval is closer to zero than that of Wald's interval (i.e., $|\tilde{p}_E - \tilde{p}_C| \leq |\hat{p}_E - \hat{p}_C|$). However, this need not be true for uneven sample sizes.

Santner et al.[36] compared the small-sample probability coverage and expected lengths of five methods for determining a 90% confidence interval for the difference of proportions. The methods include the asymptotic method of Miettinen and Nurminen,[35] which is based on the score statistic, and the exact methods of Agresti and Min,[15] Chan and Zhang,[9] Coe and Tamhane,[17] and Santner and Yamagami.[18] For seven pairs of sample sizes (three cases of balanced allocation and four cases of unbalanced allocation were examined), the average (exact) probability coverage was calculated (based on binomial distributions) across 10,000 pairs of $(p_E, p_C)$ selected evenly across the unit square. The overall sample size ranged from 20 to 70. The authors conclude that the exact method of Coe and Tamhane performed the best, and the asymptotic method of Miettinen and Nurminen performed the worst. The authors recommended the use of the Coe and Tamhane method; when that method is not available, either the method of Agresti and Min or the method of Chan and Zhang is recommended. The use of any of these five methods was strongly recommended by the authors in the abstract of the paper. However, in the conclusions of paper, use of the methods of Santner and Yamagami and Miettinen and Nurminen was discouraged.

Dann and Koch[37] reviewed the ability of various non-inferiority tests involving a risk difference to maintain the proper type I error rate for different allocation rates. The power is compared among procedures that maintain the desired type I error rate. Their results suggested that the best test to use depends on the allocation ratio. The accuracy of the corresponding sample-size formulas was also checked. The assessments were based on one-sided 97.5% confidence intervals. The methods compared were Wald's, Farrington–Manning, Agresti–Caffo, and Newcombe–Wilson intervals, with and without a continuity correction. Control success rates (for positive outcomes) between 0.60 and 0.95 were examined with non-inferiority margins ($\delta$) of 0.05, 0.075, and 0.10. The sample sizes chosen provide approximately 85% power at no difference in the control and experimental success rates. The experimental versus control allocation ratios considered were 1:2, 1:1, 3:2, 2:1, and 3:1. For each case, 100,000 simulations were performed. Simulated type I error rates were regarded as appropriate if between 0.0225 and 0.0275.

With respect to the type I error rate and power, the continuity-corrected Newcombe–Wilson method was the most conservative. We have the following observations from Figure 1 of Dann and Koch[37]:

- For all allocation ratios, the continuity corrected Newcombe–Wilson method maintained the type I error rate at or below 0.025 (i.e., between 0.01 and 0.025). As the allocation ratio increased, the distribution of type I error rates for the continuity-corrected Newcombe–Wilson method became narrower and closer to 0.025.

- For allocation ratios of 1:2 and 1:1 the Newcombe–Wilson method (without a continuity correction) maintained type I error rates below 0.0275 and within 0.0225–0.0275, respectively. As the allocation ratio increased, the type I error rate of Newcombe–Wilson method tended to increase.

- For a given allocation ratio, the distribution of type I error rates for the Farrington–Manning method was slightly smaller than that of the Newcombe–Wilson method. For allocation ratios of 1:2, 1:1, and 3:2, the Newcombe–Wilson method (without a continuity correction) maintained type I error rates below 0.0275, within 0.0225–0.0275 and 0.0225–0.0275, respectively. As the allocation ratio increased, there was a modest increasing trend in the type I error rate of the Farrington–Manning method.

- For a 1:2 allocation ratio, Wald's method produced type I error rates between 0.025 and 0.05. As the allocation ratio increased, the type I error rate of Wald's method decreased. Wald's maintained type I error rates within 0.0225–0.0275 for a 3:2 allocation ratio and type I error rates at or below 0.025 for 2:1 and 3:1 allocation ratios.

- For any given allocation ratio, the distribution of type I error rates for the Agresti–Caffo method was concentrated between 0.025 and the distribution of type I error rates of Wald's method. How the type I error rates of the Agresti–Caffo method compare also depend on the control success.

Dann and Koch[37] seem to prefer Wald's method and the Agresti–Caffo method when the allocation ratio is large (3:2, 2:1, or 3:1), and the Farrington–Manning method or the Newcombe–Wilson method for smaller allocation ratios (1:2 or 1:1). It should be noted that their results do not directly apply to control success rates (for positive outcomes) less than 0.5.

We close this section with Example 11.4, which applies the 95% confidence intervals for the difference in proportions from various methods to the results from a hypothetical clinical trial.

### Example 11.4

Suppose there are 131 successes among 150 subjects in the experimental arm and 135 successes among 150 subjects in the control arm. For various asymptotic and Bayesian methods (see Section 11.5), the corresponding 95% two-sided confidence intervals for $\hat{p}_E - \hat{p}_C$ is determined. The results are provided in Table 11.2 where the methods are listed in decreasing order with respect to the lower confidence limit. Four of the nine methods would lead to a non-inferiority conclusion if the $\delta = 0.10$ (the lower limit of the Newcombe–Wilson interval is −0.1002).

### 11.2.5 Sample Size Determination

The selection of sample size is based primarily on the desired power of the test at a reasonable possibility ("selected alternative") in the alternative hypothesis. It should be noted that many standard software packages for sample size and power calculations are not designed for non-inferiority

**TABLE 11.2**

95% Confidence Intervals for Difference in Proportions

| Method | 95% Confidence Interval |
| --- | --- |
| Wald | (−0.098, 0.045) |
| Zero prior Bayesian[a] | (−0.099, 0.045) |
| Jeffreys | (−0.099, 0.045) |
| Agresti–Caffo | (−0.099, 0.046) |
| Newcombe–Wilson | (−0.100, 0.047) |
| Newcombe–Wilson with CC | (−0.100, 0.047) |
| Farrington–Manning[b] | (−0.101, 0.047) |
| Wald with Hauck and Anderson CC | (−0.102, 0.048) |
| Wald with Yates CC | (−0.105, 0.052) |

[a] Based on the resulting posterior distributions after the prior $\alpha \to 0$ $\beta \to 0$ for each arm.
[b] Standard errors based on the null restricted estimates of the proportions.

analyses. Care is urged in interpreting the results when trying to design a non-inferiority trial.

Sample sizes for non-inferiority trials are often developed under the assumption that the control and experimental therapies have identical efficacy. This assumption will not always lead to an adequately designed trial. For indications in which the active control has a very large effect compared to placebo, the assumption of identical efficacy may be optimistic and would thus lead to an undersized, underpowered trial. In such a situation, an experimental treatment might have less efficacy than the active control but still be far superior to a placebo. For indications in which the control therapy has a small effect compared to placebo, the non-inferiority margin is small. In such situations, assuming identical efficacy of the active control and experimental therapies will lead to a large sample size. However, in this setting, the assumption of identical efficacy means that the experimental therapy has a small effect compared to placebo, and thus a superiority trial versus placebo would also require a large sample size. In such cases, it makes sense to power the non-inferiority trial at an appropriate superiority alternative compared to the control therapy. Assuming that the experimental therapy provides a modest benefit in efficacy compared to the control therapy can lead to a marked reduction in sample size.

If there is evidence that the experimental treatment may be superior to the control therapy, additional consideration is needed for determining an appropriate sample size. If the objective of demonstrating non-inferiority is sufficient, a smaller sample size can be used by assuming that the experimental treatment is slightly superior to the active control. However, if it is desirable to demonstrate that the experimental therapy is superior to the active control, the trial can be powered for superiority. The value of an additional conclusion of superiority needs to be weighed against the associated increases in the number of patients, study sites, and study time.

Since the standard deviation for the sample proportion of successes depends on the true probability of a success, sample-size formulas for proportions are more complicated than those for means. The respective sample-size formulas for the risk difference, relative risk, and odds ratios are not only functions of the selected alternative values for the risk difference, relative risk, and odds ratios, but also depend on the specific success probabilities for the two arms. In addition, when a normalized test statistic is used, the sample-size formula depends on whether the estimated standard error used for the denominator is an unrestricted estimate or an estimate restricted to the null hypothesis (or more specifically, the boundary of the null hypothesis).

Let $p_1$ and $p_2$ represent success rates for the experimental and control arms, respectively, where $\Delta_a = p_1 - p_2 > -\delta$. Then let $\sigma_a = \sqrt{\dfrac{p_1(1-p_1)}{n_E} + \dfrac{p_2(1-p_2)}{n_C}}$ and $\sigma_0 = \sqrt{\dfrac{p_1'(1-p_1')}{n_E} + \dfrac{p_2'(1-p_2')}{n_C}}$, where $(p_1', p_2')$ either represents for some criteria that pair of probabilities closest to $(p_1, p_2)$ satisfying $p_1' - p_2' = -\delta$ (when a restricted estimate of the standard error is used), or $p_1' = p_1$ and $p_2' = p_2$ (when an unrestricted estimate of the standard error is used). Then the power

is approximately $P\left(\dfrac{\hat{p}_E - \hat{p}_C + \delta}{\sigma_o} > z_{\alpha/2}\right) = P\left(\dfrac{\hat{p}_E - \hat{p}_C - \Delta_a}{\sigma_a} > \dfrac{z_{\alpha/2}\sigma_o - \delta - \Delta_a}{\sigma_a}\right)$

$\approx \Phi\left(\dfrac{\Delta_a + \delta - z_{\alpha/2}\sigma_o}{\sigma_a}\right).$

For a desired power of $1 - \beta$, the right-hand term above is set equal to $\Phi(z_\beta)$. Simplifying the equation leads to

$$z_\beta \sigma_a + z_{\alpha/2}\sigma_o = \Delta_a + \delta \tag{11.7}$$

For $k = n_E/n_C$, solving for $n_C$ yields

$$n_C = \left(\frac{z_\beta\sqrt{p_1(1-p_1)/k + p_2(1-p_2)} + z_{\alpha/2}\sqrt{p_1'(1-p_1')/k + p_2'(1-p_2')}}{\Delta_a + \delta}\right)^2. \tag{11.8}$$

For the analyses proposed in previous papers,[5,35] using $p_1' = (p_1 + p_2 - \delta)/2$ and $p_2' = (p_1 + p_2 + \delta)/2$ or using $p_1' = (kp_1 + p_2 - \delta)/(1 + k)$ and $p_2' = (kp_1 + p_2 + k\delta)/(1 + k)$ could also be appropriate, as they more closely match the analysis method. Otherwise, $(p_1', p_2')$ can be selected using some rule for determining the pair of $(p_1', p_2')$ in the null hypothesis that is the most difficult to reject when $(p_1, p_2)$ is the true pair of the probabilities of a success. The change in the estimated sample size may not be dramatically affected unless $\delta$ is very large, or $p_1$ and $p_2$ are close to 0 or 1. When $\delta = 0$, corresponding to a superiority analysis, it is common to use $p_1' = p_2' = (p_1 + p_2)/2$ or $p_1' = p_2' = (kp_1 + p_2)/(1 + k)$. Otherwise, the approaches of Farrington and Manning[5] and Miettinen and Nurminen[35] to obtain MLEs of the true success rates restricted to the null hypothesis can be adapted to determine $p_1'$ and $p_2'$. This is accomplished by treating $p_1$ and $p_2$ as the observed success rates.

The use of the sample-size formula in Equation 11.8 is illustrated in Example 11.5.

### Example 11.5

Suppose an investigational antibiotic is being compared with a standard antibiotic in uncomplicated bacterial sinusitis. The standard is expected to cure approximately 85% of all treated cases, and the non-inferiority margin has been determined to be $\delta = 0.10$. The test will be one-sided at $\alpha = 0.025$, and 90% power is desired. An obvious starting point for the sample size calculation is to assume that the investigational antibiotic also cure 85% of all treated cases. In this situation,

$p_1 = p_2 = 0.85$, $\Delta_a = 0$, $p_1' = (p_1 + p_2 - \delta)/2 = 0.80$, $p_2' = (p_1 + p_2 + \delta)/2 = 0.90$, $z_{0.025} = 1.96$, and $z_{0.10} = 1.28$. Then by 11.8, the required sample size per arm for a one-to-one randomization ($k = 1$) is calculated as $16.26^2 = 264.5$. Thus around 265 subjects should be randomized to each treatment group. For $p_1' = p_1 = 0.85$ and $p_2' = p_2 = 0.85$, the calculated sample size is 268 subjects per arm.

A new antibiotic might be developed to have a higher cure rate than currently available treatments. If the investigational antibiotic might cure 95% of all cases ($p_1 = 0.95$ $p_2 = 0.85$, $\Delta_a = 0.10$, $p_1' = (p_1 + p_2 - \delta)/2 = 0.850$, $p_2' = (p_1 + p_2 + \delta)/2 = 0.95$), then the required sample size is only 46 per treatment arm by 11.8. This sample size may be low enough to cause concerns with whether the difference in sample proportions has a normal distribution. A sample size of 200 per group would provide 90% power to show that the experimental treatment is superior with a greater power to show non-inferiority.

Alternatively, a new antibiotic may be expected to have a slightly lower cure rate than currently available treatments, but have some other advantage such as better tolerability. If the investigational antibiotic might cure 80% of all cases, the required sample size to show non-inferiority is about 1200 per treatment group ($p_1 = 0.80$ $p_2 = 0.85$, $\Delta_a = -0.05$, $p_1' = (p_1 + p_2 - \delta)/2 = 0.775$, $p_2' = (p_1 + p_2 + \delta)/2 = 0.875$).

### 11.2.5.1 Optimal Randomization Ratio

The standard deviation for a sample proportion depends on the true proportion and would thus be different for the selected null and alternative probabilities. Because of this, the randomization or allocation ratio that minimizes the overall study size will seldom equal 1, even for a superiority analysis. Hilton[38] showed that a balanced randomization allocation is optimal only for superiority trials when $\sigma_a = \sigma_o$ (where $p_1' = p_2' = (p_1 + p_2)/2$); otherwise, some imbalanced allocation is optimal. By Equation 11.8, for an experimental to control allocation ratio of $k$, the overall study size is given by

$$\left( \frac{z_\beta \sqrt{p_1(1-p_1)/k + p_2(1-p_2)} + z_{\alpha/2}\sqrt{p_1'(1-p_1')/k + p_2'(1-p_2')}}{\Delta_a + \delta} \right)^2 (1+k). \quad (11.9)$$

The optimal $k$ that minimizes Expression 11.9 can be found using calculus by taking a derivative of Expression 11.9 with respect to $k$, setting the result equal to zero and then solving for $k$, or by a "grid search" by evaluating Expression 11.9 for many candidates for $k$.

Example 11.6 determines the optimal allocation ratio using the assumptions in Example 11.5

### Example 11.6

Consider the case in Example 11.5 where $p_1 = p_2 = 0.85$, $\Delta_a = 0$, $p_1' = (p_1 + p_2 - \delta)/2 = 0.80$, $p_2' = (p_1 + p_2 + \delta)/2 = 0.90$, $z_{0.025} = 1.96$, and $z_{0.10} = 1.28$. The optimal allocation ratio under those assumptions is $k = 1.187$, where $n_E = 286$ and $n_C = 240$. If instead $p_1' = (kp_1 + p_2 - \delta)/(1 + k)$, $p_2' = (kp_1 + p_2 + k\delta)/(1 + k)$ are used, the optimal allocation ratio under those assumptions is $k = 1.145$, where $n_E = 278$ and $n_C = 242$. The values for $p_1'$ and $p_2'$ at $k = 1.145$ are 0.803 and 0.903, respectively. Whenever $p_1' = p_1 = p_2' = p_2$, the optimal allocation ratio will be $k = 1$ (and thus 268 subjects per arm as noted above).

Before choosing a particular unbalanced allocation, its impact on the evaluation of other endpoints needs to be considered. Although an unbalanced allocation may reduce the required sample size for the primary endpoint, it may not be optimal for the evaluation of secondary endpoints and/or safety endpoints.

Dann and Koch[37] found that for experimental to control allocation ratios of 1:2 and 1:1 (allocation ratios for which the Farrington–Manning method was recommended), the Farrington–Manning method for a risk difference maintained the desired type I error rate with a corresponding sample-size formula that is always conservative (the simulated power was larger than the calculated power based on normal approximation). Also, for experimental to control allocation ratios of 3:2, 2:1, and 3:1, Wald's sample-size formula was appropriate to design a clinical trial for a risk difference and was slightly conservative.

Finally, we note that the relationship in Equation 11.7 holds in general (see Figure 11.1). Consider testing $H_0: \eta \le \eta_o$ versus $H_a: \eta > \eta_o$ at a one-sided level of $\alpha/2$ based on the estimator $\hat{\eta}$. Suppose when $\eta = \eta_o$ ($\eta = \eta_a$), $\hat{\eta}$ has an approximate normal distribution with standard deviation $\sigma_o$ ($\sigma_a$). Let $1 - \beta$ denote the power of the test when $\eta = \eta_a$. Let $\eta^*$ be that value for which statistical significance is reached if and only if the value for $\hat{\eta}$ is less than $\eta^*$. Since the one-sided level is $\alpha/2$, $\eta_o + z_{\alpha/2}\sigma_o = \eta^*$, and since the power is $1 - \beta$ when $\eta = \eta_a$, $\eta^* = \eta_a - z_\beta\sigma_a$ (see Figure 11.1). Therefore, $z_\beta\sigma_a + z_{\alpha/2}\sigma_o = \eta_a - \eta_o$.

Since the standard error in the denominator of a test statistic involving proportions is not constant, there is no specific value for the estimate (i.e., $\eta^*$) on the boundary of statistical significance. However, sample-size calculations for inferences involving proportions generally reduce to approximations that are represented by Figure 11.1. We will also use this relationship for sample-size formulas involving a relative risk or an odds ratio.

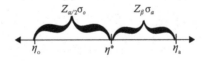

**FIGURE 11.1**
Relationship between null and selected alternative.

## 11.3 Fixed Thresholds on Ratios

In this section we will consider the situation in which a margin has been chosen on the basis of the ratio of two proportions, which is often referred to as the *relative risk* or *risk ratio*.

### 11.3.1 Hypotheses and Issues

Let $\theta = p_E/p_C$ denote the relative risk. When a success is a positive outcome, the hypotheses for testing that the experimental therapy is noninferior to the control therapy based on a prespecified threshold of $\theta_o$ ($0 < \theta_o \leq 1$) are

$$H_o: \theta \leq \theta_o \text{ vs. } H_a: \theta > \theta_o \tag{11.10}$$

That is, the null hypothesis is that the success rate of the experimental arm is smaller than a prespecified fraction, $\theta_o$, of the success rate of the control arm, whereas the alternative is that the active control is superior by a smaller amount, the two treatments are identical, or the experimental treatment is superior. When $\theta_o = 1$, the hypotheses in Expression 11.10 reduce to classical one-sided hypotheses for a superiority trial. If a "success" is a negative outcome, the roles of $p_C$ and $p_E$ in the hypotheses in Expression 11.10 would be reversed, leading to hypotheses that can expressed as $H_o: \theta = p_E/p_C \geq \theta_o$ versus $H_a: \theta = p_E/p_C < \theta_o$, where $\theta_o \geq 1$.

In the simplest application with no covariates, a confidence interval can be calculated for $p_E/p_C$ and non-inferiority is concluded (the null hypothesis in Expression 11.10 is rejected) if the lower bound of the confidence interval is greater than $\theta_o$.

### 11.3.2 Exact Methods

Many asymptotic tests of non-inferiority using the relative risk have been proposed in the literature. However, their performance in small-sample studies is uncertain and the test procedures may be too liberal owing to type I error inflation. In these situations, exact test procedures are desirable as they guarantee control of the type I error rate. In this section we will discuss two exact test procedures: one unconditional exact test based on the score test statistic and one conditional exact test based on the Poisson approximation for rare disease endpoints.

Similar to the exact unconditional test based on the $Z_1$ statistic in Equation 11.4, an exact unconditional test for Expression 11.10 has been developed on the basis of the following statistic:

$$Z_2 = Z_2(X,Y) = \frac{\hat{p}_E - \theta_o \hat{p}_C}{\left\{\tilde{p}_E\left(1-\tilde{p}_E\right)/n_E + \theta_o^2 \tilde{p}_C\left(1-\tilde{p}_C\right)/n_C\right\}^{1/2}} \tag{11.11}$$

where $\hat{p}_C$ and $\hat{p}_E$ are the observed response rates and $\tilde{p}_C$ and $\tilde{p}_E$ are the MLEs of $p_E$ and $p_C$, respectively, under the constraint $p_E = \theta_o p_C$ given in the null hypothesis in Expression 11.10. The closed form solutions for $\hat{p}_C$ and $\hat{p}_E$ are given in Farrington and Manning's study[5] and in Expression 11.18 of this text. Since large values of $Z_2$ favor the alternative hypothesis, the tail region includes tables whose $Z_2$ values are greater than or equal to $z_{obs}$, the $Z_2$ value for the observed table. Therefore, the exact $p$-value is calculated as

$$p\text{-value} = \max_{p \in D} P((x,y): Z_2 \geq z_{obs} \mid H_o, p) \tag{11.12}$$

where $p$ $(=p_C)$ is the nuisance parameter with the domain $A = [0, \theta_o]$, and the probability is evaluated using the following null probability function

$$P(x = i, y = j \mid H_o) = \binom{n_C}{i}\binom{n_E}{j} \theta_o^{-i} p^{i+j}(1-\theta_o^{-1}p)^{n_C-i}(1-p)^{n_T-j}.$$

For a nominal $\alpha$-level test, the CR and true size can be calculated in a similar fashion as for the exact unconditional test using the difference measure described in Section 11.2.2. In the special case where $\theta_o = 1$, the hypotheses in Expression 11.10 are those in Expression 11.1 with $\delta = 0$. In this special case, the $Z_1$ and $Z_2$ statistics are identical. Chan and Bohidar[39] have studied the utility of this exact unconditional test in designing clinical trials and found that the empirical performance of this exact test compares very favorably with its asymptotic counterpart ($Z_2$ test) in terms of type I error rate, power, and sample size under a wide range of true parameter values.

### Example 11.7

Chan[12] reanalyzed the data in Example 11.2 using the relative risk measure to show non-inferiority based on the criterion requiring that the tumor response rate to the chemotherapy treatment ($p_E$) be greater than 90% of the response to the radiation therapy ($p_C$). This corresponds to a threshold of $\theta_o = 0.9$ for the relative risk. Since $\hat{p}_E = 0.943$ [83/88] and $\hat{p}_C = 0.908$ [69/76], the observed relative risk is $\hat{\theta} = 1.039$. The MLEs when $p_E/p_C = 0.9$ are $\tilde{p}_C = 0.946$ and $\tilde{p}_E = 0.851$, which gives $Z_2 = 2.835$ from Equation 11.11. From Equation 11.12 the exact unconditional test based on $Z_2$ yielded a $p$-value of 0.0028, compared with the asymptotic $p$-value of 0.0024. Both tests strongly supported the conclusion of non-inferiority. At the one-sided 5% level, the size of the exact test is 4.82%, whereas the type I error rate of the asymptotic $Z_2$ test is 5.32% when $p_C = 0.9$ and the size of the test is approximately 5.59%.

### 11.3.2.1 Exact Conditional Non-Inferiority Test

For rare disease endpoints, a large-scale study is usually required to demonstrate non-inferiority or equivalence between two treatments. In such cases, it is reasonable to assume that the number of disease events in the control and experimental groups ($X$ and $Y$, respectively) are approximately distributed as independent Poisson random variables with respective means $\lambda_C = n_C p_C$ and $\lambda_E = n_E p_E$. Since the rates of interest are for negative outcomes, the hypotheses to be tested are $H_o: \theta = p_E/p_C \geq \theta_o$ versus $H_a: \theta = p_E/p_C < \theta_o$, where $\theta_o \geq 1$. Let $S = X + Y$ be the total number of disease cases observed in the study. Then conditional on $S = s$, the number of disease cases in the experimental group ($Y$) is distributed as Binomial ($s$, $\phi$), where

$$\phi = \frac{\lambda_E}{\lambda_C + \lambda_E} = \frac{n_E p_E}{n_C p_C + n_E p_E} = \frac{\theta}{\theta + u}$$

and $u = n_C/n_E$.

Since $\phi$ is increasing in $\theta$, the non-inferiority hypotheses in Expression 11.10 are equivalent to

$$H_o: \phi \geq \phi_o \text{ vs. } H_a: \phi < \phi_o \qquad (11.13)$$

where $\phi_o = \theta_o/(\theta_o + u)$. Thus, inferences can be based on a simple exact test involving a one-sample binomial distribution. Suppose $y_{obs}$ is the number of disease cases observed in the experimental group, then the exact $p$-value conditional on the total number of disease cases $S = s$ is

$$p\text{-value} = \Pr[Y < y_{obs} \mid Y \sim \text{Binomial}(s, \phi_o)] = \sum_{k=0}^{y_{obs}} \frac{s!}{k!(s-k)!} \phi_o^k (1 - \phi_o)^{s-k}. \qquad (11.14)$$

For an $\alpha$-level test, the critical value $y_\alpha$ can be determined as the largest value satisfying that $\Pr\{Y \leq y_\alpha \mid Y \sim \text{Binomial}(s, \phi_o)\}$ is as close as possible to $\alpha$ from below without exceeding it. The power conditional on $S = s$ for testing the hypotheses in Expression 11.13 against a specific alternative $\phi = \phi_1 < \phi_o$ is then calculated as $\Pr\{Y \leq y_\alpha \mid y \sim \text{Binomial}(s, \phi_1)\}$, where $\phi_1 = \theta_1/(\theta_1 + u)$.

Note that Equation 11.14 could also be evaluated via the $F$-distribution using the following relationship (see, e.g., Johnson, Kotz, and Kemp,[40] p.110):

$$\sum_{k=0}^{y} \frac{s!}{k!(s-k)!} \phi^k (1 - \phi)^{s-k} = F_{v_1, v_2}\left(\frac{v_2 \phi}{v_1 (1 - \phi)}\right)$$

where $F_{v_1, v_2}(\cdot)$ is the central $F$-distribution function with parameters $v_1 = 2(y + 1)$ and $v_2 = 2(s - y)$.

In addition, conditional on $S = s$, an exact confidence interval for $\theta$ is determined from the exact confidence interval for $\phi$. For example, using the Clopper and Pearson[13] method, a two-sided $100(1 - \alpha)\%$ test-based exact confidence interval, $(\phi_L, \phi_U)$, for $\phi$ can be constructed as

$$\phi_L = \frac{v_1 F^{-1}_{v_1,v_2}(\alpha/2)}{v_2 + v_1 F^{-1}_{v_1,v_2}(\alpha/2)},$$

where
$v_1 = 2y_{obs}$
$v_2 = 2(s - y_{obs} + 1)$,

and

$$\phi_U = \frac{v_1 F^{-1}_{v_1,v_2}(1-\alpha/2)}{v_2 + v_1 F^{-1}_{v_1,v_2}(1-\alpha/2)},$$

where
$v_1 = 2(y_{obs} + 1)$,
$v_2 = 2(s - y_{obs})$.

Then, a $100(1 - 2\alpha)\%$ exact confidence interval for the relative risk $\theta$ $(\theta_L, \theta_U)$ is given by

$$\theta_L = \frac{u\phi_L}{1 - \phi_L}, \quad \theta_U = \frac{u\phi_U}{1 - \phi_U} \tag{11.15}$$

This exact conditional method can be applied to design a study with a goal to obtain a fixed total number of events instead of running for a fixed duration. Once the desired total number of events $(S)$ is fixed, the power of the study depends on incidence rates only through the relative risk $(\theta = p_E/p_C)$; thus, one can avoid the situation potentially encountered in a fixed-duration trial where the anticipated power is not achieved at the end of the trial because the number of events is too few owing to unexpectedly low incidence rates. Since the unconditional expected value of $S$ is $(n_C p_C + n_E p_E)$, the expected number of subjects required for the study can be estimated on the basis of the incidence rate in the control group $(p_C)$ and the relative risk $(\theta = \theta_1)$ under the alternative hypothesis:

$$n_E \approx \frac{S}{(u + \theta_1)p_C} \tag{11.16}$$

In addition, the Poisson approximation allows the exact conditional method to easily handle person-time or exposure-time data and account for potential differential follow-up between treatment groups. The only modification required is to change the constant $u$ in Equations 11.15 and 11.16 to be the ratio of the respective total person-times instead of the respective sample sizes. This flexibility is important in long-term studies where differential follow-up between treatment groups often occurs, and as a result the asymptotic $Z_2$ test can no longer be used appropriately.

## Example 11.8

Chan[12] used the exact conditional method to design a non-inferiority trial comparing a new hepatitis A vaccine with immune globulin (IG, standard treatment C) in postexposure prophylaxis of hepatitis A disease. IG is believed to have approximately 90% efficacy for postexposure prophylaxis. However, IG is a derived blood product, and thus there are concerns about its safety and purity in addition to its short-lived protection. In contrast, the hepatitis A vaccine has been demonstrated to be safe and highly efficacious ($\approx$100%) in preexposure prophylaxis,[41] and capable of inducing long-term protective immunity against hepatitis A in healthy subjects.[42] Recognizing the potential long-term benefit of the vaccine, investigators of this study intended to show that the vaccine is noninferior to IG in terms of postexposure efficacy. Since we are dealing with negative outcomes (have the disease), the hypotheses to be tested are $H_0$: $\theta = p_E/p_C \geq \theta_o$ versus $H_a$: $\theta = p_E/p_C < \theta_o$, where $\theta_o \geq 1$. If non-inferiority is established ($\theta < \theta_o$), one can infer that the new vaccine has reasonable efficacy ($\pi_E$) for postexposure prophylaxis on the basis of the following indirect argument:

$$\pi_E = 1 - \frac{p_E}{p_U} = 1 - \frac{p_E}{p_C} \cdot \frac{p_C}{p_U} = 1 - \theta(1 - \pi_C) > 1 - \theta_o(1 - \pi_C) \quad (11.17)$$

where $p_U$ is the assumed disease incidence rate in unvaccinated (or untreated) subjects. The choice of the non-inferiority threshold ($\theta_o$) was determined by considering the lower bound of the "estimated vaccine efficacy" given the assumed efficacy of IG based on Equation 11.17. For example, a non-inferiority threshold of $\theta_o$ = 3.0 will correspond to the estimated vaccine efficacy lower bound of 70%, which will imply a preservation of at least 78% of IG efficacy. Likewise, a choice of $\theta_o$ = 2.0 will indicate that the new vaccine will preserve (with lower bound) at least 89% of the IG efficacy. Although an equivalence margin that preserves 50% of the treatment effect has been proposed for evaluation of drug treatments,[43-45] there is a general perception that a narrower margin should be used in vaccine trials because the vaccine will be given to healthy subjects for prophylactic purpose.

For a relative risk threshold of 2.0, the study requires a total of 94 disease cases to achieve 91.0% power for demonstrating non-inferiority of the vaccine to IG when the true efficacy of the vaccine is identical to that of IG. In this case, the projected number of subjects needed in this study will be approximately 4700 subjects per treatment group if the disease incidence rate in the IG group is 1%.

## 11.3.3 Asymptotic Methods

When the sample size allows, use of asymptotic methods will generally be preferred. For some prespecified $0 < \theta_o < 1$, the alternative hypothesis of $p_E/p_C > \theta_o$ can be reexpressed as $p_E - \theta_o p_C > 0$. Inference is then based on the estimator $\hat{p}_E - \theta_o \hat{p}_C$. If $Z^* = \dfrac{\hat{p}_E - \theta_o \hat{p}_C}{se(\hat{p}_E - \theta_o \hat{p}_C)}$ exceeds the critical value or equivalently every value in the corresponding confidence interval for $p_E - \theta_o p_C$ is positive, then non-inferiority is concluded.

For some $0 < \alpha < 1$, a two-sided $100(1 - \alpha)\%$ Fieller confidence interval for $p_E/p_C$ equals $\{k: \left\{ \left| \dfrac{\hat{p}_E - k\hat{p}_C}{se(\hat{p}_E - k\hat{p}_C)} \right| \right\} < z_{\alpha/2}\}$ or those values $k$ that can replace $\theta_o$ for which statistical significance is not reached. The approach used by Farrington and Manning[5] and Miettinen and Nurminen[35] modifies this procedure by using the MLE of $se(\hat{p}_E - \theta_o \hat{p}_C)$ restricted to $p_E - \theta_o p_C = 0$. The MLEs, $\tilde{p}_E$ and $\tilde{p}_C$, of $\hat{p}_E$ and $\hat{p}_C$ restricted to $p_E - \theta_o p_C = 0$ are given by

$$\tilde{p}_E = \frac{\theta_o(n_E + x) + y + n_C - \sqrt{\{\theta_o(n_E + x) + y + n_C\}^2 - 4\theta_o(n_E + n_C)(x + y)}}{2(n_E + n_C)} \tag{11.18}$$

and $\tilde{p}_C = \tilde{p}_E/\theta_o$, where $x$ and $y$ are the observed number of successes in the control and experimental arms, respectively, given in Table 11.1.

An alternative would be to base the inference on the log-relative risk, $\log(p_E/p_C) = \log p_E - \log p_C$. Katz et al.[46] proposed the use of the standard Taylor series method for determining a one-sided or two-sided confidence interval for a relative risk.

By using the asymptotic standard error of $\log(\hat{p}_E) - \log(\hat{p}_C)$, a $100\,(1 - \alpha)\%$ confidence interval for $\log p_E - \log p_C$ can be calculated as

$$\log(\hat{p}_E) - \log(\hat{p}_C) \pm z_{\alpha/2} \sqrt{\frac{(1 - \hat{p}_E)}{n_E \hat{p}_E} + \frac{(1 - \hat{p}_C)}{n_C \hat{p}_C}}\,.$$

Non-inferiority is concluded if the confidence interval contains only values greater than $\log(\theta_o)$. It should be understood that this standard error is an approximation and not the actual standard error, and that $\log(\hat{p}_E) - \log(\hat{p}_C)$ may not lie within $z_{\alpha/2}\sqrt{\dfrac{(1 - \hat{p}_E)}{n_E \hat{p}_E} + \dfrac{(1 - \hat{p}_C)}{n_C \hat{p}_C}}$ of $\log p_E - \log p_C$, $100(1 - \alpha)\%$ of the time.

Section 11.3.4 introduces some other confidence interval methods that can be used to perform a test of non-inferiority using a relative risk. The results of comparisons of methods found in the literature are also discussed.

### 11.3.4 Comparisons of Methods

For a relative risk of success rates, we will discuss the results and conclusions of many papers that compare the probability coverage, type I error rates, and power of various methods. Some research involving two-sided testing on a relative risk may not directly apply to non-inferiority testing, which is one-sided.

Katz et al.[46] compared the coverage probabilities of the standard Taylor series method, the standard quadratic method, and the method of Thomas and Gart.[47] The standard quadratic method for determining a one-sided or two-sided confidence interval is based on testing $H_0$: $\theta = p_E / p_C = \theta_0$ with the normalized test statistic

$$\frac{\hat{p}_E - \theta_0 \hat{p}_C}{\sqrt{\hat{p}_E(1-\hat{p}_E)/n_E + \theta_0^2 \hat{p}_C(1-\hat{p}_C)/n_C}}$$

A Fieller method can be used to determine the corresponding one-sided or two-sided $100(1 - \alpha)\%$ confidence interval for the relative risk, which will consist of those values that can be specified as $\theta_0$ for which the respective one-sided or two-sided test of level $\alpha$ fails to reject the null hypothesis.

Katz et al.[46] provided detailed results for the coverage probabilities of one-sided lower confidence intervals for the possible combinations where $n_E = n_C = 100$; $p_C = 0.1$, 0.2, or 0.4; and $\theta_0 = 0.25$, 0.5, 0.667, 1, 1.5, 2, and 4. For each case, 5000 simulations were performed. Katz et al.[46] concluded that the Taylor series method and the Thomas and Gart method produced "similar and appropriate results," and that the quadratic method could be erratic and was not recommended. However, for cases that represent non-inferiority testing, the standard quadratic method and the Thomas and Gart method had simulated type I error rates (=1 − simulated coverage probability) of less than 0.025, whereas the standard Taylor series method usually had an inflated type I error rate. For a superiority test ($\theta_0 = 1$), the Thomas and Gart method had simulated significance levels always less than 0.025 in every case, whereas the standard quadratic method had type I error rates slightly greater than 0.025 and the Taylor series method had type I error rates either slightly greater than or slightly less than 0.025. Also, the simulated type I error rates for the standard quadratic method increased as $\theta_0$ increased, whereas the simulated type I error rates for the Taylor series method decreased as $\theta_0$ increased.

Koopman[48] considered confidence intervals based on testing $H_0$: $\theta = p_E/p_C = \theta_0$ using a Pearson's $\chi^2$ goodness-of-fit statistic. The test statistic is given by

$$\frac{(y - n_E \tilde{p}_E)^2}{n_E \tilde{p}_E(1-\tilde{p}_E)} + \frac{(x - n_C \tilde{p}_C)^2}{n_E \tilde{p}_C(1-\tilde{p}_C)}$$

where $\tilde{p}_E$ and $\tilde{p}_C$ are the MLEs restricted to $p_E/p_C = \theta_0$ (see Equation 11.18 for the expressions for $\tilde{p}_E$ and $\tilde{p}_C$). The test statistic is compared with the appropriate upper percentile of a $\chi^2$ distribution with 1 degree of freedom. A two-sided $100(1 - \alpha)\%$ confidence interval for the relative risk consists of values that can be specified as $\theta_0$ for which the respective one-sided or two-sided test of level $\alpha$ fails to reject the null hypothesis. A one-sided confidence interval would extend the appropriate side of a two-sided confidence interval.

Koopman[48] compared the coverage probabilities of the Taylor series method and Pearson's maximum likelihood method. Koopman provided detailed results for the coverage probabilities of two-sided 95% confidence intervals and one-sided 97.5% confidence intervals for various combinations where $n_E = n_C = 100$, $n_E = 50$ and $n_C = 150$, or $n_E = 150$ and $n_C = 50$; $p_E = 0.05$, 0.2, 0.35, 0.5, 0.65, 0.8 or 0.95; and $\theta_0 = 1$, 1.5, 2, or 4. Pearson's method maintained the desired two-sided coverage probability notably better than the Taylor series method. For non-inferiority testing on an undesirable outcome targeting a one-sided type I error rate of 0.025 when $n_E = n_C = 100$, the type I error rate ranged from 1.26% to 3.55% and from 2.39% to 4.10% for the Taylor series method and Pearson's method, respectively. When $n_E = n_C = 100$ as the probability of a "success" increased, the type I error rate for the Taylor series method tended to increase, whereas the type I error rate for the Pearson's method tended to decrease. When $n_E = 150$ and $n_C = 50$, the type I error rate ranged from 3.51% to 3.78% and from 2.97% to 4.91% for the Taylor series method and Pearson's method, respectively. Results also apply to non-inferiority testing on a desirable outcome with the roles of the experimental and control arms reversed (for relative risks of 0.25, 0.5, and 0.667).

Bailey[49] modified the standard quadratic method to reduce the skewness of the confidence interval so as to improve the coverage probability. Bailey's method determines a one-sided or two-sided $100(1 - \alpha)\%$ confidence interval for the relative risk based on the asymptotic normalized test statistic

$$\frac{\hat{p}_E^{1/3} - \theta_0^{1/3}\hat{p}_C^{1/3}}{(1/3)\sqrt{\hat{p}_E^{-1/3}(1-\hat{p}_E)/n_E + \theta_0^{2/3}\hat{p}_C^{-1/3}(1-\hat{p}_C)/n_C}}.$$

The resulting two-sided $100(1 - \alpha)\%$ confidence interval is

$$\left(\frac{\hat{p}_E}{\hat{p}_C}\right)\left[\frac{1 \mp z_{\alpha/2}\sqrt{(1-\hat{p}_E)/y + (1-\hat{p}_C)/x - z_{\alpha/2}^2(1-\hat{p}_E)(1-\hat{p}_C)/(9xy)/3}}{1 - z_{\alpha/2}^2(1-\hat{p}_C)/(9x)}\right]^3.$$

The continuity-corrected version replaces $y$ with $y + 0.5$ when determining the lower limit of the confidence interval and replaces $x$ with $x + 0.5$ when determining the upper limit of the confidence interval.

Bailey[49] compared this new method with and without a continuity correction to the standard Taylor series method. Bailey's method had exact coverage probabilities closer to the desired level than the standard Taylor series method. Bailey provided detailed results for the exact coverage probabilities of one-sided lower 95% confidence intervals for the same possible combinations as in Katz et al.'s[46] study ($n_E = n_C = 100$; $p_C = 0.1, 0.2,$ or $0.4$; and $\theta_o = 0.25, 0.5, 0.667, 1, 1.5, 2,$ and 4). Bailey's method had coverage probabilities much closer to the desired level than the Taylor series method and Pearson's method. For non-inferiority testing targeting a one-sided type I error rate of 5%, the corresponding (one-sided) type I error rates ranged from 4.6% to 5.2% for Bailey's method, from 4.3% to 5.8% for the Taylor series method, and from 3.8% to 4.6% for Bailey's method with a continuity correction.

Dann and Koch[1] evaluated and compared several methods of constructing a confidence interval for the relative risk based on the calculated limits, power, type I error rate, and agreement/disagreement with other methods. The methods included were classified into three categories: Taylor series methods, solution to quadratic equation methods (Fieller-based methods applied to normalized test statistics), and maximum likelihood–based methods (asymptotic likelihood ratio test and Pearson's $\chi^2$ test).

The Taylor series methods are the standard Taylor series method, the modified Taylor series method of Gart and Nam[50] that adds 0.5 to the number of successes in each group and to the common sample size, and another modified method (adapted Agresti–Caffo) that adds $4\theta_o/(1 + \theta_o)$ successes and $4/(1 + \theta_o)$ failures to the experimental group and $4/(1 + \theta_o)$ successes and $4\theta_o/(1 + \theta_o)$ failures to the control group. In addition, a Taylor series adjusted alpha method that uses $z_{0.0225} = 2.005$ instead of $z_{0.025} = 1.96$ for the standard Taylor series method is investigated.

The quadratic methods studied included the standard quadratic method (referred to as "F-M 1" by Dann and Koch[1]), an adapted version that divides by one less the sample size when determining the standard error (referred to as "the quadratic method" in the paper of Dann and Koch[1]), Bailey's method, and two variations of the quadratic method provided by Farrington and Manning.[5] One of the variations in Farrington and Manning's study[5] uses the MLEs in Equation 11.18 in determining the standard error. The other variation uses the approach of Dunnet and Gent[8] in obtaining estimates of the success rates for determining the standard error.

The maximum likelihood methods include the Pearson's method proposed by Koopman[48] and the generalized likelihood ratio test (referred to as the "deviance method" by Dann and Koch[1]). The deviance statistic is $2\ln[L(\hat{p}_E, \hat{p}_C)/L(\tilde{p}_E, \tilde{p}_C)]$, where $\tilde{p}_E$ and $\tilde{p}_C$ are the MLEs restricted to $p_E/p_C = \theta_o$ provided in Section 11.3.3. The test statistic is compared with the appropriate upper percentile of a $\chi^2$ distribution with 1 degree of freedom. A two-sided $100(1 - \alpha)\%$ confidence interval for the relative risk consists of those values that can be specified as $\theta_o$ for which the respective one-sided or two-sided test of level $\alpha$ fails to reject the null hypothesis. A one-sided confidence interval would extend the appropriate side of a two-sided confidence interval to zero or infinity.

The assessments were based on one-sided 97.5% confidence intervals. The trial sizes used were 100, 140, and 200 patients per group. Control success rates (for undesirable outcomes) of 0.10, 0.15, 0.20, and 0.25 were examined with null relative risks (experimental/control) of 0.667, 0.8, 1, 1.25, 1.5, and 2. For each combination of sample size, control success rate, and relative risk, 100,000 simulations were performed. For the majority of methods, when the experimental or control number of "successes" was three or fewer, the confidence interval was replaced by the corresponding exact confidence interval for the odds ratio. When there were no successes in the control group, the upper limit was assigned the value of 100.

The simulated type I error rates were provided in the paper of Dann and Koch[1] for a null relative risk of 2. The quadratic method had the smallest simulated type I error rate in every case. In all 12 of these cases, the Taylor series adjusted alpha method and the deviance methods maintained the type I error rate between 0.023 and the desired 0.025. Bailey's method, maintained a type I error rate between 0.0225 and 0.0275 in all cases. The adapted Agresti-Caffo method had simulated type I error rates between 0.020 and 0.027. The other methods did not perform consistently as well in maintaining the targeted type I error rate as the Taylor series adjusted alpha, deviance, adapted Agresti-Caffo, and Bailey methods. In cases having a null relative risk of 0.667, 0.8, 1, 1.25, and 1.5, the power was determined for each method for a true relative risk of 2. In those 60 cases, the Taylor series adjusted alpha, deviance, and Bailey's methods all produced extraordinarily similar simulated power.

We close with Example 11.9, which applies the 95% confidence intervals for the relative risk from various methods to the results from a hypothetical clinical trial.

### Example 11.9

Suppose there are 131 successes among 150 subjects in the experimental arm and 135 successes among 150 subjects in the control arm. For various asymptotic and Bayesian methods (see Section 11.5), the corresponding 95% two-sided confidence intervals for $\hat{p}_E/\hat{p}_C$ is determined. The results are provided in Table 11.3

**TABLE 11.3**

95% Confidence Intervals for Relative Risk from Various Methods

| Method | 95% Confidence Interval |
|---|---|
| Taylor series (Katz) | (0.895, 1.052) |
| Bailey | (0.895, 1.052) |
| Standard quadratic | (0.894, 1.052) |
| Zero prior Bayesian[a] | (0.893, 1.052) |
| Taylor series adjusted alpha | (0.893, 1.054) |
| Jeffreys | (0.893, 1.053) |
| Deviance | (0.892, 1.053) |
| Farrington–Manning | (0.890, 1.055) |
| Koopman–Pearson | (0.890, 1.055) |

[a] Based on the resulting posterior distributions after the prior $\alpha \to 0$ $\beta \to 0$ for each arm.

where the methods are listed in decreasing order with respect to the lower confidence limit. The order of the lower limits can be fairly arbitrary and depends on the general success rate, which of the arms had the greater rate, the sample size, and the allocation ratio.

## 11.3.5 Sample-Size Determination

We will see that simulations may often be necessary to determine or assist in determining the appropriate sample size when an inference is based on a relative risk. Dann and Koch[1] compared the difference between calculated and simulated power for some of the methods. For those methods compared, the calculated power was less than the simulated power, indicating that sample-size calculations are conservative.

For the relative risk, let $p_1$ and $p_2$ represent success rates for the experimental and control arms, respectively, where $\theta_a = p_1/p_2 > \theta_o$. There are multiple ways of determining the sample size for a non-inferiority analysis based on the relative risk. We will first consider basing the sample-size calculations using the estimator of the log-relative risk ($\log(\hat{p}_E/\hat{p}_C)$) and its asymptotic distribution. The sample-size formula will depend on whether the estimated standard error used for the denominator is an unrestricted estimate or an estimate restricted to the null hypothesis. Although it is typical in this setting to base the inference on a test statistic that uses the unrestricted estimate of the standard error, we will consider both cases. Let

$$\sigma_a = \sqrt{\frac{1-p_1}{n_E p_1} + \frac{1-p_2}{n_C p_2}} \text{ and } \sigma_o = \sqrt{\frac{1-p_1'}{n_E p_1'} + \frac{1-p_2'}{n_C p_2'}} \text{, where } (p_1', p_2') \text{ either repre-}$$

sents for some criteria that pair of probabilities closest to $(p_1, p_2)$ satisfying $p_1'/p_2' = \theta_o$ (when a restricted estimate of the standard error is used), or $p_1' = p_1$ and $p_2' = p_2$ (when an unrestricted estimate of the standard error is used). When the sample sizes are large and the true success probabilities of $p_1$ and $p_2$ ($p_1'$ and $p_2'$) are not close to 0 or 1, the log-relative risk will have an approximate normal distribution with standard deviation of $\sigma_a$ ($\sigma_o$). Analogous to Expression 11.7, we have for a desired power of $1 - \beta$ that $z_\beta \sigma_a + z_{\alpha/2} \sigma_o = \log \theta_a - \log \theta_o$. For $k = n_C/n_E$, solving for $n_C$ yields

$$n_C = \left( \frac{z_\beta \sqrt{(1-p_1)/(kp_1) + (1-p_2)/p_2} + z_{\alpha/2} \sqrt{(1-p_1')/(kp_1') + (1-p_2')/p_2'}}{\log \theta_a - \log \theta_o} \right)^2. \quad (11.19)$$

Simulations should be used when it is believed that the distribution for $\hat{\theta}$ will not closely be approximated by a normal distribution.

Another way to compute the sample size is based on the distribution for $\hat{p}_E - \theta_o \hat{p}_C$. When the true success probabilities are $p_1$ and $p_2$, the respective standard deviation for $\hat{p}_E - \theta_o \hat{p}_C$ is $\sigma_a = \sqrt{\frac{p_1(1-p_1)}{n_E} + \theta_o^2 \frac{p_2(1-p_2)}{n_C}}$. Let

$$\sigma_o = \sqrt{\frac{p_1'(1-p_1')}{n_E} + \theta_o^2 \frac{p_2'(1-p_2')}{n_C}} \text{, where } (p_1', p_2') \text{ either represents for some crite-}$$

ria that pair of probabilities closest to $(p_1, p_2)$ satisfying $p_1' - \theta_o p_2' = 0$ (when a restricted estimate of the standard error is used), or $p_1' = p_1$ and $p_2' = p_2$ (when an unrestricted estimate of the standard error is used). Then analogous to Expression 11.7, we have for a desired power of $1 - \beta$ that $z_\beta \sigma_a + z_{\alpha/2} \sigma_o = p_1 - \theta_o p_2$. For $k = n_C / n_E$, solving for $n_C$ yields

$$n_C = \left( \frac{z_\beta \sqrt{p_1(1-p_1)/k + \theta_o^2 p_2(1-p_2)} + z_{\alpha/2} \sqrt{p_1'(1-p_1')/k + \theta_o^2 p_2'(1-p_2')}}{p_1 - \theta_o p_2} \right)^2. \quad (11.20)$$

Using $p_1' = \theta_o(p_1 + p_2)/(1 + \theta_o)$ and $p_2' = (p_1 + p_2)/(1 + \theta_o)$ or using $p_1' = \theta_o(kp_1 + p_2)/(1 + k\theta_o)$ and $p_2' = (kp_1 + p_2)/(1 + k\theta_o)$ could be appropriate in many cases. Otherwise, $(p_1', p_2')$ can be selected using some rule for determining the pair of $(p_1', p_2')$ in the null hypothesis that is the most difficult to reject when $(p_1, p_2)$ is the true pair of the probabilities of a success. The change in the estimated sample size may not be dramatically affected unless $\theta_o$ is very small, or $p_1$ and $p_2$ are close to 0 or 1. It should be understood that $p_1' = \theta_o(kp_1 + p_2)/(1 + k\theta_o)$ and $p_2' = (kp_1 + p_2)/(1 + k\theta_o)$ need not both be between 0 and 1 (e.g., when $p_1 = p_2 = 0.9$, $\theta_o = 0.5$ and $k = 1$, $p_2' = 1.2$).

Example 11.10 compares and contrasts the sample-size formulas in Equations 11.19 and 11.20.

## Example 11.10

We will first compare and contrast the sample-size formulas in Equations 11.19 and 11.20 at both 80% and 90% power for three cases based on a one-to-one randomization ($k = 1$). The values for $\theta_o$, $p_1$, and $p_2$ are provided for each case below.

- Case 1: $\theta_o = 0.7$ and $p_1 = p_2 = 0.4$
- Case 2: $\theta_o = 0.3$ and $p_1 = p_2 = 0.04$
- Case 3: $\theta_o = 0.1$ and $p_1 = p_2 = 0.04$.

The values chosen for $p_1'$ and $p_2'$ will be based on the formulas $p_1' = \theta_o(p_1 + p_2)/(1 + \theta_o)$ and $p_2' = (p_1 + p_2)/(1 + \theta_o)$. The results are summarized in Table 11.4.

In all cases examined, the sample size was smaller using formula 11.20 than formula 11.19. In each case, the respective calculated sample sizes from the formulas are closer for 90% power than for 80% power. The calculated sample sizes using formula 11.19 grew at a faster rate or at least a faster relative rate as the power increased from 80% to 90%. This occurs because when the inference is based on the distribution for $\hat{p}_E - \theta_o \hat{p}_C$, $\sigma_a/\sigma_o$ is larger than when the inference is based on the distribution of the estimator of the log-relative risk. For case 1, there was very little difference in the sample-size calculation. Case 2 had a smaller value for $\theta_o$

**TABLE 11.4**

Calculated Sample Sizes for 80% and 90% Power for Cases 1 through 3

| Power (%) | $\theta_o$ | $(p_1, p_2)$ | $(p_1', p_2')$ | $\log(\hat{p}_E/\hat{p}_C)$[a] | $\hat{p}_E - \theta_o\hat{p}_C$[b] |
|---|---|---|---|---|---|
| | | | | **Sample Size per Arm** | |
| 80 | 0.7 | (0.4, 0.4) | (0.329, 0.471) | 192 | 190 |
| 90 | 0.7 | (0.4, 0.4) | (0.329, 0.471) | 256 | 255 |
| 80 | 0.3 | (0.04, 0.04) | (0.018, 0.062) | 336 | 284 |
| 90 | 0.3 | (0.04, 0.04) | (0.018, 0.062) | 435 | 403 |
| 80 | 0.1 | (0.04, 0.04) | (0.007, 0.073) | 168 | 90 |
| 90 | 0.1 | (0.04, 0.04) | (0.007, 0.073) | 204 | 141 |

[a] Calculations based on Equation 11.19.
[b] Calculations based on Equation 11.20.

and values for $p_1$ and $p_2$ close to zero. The calculated sample size begins to deviate between the two formulas. Deviation is larger despite the sample sizes being smaller for case 3, which had an even smaller value for $\theta_o$ than case 2. When $p_1' = p_1$ and $p_2' = p_2$, the results are summarized in Table 11.5.

For case 1, comparing Tables 11.4 and 11.5, there was only moderate change in the calculated sample sizes using $p_1' = p_1 = 0.4$ and $p_2' = p_2 = 0.4$ instead of $p_1' = \theta_o(p_1 + p_2)/(1 + \theta_o) = 0.329$ and $p_2' = (p_1 + p_2)/(1 + \theta_o) = 0.471$. For cases 2 and 3, there were dramatic changes in the calculated sample sizes. Although in Table 11.4, for all cases examined, the sample size was larger using formula 11.20 than formula 11.19, the reverse is seen in Table 11.5. As with Table 11.4, for cases 2 and 3, there were different calculated sample sizes between formulas 11.19 and 11.20. For each case when the inference is based on the log-relative risk estimator, the calculated sample size decreases when $p_1' = p_1$ and $p_2' = p_2$ is used instead of $p_1' = \theta_o(p_1 + p_2)/(1 + \theta_o)$ and $p_2' = (p_1 + p_2)/(1 + \theta_o)$. This is because when $p_1' + p_2'$ is fixed, $\sqrt{(1 - p_1')/p_1' + (1 - p_2')/p_2'} = \sqrt{1/p_1' + 1/p_2' - 2}$ becomes smaller when the probabilities $p_1'$ and $p_2'$ become more similar.

Conversely, for an inference based on $\hat{p}_E - \theta_o\hat{p}_C$, the calculated sample size increases ($\sqrt{p_1'(1 - p_1') + \theta_o^2 p_2'(1 - p_2')}$ increases) when $p_1' = p_1$ and $p_2' = p_2$ is used

**TABLE 11.5**

Calculated Sample Sizes when $p_1' = p_1$ and $p_2' = p_2$

| Power (%) | $\theta_o$ | $(p_1, p_2)$ | $(p_1', p_2')$ | $\log(\hat{p}_E/\hat{p}_C)$[a] | $\hat{p}_E - \theta_o\hat{p}_C$[b] |
|---|---|---|---|---|---|
| | | | | **Sample Size per Arm** | |
| 80 | 0.7 | (0.4, 0.4) | (0.4, 0.4) | 186 | 195 |
| 90 | 0.7 | (0.4, 0.4) | (0.4, 0.4) | 248 | 261 |
| 80 | 0.3 | (0.04, 0.04) | (0.04, 0.04) | 260 | 420 |
| 90 | 0.3 | (0.04, 0.04) | (0.04, 0.04) | 348 | 561 |
| 80 | 0.1 | (0.04, 0.04) | (0.04, 0.04) | 72 | 235 |
| 90 | 0.1 | (0.04, 0.04) | (0.04, 0.04) | 96 | 315 |

[a] Calculations based on Equation 11.19.
[b] Calculations based on Equation 11.20.

instead of $p_1' = \theta_0(p_1 + p_2)/(1 + \theta_0)$ and $p_2' = (p_1 + p_2)/(1 + \theta_0)$. When $p_1' + p_2' = s$ for fixed $s$ and $0 \le \theta_0 \le 1$, the maximum value for $\sqrt{p_1'(1-p_1') + \theta_0^2 p_2'(1-p_2')}$ occurs when $p_1' = \min \{s, (1 + 2s\theta_0 - \theta_0)/(2\theta_0 + 2)\}$.

Example 11.10 illustrates that the choice for $p_1'$ and $p_2'$ can have a small or rather large effect on the calculated sample size depending on the value for $\theta_0$ and the expected success probabilities. Whenever the calculated sample size changes greatly as the choices for $p_1'$ and $p_2'$ change, simulations should be used to find the appropriate sample size or validate a calculated sample size.

### 11.3.5.1 Optimal Randomization Ratio

By Equation 11.19, the overall study size when the inference is based on the distribution of $\log(\hat{p}_E/\hat{p}_C)$ for an experimental to control allocation ratio of $k$ is given by

$$\left( \frac{z_\beta\sqrt{(1-p_1)/(kp_1)+(1-p_2)/p_2} + z_{\alpha/2}\sqrt{_1(1-p_1')/(kp')+(1-p_2')/p_2'}}{\log\theta_a - \log\theta_0} \right)^2 (1+k). \quad (11.21)$$

When the inference is based on the distribution for $\hat{p}_E - \theta_0\hat{p}_C$, it follows from Equation 11.20 that the overall study size is given by

$$\left( \frac{z_\beta\sqrt{p_1(1-p_1)/k+\theta_0^2 p_2(1-p_2)} + z_{\alpha/2}\sqrt{p_1'(1-p_1')/k+\theta_0^2 p_2'(1-p_2')}}{p_1 - \theta_0 p_2} \right)^2 (1+k). \quad (11.22)$$

In either case, the optimal $k$ that minimizes Equation 11.21 or 11.22 can be found by using calculus or by a "grid search." Example 11.11 compares and contrasts the sample-size formulas in Equations 11.21 and 11.22.

### Example 11.11

We will compare and contrast the results for the optimal overall study size based on formulas 11.21 and 11.22 for cases 1 through 3 in Example 11.10 at both 80% and 90% power when $p_1' = \theta_0(p_1 + p_2)/(1 + \theta_0)$ and $p_2' = (p_1 + p_2)/(1 + \theta_0)$. The results are summarized in Tables 11.6 and 11.7.

We see that the study size reduction is more prominent when the inference is based on $\hat{p}_E - \theta_0\hat{p}_C$. Within a case, the optimal $k$ for 90% power is smaller than that for 80% when the inference is based on the distribution of $\log(\hat{p}_E/\hat{p}_C)$, but larger when the inference is based on the distribution of $\hat{p}_E - \theta_0\hat{p}_C$.

Tables 11.8 and 11.9 provide analogous results on the calculation of the optimal allocation ratio, $k$, and the corresponding study size when $p_1' = \theta_0(kp_1 + p_2)/(1 + k\theta_0)$ and $p_2' = (kp_1 + p_2)/(1 + k\theta_0)$.

**TABLE 11.6**

Sample Sizes for Log-Relative Risk Based on Optimal Allocation

| Case | Power (%) | Ratio | $n_E$ | $n_C$ | $n$ | Reduction in Study Size[a] |
|------|-----------|-------|-------|-------|-----|---------------------------|
| | | | $\log(\hat{p}_E/\hat{p}_C)$ | | | |
| 1 | 80 | 1.23 | 210 | 170 | 380 | 4 (1%) |
| 1 | 90 | 1.20 | 277 | 231 | 508 | 4 (1%) |
| 2 | 80 | 1.55 | 390 | 251 | 641 | 31 (5%) |
| 2 | 90 | 1.47 | 499 | 341 | 840 | 30 (3%) |
| 3 | 80 | 2.34 | 203 | 87 | 290 | 46 (13%) |
| 3 | 90 | 2.11 | 246 | 117 | 363 | 45 (11%) |

[a] Reduction is relative to the sample-size calculation in Table 11.4 using Equation 11.19.

For this scenario, the calculated samples are more similar between formulas 11.19 and 11.20 than in the earlier scenarios. Although earlier when $p'_1 = \theta_o(p_1 + p_2)/(1 + \theta_o)$ and $p'_2 = (p_1 + p_2)/(1 + \theta_o)$, the study reduction was more prominent using an optimal allocation ratio when the inference is based on $\hat{p}_E - \theta_o\hat{p}_C$, we see that when $p'_1 = \theta_o(kp_1 + p_2)/(1 + k\theta_o)$ and $p'_2 = (kp_1 + p_2)/(1 + k\theta_o)$, the study size reduction is more prominent using an allocation ratio when the inference is based on $\log(\hat{p}_E/\hat{p}_C)$. As before, within a case, the optimal $k$ for 90% power is smaller than that for 80% when the inference is based on the distribution of $\log(\hat{p}_E/\hat{p}_C)$, but larger when the inference is based on the distribution of $\hat{p}_E - \theta_o\hat{p}_C$. Compared with when $p'_1 = \theta_o(p_1 + p_2)/(1 + \theta_o)$ and $p'_2 = (p_1 + p_2)/(1 + \theta_o)$, the calculated study size for the optimal allocation ratio is smaller when the inference is based on the distribution of $\log(\hat{p}_E/\hat{p}_C)$ when $p'_1 = \theta_o(kp_1 + p_2)/(1 + k\theta_o)$ and $p'_2 = (kp_1 + p_2)/(1 + k\theta_o)$, but larger when the inference is based on the distribution of $\hat{p}_E - \theta_o\hat{p}_C$. For cases 1 through 3, the optimal allocation ratios when $p'_1 = \theta_o(p_1 + p_2)/(1 + \theta_o)$ and $p'_2 = (p_1 + p_2)/(1 + \theta_o)$ and the inference based on the distribution of $\log(\hat{p}_E/\hat{p}_C)$ were similar to the optimal allocation ratios when $p'_1 = \theta_o(kp_1 + p_2)/(1 + k\theta_o)$ and $p'_2 = (kp_1 + p_2)/(1 + k\theta_o)$ and the inference based on the distribution of $\hat{p}_E - \theta_o\hat{p}_C$. Likewise, the optimal allocation ratios when $p'_1 = \theta_o(kp_1 + p_2)/(1 + k\theta_o)$ and $p'_2 = (kp_1 + p_2)/(1 + k\theta_o)$ with the inference is based on the distribution of $\log(\hat{p}_E/\hat{p}_C)$

**TABLE 11.7**

Sample Sizes for $\hat{p}_E - \theta_o\hat{p}_C$ Based on Optimal Allocation

| Case | Power (%) | Ratio | $n_E$ | $n_C$ | $n$ | Reduction in Study Size[a] |
|------|-----------|-------|-------|-------|-----|---------------------------|
| | | | $\hat{p}_E - \theta_o\hat{p}_C$ | | | |
| 1 | 80 | 1.37 | 214 | 156 | 370 | 10 (3%) |
| 1 | 90 | 1.38 | 288 | 209 | 497 | 13 (3%) |
| 2 | 80 | 2.10 | 337 | 161 | 498 | 70 (12%) |
| 2 | 90 | 2.34 | 486 | 208 | 694 | 112 (14%) |
| 3 | 80 | 4.53 | 105 | 23 | 128 | 52 (29%) |
| 3 | 90 | 5.00 | 163 | 33 | 196 | 86 (30%) |

[a] Reduction is relative to the sample-size calculation in Table 11.4 using Equation 11.20.

**TABLE 11.8**

Sample Sizes for Log-Relative Risk Based on Optimal Allocation

| Case | Power (%) | $(p_1', p_2')$ | Ratio | $n_E$ | $n_C$ | $n$ | Reduction in Study Size[a] |
|------|-----------|----------------|-------|-------|-------|-----|---------------------------|
| 1 | 80 | (0.342, 0.489) | 1.53 | 222 | 146 | 368 | 16 (4%) |
| 1 | 90 | (0.340, 0.486) | 1.44 | 293 | 203 | 496 | 16 (3%) |
| 2 | 80 | (0.024, 0.079) | 2.40 | 394 | 165 | 559 | 113 (17%) |
| 2 | 90 | (0.023, 0.077) | 2.13 | 515 | 242 | 757 | 113 (13%) |
| 3 | 80 | (0.016, 0.164) | 5.24 | 154 | 30 | 184 | 152 (45%) |
| 3 | 90 | (0.015, 0.146) | 4.19 | 206 | 50 | 256 | 152 (37%) |

[a] Reduction is relative to the sample-size calculation in Table 11.4 using Equation 11.19.

were similar to the optimal allocation ratios when $p_1' = \theta_o(p_1 + p_2)/(1 + \theta_o)$ and $p_2' = (p_1 + p_2)/(1 + \theta_o)$ and the inference is based on the distribution of $\hat{p}_E - \theta_o\hat{p}_C$.

These examples illustrate how the "optimal" allocation ratio depends on the selection of $p_1'$ and $p_2'$. Therefore, for the relative risk, the "optimal" allocation ratio should be interpreted with caution. A moderate or even large change in the allocation ratio often provides only a small change in the power (for a fixed sample size) or sample size (for fixed power). Also, the allocation ratio selected to maximize the power for the analysis of the primary efficacy endpoint may not be optimal or appropriate for the evaluation of secondary efficacy endpoints and/or safety endpoints.

When the inference is based on $\log(\hat{p}_E/\hat{p}_C)$ and $p_1' = p_1 = p_2' = p_2$, the optimal allocation ratio will be $k = 1$ (and thus the sample sizes are those provided in Table 11.5). When the inference is based on $\hat{p}_E - \theta_o\hat{p}_C$ and $p_1' = p_1 = p_2' = p_2$, the optimal allocation ratio will be $k = 1/\theta_o$. The sample sizes are provided in Table 11.10 for cases 1 through 3. Compared with Table 11.9 when $p_1' = \theta_o(kp_1 + p_2)/(1 + k\theta_o)$ and $p_2' = (kp_1 + p_2)/(1 + k\theta_o)$, the sample sizes for the optimal allocation ratio are larger when $p_1' = p_1$ and $p_2' = p_2$, as are the corresponding calculated optimal allocation ratios.

**TABLE 11.9**

Sample Sizes for $\hat{p}_E - \theta_o\hat{p}_C$ Based on Optimal Allocation

| Case | Power (%) | $(p_1', p_2')$ | Ratio | $n_E$ | $n_C$ | $n$ | Reduction in Study Size[a] |
|------|-----------|----------------|-------|-------|-------|-----|---------------------------|
| 1 | 80 | (0.338, 0.482) | 1.32 | 212 | 160 | 372 | 8 (2%) |
| 1 | 90 | (0.338, 0.483) | 1.34 | 286 | 214 | 500 | 10 (2%) |
| 2 | 80 | (0.020, 0.068) | 1.43 | 325 | 226 | 551 | 17 (3%) |
| 2 | 90 | (0.021, 0.070) | 1.59 | 471 | 296 | 767 | 39 (5%) |
| 3 | 80 | (0.011, 0.107) | 2.27 | 106 | 47 | 153 | 27 (15%) |
| 3 | 90 | (0.012, 0.117) | 2.71 | 169 | 63 | 232 | 50 (18%) |

[a] Reduction is relative to the sample-size calculation in Table 11.4 using Equation 11.20.

**TABLE 11.10**

Sample Sizes for $\hat{p}_E - \theta_o \hat{p}_C$ Based on Optimal Allocation

| Case | Power (%) | $(p_1', p_2')$ | Ratio | $n_E$ | $n_C$ | $n$ | Reduction in Study Size[a,b] |
|------|-----------|----------------|-------|-------|-------|-----|------------------------------|
| 1 | 80 | (0.4, 0.4) | 1.43 | 223 | 156 | 379 | 1 (0%) |
| 1 | 90 | (0.4, 0.4) | 1.43 | 298 | 209 | 507 | 3 (1%) |
| 2 | 80 | (0.04, 0.04) | 3.33 | 500 | 150 | 650 | −82 (−14%) |
| 2 | 90 | (0.04, 0.04) | 3.33 | 669 | 201 | 870 | −64 (−8%) |
| 3 | 80 | (0.04, 0.04) | 10 | 256 | 26 | 282 | −102 (−57%) |
| 3 | 90 | (0.04, 0.04) | 10 | 343 | 34 | 377 | −95 (−34%) |

[a] Reduction is relative to the sample-size calculation in Table 11.4 using Equation 11.20.
[b] Reductions in sample sizes relative to Table 11.5, using Equation 11.20, are 11 (3%), 15 (3%), 190 (24%), 252 (22%), 188 (40%), 253 (40%), respectively.

---

## 11.4 Fixed Thresholds on Odds Ratios

### 11.4.1 Hypotheses

The odds ratio of two binomial proportions is commonly used in case–control and other retrospective studies. Although it is less frequently used in prospective clinical trials, we will briefly describe the methods used to prove non-inferiority using the odds ratio. For a probability of success, $p$, the odds are defined as

$$\omega = p/(1 - p).$$

The odds are a nonnegative number, representing the likelihood of a success occurring relative to the likelihood of a failure. For example, when $p = 0.8$, the odds are $\omega = 0.8/0.2 = 4.0$, indicating a success is four times as likely as a failure. Inversely,

$$p = \omega/(1 + \omega)$$

When the odds $\omega = 2$, $p = 2/3$, indicating 2 successes for every failure. For a comparative study with two binomial parameters, $p_E$ and $p_C$, the odds ratio between the new treatment (E) and the control (C) is

$$\psi = \omega_E/\omega_C = p_E(1 - p_C)/(p_C(1 - p_E)).$$

It can be seen that the odds ratio = relative risk of a success ÷ relative risk of a failure. When the probabilities of a success are very small (the relative risk of a failure ≈ 1), the odds ratio is approximately equal to the relative risk of a success.

When a success is a desirable outcome, the hypotheses for testing that the experimental therapy is noninferior to the control therapy based on a pre-specified threshold of $\psi_0$ $(0 < \psi_0 \leq 1)$ are

$$H_o: \psi \leq \psi_0 \text{ vs. } H_a: \psi > \psi_0 \tag{11.23}$$

### 11.4.2 Exact Methods

Let $X$ and $Y$ be independent binomial random variables with parameters $(n_C, p_C)$ and $(n_E, p_E)$, respectively. Conditional on the marginal sum $(S = X + Y = s)$, the random variable $Y$ is a sufficient statistic for $\psi$ with its conditional probability distribution given by (Fisher)

$$P(Y = y \mid s, \psi) = \frac{\binom{n_E}{y}\binom{n_C}{s-y}\psi^y}{\sum\limits_{k}\binom{n_E}{k}\binom{n_C}{s-k}\psi^k} \tag{11.24}$$

where the permissible values of $y$ and $k$ consist of all integers within the range $\max(0, s - n_C)$ to $\min(n_E, s)$. This is called the *extended hypergeometric distribution*, and more details can be found in the papers of Zelterman[51] and Johnson et al.[40] Note that for a classical null hypothesis of unity odds ratio $(H_o: \psi = 1)$, the probability function in Equation 11.24 will reduce to the hypergeometric distribution under the null hypothesis.

Suppose $y_{obs}$ is the observed number of positive responses in the new treatment group, then the exact $p$-value for testing hypothesis in Expression 11.23 is given by

$$p\text{-value} = P(Y \geq y_{obs} \mid s, \psi_0).$$

As noted in previous papers,[8,52,53] a $100(1 - \alpha)\%$ exact confidence interval for $\psi$, $(\psi_L, \psi_U)$ can be constructed by inverting two one-sided tests based on the conditional probability distribution given in Equation 11.24:

$$\sum_{i=A}^{y_{obs}} P(Y = i \mid s, \psi_U) = \alpha/2, \text{ and } \sum_{i=y_{obs}}^{B} P(Y = i \mid s, \psi_L) = \alpha/2$$

where $A = \max(0, s - n_C)$ and $B = \min(n_E, s)$. Agresti and Min[15] have also discussed the construction of confidence interval for the odds ratio by inverting a two-sided test.

The above method of analyzing the odds ratio in a single $2 \times 2$ table has been extended to analyze a common odds ratio in a series of $2 \times 2$ tables.[54–57]

### 11.4.3 Asymptotic Methods

The sample odds ratio is defined as

$$\hat{\psi} = \frac{y(n_C - x)}{x(n_E - y)} = \frac{\hat{p}_E(1 - \hat{p}_C)}{\hat{p}_C(1 - \hat{p}_E)}.$$

If there are zero counts, the following amended estimator has been shown to have good large sample behavior[58]:

$$\tilde{\psi} = \frac{(y + 0.5)(n_C - x + 0.5)}{(x + 0.5)(n_E - y + 0.5)}.$$

Both estimators have the same asymptotic normal distribution around $\psi$. However, the convergence is more rapid with the log transformation, and thus the normal approximation is better for a log-odds ratio. An estimator of the standard error for $\log \hat{\psi}$ is given by

$$\hat{\sigma} = \left( \frac{1}{Y} + \frac{1}{n_E - Y} + \frac{1}{X} + \frac{1}{n_C - X} \right)^{1/2} = \left( \frac{1}{n_E \hat{p}_E (1 - \hat{p}_E)} + \frac{1}{n_C \hat{p}_C (1 - \hat{p}_C)} \right)^{1/2}.$$

To protect against having a zero cell count, 1 or 0.5 can be added to each cell count in the estimator of standard error. On the basis of the asymptotic normality of $\log \hat{\psi}$, a two-sided $100(1 - \alpha)\%$ Wald's confidence interval for $\log \psi$ is given by

$$\log \hat{\psi} \pm z_{\alpha/2} \hat{\sigma}$$

where $z_{\alpha/2}$ is the upper $\alpha/2$ percentile of the standard normal distribution. Therefore, a confidence interval $(\psi_L, \psi_U)$ for the odds ratio $(\psi)$ can be obtained by exponentiating the above limits.

For testing the non-inferiority hypothesis in Expression 11.23, the test statistic is

$$Z = \frac{\log \hat{\psi} - \log \psi_o}{\hat{\sigma}},$$

and the $p$-value can be calculated asymptotically as

$$p\text{-value} = P \left\{ Z > \frac{\log \hat{\psi} - \log \psi_o}{\hat{\sigma}} \right\} = 1 - \Phi \left( \frac{\log \hat{\psi} - \log \psi_o}{\hat{\sigma}} \right)$$

where $Z$ follows a standard normal distribution and $\Phi(\bullet)$ is the standard normal distribution function. The hypothesis will be rejected if $Z > z_{\alpha/2}$ or if the $p$-value is $<\alpha/2$. Equivalently, one can compare the lower limit $(\psi_L)$ of the $100(1-\alpha)\%$ confidence interval for $\psi$ with $\psi_0$. The non-inferiority hypothesis will be rejected at the one-sided $\alpha$ level if $\psi_L > \psi_0$.

If there are covariates to be adjusted in the analysis, one can consider performing a logistic regression with covariates or a log-linear model if all covariates are categorical. Then the odds ratio between the treatments can be estimated from the regression parameter.

## 11.4.4 Sample Size Determination

For a design based on the log-odds ratio, let $p_1$ and $p_2$ represent the selected success rates for the experimental and control arms, respectively, where $\psi_a = p_1(1-p_2)/(p_2(1-p_1)) > \psi_0$. We will base the sample-size calculations on the estimator of the log-odds ratio $(\log(\hat{p}_E(1-\hat{p}_C)/(\hat{p}_C(1-\hat{p}_E))))$

and its asymptotic distribution. Let $\sigma_a = \sqrt{\dfrac{1}{n_E p_1(1-p_1)} + \dfrac{1}{n_C p_2(1-p_2)}}$ and

$\sigma_0 = \sqrt{\dfrac{1}{n_E p_1'(1-p_1')} + \dfrac{1}{n_C p_2'(1-p_2')}}$, where $(p_1', p_2')$ either represents for some

criteria that pair of probabilities closest to $(p_1, p_2)$ satisfying $p_1'(1-p_2')/(p_2'(1-p_1')) = \psi_0$ (when a restricted estimate of the standard error is used), or $p_1' = p_1$ and $p_2' = p_2$ (when an unrestricted estimate of the standard error is used). When the sample sizes are large and the true success probabilities of $p_1$ and $p_2$ ($p_1'$ and $p_2'$) are not close to 0 or 1, the estimator of the log-odds ratio will have an approximate normal distribution with standard deviation of $\sigma_a$ ($\sigma_0$). Analogous to Equation 11.7, we have for a desired power of $1-\beta$ that $z_\beta \sigma_a + z_{\alpha/2}\sigma_0 = \log\psi_a - \log\psi_0$. For $k = n_E/n_C$, solving for $n_C$ yields

$$\left( \frac{z_\beta \sqrt{(kp_1(1-p_1))^{-1} + (p_2(1-p_2))^{-1}} + z_{\alpha/2}\sqrt{(kp_1'(1-p_1'))^{-1} + (p_2'(1-p_2'))^{-1}}}{\log\psi_a - \log\psi_0} \right)^2. \quad (11.25)$$

Alternatively, simulations can be performed to determine the required sample size. Simulations should be used when it is believed that the distribution for $\hat{\psi}$ will not closely be approximated by a normal distribution.

By Equation 11.25, the overall study size for an experimental to control allocation ratio of $k$, is given by

$$\left( \frac{z_\beta \sqrt{(kp_1(1-p_1))^{-1} + (p_2(1-p_2))^{-1}} + z_{\alpha/2}\sqrt{(kp_1'(1-p_1'))^{-1} + (p_2'(1-p_2'))^{-1}}}{\log\psi_a - \log\psi_0} \right)^2 (1+k).$$

As before, the optimal $k$ that minimizes the above expression can found by using calculus or by a "grid search."

---

## 11.5 Bayesian Methods

In this section we will consider non-inferiority testing based on a credible interval when independent beta prior distributions are used. The beta distributions are a conjugate family of prior distributions for a random sample from a Bernoulli distribution. For a random sample of $n$ binary observations where $x$ are successes, a beta prior distribution with parameters $\alpha$ and $\beta$ for $p$ (the probability of success) leads to a beta posterior distribution for $p$ with parameters $\alpha + x$ and $\beta + n - x$. The mean of the posterior distribution, $(\alpha + x)/(\alpha + \beta + n)$ is the most frequently used estimate of $p$. This fraction can be regarded as a fraction of "successes," where the prior distribution contributes $\alpha$ "successes" and $\beta$ "failures" among $\alpha + \beta$ Bernoulli trials. Thus, in essence, a beta prior distribution establishes a beta posterior distribution to base inferences while also contributing additional successes and failures. When $\alpha$ and $\beta$ for the prior distribution are relatively small compared to the sample size, the specific choice of $\alpha$ and $\beta$ have little impact. For a Jeffreys prior, $\alpha = \beta = 0.5$, the 95% credible interval for $p$ is similar to the 95% exact confidence interval for $p$.

For a superiority comparison, a Bayesian analysis gives the same results whether the inference is based on a difference in proportions, the relative risk, or an odds ratio (i.e., $P(p_E - p_C > 0) = P(p_E/p_C > 1) = P((p_E/(1 - p_E))/(p_C/(1 - p_C)) > 1))$. The results will depend on the chosen prior distributions. The results may also differ among frequentist analyses based on different asymptotic methods. For example, study GAN040 randomized 304 subjects receiving orthotopic liver transplants between a Cytovene and a placebo arm.[59] Forty-six R$^-$ subjects (seronegative recipients), 21 in the Cytovene arm and 25 in the placebo arm, were transplanted with D$^+$ (seropositive donors) livers. Three of these 21 subjects on the Cytovene arm and 11 of these 25 subjects on the placebo arm had CMV disease at 6 months. For the subgroup of R$^-$ subjects, the respective 95% confidence intervals based on asymptotic results (use of Taylor series/delta methods for the relative risk and the odds ratio) for the difference in incidence of CMV disease at 6 months, the relative risk, and the odds ratio are (−0.54, −0.05), (0.10, 1.01), and (0.05, 0.91), respectively. If the 95% confidence interval for either the difference in proportions or the odds ratios were used to make inferences, the conclusion would be that Cytovene reduces the incidence of CMV disease at 6 months for R$^-$ subjects transplanted with D$^+$ livers. However, if the 95% confidence interval for the relative risk were used, the conclusion that Cytovene reduces the incidence of CMV disease at 6 months for R$^-$ subjects transplanted with D$^+$ livers cannot be made.

**TABLE 11.11**

Integrals Representing Posterior Probability of Non-Inferiority

| Characteristic | Determination of the Posterior Probability |
|---|---|
| Difference in proportions | For any $0 \leq k \leq 1$, $P(p_E - p_C > -k) =$ $$1 - \int_k^1 \int_0^{v-k} g_E(u\|y)g_C(v\|x)\mathrm{d}u\mathrm{d}v.$$ |
| Relative risk | For any $0 < k \leq 1$, $P(p_E/p_C > k) =$ $$1 - \int_0^1 \int_0^{kv} g_E(u\|y)g_C(v\|x)\mathrm{d}u\mathrm{d}v.$$ |
| Odds ratio | For any $0 < k < 1$, $P((p_E/(1-p_E))/(p_C/(1-p_C)) > k) =$ $$\int_0^1 \int_{kv/(1+(k-1)v)}^1 g_E(u\|y)g_C(v\|x)\mathrm{d}u\mathrm{d}v.$$ |

In the above Cytovene example, if Jeffreys prior distributions are chosen for the respective probabilities of CMV disease at 6 months, the 95% credible intervals for difference in proportions, relative risk, and odds ratio are (−0.51, −0.04), (0.09, 0.88), and (0.05, 0.84), respectively. For the Bayesian analysis, the choice between using a difference in proportions, the relative risk, or the odds ratio does not affect the decision on whether Cytovene reduces the incidence of CMV disease at 6 months for R⁻ subjects transplanted with D⁺ livers. The choice of a prior distribution will influence whether that conclusion can be made. If the parameters for the beta prior distribution are very small relative to the sample size, the prior distribution will have little influence on the results of the analysis.

For a difference in proportions, relative risk, or odds ratio, a credible interval and the posterior probability of non-inferiority (i.e., $P(p_E - p_C > -k)$, $P(p_E/p_C > k)$, or $P((p_E/(1-p_E))/(p_C/(1-p_C)) < k)$ for appropriate $k$) can be determined from either the joint distribution for $(p_E, p_C)$; the respective posterior distribution for $p_C - p_E, p_C/p_E$, or $(p_E/(1-p_E))/(p_C/(1-p_C))$; or from simulations. Let $g_E(p_E\|y)$ and $g_C(p_C\|x)$ denote the posterior densities for $p_E$ and $p_C$, respectively. Table 11.11 provides integral forms for determining the posterior probabilities of non-inferiority for the different metrics used to compare proportions. The relationship that $P((p_E/(1-p_E))/(p_C/(1-p_C)) < k) = P(p_E < kp_C/(1+(k-1)p_C))$ is used to construct the posterior probability for an odds ratio.

Example 11.12 illustrates the impact of the choice of the prior distributions for the probability of a success on the non-inferiority inference.

### Example 11.12

Suppose for a randomized clinical trial, 80 of 100 subjects in the experimental arm and 85 of 100 subjects in the control arm have a "response." We will examine the equal-tailed 95% credible intervals for the difference, relative risk, and odds ratio of the probabilities of a response under a variety of choices for the prior distributions for the probability of a response. For each case, independent beta distributions are selected as the prior distribution for the probability of a response for the control and the experimental arms. Table 11.12 summarizes the equal-tailed 95% credible intervals for the difference, relative risk, and odds ratio of the probabilities of a response under seven different pairs of the prior distributions. Each credible interval was based on 1 million simulations from the corresponding beta posterior distributions. The first case uses the limiting posterior distributions, as the parameters ($\alpha$ and $\beta$) for each prior distribution tend toward zero. This establishes a limiting beta posterior distribution where the inference is essentially based entirely on the data (the mean of each posterior distribution is the respective observed proportion of responders). The second case has a Jeffreys prior distribution for each probability of a response. The remaining cases are based on having prior information on each probability of a response that is essentially equivalent to having response data on 40 subjects. The third case is essentially equivalent to beginning with 20 responders and 20 nonresponders for each arm. When compared with case 1, this choice of a common prior distribution makes the proportion of responders between arms more similar and closer to 0.5, while not having a great impact on the variance of the posterior distributions. The fourth case represents starting with 34 responders out of 40 subjects (85%, the same as the observed proportion in the control arm) in each arm. When compared with case 1, this choice of a common prior distribution makes the proportion of responders between arms more similar and reduces the variance of each posterior distribution. The fifth case represents starting with 34 responders out of 40 subjects in the control arm and 32 responders out of 40 subjects in the experimental arm. When compared with case 1, this choice of prior distributions does not change the proportion of responders in the respective arms while reducing the variance of each posterior distribution.

The sixth case represents starting with 34 responders out of 40 subjects in the control arm and 28 responders out of 40 subjects in the experimental arm. For a non-inferiority margin of 15%, this case starts with observed proportions whose

**TABLE 11.12**

Equal-Tailed 95% Credible Intervals for a Difference, Relative Risk, and Odds Ratio

| Case | Prior Parameters | $p_E - p_C$ | $p_E/p_C$ | Odds Ratio |
|------|------------------|-------------|-----------|------------|
| 1 | C: $\alpha \to 0$ $\beta \to 0$<br>E: $\alpha \to 0$ $\beta \to 0$ | (−0.155, 0.055) | (0.825, 1.070) | (0.329, 1.468) |
| 2 | C: $\alpha = 0.5$ $\beta = 0.5$<br>E: $\alpha = 0.5$ $\beta = 0.5$ | (−0.155, 0.055) | (0.824, 1.070) | (0.336, 1.464) |
| 3 | C: $\alpha = 20$ $\beta = 20$<br>E: $\alpha = 20$ $\beta = 20$ | (−0.139, 0.068) | (0.825, 1.098) | (0.487, 1.418) |
| 4 | C: $\alpha = 34$ $\beta = 6$<br>E: $\alpha = 34$ $\beta = 6$ | (−0.123, 0.051) | (0.861, 1.064) | (0.406, 1.449) |
| 5 | C: $\alpha = 34$ $\beta = 6$<br>E: $\alpha = 32$ $\beta = 8$ | (−0.139, 0.039) | (0.842, 1.048) | (0.372, 1.309) |
| 6 | C: $\alpha = 34$ $\beta = 6$<br>E: $\alpha = 28$ $\beta = 12$ | (−0.170, 0.012) | (0.807, 1.015) | (0.317, 1.085) |
| 7 | C: $\alpha = 34$ $\beta = 6$<br>E: $\alpha = 20$ $\beta = 20$ | (−0.231, −0.040) | (0.737, 0.950) | (0.238, 0.785) |

difference (ignoring any associated variability) would be on a boundary of the null hypothesis for testing for non-inferiority. When compared with case 5, this choice of prior distributions increases the difference in the proportion of responders in the respective arms while not greatly affecting the variance of the posterior distributions.

The seventh case represents starting with 34 responders out of 40 subjects in the control arm and 20 responders out of 40 subjects in the experimental arm. For a non-inferiority margin of 15%, this case starts with observed proportions whose difference (ignoring any associated variability) would be in the interior of a boundary of the null hypothesis for testing for non-inferiority. When compared with case 6, this choice of prior distributions increases the difference in the proportion of responders in the respective arms while not greatly affecting the variance of the posterior distributions.

For cases 1 and 2, since the corresponding posterior distributions of the probability of a responder for the control and experimental arms, $p_C$ and $p_E$, are similar, there was very little difference in the respective 95% equal-tailed credible intervals.

Compared with cases 1 and 2, using a common beta prior distribution with $\alpha = \beta = 20$ (case 3) had a different impact on the respective 95% credible intervals for $p_E - p_C$, $p_E/p_C$, and the odds ratio. For case 3, the 95% equal-tailed credible interval for $p_E - p_C$ has greater upper and lower limits than the corresponding credible interval for cases 1 and 2. The 95% equal-tailed credible interval for $p_E/p_C$ for case 3 is similar to that for cases 1 and 2. The 95% equal-tailed credible interval for the odds ratio for case 3 is noticeably narrower than that for cases 1 and 2—in particular, the lower limit is much greater.

Although using rather informative prior distributions may not be appropriate (we will later elaborate further), it should be understood that there can be a different impact of using such prior distributions for different functions of $(p_C, p_E)$.

Compared with cases 1 and 2, the upper limits of the respective 95% equal-tailed credible intervals for case 4 are about the same, whereas the lower limits are noticeably greater. Since case 5 has the same means for the posterior distributions as case 1 but with smaller variances, the corresponding 95% equal-tailed credible intervals are narrower versions (from both ends) of those for case 1. Additionally, since the prior distribution for the experimental arm in case 5 moves probability to the left compared with the corresponding prior distribution in case 4, the corresponding 95% equal-tailed credible intervals for case 5 has limits to the left of those for case 4.

Since cases 6 and 7 involve prior distributions for the experimental arm that further moves probability to the left, the corresponding 95% equal-tailed credible intervals have limits further to the left. Note for case 7, the conclusion is that the control arm is superior to the experimental arm. For cases 6 and 7, usage of these pairs of prior distributions make it more difficult to conclude non-inferiority based on a difference in the probability of a responder using a margin of 15%.

If the results of a clinical trial are to stand alone, the $\alpha$'s and $\beta$'s for the prior distributions should be relatively small when compared to the sample size. Otherwise, as would be done in a meta-analysis, the use of a beta prior distribution for each arm involves integrating prior successes and failures with the successes and failures in the present clinical trial.

We note again that it is the size of the parameters for the prior distributions that can be influential rather than the prior probability of non-inferiority, inferiority, or superiority. Suppose for the experimental and control arms, the prior distributions for the probability of a response are beta distributions with $\alpha = 5 \times 10^{-10}$ and $\beta = 9.5 \times 10^{-9}$ for the experimental arm and $\alpha = 9.5 \times 10^{-9}$ and $\beta = 5 \times 10^{-10}$ for the control arm. Then the prior probability that $p_E > p_C$ is greater than 0.9 (90%). However, these prior distributions lose any real impact once the response status is known for at least one patient in each arm.

In addition to using posterior probabilities for testing non-inferiority, Kim and Xue[60] discussed two other Bayesian approaches for non-inferiority testing. The first alternative approach determines the 5% contour region defined as those possible $(p_E, p_C)$ whose joint posterior density is 5% of the joint density of the mode. When the 5% contour region lies entirely within the non-inferiority region (the alternative hypothesis), non-inferiority is concluded. The second alternative approach concludes non-inferiority whenever the 95% credible set of highest posterior probability for $(p_E, p_C)$ lies entirely within the non-inferiority region (the alternative hypothesis).

---

## 11.6 Stratified and Adjusted Analyses

A stratified analysis yields a single comparison between arms by combining the stratum-level comparisons. An analysis that adjusts for baseline covariates will yield a comparison between arms that is not obscured by differences between arms in the distributions of known baseline covariates. Other types of adjusted analyses make adjustments over some characteristic (strata levels or combination of levels) so that the estimators and/or some comparison are reflective of some target population. This provides for unbiased estimation for that target population. For a non-inferiority comparison, it is important that the method used to compare the proportions in the non-inferiority trial is consistent with that used in the determination of the non-inferiority margin. This is particularly true when the value of the parameter of interest is dependent on the analysis. For example, for an odds ratio, unadjusted and adjusted estimators will be estimating different parameters/values. When the factors are prognostic, the mean value of an unadjusted estimator will be different from the mean value of an adjusted estimator. Consider the example provided in Table 11.13.

For each stratum, the odds ratio is 2. Therefore, the estimate adjusting for strata is 2. The unadjusted or overall odds ratio is 1.96. In general (not always), for relative measures, when the factors or covariates are prognostic and the size of the effect or relative effect is similar across factors/covariates, the adjusted estimator will tend to be further away from equality than the unadjusted estimator. For relative measures, it is important that the analysis

**TABLE 11.13**

Distribution of Successes and Failures between Arms across Strata

| | Experimental Arm | | Control Arm | |
|---|---|---|---|---|
| Strata | Success | Failure | Success | Failure |
| 1 | 40 | 20 | 30 | 30 |
| 2 | 30 | 30 | 20 | 40 |
| Total | 70 | 50 | 50 | 70 |

method used in determining the effect of the control therapy and the non-inferiority margin also be used in comparing the experimental and control arms in the non-inferiority trial. If not, some "adjustment" may be needed to the non-inferiority margin.

A Cochran–Mantel–Haenszel procedure is a very common stratified analysis when testing for the difference (or superiority or inferiority) between two treatment arms on a binary endpoint. Essentially, the test statistic has for its numerator the sum across strata within one of the arms of the difference in the observed number of successes and the expected number of successes (assuming no difference in success rates between arms). The denominator is an estimate of the corresponding standard deviation under the assumption that the success rate is equal between arms within each stratum. This is one of the primary ways of performing a stratified analysis.

Another type of stratified or adjusted analysis adjusts with respect to some preset relative frequency of some characteristic or combinations of characteristics in a target population. A particular subpopulation or stratum would consist of subjects that have the same level for that chosen characteristic or the same combination of levels of many characteristics. For valid comparisons, the same relative weights should be used for each arm.

### 11.6.1 Adjusted Rates

Suppose a target population is partitioned into $k$ subpopulations. Let $a_i$ denote the proportion of the target population that belongs to the $i$th subpopulation (all $a_i > 0$, $\Sigma a_i = 1$). Then an adjusted rate of success for the target population is given by $\hat{p}_{\text{Target}} = \sum_{i=1}^{k} a_i \hat{p}_i$, where $\hat{p}_i$ is some estimator of $p_i$, the success rate for the $i$th subpopulation. If $\hat{p}_i$ is an unbiased estimator of $p_i$, then $\hat{p}_{\text{Target}}$ will be an unbiased estimator of the true success rate for the target population.

Each $\hat{p}_i$ can be modeled as having a normal distribution with mean $p_i$ and variance $p_i(1 - p_i)/b_i$, where $b_i$ is the number of Bernoulli trials observed for the $i$th subpopulation. Then $\hat{p}_{\text{Target}}$ can be modeled as having a normal distribution with mean $\sum_{i=1}^{k} a_i p_i$ and variance $\sum_{i=1}^{k} a_i^2 p_i(1-p_i)/b_i$.

For a clinical trial, let $\hat{p}_{E,i}$ and $\hat{p}_{C,i}$ denote the observed proportion of "successes" in the $i$th stratum or subpopulation for the experimental and control arms, respectively. Then the respective estimators for the target population are given by $\hat{p}_E = \sum_{i=1}^{k} a_i \hat{p}_{E,i}$ and $\hat{p}_C = \sum_{i=1}^{k} a_i \hat{p}_{C,i}$. The difference is given by $\hat{p}_E - \hat{p}_C = \sum_{i=1}^{k} a_i(\hat{p}_{E,i} - \hat{p}_{C,i})$. Thus, the difference in the overall estimated rates is a weighted average (same weights) of the difference in the observed rates within each stratum or subpopulation. The overall relative risk is given by

$$\hat{p}_E/\hat{p}_C = \frac{\sum_{i=1}^{k} a_i \hat{p}_{E,i}}{\sum_{i=1}^{k} a_i \hat{p}_{C,i}} = \frac{\sum_{i=1}^{k} (a_i \hat{p}_{C,i})(\hat{p}_{E,i}/\hat{p}_{C,i})}{\sum_{i=1}^{k} a_i \hat{p}_{C,i}}.$$ Thus, the overall relative risk

for the target population can be expressed as a weighted average, with random weights, of the relative risks within the strata. The overall odds ratio is given by

$$\hat{p}_E(1-\hat{p}_C)/(\hat{p}_C(1-\hat{p}_E)) = \frac{\sum_{i=1}^{k} \sum_{j=1}^{k} a_i a_j \hat{p}_{E,i}(1-\hat{p}_{C,j})}{\sum_{i=1}^{k} \sum_{j=1}^{k} a_i a_j (1-\hat{p}_{E,i})\hat{p}_{C,j}}.$$ Since this

expression involves the product of terms calculated from different strata, this odds ratio estimator cannot be expressed as a weighted average of the within-stratum odds ratios. Discussion and inferences about odds ratio provided later will instead be based on a common or "average" odds ratio.

There are many choices for how to do a stratified or adjusted non-inferiority analysis of a binary endpoint. This goes beyond whether a difference in proportions, a relative risk, or an odds ratio is chosen as the basis for making an inference. When comparing two proportions or probabilities in a randomized, stratified clinical trial, one method for comparing the difference in proportions uses the overall strata sizes as the common weights for each arm. This allows for a comparison of the two proportions with respect to a target population that has the same breakdown for the strata levels as observed in the study. This is also consistent with the one proportion problem in estimating a common proportion across strata (or studies). When it is assumed that the true probability of a success for an arm is constant across strata (or studies), the MLE of the common probability of a success for that arm uses the total number of subjects in that stratum for just that arm (or the study size for that arm) as weights. This will lead to the overall proportion of successes as the estimate of the common probability of a success. However, it may not be reasonable to assume for a given arm that the true probability of a success is constant across strata. It should be noted that in a clinical trial, the most prognostic factors are selected as stratification factors. Thus, the expectation is that the success rate will vary greatly across stratification levels of the same factor.

Another adjusted analysis uses the harmonic mean of the number of subjects in the experimental and control arms within a stratum as the stratum weight. This is the adjustment that is used in determining the Mantel–Haenszel estimators of the difference in proportions and the relative risk, which will be introduced later. When the ratio of the number of subjects in the experimental arm to that of the control arm is constant across strata, using the harmonic means as the strata weights leads to the same estimates as using the overall strata sizes as the strata weights.

When making an inference on the difference in means for two arms, where the data are assumed to be normally distributed and the difference in means is constant across strata or studies, the MLE of the common difference in means of an arm uses the inverse of the variance (or mean square error) of the strata sample mean difference as the weights. If there is a common variance of the observations across strata and arms, then this approach is equivalent to using the harmonic mean of the number of subjects in the experimental and control arms within a stratum as the stratum weight.

If the non-inferiority hypothesis is based on a relative risk (or on an odds ratio), a decision needs to be made on whether to use the ratio of the two overall proportions (or an odds ratio based on the overall proportions) or to estimate a common or "average" relative risk by integrating the estimated relative risk (estimated odds ratio) across strata. How the results from other trials were used to determine the non-inferiority margin/criterion needs to be considered in making the choice.

## 11.6.2 Adjusted Estimators

For $i = 1, 2, \ldots, k$, let $X_{E,i}$ and $X_{C,i}$ have a binomial distributions with respective parameters $n_{E,i}$ and $p_{E,i}$, and $n_{C,i}$ and $p_{C,i}$. Further assume that all $X_{j,i}$ are independent for $i = 1, 2, \ldots, k$ and $j = E, C$. For the $i$th strata or subpopulation, Table 11.14 provides the notation for the cell counts.

The observed proportion of successes within strata will be denoted by $\hat{p}_{j,i}$ for $i = 1, 2, \ldots, k$ and $j = E, C$.

*Risk Difference.* Let $\Delta_i$ denote the true risk difference for the $i$th strata or subpopulation. The observed risk difference in the $i$th stratum or subpopulation is given by $\hat{\Delta}_i = x_{E,i}/n_{E,i} - x_{C,i}/n_{C,i}$. The weighted least squares estimator of a common overall risk difference ($\Delta_i \equiv \Delta$) is given by $\hat{\Delta}_w = \sum_{i=1}^{k} w_i \hat{\Delta}_i / \sum_{i=1}^{k} w_i$, where $w_i = 1/[x_{E,i}(n_{E,i} - x_{E,i})/n_{E,i} + x_{C,i}(n_{C,i} - x_{C,i})/n_{C,i}]$. Relative to the proportion of all observations that are in a stratum, this estimator downweights those strata where the observed success rates are near 0.5, and overweights those strata where the observed success rates are close to 0 or 1. For most clinical trials, the risk difference will not be constant or approximately constant across strata.

The Mantel–Haenszel estimator of the common risk difference across strata is given by

**TABLE 11.14**

Notation for the Cell Counts

| | Experimental | Control | |
|---|---|---|---|
| Success | $x_{E,i}$ | $x_{C,i}$ | |
| Failure | $n_{E,i} - x_{E,i}$ | $n_{C,i} - x_{C,i}$ | |
| | $n_{E,i}$ | $n_{C,i}$ | $n_i$ |

$$\hat{\Delta}_{MH} = \frac{\sum_{i=1}^{k} (x_{E,i} n_{C,i} - x_{C,i} n_{E,i})/N_i}{\sum_{i=1}^{k} n_{C,i} n_{E,i}/N_i} = \sum_{i=1}^{k} w_i \hat{\Delta}_i \bigg/ \sum_{i=1}^{k} w_i$$

where $w_i = n_{C,i} n_{E,i}/N_i = 1 \bigg/ \left( \dfrac{1}{n_{C,i}} + \dfrac{1}{n_{E,i}} \right)$ and $\hat{\Delta}_i = x_{E,i}/n_{E,i} - x_{C,i}/n_{C,i}$. Equivalently, the weight for a given stratum can be considered as the harmonic mean of the within-stratum sizes for the experimental and control arms.

The estimators

$$\hat{p}_{E,MH} = \frac{\sum_{i=1}^{k} \hat{p}_{E,i}(1/n_{C,i} + 1/n_{E,i})^{-1}}{\sum_{i=1}^{k} (1/n_{C,i} + 1/n_{E,i})^{-1}} \quad \text{and} \quad \hat{p}_{C,MH} = \frac{\sum_{i=1}^{k} \hat{p}_{C,i}(1/n_{C,i} + 1/n_{E,i})^{-1}}{\sum_{i=1}^{k} (1/n_{C,i} + 1/n_{E,i})^{-1}} \quad (11.26)$$

will be regarded as the Mantel–Haenszel estimators of $p_E$ and $p_C$, respectively. In this case, the parameters $p_E$ and $p_C$ are analogous weighted averages of the respective strata probabilities of success. These estimators weigh the within-stratum proportions (estimators) by the harmonic mean of the number of subjects in the experimental and control arms within the stratum. Here, we have $\hat{\Delta}_{MH} = \hat{p}_{E,MH} - \hat{p}_{C,MH}$. When $\hat{p}_{E,MH}$ and $\hat{p}_{C,MH}$ have approximate normal distributions, confidence intervals can be found for the Mantel–Haenszel average risk difference.

*Relative Risk.* Let $\theta_i$ denote the true relative risk for the $i$th strata or subpopulation. The observed relative risk in the $i$th strata or subpopulation is given by $\hat{\theta}_i = x_{E,i} n_{C,i}/x_{C,i} n_{E,i}$. Grizzle, Starmer, and Koch[61] provided the asymptotically weighted least squares estimator of a common log-relative risk (log $\theta_i \equiv$ log $\theta$), which is given by $\log \hat{\theta}_w = \sum_{i=1}^{k} w_i \log \hat{\theta}_i \bigg/ \sum_{i=1}^{k} w_i$, where $w_i = 1/x_{E,i} - 1/n_{E,i} + 1/x_{C,i} - 1/n_{C,i}$. The weight for a stratum is the inverse of the asymptotic variance of the respective log-relative risk estimators.

The Mantel–Haenszel estimator of the common or average relative risk across strata (or the relative risk for the appropriate target population) is given by

$$\hat{\theta}_{MH} = \frac{\sum\limits_{i=1}^{k} x_{E,i} n_{C,i} \Big/ N_i}{\sum\limits_{i=1}^{k} x_{C,i} n_{E,i} \Big/ N_i} = \sum_{i=1}^{k} w_i \hat{\theta}_i \Big/ \sum_{i=1}^{k} w_i \qquad (11.27)$$

where $w_i = x_{C,i} n_{E,i} / N_i = (x_{C,i}/n_{C,i})/(1/n_{C,i} + 1/n_{E,i})$. We see that the Mantel–Haenszel estimator of the relative risk equals the ratio of the Mantel–Haenszel estimators for $p_E$ and $p_C$ (i.e., $\hat{\theta}_{MH} = \hat{p}_{E,MH}/\hat{p}_{C,MH}$ ). In an example to follow, we will use Fieller's method and the approximate normal distributions of $\hat{p}_{E,MH}$ and $\hat{p}_{C,MH}$ to find a 95% confidence interval for the true Mantel–Haenszel relative risk.

*Odds Ratio.* Let $\psi_i$ denote the true odds ratio for the $i$th strata or subpopulation. The odds ratio in the $i$th strata or subpopulation is given by $\hat{\psi}_i = x_{E,i}(n_{C,i} - x_{C,i})/(x_{C,i}(n_{E,i} - x_{E,i}))$. The asymptotically weighted least squares estimator of a common log-odds ratio (log $\psi_i \equiv$ log $\psi$) is given by $\log \hat{\psi}_w = \sum\limits_{i=1}^{k} w_i \log \hat{\psi}_i \Big/ \sum\limits_{i=1}^{k} w_i$, where $w_i = 1/x_{E,i} + 1/(n_{E,i} - x_{E,i}) + 1/x_{C,i} + 1/(n_{C,i} - x_{C,i})$. This estimator is known as the Woolf estimator. The weights are the inverse of the asymptotic variance of the respective log-odds ratio estimators within stratum. Gart[62] showed that the Woolf estimator and the MLE have the same asymptotic distribution and that the Woolf estimator is asymptotically efficient.

The Mantel–Haenszel estimator of the common or average odds ratio across strata is given by

$$\hat{\psi}_{MH} = \frac{\sum\limits_{i=1}^{k} x_{E,i}(n_{C,i} - x_{C,i}) \Big/ N_i}{\sum\limits_{i=1}^{k} x_{C,i}(n_{E,i} - x_{E,i}) \Big/ N_i} = \sum_{i=1}^{k} w_i \hat{\psi}_i \Big/ \sum_{i=1}^{k} w_i \qquad (11.28)$$

where $w_i = x_{C,i}(n_{E,i} - x_{E,i})/N_i = (x_{C,i}/n_{C,i})(1 - x_{E,i}/n_{E,i})/(1/n_{C,i} + 1/n_{E,i})$. Unlike with the risk difference and the relative risk, the Mantel–Haenszel estimator of the common odds ratio does not equal the odds ratio based on the Mantel–Haenszel estimators for $p_E$ and $p_C$ ($\hat{\psi}_{MH} \neq (\hat{p}_{E,MH}(1 - \hat{p}_{C,MH})/(\hat{p}_{C,MH}(1 - \hat{p}_{E,MH})))$).

Robins, Breslow, and Greenland[63] provided an estimate of the variance of $\log(\hat{\psi}_{MH})$, which works well for both large strata and sparse strata data, that is given by

$$\sum_{i=1}^{k} P_i R_i /(2R_+^2) + \sum_{i=1}^{k} (P_i S_i + Q_i R_i)/(2R_+ S_+) + \sum_{i=1}^{k} Q_i S_i /(2S_+^2) \qquad (11.29)$$

where $P_i = (x_{E,i} + n_{C,i} - x_{C,i})/N_i$, $Q_i = x_{C,i} + n_{E,i} - x_{E,i}$, $R_i = x_{E,i}(n_{C,i} - x_{C,i})/N_i$, $S_i = x_{C,i}(n_{E,i} - x_{E,i})/N_i$, $R_+ = \sum_{i=1}^{k} R_i$, and $S_+ = \sum_{i=1}^{k} S_i$. We will use this standard error estimate in an example to construct confidence intervals for the common or average odds ratio.

Mantel and Haenszel[64] indicated their disbelief that the relative risk for an exposure factor would be constant across strata and suggested the use instead of an average relative risk or, rather, an average odds ratio.

*Logistic Regression.* A logistic regression model can be used to estimate a common log-odds ratio across all possibilities for a collection of covariates. For a logistic regression model, the log-odds of a success is a linear function of the covariate values of the given patient. A patient having baseline covariates values $x_1, \ldots, x_k$ (one or more of these covariates used to identify the treatment arm) has a log-odds of success of $\alpha + \sum_{i=1}^{k} \beta_i x_i$. When the sample size is large, the MLEs will have approximate normal distributions. For fixed values of the baseline covariates, the *treatment effect* represents the common log-odds ratio between the experimental and control arms.

Example 11.13 illustrates the use of these methods.

## Example 11.13

To illustrate these methods, data were simulated for a two-arm study having 200 subjects. One hundred subjects were randomized to each arm according to two stratification factors having two levels each. The endpoint is a binary response. Table 11.15 gives the subject breakdown according to treatment arm, stratification factors, and response status.

From Equation 11.26, the Mantel–Haenszel estimates of $p_E$ and $p_C$ are 0.631 and 0.429, respectively. The respective estimates of the corresponding standard deviations are 0.0479 and 0.0480. The approximate 95% confidence interval for $p_E - p_C$ is 0.069–0.335. From Equation 11.27, the Mantel–Haenszel estimate of the relative risk $p_E/p_C$ is 1.47. On the basis of Fieller's method, the approximate 95% confidence interval of 1.14–1.95 is found by solving for the values of $x$ that satisfy $-1.96 < (0.631 - 0.429x)/((0.0479)^2 + (0.0480x)^2)^{0.5} < 1.96$.

From Equation 11.28, the Mantel–Haenszel estimate of the common odds ratio is 2.31 with approximate 95% confidence interval of 1.30–4.09 based on the standard error estimate of the log-odds ratio of Robins, Breslow, and Greenland in Equation 11.29.

For a logistic regression model using treatment arm and the two stratification factors as factors in the model, the estimate of the common odds ratio is 2.31 with corresponding approximate 95% confidence interval of 1.30–4.09, identical to the Mantel–Haenszel estimate and the corresponding confidence interval. When an interaction term for the stratification factors is added, the estimate of the common odds ratio is 2.33 with a corresponding approximate 95% confidence interval of 1.31–4.15.

**TABLE 11.15**

Breakdown by Treatment Arm, Stratification Factors, and Response Status

| Arm | Factor 1 | Factor 2 | $n$ | Number of Responses |
|------|------|------|------|------|
| Experimental | 0 | 0 | 30 | 21 |
| | 0 | 1 | 25 | 15 |
| | 1 | 0 | 23 | 15 |
| | 1 | 1 | 22 | 12 |
| Control | 0 | 0 | 32 | 18 |
| | 0 | 1 | 25 | 6 |
| | 1 | 0 | 22 | 9 |
| | 1 | 1 | 21 | 10 |

## 11.7 Variable Margins

There are non-inferiority testing procedures and real cases involving proportions that are not based on a fixed margin for a risk difference, a relative risk, or an odds ratio. These other procedures generally prespecify which pairs of $(p_E, p_C)$ satisfy that the experimental therapy is noninferior to the control therapy. Whether a difference, $p_E - p_C$, can be regarded as acceptable (i.e., that the experimental therapy is noninferior to the control therapy) depends on the control success rate. The non-inferiority criteria can be expressed by a "variable margin" on the risk difference. That is, a function $\delta$ is prespecified so that the hypotheses for non-inferiority testing are given by (for positive outcomes)

$$H_o: p_E - p_C \leq -\delta(p_C) \text{ vs. } H_a: p_E - p_C > -\delta(p_C) \qquad (11.30)$$

For a given value of $0 \leq p_C \leq 1$, the value of $\delta(p_C)$ satisfies $0 \leq \delta(p_C) \leq p_C$ and is prespecified. If a success is an undesirable outcome, the roles of $p_C$ and $p_E$ in the hypotheses in Expression 11.30 would be reversed (i.e., test $H_o: p_C - p_E \leq -\delta(p_C)$ vs. $H_a: p_C - p_E > -\delta(p_C)$).

Note that all previous cases of non-inferiority testing of proportions involve hypotheses that are specific cases of those given in Expression 11.30. For a fixed margin on a risk difference, we have $\delta(p_C) = \delta$. For a fixed margin on a relative risk, we have $\delta(p_C) = (1 - \theta_o)p_C$. For a fixed margin on an odds ratio, we have $\delta(p_C) = (1 - \psi_o)p_C(1 - p_C)/[1 - (1 - \psi_o)p_C]$.

For testing the hypotheses in Expression 11.30, Zhang[65] provided the delta-method large sample normal statistic of

$$Z_S = \frac{\hat{p}_E - \hat{p}_C + \delta(p_C)}{\sqrt{\tilde{p}_E(1 - \tilde{p}_E)/n_E + \tilde{p}_C(1 - \tilde{p}_C)(1 - \delta'(\tilde{p}_C))^2/n_C}}$$

where $\delta'$ is the first derivate of $\delta$. The values for $\tilde{p}_E$ and $\tilde{p}_C$ may correspond to the sample proportions or to the MLEs of the proportions under the null hypothesis. The hypotheses in Expression 11.30 can also be tested on the basis of the posterior probability that $p_E - p_C > -\delta(p_C)$. Alternatively, the specific form for $\delta(p_C)$ may dictate an appropriate method of analysis.

*Various Variable Margins.* For the risk difference and the relative risk, the corresponding variable margin is a linear function of $p_C$. Phillips[66] proposed the use of a linear function in $p_C$, $\delta(p_C) = a + bp_C$. A motivation was to "fit" a line to the random margin provided in the U.S. Food and Drug Administration (FDA) guidelines for anti-infective products[67] by having that margin based on $p_C$. The value for $b < 0$ in that fit is indicative of a margin that increases as $p_C$ decreases. Thus, when the success rate (and possibly the effect) of the active control appears smaller, the acceptable amount of inferiority becomes larger. This seems counterintuitive. It would make more sense to either maintain the same or smaller amount of the effect of the control therapy as the perceived effect becomes smaller. When $b > 0$, the variable margin of $\delta(p_C) = a + bp_C$ appropriately decreases as $p_C$ decreases. For $a > 0$ and $0 < b < 1$, $p_C - (a + bp_C) < 0$ whenever $p_C < a/(1 - b)$. Thus, such a variable margin should be avoided whenever $p_C$ may be less than $a/(1 - b)$.

Röhmel[68,69] proposed various functions for $\delta(\bullet)$. In one study, Röhmel[69] proposed to use $\delta(p_C) = 0.223\sqrt{p_C(1-p_C)}$ and $\delta(p_C) = 0.333\sqrt{p_C(1-p_C)}$ for the purpose of stabilizing the desired power and providing a variable margin fairly consistent with those provided in the FDA guidelines for anti-infective products.[67] The power should be fairly stable among possibilities for $p_C$ that are not close to 0 or 1. Röhmel appears to have been recommending such a variable margin for real situations—for example, antibiotics or anti-infective products—when the anticipated success rate is greater than 50%. For $0.50 < p_C < 1$, the margin of $\delta(p_C) = c\sqrt{p_C(1-p_C)}$ increases as $p_C$ decreases. When it is anticipated that $0.50 < p_C < 1$, such a function for the margin is probably no longer appropriate for a non-inferiority registration trial. For $0 < p_C < 0.50$, the margin of $\delta(p_C) = c\sqrt{p_C(1-p_C)}$ decreases as $p_C$ decreases. This is more appropriate. However, $p_C - c\sqrt{p_C(1-p_C)} < 0$ when $p_C$ is close to zero. Thus, such a variable margin should be avoided when $p_C$ may be close to zero.

By stabilizing the power for a given sample size, such a variable margin may be appropriate for a randomized phase 2 study to assist in making a go/no go decision to phase 3. For a one-sided significance level of $\alpha/2$ and power of $1 - \beta$ at $p_E - p_C = p$, where $\delta(p_C) = c\sqrt{p_C(1-p_C)}$ with $c > 0$, a crude sample-size calculation of the number of patients per arm is given by $2(z_\beta + z_{\alpha/2})^2/c^2$ (derived from Equation 11.7 with $k = 1$, $\delta = c\sqrt{p_C(1-p_C)}$, $\Delta_a = 0$, and $p_1' = p_1 = p_2 = p_2'$).

In another study, Röhmel[68] proposed possibly defining $\delta$ ($\bullet$), so that $\Phi^{-1}(p_C) - \Phi^{-1}(p_C - \delta(p_C)) = d$, where $d$ is a prespecified constant and $\Phi$ is the standard normal distribution function. This leads to $\delta(p_C) = p_C - \Phi(\Phi^{-1}(p_C) - d)$. Use of this function of $\delta(\bullet)$ corresponds to testing the hypotheses $H_o$: $\Phi^{-1}(p_E) - \Phi^{-1}(p_C) \leq -d$ versus $H_a$: $\Phi^{-1}(p_E) - \Phi^{-1}(p_C) > -d$. Zhang[65] noted that Röhmel's proposal could be extended to any distribution function $F$. For $F(x) = (1 + e^{-x})^{-1}$, the hypotheses reduce to the standard non-inferiority hypotheses of an odds ratio given in Expression 11.23.

Some authors, including Tsou et al.,[70] have discussed the use of a mixed margin. When the control success rate lies in particular range (e.g., above 0.2), a fixed margin on the difference is used. When the control success rate lies outside this range, a fixed margin on the relative risk of success (e.g., control success rates below 0.2) is used. Tsou et al. proposed a synthesis method for evaluating such cases. A similar approach can also be considered when using a prespecified variable margin.

*A Comparison of Methods.* Kim and Xue[60] compared the type I error rate and power of various methods based on a piece-wise linear model for the variable margin for a non-inferiority evaluation of the Safety of Estrogens in Systemic Lupus Erythematosus National Assessment (SELENA) study. The desired one-sided type I error rate is 5%. For various selected placebo rates, eight experts were independently queried on the appropriate non-inferiority margin. After averaging, discussion, and consensus, the variable margin was defined by line segments connecting the choices for the non-inferiority margin for those selected control rates.

The methods compared by Kim and Xue[60] include an "observed event rate" approach, two Bayesian approaches, and an exact likelihood ratio test. The observed event rate approach concludes non-inferiority (involving negative/ undesirable outcomes) when the upper bound of a one-sided 95% confidence interval for $p_E - p_C$ is less than $\delta(\hat{p}_C)$. The confidence interval is based on the Farrington and Manning approach, where the standard error is based on the null restricted MLEs of $p_E$ and $p_C$.

The first of two Bayesian approaches determines the posterior probability that $p_E$ and $p_C$ lie in the alternative hypothesis based on independent uniform prior distributions. When the posterior probability is greater than 0.95, non-inferiority is concluded.

The second Bayesian approach uses a uniform prior distribution for $p_E$ and an informative prior distribution for $p_C$. The informative prior distribution for $p_C$ is determined after the non-inferiority study is completed and equals the posterior distribution that arises from a uniform prior distribution and observing a number of successes in 100 Bernoulli trials that equals to 100× the observed control rate from the non-inferiority study. This case appears to represent having prior information about the control rate.

Control rates between 5% and 40% were used to evaluate the type I error rate and power of these procedures. For the type I error evaluations, $p_E = p_C - \delta(p_C)$; for the power evaluations, $p_E = p_C$. For each combination of $(p_E, p_C)$

and sample size per arm of 150 and 300 subjects, 1000 trials were simulated and the proportion in which non-inferiority was demonstrated with each approach was determined.

The second Bayesian approach (using a retrospective prior for the control rate) and the exact likelihood ratio test maintained a type I error near 0.05 in all cases. The Bayesian approach using independent uniform prior distributions slightly inflates the type I error rate in all cases. The observed event rate procedure had an inflated type I error rate that was as high as 0.10 when the control success rate was 15%. The inflation at all time points is mostly due to the statistic ignoring the variability in $\delta(\hat{p}_C)$. The particularly high inflated type I error rate when the control rate was 15% appears to be due to the value of $\delta(0.15)$ being smaller than what would be consistent with values of $\delta(p_C)$ for $p_C$ near 0.15. The value of $\delta$ (0.15) is 0.09, whereas from our interpolations $\delta(p_C) \approx 0.04 + 0.4p_C$ for $p_C$ near 0.15, which would lead to $\delta(0.15) \approx 0.10$. Thus, although $(p_E, p_C) = (0.24, 0.15)$ is on the boundary of the null hypothesis, the general behavior of $\delta(p_C)$ is such that $(p_E, p_C) = (0.25, 0.15)$ is "expected" to be on the boundary of the null hypothesis and $(p_E, p_C) = (0.24, 0.15)$ is expected to be just in the alternative hypothesis of non-inferiority. Hence, the added inflation of the type I error rate when the control rate is 15%.

The second Bayesian approach tended to have the greatest power followed by the observed event rate approach, and then followed by the first Bayesian approach.

A type I error rate evaluation of using the testing procedure described in the 1992 FDA Guidance[67] is given in Example 11.14.

### Example 11.14

Much of the work on variable margins for proportions has been motivated by experience in testing the efficacy of anti-infective products. The observed margin is given by

$$\hat{\delta} = \begin{cases} 0.10, \text{if } \hat{p}_{max} > 0.90 \\ 0.15, \text{if } 0.80 < \hat{p}_{max} \le 0.90 \\ 0.20, \text{if } \hat{p}_{max} \le 0.80 \end{cases}$$

where $\hat{p}_{max} = \max\{\hat{p}_E, \hat{p}_C\}$. Non-inferiority would be concluded if the lower limit of the 95% two-sided confidence interval for the experimental versus control difference in the cure rates is greater than the negative of the observed margin. Formally, this test procedure does not perfectly correspond to a test of two specific statistical hypotheses. Statistical hypotheses can be specified for a test that is approximately equal to this non-inferiority test. The alternative hypothesis would be defined as the union of the sets $\{(p_E, p_C):p_C < 0.8, p_E > p_C - 0.2\}$, $\{(p_E, p_C):0.8 \le p_C < 0.9, p_E > p_C - 0.15\}$, and $\{(p_E, p_C):p_C \ge 0.9, p_E > p_C - 0.1\}$. The null hypothesis is the complement. The variable margin is given by $\delta(p_C) = 0.2$, if $p_C < 0.8$; $\delta(p_C) = 0.15$, if $0.8 \leq p_C < 0.9$; and $\delta(p_C) = 0.1$, if $p_C \geq 0.9$.

For three possibilities in the null hypothesis and sample sizes of 150 and 300 per arm, the simulated probabilities of rejecting the null hypothesis (the type I error rate) for testing these hypotheses by using the testing procedure described in the 1992 FDA Guidance are provided in Table 11.16. The two-sided 95% Wald's confidence interval for the difference in proportions was used. Two possibilities are located at or near where the variable margin changes. One million simulations were used in each case. For the case where $(p_E, p_C) = (0.65, 0.8)$ and the sample size is 150 per arm, only 124 simulations (about 1 in every 8000 simulations) had $\delta(\hat{p}_C) \neq \hat{\delta} = \delta(\max\{\hat{p}_E, \hat{p}_C\})$. In all of these 124 simulations, non-inferiority was concluded using the smaller margin of $\hat{\delta} = \delta(\max\{\hat{p}_E, \hat{p}_C\})$. In all other studied cases for $(p_E, p_C)$ and the sample size, observing $\delta(\hat{p}_C) \neq \hat{\delta} = \delta(\max\{\hat{p}_E, \hat{p}_C\})$ was more rare and never influenced the conclusion on non-inferiority. Unless the sample size is small, non-inferiority will be demonstrated with respect to $\hat{\delta} = \delta(\max\{\hat{p}_E, \hat{p}_C\})$ whenever $\hat{p}_E > \hat{p}_C$. Thus, for all practical purposes, the observed margin could have been regarded as $\delta(\hat{p}_C)$. When $(p_E, p_C) = (0.65, 0.8)$, the type I error rate is greatly inflated and tends to increase toward some value slightly larger than 0.5 as the common sample size increases without bound. When $(p_E, p_C) = (0.599, 0.799)$, the type I error rate is slightly deflated and tends to increase toward 0.025 as the common sample size increases without bound. When $(p_E, p_C) = (0.70, 0.85)$, a value on the boundary of the null hypothesis not very near a change point in the variable margin, the type I error rate is inflated and tends to decrease toward 0.025 as the common sample size increases without bound.

## 11.8 Matched-Pair Designs

Matched-pair designs are frequently used in assessing the performance of diagnostic or laboratory tests. It is also often used in clinical trials to increase the efficiency of treatment comparison. For example, subjects are asked to take both an experimental and a standard treatment in a crossover clinical trial to reduce the variability of treatment comparison. In many laboratory studies of new assays or diagnostic tests, a sample is split into

**TABLE 11.16**

Type I Error Rates Consistent with Old FDA Guidelines on Anti-Infective Products

| | Sample Size per Arm | | |
|---|---|---|---|
| $(p_E, p_C)$ | 150 | 300 | $\infty$ |
| (0.599, 0.799) | 0.0234 | 0.0235 | 0.025 |
| (0.65, 0.8) | 0.1385 | 0.2201 | $\approx 0.5007$ |
| (0.70, 0.85) | 0.0438 | 0.0293 | 0.025 |

two with one tested by a new assay (or diagnostic test) and the other by the standard method. When the outcome measure is dichotomous, risk difference and risk ratios are often used to compare treatments. In this section we describe the statistical methods for evaluating non-inferiority of the difference and the rate ratio of two proportions in a matched-pair design. Methods appropriate for large samples and small to moderate samples will be discussed.

Consider the matched-pair design in which two treatments (e.g., experimental and control) are performed on the same $n$ subjects. A "response" to a treatment will be denoted by a "1," whereas a "2" will denote "no response" to the treatment. For any subject the possible outcomes are denoted by (1, 1), (1, 2), (2, 1), and (2, 2), where the first (second) entry is the outcome to the experiment (control) treatment. Let $q_{11}$, $q_{12}$, $q_{21}$, and $q_{22}$ be the corresponding probabilities of the pairs and let $a$, $b$, $c$, and $d$ $(a + b + c + d = n)$ be the observed numbers for the pairs (1, 1), (1, 2), (2, 1), and (2, 2), respectively. The observed vector $(a, b, c, d)$ is assumed to come from the usual *multinomial distribution model*:

$$P((a,b,c,d) \mid n,(q_{11},q_{12},q_{11},q_{12})) = \frac{n!}{a!b!c!d!} q_{11}^a\, q_{12}^b\, q_{21}^c\, q_{22}^d.$$

Then $p_E = q_{11} + q_{12}$ and $p_C = q_{11} + q_{21}$ are the probability of a response to the experimental and control treatments, respectively. For a classical hypothesis test of no difference between the new and standard treatments,

$$H_o\colon p_E = p_C \text{ (i.e., } q_{12} = q_{21}) \text{ vs. } H_a\colon p_E \neq p_C \text{ (i.e., } q_{12} \neq q_{21}),$$

the well-known McNemar[71] test is

$$Z_M = \frac{b-c}{\sqrt{b+c}}. \tag{11.31}$$

The square of $Z_M$ is a $\chi^2$ statistic with 1 degree of freedom. The McNemar statistic depends only on the off-diagonal cells (discordant pairs) for the inference about whether the two treatments are different.

### 11.8.1 Difference in Two Correlated Proportions

The difference of two correlated proportions is frequently used in comparing two treatments or diagnostic procedures in a matched-pair design. For a non-inferiority test for some $\delta > 0$, the hypotheses are

$$H_o : p_E - p_C \leq -\delta \quad \text{vs.} \quad H_a : p_E - p_C > -\delta. \tag{11.32}$$

Rejection of the null hypothesis will lead to a conclusion of non-inferiority. The definition of a positive response will depend on the set up of the study. For example, the positive response may refer to the ability of an assay to detect an antigen of interest in assay development or it may represent a successful outcome following a treatment in a crossover study.

For the non-inferiority hypotheses in Expression 11.32, Tango[72] proposed a score-type test

$$Z_D = \frac{b - c + n\delta}{\left\{ n\left(2\hat{q}_{21} - \delta(\delta + 1)\right)\right\}^{1/2}}  \tag{11.33}$$

where the estimator $\hat{q}_{21}$ is the constrained MLE of $q_{21}$ under $p_E - p_C = -\delta$. Specifically, Tango[72] showed that

$$\hat{q}_{21} = \frac{\sqrt{(B^2 - 4AC)} - B}{2A}  \tag{11.34}$$

where

$$A = 2n, \ B = -b - c - (2n - b + c)\delta, \ C = c\delta(\delta + 1).$$

The null hypothesis will be rejected at the one-sided $\alpha/2$ level if $Z_D > z_{\alpha/2}$, where $z_{\alpha/2}$ is the upper $\alpha$-percentile of the standard normal distribution.

Tango provided special cases for this test. From Equation 11.34, when $\delta = 0$, corresponding to the test of no difference between the test and control, $\hat{q}_{21} = (b + c)/2n$. Thus, the test statistic $Z_D$ in Equation 11.33 simplifies to the McNemar test statistic in Equation 11.31. When the off-diagonal cells are all zero ($b = c = 0$), $\hat{q}_{21} = \delta$ and the test statistics reduce to

$$Z_D = \left(\frac{n\delta}{1 - \delta}\right)^{1/2}.$$

In other words, if there are no discordant pairs observed from the study, the non-inferiority hypothesis will be rejected at the one-sided $\alpha$ level if the sample size ($n$) is large enough:

$$n > \frac{1 - \delta}{\delta} z_{\alpha/2}^2.$$

On the basis of the score test statistic $Z_D$ in Equation 11.33, a $100(1 - \alpha)\%$ confidence interval for the difference in proportions $\Delta = p_E - p_C = q_{12} - q_{21}$ can be constructed by solving for $\delta$ in the equations: $Z_D = \pm z_{\alpha/2}$. Example 11.15

illustrates the use of non-inferiority testing based on the difference in proportions in a matched-pair design.

### Example 11.15

Miyanaga[73] conducted a crossover clinical trial of 44 patients comparing a chemical disinfection system with a thermal disinfection system. The outcome of the trial is summarized in Table 11.17. It was concluded that the two methods are equivalent based on a one-sided $p$-value of 0.5 from Fisher's exact test. However, as previously mentioned in this text, failure to demonstrate a difference does not mean there is no difference.

The data were reanalyzed by Tango[72] using the non-inferiority test framework with $\delta = 0.1$. From Equation 11.33, $Z_D = 1.709$ with a one-sided $p$-value of .044, based on a standard normal distribution. The lower limit of the corresponding 90% confidence interval is $-0.096$. As a result, it was concluded that the chemical disinfection system is noninferior to the thermal disinfection system at the one-sided 0.05 level, but not at the one-sided 0.025 level. We note that as all observations but one were the same, $Z_D$ may not have a normal distribution.

It has been shown (see Tango[72]) that the score statistic $Z_D$ performs better in small samples than in two other asymptotic tests proposed by Lu and Bean[74] and Morikawa and Yanagawa.[75] In particular, the type I error rate of the $Z_D$ test statistic is much closer to the nominal level than the other two statistics, while maintaining similar power. For a matched-pair design with small sample sizes, an exact test of non-inferiority proposed by Hsueh, Liu, and Chen[76] can be used to guarantee control of the type I error rates. Sample-size and power calculation methods have been developed in Lu and Bean[74] and Nam.[77] Nam[77] showed that a method based on the score-type statistic performed better than the method of Lu and Bean.[74]

### 11.8.2 Ratio of Two Correlated Proportions

When an outcome measure is dichotomous, the relative risk is also often used to compare treatments. A few publications have focused on the relative

**TABLE 11.17**

Outcomes of Disinfection Systems for Soft Contact Lenses

| Chemical Disinfection | Thermal Disinfection | | |
|---|---|---|---|
| | Effective | Ineffective | Total |
| Effective | 43 | 0 | 43 |
| Ineffective | 1 | 0 | 1 |
| Total | 44 | 0 | 44 |

risk measure in matched-pair designs. To show that the experimental treatment is noninferior to the standard treatment based on the relative risk measure, we formulate the hypothesis as follows

$$H_0: \theta = p_E/p_C \le \theta_0 \quad \text{vs.} \quad H_a: \theta = p_E/p_C > \theta_0 \tag{11.35}$$

where $0 < \theta_0 < 1$ is a prespecified acceptable threshold for the ratio of the two proportions. Rejection of the null hypothesis will lead to a conclusion of non-inferiority in the sense that the experimental treatment will have similar positive response rate compared with the standard treatment based on the marginal response rates. Extending the work of Tango[72], a score test was derived by Tang, Tang, and Chan[78] for testing the non-inferiority hypothesis in Equation 11.35

$$Z_R = \frac{a+b-(a+c)\theta_0}{\sqrt{n\{(1+\theta_0)\hat{q}_{21}+(a+b+c)(\theta_0-1)/n\}}}, \tag{11.36}$$

where $\hat{q}_{21}$ is the constrained MLE of $q_{21}$ under the null hypothesis, given by the larger root of the following quadratic equation

$$f(x) = n(1+\theta_0)x^2 + [(a+c)\theta_0^2 - (a+b+2c)]x + c(1-\theta_0)(a+b+c)/n = 0.$$

That is,

$$\hat{q}_{21} = \left[\sqrt{B^2-4AC} - B\right]/(2A) \tag{11.37}$$

where

$$A = n(1+\theta_0), B = [(a+c)\theta_0^2 - (a+b+2c)], \text{ and } C = c(1-\theta_0)(a+b+c)/n.$$

The $Z_R$ statistic has an asymptotic standard normal distribution under the null hypothesis, and the test rejects the null hypothesis at the one-sided $\alpha/2$ level if $Z_R \ge z_{\alpha/2}$.

From Equation 11.37, $\hat{q}_{21} = (b+c)/(2n)$ for the special case of $\theta_0 = 1$, and thus $Z_R$ in Equation 11.36 reduces to the McNemar's test statistic in Equation 11.31. When $a > 0$ and $b = c = 0$ (complete concordance), $Z_R = \{a(1-\theta_0)/\theta_0\}^{1/2}$; thus, $Z_R$ increases with the concordant positive response ($a$). In the rare case where both new and standard treatments have no observed responses ($a = b = c = 0$), $Z_R$ is undefined.

Non-inferiority requires that the lower limit of the confidence interval for the risk ratio, $\theta = p_E/p_C$, is greater than $\theta_0$. A $100(1-\alpha)\%$ confidence interval for the risk ratio $\theta = p_E/p_C$ can be obtained on the basis of the $Z_R$ statistic using iterative procedures. Since $Z_R$ has an asymptotic standard normal distribution, the lower and upper limits of the $100(1 - \alpha)\%$ confidence interval can be obtained as the two roots to the equation of $Z_R^2 = \chi_{1,\alpha}^2$, where $\chi_{1,\alpha}^2$ is the upper $\alpha$th percentile of the central $\chi^2$ distribution with 1 degree of freedom.

Tang, Tang, and Chan[78] compared the performance of $Z_R$ in Equation 11.36 with several other potential test statistics, including one proposed by Lachenbruch and Lynch.[79] The empirical comparison showed that $Z_R$ was the only test statistic that behaved satisfactorily in the sense that its empirical type I error rate was much closer to the desired, nominal level than those for the other tests. The $Z_R$ statistic tends to be slightly conservative in cases where $p_E$ and $p_C$ are large and the probability of discordance ($q_{21}$) is low. In addition, the empirical coverage probabilities of the confidence intervals based on $Z_R$ were close to the nominal level, and the error rates of both tails were generally similar. Sample sizes and power calculation formulas based on the $Z_R$ statistic for both hypothesis testing and confidence interval estimation were given by Tang et al.[80]

Example 11.16 illustrates the use of non-inferiority testing based on a relative risk in a matched-pair design.

## Example 11.16

Tang, Tang, and Chan[78] revisited the crossover clinical trial described in Example 11.15, where a chemical disinfection system was compared with a thermal disinfection system for soft contact lenses. Here suppose the interest is to assess non-inferiority using the relative risk with a margin of 0.9 (requiring the response of chemical method be at least 90% of the thermal method). The observed risk ratio (chemical/thermal) is 0.977 with the 90% confidence interval based on $Z_R$ in Equation 11.36 of (0.904–1.038). The p-value for testing the null hypothesis in Equation 11.35 is .044. The results indicate that the chemical method is noninferior to the thermal method at a one-sided 0.05 level, but not at the one-sided level of 0.025.

For small studies, the score test sometimes may be anticonservative—that is, its type I error rates may be inflated. For this reason, Chan et al.[81] studied the performance of an exact unconditional test and an "approximate" exact test for testing non-inferiority hypothesis based on risk ratios in matched-pair designs. Their research showed that the exact unconditional test guarantees the type I error rate and should be recommended if strict control of type I error (protection against any inflated risk of accepting inferior treatments) is required. However, the exact method tends to be overly conservative and computationally demanding. Through empirical studies, it was demonstrated that an "approximate" exact test, which is computationally simple to implement, controls the type I error rate reasonably well while maintaining high power for the hypothesis testing.

# References

1. Dann, R.S. and Koch, G.G., Review and evaluation of methods for computing confidence intervals for the ratio of two proportions and considerations for non-inferiority clinical trials, *J. Biopharm. Stat.*, 15, 85–107, 2005.

2. Barnard, G.A., Significance tests for 2 × 2 tables, *Biometrika*, 34, 123–138, 1947.

3. Basu, D., On the elimination of nuisance parameters, *J. Am. Stat. Assoc.*, 72, 355, 1977.

4. Chan, I.S.F., Exact tests of equivalence and efficacy with a non-zero lower bound for comparative studies, *Stat. Med.*, 17, 1403–1413, 1998.

5. Farrington, C.P. and Manning, G., Test statistics and sample size formulae for comparative binomial trials with null hypothesis of non-zero risk difference or non-unity relative risk, *Stat. Med.*, 9, 1447–1454, 1990.

6. Suissa, S. and Shuster, J.J., Exact unconditional sample sizes for the 2 × 2 binomial trial, *J. R. Stat. Soc. A*, 148, 317–327, 1985.

7. Haber, M., An exact unconditional test for the 2 × 2 comparative trials, *Psychol. Bull.*, 99, 129–132, 1986.

8. Dunnet, C.W. and Gent, M., Significance testing to establish equivalence between treatments with special reference to data in the form of 2 × 2 tables, *Biometrics*, 33, 593–602, 1977.

9. Chan, I.S.F. and Zhang, Z., Test-based exact confidence intervals for the difference of two binomial proportions, *Biometrics*, 55, 1201–1209, 1999.

10. Röhmel, J. and Mansmann, U., Unconditional non-asymptotic one-sided tests for independent binomial proportions when the interest lies in showing non-inferiority and/or superiority, *Biom. J.*, 41, 149–170, 1999.

11. Andres, A.M. and Mato, A.S., Choosing the optimal unconditional test for comparing two independent proportions, *Comput. Stat. Data Anal.*, 17, 555–574, 1994.

12. Chan, I.S.F., Providing non-inferiority or equivalence of two treatments with dichotomous endpoints using exact methods, *Stat. Method. Med. Res.*, 12, 37–58, 2003.

13. Clopper, C.J. and Pearson, E.S., The use of confidence or fiducial limits illustrated in the case of the binomial, *Biometrika*, 26, 404–413, 1934.

14. Santner, T.J. and Snell, M.K., Small-sample confidence intervals for $p_1 - p_2$ and $p_1/p_2$ in 2 × 2 contingency tables, *J. Am. Stat. Assoc.*, 75, 386–394, 1980.

15. Agresti, A. and Min, Y., On small-sample confidence intervals for parameters in discrete distributions, *Biometrics*, 57, 963–971, 2001.

16. Chen, X., A quasi-exact method for the confidence intervals of the difference of two independent binomial proportions in small sample cases, *Stat. Med.*, 21, 943–956, 2002.

17. Coe, P.R. and Tamhane, A.C., Small sample confidence intervals for the difference, ratio, and odds ratio of two success probabilities, *Commun. Stat. B Simul.*, 22, 925–938, 1993.

18. Santner, T.J. and Yamagami, S., Invariant small sample confidence intervals for the difference of two success probabilities, *Commun. Stat. B Simul.*, 22, 33–59, 1993.

19. Fries, L.F. et al., Safety and immunogenicity of a recombinant protein influenza A vaccine in adult human volunteers and protective efficacy against wild-type H1N1 virus challenge, *J. Infect. Dis.*, 167, 593–601, 1993.

20. Boschloo, R.D., Raised conditional level of significance for the 2×2-table when testing the equality of two probabilities, *Stat. Neerl.*, 24, 1–35, 1970.

21. Rodary, C., Com-Nougue, C., and Tournade, M.F., How to establish equivalence between treatments: A one-sided clinical trial in paediatric oncology, *Stat. Med.*, 8, 593–598, 1989.

22. Hauck, W.W. and Anderson, S., A comparison of large sample confidence interval methods for the differences of two binomial probabilities, *Am. Stat.*, 40, 318–322, 1986.

23. Ghosh, B.K., A comparison of some approximate confidence intervals for the binomial parameter, *J. Am. Stat. Assoc.*, 74, 894–900, 1979.

24. Vollset, S.E., Confidence intervals for a binomial proportion, *Stat. Med.*, 12, 809–824, 1993.

25. Agresti, A. and Coull, B.A., Approximate is better than 'exact' for interval estimation of binomial proportions, *Am. Stat.*, 52, 119–126, 1998.

26. Agresti, A. and Caffo, B., Simple and effective confidence intervals for proportions and differences of proportions result from adding two successes and two failures, *Am. Stat.*, 54, 280–288, 2000.

27. Newcombe, R.G., Two-sided confidence intervals for the single proportion: comparison of seven methods, *Stat. Med.*, 17, 857–872, 1998.

28. Newcombe, R.G., Interval estimation for the difference between independent proportions: Comparison of seven methods, *Stat. Med.*, 17, 873–890, 1998.

29. Brown, L.D., Cai, T., and Dasgupta, A., Interval estimation for a binomial proportion (with discussion), *Stat. Sci.*, 16, 101–133, 2001.

30. Wilson, E.B., Probable inference, the law of succession, and statistical inference. *J. Am. Stat. Assoc.*, 22, 209–212, 1927.

31. Schouten, H.J.A. et al., Comparing two independent binomial proportions by a modified chi-square test, *Biom. J.*, 22, 241–248, 1980.

32. Tu, D., A comparative study of some statistical procedures in establishing therapeutic equivalence of nonsystemic drugs with binary endpoints, *Drug Inf. J.*, 31, 1291–1300, 1997.

33. Li, Z. and Chuang-Stein, C., A note on comparing two binomial proportions in confirmatory non-inferiority trials, *Drug Inf. J.*, 40, 203–208, 2006.

34. Mee, R.W., Confidence bounds for the difference between two probabilities, *Biometrics*, 40, 1175–1176, 1984.

35. Miettinen, O.S. and Nurminen, M., Comparative analysis of two rates, *Stat. Med.*, 4, 213–226, 1985.

36. Santner, T.J. et al., Small-sample comparisons of confidence intervals for the difference of two independent binomial proportions, *Comput. Stat. Data Anal.*, 51, 5791–5799, 2007.

37. Dann, R.S. and Koch, G.G., Methods for one-sided testing of the difference between proportions and sample size considerations related to non-inferiority clinical trials, *Pharm. Stat.*, 7, 130–141, 2008.

38. Hilton, J.F., Designs of superiority and non-inferiority trials for binary responses are noninterchangeable, *Biom. J.*, 48, 934–947, 2006.

39. Chan, I.S.F. and Bohidar, N.R., Exact power and sample size for vaccine efficacy studies, *Commun. Stat. Theory*, 27, 1305–1322, 1998.

40. Johnson, N.L., Kotz, S., and Kemp, A.W., *Univariate Discrete Distributions*, Wiley, New York, NY, 1992.
41. Werzberger, A. et al., A controlled trial of a formalin-inactivated hepatitis A vaccine in healthy children, *New Engl. J. Med.*, 327, 453–457, 1992.
42. Wiens, B.L. et al., Duration of protection from clinical hepatitis A disease after vaccination with VAQTA®, *J. Med. Virol.*, 49, 235–241, 1996.
43. Temple, R., Problems in interpreting active control equivalence trials, *Acct. Res.*, 4, 267–275, 1996.
44. Jones, B. et al., Trials to assess equivalence: The importance of rigorous methods, *Br. Med. J.*, 313: 36–39, 1996.
45. Ebbutt, A.F. and Frith, L., Practical issues in equivalence trials, *Stat. Med.*, 17, 1691–1701, 1998.
46. Katz, D. et al., Obtaining confidence intervals for the risk ratio in cohort studies, *Biometrics*, 34, 469–474, 1978.
47. Thomas, D.G. and Gart, J.J., A table of exact confidence limits for differences and ratios of two proportions and their odd ratios, *J. Am. Stat. Assoc.*, 72, 73–76, 1977.
48. Koopman, P.A.R., Confidence intervals for the ratio of two binomial proportions, *Biometrics*, 40, 513–517, 1984.
49. Bailey, B.J.R., Confidence limits to the risk ratio, *Biometrics*, 43, 201–205, 1987.
50. Gart, J.J. and Nam, J., Approximate interval estimation of the ratio of binomial parameters: A review and corrections for skewness, *Biometrics*, 44, 323–338, 1988.
51. Zelterman, D., *Models for Discrete Data*, Oxford University Press, Oxford, 1999.
52. Cornfield, J., A statistical problem arising from retrospective studies. *Proceedings of the Third Berkeley Symposium on Mathematical Statistics and Probability IV*, J. Neyman (ed.). 135–148, California Press, Berkeley, CA, 1956.
53. Gart, J.J., The comparison of proportions: A review of significance tests, confidence intervals and adjustments for stratification, *Rev. Inst. Int. Stat.*, 39, 148–169, 1971.
54. Mehta, C.R., Patel, N.R., and Gray, R., Computing an exact confidence interval for the common odds ratio in several 2 by 2 contingency tables, *J. Am. Stat. Assoc.*, 80, 969–973, 1985.
55. Vollset, S.E., Hirji, K.F., and Elashoff, R.M., Fast computation of exact confidence limits for the common odds ratio in a series of 2×2 tables, *J. Am. Stat. Assoc.*, 86, 404–409, 1991.
56. Mehta, C.R. and Walsh, S.J., Comparison of exact, mid-p, and Mantel–Haenszel confidence intervals for the common odds ratio across several 2×2 contingency tables, *Am. Stat.*, 46, 146–150, 1992.
57. Emerson, J.D., Combining estimates of the odds ratio: The state of the art, *Stat. Methods Med. Res.*, 3, 157–178, 1994.
58. Gart, J.J. and Zweifel, J.R., On the bias of various estimators of the logit and its variance with application to quantal bioassay, *Biometrika*, 54, 181–187, 1967.
59. Cytovene product labeling available at www.fda.gov/cder/foi/label/2000/20460s10lbl.pdf.
60. Kim, M.Y. and Xue, X., Likelihood ratio and a Bayesian approach were superior to standard non-inferiority analysis when the non-inferiority margin varied with the control event rate, *J. Clin. Epidemiol.*, 57, 1253–1261, 2004.

61. Grizzle, J.E., Starmer, C.F., and Koch, G.G., Analysis of categorical data by linear models, *Biometrics*, 25, 489–504, 1969.
62. Gart, J.J., On the combination of relative risks, *Biometrics*, 18, 601–610, 1962.
63. Robins, J., Breslow, N., and Greenland, S., Estimators of the Mantel–Haenszel variance consistent in both sparse data and large-strata limiting models, *Biometrics*, 42, 311–323, 1986.
64. Mantel, N. and Haenszel, W., Statistical aspects of the analysis of data from retrospective studies of disease, *J. Natl. Cancer I.*, 22, 71 9–748, 1959.
65. Zhang, Z., Non-inferiority testing with a variable margin, *Biom. J.*, 48, 948–965, 2006.
66. Phillips, K.F., A new test of non-inferiority for anti-infective trials, *Stat. Med.*, 22, 201–212, 2003.
67. U.S. Food and Drug Administration, Division of Anti-infective Drug Products, Clinical Development and Labeling of Anti-Infective Drug Products. Points-to-consider. U.S. Food and Drug Administration, Washington, DC, 1992.
68. Röhmel, J., Therapeutic equivalence investigations: Statistical considerations, *Stat. Med.*, 17, 1703–1714, 1998.
69. Röhmel, J., Statistical considerations of FDA and CPMP rules for the investigation of new antibacterial products, *Stat. Med.*, 20, 2561–2571, 2001.
70. Tsou, H.H. et al., Mixed non-inferiority margin and statistical tests in active controlled trials, *J. Biopharm. Stat.*, 17, 339–357, 2007.
71. McNemar, Q., Note on the sampling error of the difference between correlated proportions or percentages, *Psychometrika* 12, 153–157, 1947.
72. Tango, T., Equivalence test and confidence interval for the difference in proportions for the paired-sample design, *Stat. Med.*, 17, 891–908, 1998.
73. Miyanaga, Y., Clinical evaluation of the hydrogen peroxide SCL disinfection system (SCL-D), *Jpn. J. Soft Contact Lenses*, 36, 163–173, 1994.
74. Lu, Y. and Bean, J.A., On the sample size for one-sided equivalence of sensitivities based upon McNemar's test, *Stat. Med.*, 14, 1831–1839, 1995.
75. Morikawa, T. and Yanagawa, T., Taiounoaru 2chi data ni taisuru doutousei kentei (Equivalence testing for paired dichotomous data), *P. Ann. Conf. Biometric Soc. Jpn.*, 123–126, 1995.
76. Hsueh, H.M., Liu, J.P., and Chen, J.J., Unconditional exact tests for equivalence or non-inferiority for paired binary endpoints, *Biometrics*, 57, 478–483, 2001.
77. Nam, J., Establishing equivalence of two treatments and sample size requirements in matched-pairs design, *Biometrics*, 53, 1422–1430, 1997.
78. Tang, N.S., Tang, M.L., and Chan, I.S.F., On tests of equivalence via non-unity relative risk for matched-pairs design, *Stat. Med.*, 22, 1217–1233, 2003.
79. Lachenbruch, P.A. and Lynch, C.J., Assessing screening tests: Extensions of McNemar's Test, *Stat. Med.*, 17, 2207–2217, 1998.
80. Tang, N.S. et al., Sample size determination for establishing equivalence/non-inferiority via ratio of two proportions in matched-pair design, *Biometrics*, 58, 957–963, 2002.
81. Chan, I.S.F. et al., Statistical analysis of non-inferiority trials with a rate ratio in small-sample matched-pair designs, *Biometrics*, 59, 1170–1177, 2003.

# 12

## Inferences on Means and Medians

### 12.1 Introduction

This chapter discusses non-inferiority based on the underlying means or medians when there are no missing data or censored observations. Means and medians are often used to describe the typical value or the central location of a distribution. Medians are preferred when the data are skewed. The outcomes may be continuous or discrete. For continuous outcomes where larger outcomes are more desirable and differences between outcomes have meaning (i.e., the data have an interval or ratio scale), the difference in the means of the experimental and control arms in a randomized trial represents the average benefit across trial subjects from being randomized to the experimental arm instead of the control arm. A difference in the medians does not have any analogous interpretation unless additional assumptions are made on the underlying distributions (e.g., that the shapes of the underlying distributions are equal).

For discrete outcomes, such as scores, the value for the mean will probably not be a possible value and may not be interpretable. In such a case, inferences based on means may be difficult to interpret without additional assumptions (e.g., the distributions, when different, are ordered). In these situations, testing should not be based on the mean. For binary data, the mean is the proportion of 1s or successes, which is interpretable.

When the difference in means (medians) defines the benefit or loss of benefit, non-inferiority testing should be based on the difference in means (medians). When the data are positive and relative changes are most important, it may be more appropriate to base non-inferiority testing on the ratio of the means (medians). The mean for the control group may be needed to understand and interpret a ratio of means.

A normal model is frequently used for inferences on the mean when the sample size is large. The sample mean is assumed to be a random value from an approximate normal distribution, with mean equal to the true mean or population mean and variance equal to $\sigma^2/n$, where $\sigma^2$ is the population variance and $n$ is the sample size. Inferences on a median are often based on the behavior of the order statistics (see Section 12.4).

The experimental therapy is noninferior to the control therapy if the mean (median) for the experimental arm is better than or not too much worse than that of the control arm. When larger outcomes are more desirable, non-inferiority corresponds to the mean (median) for the experimental arm either being greater than or not too much less than that for the control arm. When smaller outcomes are more desirable, non-inferiority corresponds to the mean (median) for the experimental arm either being less than or not too much greater than that for the control arm.

In this chapter we will discuss non-inferiority analyses based on the differences in means (Section 12.2), the ratio of means (Section 12.3), and the differences in medians (Section 12.4). The chapter ends with a brief discussion of non-inferiority testing involving ordinal data (Section 12.5). The method discussed in Section 12.5 can also be adapted for use in continuous data.

## 12.2 Fixed Thresholds on Differences of Means

Many non-inferiority trials use a continuous measure for the primary outcome. The mean is a natural summary for the typical or central value. In this section we will discuss non-inferiority analysis methods for the difference in means.

The distribution for continuous data generally cannot be solely described by the mean. Another parameter (or more than one) may also need to be assessed. For normally distributed data, the standard deviation (or alternatively the variance) is used in conjunction with the mean to completely summarize the distribution. The need for a second parameter to define the distribution adds complexity. This second parameter is often treated as a nuisance parameter. For normal data, the standard deviations of responses from the two treatment groups are often assumed to be identical, negating the need to base inference on differences in distribution other than the difference in means.

### 12.2.1 Hypotheses and Issues

For this discussion, consider the situation where larger outcomes are more desired than smaller outcomes and the margin for the difference in means, $\delta > 0$, is prespecified. The null and alternative hypotheses to be considered are

$$H_o: \mu_C - \mu_E \geq \delta \text{ and } H_a: \mu_C - \mu_E < \delta \tag{12.1}$$

That is, the null hypothesis is that the mean in the active control group is superior to the mean in the experimental treatment group by at least a quantity of $\delta$, whereas the alternative is that the active control is superior by a smaller amount, or the two treatments are identical, or the experimental treatment is superior.

In the simplest application with no covariates, a confidence interval can be determined for $\mu_C - \mu_E$, and non-inferiority is concluded (the null hypothesis rejected) if the upper bound of the confidence interval is less than $\delta$. For situations that include covariates, the methods will become more complex.

## 12.2.2 Exact and Distribution-Free Methods

With inference on means, the "exact" distribution will have a different meaning than with inference on proportions discussed in the previous chapter. For continuous data, we will use the term "distribution-free" to signify the lack of reliance on a normal distribution of the data or even asymptotic normality of the parameter estimates.

One proposed method of distribution-free non-inferiority analysis is to use a permutation test. With a permutation test, the analysis occurs by estimating a sampling distribution of the test statistic, which might be a $z$-statistic, a $p$-value, a point estimate, or another statistic. The sampling distribution is estimated by multiple rerandomizations of treatments to study subjects, using the same mechanism as was used for the real allocation of treatments to subjects in the clinical trial (e.g., using permuted blocks of the same block size, with the same stratification) and determining the test statistic from the rerandomization. This process is completed many times (say, 10,000) and a distribution of the test statistics is estimated. This process is well known for tests of superiority; however, for tests of non-inferiority, one more adjustment is required. Since the null hypothesis represents a nonzero difference of $\delta$, we start by assuming that subjects on the control therapy will have on average a response that exceeds that of subjects on the experimental treatment by $\delta$. This leads to an analogous modification of the test statistic.

An easy implementation of this is to use the actual outcomes and add or subtract $\delta$ to the actual observed values as necessary. For example, if a subject was assigned to the control arm in the real study but to the experimental arm in the rerandomization, $\delta$ is subtracted from that subject's actual observed value for the rerandomization calculations; if a subject was assigned to the experimental treatment in the real study but to the active control in the rerandomization, $\delta$ is added to that subject's actual observed value for the rerandomization calculations; and if a subject receives the same allocation in the rerandomization as in the real trial and, the actual observed value is used. In such an approach, we are assuming the shapes of the distributions for the outcomes are identical between the control and experimental arms.

Hollander and Wolfe[1] proposed a modification of the usual distribution-free testing procedure for testing a null hypothesis of a nonzero difference of $\delta$, a modification that is operationally identical to that described in the previous paragraph. To test a null hypothesis of a $\delta$ difference in mean, each observation from the active control group is transformed by subtracting $\delta$ from the observed value and each observation from the experimental treatment is unchanged. With the resulting transformed values, a test for superiority of the investigational treatment over the active control by any method, including a parametric test, will be equivalent to a test of non-inferiority of the original values. A permutation test for superiority can easily be used to test the null hypotheses after transformation.[2]

The permutation test being valid means that the residuals are exchangeable. That is, if the distribution of $X_i - \mu_i$ (observed difference in a value minus the treatment group mean value) is identical for the two treatment groups, the permutation test is valid. This requirement is often assumed to be correct (at least mostly correct) but rarely checked in a detailed manner. Obvious examples of situations where the residuals are not exchangeable include when one treatment produces a unimodal distribution and the other produces a bimodal distribution, or when one treatment produces responses with a larger variance than those produced by the other treatment. In such cases, the permutation test will not be appropriate.[3]

A sufficient condition for the permutation test to be valid is that each subject would have a response, if assigned to receive the active control, that exceeds that subject's response, if assigned to receive the experimental treatment, by exactly $\delta$. This condition guarantees that the necessary condition from the previous paragraph is met, but this condition is not in itself necessary.

### 12.2.3 Normalized Methods

When the sample size allows, use of normalized methods are generally preferred. Like binary data, this has historically been due to computational ease compared with other methods (e.g., exact methods). However, as discussed in the previous section, distribution-free methods are not easily applied in each situation. In addition, if the data are normally or approximately normally distributed, tests based on normal theory will be more powerful than tests that ignore the underlying distribution.

A confidence interval for $\mu_C - \mu_E$ is needed to test whether the difference in means is less than $\delta$. The standard formula of the confidence interval is $\bar{x}_C - \bar{x}_E \pm k \times \text{se}(\bar{X}_C - \bar{X}_E)$, where $\bar{X}_C$ and $\bar{X}_E$ are the sample means for the control and experimental groups respectively, with $\bar{X}_C$ and $\bar{X}_E$ denoting their corresponding observed values. With normally distributed data, the multiplier or critical value, $k$, is either $z_{\alpha/2}$, the upper $\alpha/2$ percentile of the standard normal distribution or the upper $\alpha/2$ percentile of some $t$ distribution (which are almost equal for large sample sizes). When the data are not normally distributed but the sample size is large, the Central Limit Theory supports using $z_{\alpha/2}$ as the critical value. The standard error of the difference $\bar{X}_C - \bar{X}_E$ is generally estimated using either the empirical (method of moments) esti-

mator, $\sqrt{\dfrac{\hat{\sigma}_C^2}{n_C} + \dfrac{\hat{\sigma}_E^2}{n_E}}$, or the unbiased estimator, $\sqrt{\dfrac{\hat{\sigma}_C^2}{n_C - 1} + \dfrac{\hat{\sigma}_E^2}{n_E - 1}}$, where $\hat{\sigma}_C^2$ and $\hat{\sigma}_E^2$ are the empirical (method of moments) estimators of the variances for the control and experimental groups, respectively. Again, with large sample sizes, the relative difference in the two estimates will be negligible.

As an equivalent alternative to the confidence interval methodology, a test statistic can be calculated. If $\dfrac{\bar{x}_C - \bar{x}_E - \delta}{\mathrm{se}(\bar{X}_C - \bar{X}_E)}$ is less than the critical value (e.g., less than $-z_{\alpha/2}$), non-inferiority is concluded. Alternatively, non-inferiority is concluded when the appropriate-level confidence interval for $\mu_C - \mu_E$ contains only values less than $\delta$. Using a test statistic has the advantage of being able to calculate a *p*-value for the test of the null hypothesis. However, *p*-values are not often calculated for such non-inferiority tests and, when they are calculated, they are prone to misinterpretation as an indication of the existence of differences, not the rejection of the null hypothesis of a specific nonzero difference. We will later compare the results of different analysis methods based on both the calculated *p*-values (and an analogous posterior probability) for given various margins and compare the calculated 95% confidence/credible intervals.

### 12.2.3.1 Test Statistics

Let $X_1, X_2, \ldots, X_{n_C}$ and $Y_1, Y_2, \ldots, Y_{n_E}$ denote independent random samples for the control and experimental arms, respectively with corresponding sample means $\bar{X}$ and $\bar{Y}$. Let $\mu_C$ and $\mu_E$ denote the means of the underlying distributions for the control and experimental arms and let $\sigma_C^2$ and $\sigma_E^2$ denote the respective variances. Let $S_C^2$ and $S_E^2$ denote the respective sample variances given by $S_C^2 = \sum_{i=1}^{n_C} (X_i - \bar{X})^2 / (n_C - 1)$ and $S_E^2 = \sum_{j=1}^{n_E} (Y_j - \bar{Y})^2 / (n_E - 1)$. We will consider three cases for testing the hypotheses in Expression 12.1: (1) large sample normal–based inference, (2) using Satterthwaite degrees of freedom,[4] and (3) using a *t* statistic under the assumption of unknown but equal variances. Procedures (1) and (2) are just two cases to address the *Behrens–Fisher problem*—making statistical inferences on the difference in the means of two normal distributions having unknown variances that are not assumed to be equal.

*Large Sample Normal Inference.* For large sample sizes, it follows from the central limit theorem that the test statistic

$$Z = \frac{\bar{X} - \bar{Y} - \delta}{\sqrt{S_C^2/n_C + S_E^2/n_E}} \tag{12.2}$$

will have an approximate standard normal distribution when $\mu_C - \mu_E = \delta$. The null hypothesis in Expression 12.1 is rejected and non-inferiority is concluded at a one-sided level of $\alpha/2$ when the observed value for $Z$ is less than $-z_{\alpha/2}$. An approximate $100(1 - \alpha)\%$ confidence interval is given by $\bar{x} - \bar{y} \pm z_{\alpha/2}\sqrt{s_C^2/n_C + s_E^2/n_E}$, where $\bar{x}$ and $\bar{y}$ are the respective observed sample means and $s_C^2$ and $s_E^2$ are the respective observed sample variances.

*Using Satterthwaite Degrees of Freedom.* The hypotheses in Expression 12.1 are still tested using the test statistic in Equation 12.2. However, the observed value for $Z$ is compared with $-t_{\alpha/2,\nu}$ instead of $-z_{\alpha/2}$, where, from Satterthwaite,[4] the degrees of freedom $\nu$ is the greatest integer less than or equal to

$$\frac{\left(\dfrac{s_C^2}{n_C}+\dfrac{s_E^2}{n_E}\right)^2}{\dfrac{s_C^4}{n_C^2(n_C-1)}+\dfrac{s_E^4}{n_E^2(n_E-1)}} \tag{12.3}$$

The approximate $100(1-\alpha)\%$ confidence interval is then given by $\bar{x}-\bar{y}\pm$ $t_{\alpha/2,\nu}\sqrt{s_C^2/n_C+s_E^2/n_E}$. The value $t_{\alpha/2,\nu}$ is the upper $\alpha/2$ percentile of a $t$ distribution with $\nu$ degrees of freedom. The Satterthwaite degrees of freedom are random and can only be determined after calculating the observed sample variances. Because $t_{\alpha/2,\nu}>z_{\alpha/2}$, the confidence intervals using $t_{\alpha/2,\nu}$ will be wider than those using $z_{\alpha/2}$ and the one-sided $p$-values based on the $t$ distribution with $\nu$ degrees of freedom will be closer to 0.5 than the one-sided $p$-values based on the standard normal distribution.

*t Statistic Assuming Unknown but Equal Variances.* The test statistic is based on the assumptions that the underlying distributions are normal distributions with equal variances. The test statistic is given by

$$T=\frac{\bar{X}-\bar{Y}-\delta}{\sqrt{S^2(1/n_C+1/n_E)}} \tag{12.4}$$

where $S^2=[(n_C-1)S_C^2+(n_E-1)S_E^2]/(n_C+n_E-2)$ is a pooled estimator of the common underlying variance. The null hypothesis in Expression 12.1 is rejected and non-inferiority is concluded at a one-sided level of $\alpha/2$ when the observed value for $T$ is less than $-t_{\alpha/2,n_C+n_E-2}$. The multiplier or critical values for the three methods are ordered $t_{\alpha/2,\nu}>t_{\alpha/2,n_C+n_E-2}>z_{\alpha/2}$. The following examples compare the results based on $p$-values and confidence interval from applying these three methods. They also provide insight on how the two estimated standard errors for $\bar{X}-\bar{Y}$ in Equations 12.2 and 12.4 compare.

### Example 12.1

Consider testing the hypotheses in Expression 12.1 with $\delta=4$ based on a randomized trial having 25 subjects in the control arm and 30 subjects in the experimental arm. For the control arm the observed sample mean is $\bar{x}=40.5$ and the observed sample variance is $s_C^2=4$. For the experimental arm the observed sample mean is $\bar{y}=39.1$ and the observed sample variance is $s_E^2=49$. The one-sided $p$-values and two-sided 95% confidence intervals are provided in Table 12.1. From Equation 12.3 the degrees of freedom for the Satterthwaite method equal 34.

**TABLE 12.1**

Summary of One-Sided *p*-Values and 95% Confidence Intervals

|  | Large Sample Normal | Satterthwaite | Equal Variance |
|---|---|---|---|
| *p*-Value | 0.026 | 0.030 | 0.039 |
| 95% CI | (−1.23, 4.03) | (−1.32, 4.12) | (−1.51, 4.31) |

For each method, the one-sided *p*-values are greater than 0.025 and each 95% confidence interval contains the non-inferiority margin of 4. Therefore non-inferiority cannot be concluded. The upper limits of the 95% confidence intervals represent the smallest margin that could have been prespecified for which non-inferiority would have been concluded. The equal variance method has both the largest confidence interval upper limit of 4.31 and the largest *p*-value of 0.039. This is primarily due to the larger estimated standard error for $\bar{X} - \bar{Y}$ used by the equal variance method. For the equal variance method, the estimated standard error for $\bar{X} - \bar{Y}$ equals 1.45, whereas the estimated standard error for $\bar{X} - \bar{Y}$ equals 1.34 for the large sample normal and Satterthwaite methods. When the treatment group having the larger sample size has the larger (smaller) observed sample variance, the estimated standard error for $\bar{X} - \bar{Y}$ used by the equal variance method will be larger (smaller) than the estimated standard error for $\bar{X} - \bar{Y}$ used by the large sample normal and Satterthwaite methods. When the sample sizes are equal, the same standard error for $\bar{X} - \bar{Y}$ is used in all three methods. Note that the multipliers (i.e., the absolute values of the critical values) used for the confidence intervals were 1.960, 2.032, and 2.006.

In Example 12.2 the observed sample variances in Example 12.1 are reversed, and again the hypotheses in Expression 12.1 are tested with $\delta = 4$.

## Example 12.2

We still have $n_C = 25$, $n_E = 30$, $\bar{x} = 40.5$, and $\bar{y} = 39.1$. However, now $s_C^2 = 49$ and $s_E^2 = 4$. The one-sided *p*-values and two-sided 95% confidence intervals are provided in Table 12.2. From Equation 12.3 the degrees of freedom for the Satterthwaite method equal 27.

For each method, the one-sided *p*-values are greater than 0.025 and each 95% confidence interval contains the non-inferiority margin of 4. Therefore, non-inferiority cannot be concluded. The Satterthwaite method has both the largest confidence interval upper limit of 4.37 and the largest *p*-value of 0.042. The estimated standard errors used for $\bar{X} - \bar{Y}$ are approximately reversed from Example 12.1. For

**TABLE 12.2**

Summary of One-Sided *p*-Values and 95% Confidence Intervals

|  | Large Sample Normal | Satterthwaite | Equal Variance |
|---|---|---|---|
| *p*-Value | 0.036 | 0.042 | 0.029 |
| 95% CI | (−1.44, 4.24) | (−1.57, 4.37) | (−1.28, 4.08) |

**TABLE 12.3**

Summary of One-Sided $p$-Values and 95% Confidence Intervals

|  | Large Sample Normal | Satterthwaite | Equal Variance |
|---|---|---|---|
| $p$-Value | 0.015 | 0.016 | 0.017 |
| 95% CI | (–1.76, 2.75) | (–1.79, 2.78) | (–1.83, 2.82) |

the equal variance method the estimated standard error for $\bar{X} - \bar{Y}$ equals 1.34, whereas the estimated standard error for $\bar{X} - \bar{Y}$ equals 1.45 for the large sample normal and Satterthwaite methods. The multipliers used for the confidence intervals were 1.960, 2.052, and 2.006.

In Example 12.3 we consider a case where the observed sample variances are similar.

**Example 12.3**

Consider testing the hypotheses in Expression 12.1, where $\delta = 3$ based on a randomized trial having 50 subjects in the control arm and 55 subjects in the experimental arm. For this example the data were randomly generated. For the control arm the observed sample mean is $\bar{x} = 29.80$ and the observed sample variance is $s_C^2 = 23.27$. For the experimental arm the observed sample mean is $\bar{y} = 29.31$ and the observed sample variance is $s_E^2 = 47.22$. The one-sided $p$-values and two-sided 95% confidence intervals are provided in Table 12.3. From Equation 12.3 the degrees of freedom for the Satterthwaite method equal 97.

For each method the one-sided $p$-values are less than 0.025 and each 95% confidence interval contains only values less than the non-inferiority margin of 3. Therefore, non-inferiority is concluded. The $p$-values and confidence intervals are fairly similar across the trials. For the equal variance method the estimated standard error for $\bar{X} - \bar{Y}$ equals 1.17, whereas the estimated standard error for $\bar{X} - \bar{Y}$ equals 1.15 for the large sample normal and Satterthwaite methods. The multipliers used for the confidence intervals were similar: 1.960, 1.983, and 1.985.

When the observed sample variances are not greatly dissimilar and the sample sizes are large, the results from these three analysis methods should be similar.

We will revisit these examples in greater detail with addition of a Bayesian method in Section 12.2.4.

### 12.2.4 Bayesian Methods

In this section we will consider non-inferiority analysis of the difference in means using Bayesian approaches. For ease, conditional on the values of the parameters, we will consider the samples from the experimental and control arms as being independent random samples from normal distributions. For these pairs of normal distributions, we will consider the cases where (1) the population variances are known and (2) the population variances are unknown. Both cases are provided to illustrate the similarities and differences in applying the methods. As in Section 12.2.3.1, only the case where the population variances are unknown will be carried forward and compared in revisited examples with the methods in Section 12.2.3.1.

*Variances Known.* For a random sample of $n$ from a normal distribution with mean $\mu$ and known variance $\sigma^2$, a normal prior distribution for $\mu$ that has mean $v$ and variance $\tau^2$ leads to normal posterior distribution for $\mu$ that has mean

$$\left(\left(\frac{n}{\sigma^2}\right)\bar{x} + \left(\frac{1}{\tau^2}\right)v\right) \Big/ \left(\left(\frac{n}{\sigma^2}\right) + \left(\frac{1}{\tau^2}\right)\right)$$

and variance

$$1 \Big/ \left(\left(\frac{n}{\sigma^2}\right) + \left(\frac{1}{\tau^2}\right)\right)$$

When $\tau^2$ is relatively large compared to $\sigma^2/n$, the specific choice of $\tau^2$ will have little impact. The Jeffreys prior has density $h(\mu) \propto \sqrt{I(\mu)} = 1/\sigma$ for $-\infty < \mu < \infty$, which is not a proper density and is a noninformative prior for $\mu$. When $h(\mu) = 1/\sigma$ is used, the resulting posterior density is the density for a normal distribution having mean equal to $\bar{x}$ and variance $\sigma^2/n$.

The parameters $\mu_C$ and $\mu_E$ can be regarded as independent. Therefore, the posterior distribution for $\mu_C - \mu_E$ is a normal distribution with mean equal to the difference in the posterior means for $\mu_C$ and $\mu_E$ and variance equal to the sum of the posterior variances for $\mu_C$ and $\mu_E$.

When testing $H_0$: $\mu_C - \mu_E \geq \delta$ versus $H_a$: $\mu_C - \mu_E < \delta$, the null hypothesis is rejected and non-inferiority is concluded when the posterior probability of $\mu_C - \mu_E < \delta$ exceeds some threshold (e.g., exceeds $1 - \alpha/2$) or alternatively when the appropriate level credible interval for $\mu_C - \mu_E$ contains only values less than $\delta$.

*Variances Unknown.* In the frequentist setting, the analysis simplifies when the additional assumption is made that the two underlying normal distributions have the same variance. In the Bayesian setting this additional assumption complicates the analysis by leading to a joint posterior distribution where $\mu_C$ and $\mu_E$ are not independent ($\mu_C$ and $\mu_E$ are conditionally independent given $\sigma$). We will discuss two Bayesian procedures, which will be referred to as the Bayesian-$\gamma$ and Bayesian-$T$ procedures. These procedures were introduced in Section 6.3 for three-arm non-inferiority trials. We repeat the explanations of the procedures below.

*Bayesian-$\gamma$ Procedure.* This procedure is similar to a procedure provided by Ghosh et al.[5] There are different choices on what function of the variance (e.g., $\sigma$, $\sigma^2$, $1/\sigma$, or $1/\sigma^2$) to model. Applying the joint Jeffreys prior in each case leads to joint posterior distributions that provide different posterior probabilities. For this discussion, the variance will be modeled with $\sigma^2$, as this will lead to a more convenient form for the joint posterior distribution. For $\theta = \sigma^2$, the density of the Jeffreys prior, $h$, satisfies $h(\theta) \propto \theta^{-3/2}$ for $-\infty < \mu < \infty$, and $\theta > 0$. Then for $X_1$, $X_2$ ..., $X_n$, a random sample from a normal distribution with mean $\mu$ and variance $\theta$, where the prior density satisfies $h(\theta) \propto \theta^{-3/2}$, the joint posterior density satisfies

$$g(\mu,\theta \mid x_1, x_2, \ldots, x_n) \propto \theta^{-1/2} \exp\left\{ -\frac{(\mu - \overline{x})^2}{2\theta/n} \right\} \times \theta^{-n/2-1} \exp\left\{ -\frac{1}{2}\sum_{i=1}^{n}(x_i - \overline{x})^2/\theta \right\}$$

$$(12.5)$$

We see from Expression 12.5 that the joint density factors into the product of an inverse gamma marginal distribution for $\theta$ and a normal conditional distribution for $\mu$ given $\theta$. The inverse gamma distribution has shape and scale parameters equal to $n/2$ and $\sum_{i=1}^{n}(x_i - \overline{x})^2/2$, respectively, with mean equal to $\sum_{i=1}^{n}(x_i - \overline{x})^2/(n-2)$ and variance equal to $2\left(\sum_{i=1}^{n}(x_i - \overline{x})^2\right)^2/[(n-2)^2(n-4)]$. Note that $\theta$ has an inverse gamma distribution with parameters $n/2$ and $\sum_{i=1}^{n}(x_i - \overline{x})^2/2$, if and only if $1/\theta$ has a gamma distribution with parameters $n/2$ and $2/\sum_{i=1}^{n}(x_i - \overline{x})^2$ with mean equal to $n/\sum_{i=1}^{n}(x_i - \overline{x})^2$. Given $\theta$, $\mu$ has a normal distribution with mean equal to $\overline{x}$ and variance equal to $\theta/n$. Therefore, to simulate probabilities involving $\mu$, a random value for $1/\theta$ can be taken from the gamma distribution with parameters $n/2$ and $2/\sum_{i=1}^{n}(x_i - \overline{x})^2$, and then a random value for $\mu$ can be taken from a normal distribution having mean $\overline{x}$ and variance $\theta/n$.

*Bayesian-T Procedure.* Another approach consistent with Gamalo et al.[6] to address the problem of unknown variances uses translated $t$ distributions for the posterior distributions. The mean of the control arm, $\mu_C$, has a posterior distribution equal to the distribution of

$$\overline{x} + T_C s_C / \sqrt{n_C}$$

where $T_C$ has a $t$ distribution with $n_C - 1$ degrees of freedom.

The mean of the experimental arm, $\mu_E$, has a posterior distribution equal to the distribution of

$$\overline{y} + T_E s_E / \sqrt{n_E}$$

where $T_E$ has a $t$ distribution with $n_E - 1$ degrees of freedom.

Example 12.4 revisits the examples in Section 12.2.3.1. We compare the posterior probabilities of the null hypothesis for various margins and the corresponding 95% credible interval for the above Bayesian procedures with the methods in Section 12.2.3.1.

## Example 12.4

We can determine from the information provided in Example 12.1 that $2/\sum_{i=1}^{25}(x_i - \bar{x})^2 = 0.02083$ and $2/\sum_{j=1}^{30}(y_j - \bar{y})^2 = 0.001408$. One hundred thousand values for $\mu_C - \mu_E$ were simulated for each Bayesian method.

For the Bayesian-$\gamma$ procedure, a value for $\mu_C$ was simulated by first randomly selecting a value for $1/\sigma_C^2$ from a gamma distribution with parameters 12.5 and 0.02083. Then for that randomly selected value that will be denoted by $1/\theta_C^*$, a value for $\mu_C$ was randomly taken from a normal distribution having mean $\bar{x} = 40.5$ and variance equal to $\theta_C^*/25$. A value for $\mu_E$ was simulated by first randomly selecting a value for $1/\sigma_E^2$ from a gamma distribution with parameters 15 and 0.001408. Then for that randomly selected value that will be denoted by $1/\theta_E^*$, a value for $\mu_E$ was randomly taken from a normal distribution having mean $\bar{y} = 39.1$ and variance equal to $\theta_E^*/30$. The simulated probability that $\mu_C - \mu_E < 4$ is 0.971 (=1 − 0.029), which is less than 0.975; thus, non-inferiority cannot be concluded for a non-inferiority margin of 4.

For the Bayesian-$T$ procedure, a value for $\mu_C$ was selected at random from the distribution for $40.5 + 2T_C/5$, where $T_C$ has a $t$ distribution with 24 degrees of freedom. A value for $\mu_E$ was selected at random from the distribution for $39.1 + 7T_E/\sqrt{30}$, where $T_E$ has a $t$ distribution with 29 degrees of freedom.

Table 12.4 provides a summary of the 95% confidence/credible intervals for the Bayesian methods and the three methods discussed in Section 12.2.3.1 along with the calculated $p$-values or simulated probabilities of the null hypothesis in Expression 12.1 for various choices of a non-inferiority margin, $\delta$. In this example the posterior probabilities and 95% credible interval for the Bayesian methods are respectively similar to $p$-values and the 95% confidence intervals from the large sample normal and Satterthwaite methods.

**TABLE 12.4**

Summary of $p$-Values, Posterior Probabilities, and 95% Confidence/Credible Intervals

| Non-Inferiority Margin, $\delta$ | Bayesian-$\gamma$ | Bayesian-$T$ | Large Sample Normal | Satterthwaite | Equal Variance |
|---|---|---|---|---|---|
| 0 | 0.850 | 0.846 | 0.865 | 0.861 | 0.831 |
| 1 | 0.618 | 0.618 | 0.617 | 0.617 | 0.608 |
| 2 | 0.327 | 0.330 | 0.327 | 0.329 | 0.340 |
| 3 | 0.117 | 0.121 | 0.116 | 0.120 | 0.137 |
| 4 | 0.029 | 0.030 | 0.026 | 0.030 | 0.039 |
| 5 | 0.005 | 0.005 | 0.004 | 0.006 | 0.008 |
| 95% CI | (−1.38, 4.08) | (−1.34, 4.12) | (−1.23, 4.03) | (−1.32, 4.12) | (−1.51, 4.31) |

For Example 12.2, where the observed sample variances are reversed, we have that $2/\sum_{j=1}^{25}(x_i - \bar{x})^2 = 0.001701$ and $2/\sum_{j=1}^{30}(y_j - \bar{y})^2 = 0.01724$. One hundred thousand values for $\mu_C - \mu_E$ were simulated for each Bayesian method.

For the Bayesian-$\gamma$ procedure, the first step in simulating a value for $\mu_C$ ($\mu_E$) is randomly selecting a value for $1/\sigma_C^2$ ($1/\sigma_E^2$) from a gamma distribution with parameters 12.5 and 0.001701 (15 and 0.01724). The second steps are the same as before. For the Bayesian-$T$ procedure, a value for $\mu_C(\mu_E)$ was selected at random from the distribution for $40.5 + 7T_C/5$ ($39.1 + 2T_E/\sqrt{30}$), where $T_C$ ($T_E$) has a $t$ distribution with 24 (29) degrees of freedom.

Table 12.5 provides a summary of the 95% confidence/credible intervals for these Bayesian methods and the three methods discussed in Section 12.2.3.1, along with the calculated $p$-values or simulated probabilities of the null hypothesis in Expression 12.1 for various choices of a non-inferiority margin, $\delta$. In this example the posterior probabilities and 95% credible interval for the Bayesian methods are respectively similar to $p$-values and the 95% confidence intervals from the large sample normal and Satterthwaite methods.

For Example 12.3, we have that $2/\sum_{j=1}^{50}(x_i - \bar{x})^2 = 0.001748$ and $2/\sum_{j=1}^{55}(y_j - \bar{y})^2 = 0.000784$. These two values are the values of the scale parameters for simulating values for $1/\sigma_C^2$ and $1/\sigma_E^2$ from respective gamma distributions with values for the shape parameters of 25 and 27.5. For both Bayesian methods, a value for $\mu_C - \mu_E$ is simulated in analogous fashion as above. Again, 100,000 values for $\mu_C - \mu_E$ were simulated for each method. Table 12.6 provides a summary of the 95% confidence/credible intervals for the Bayesian methods and the three methods discussed in Section 12.2.3.1 along with the calculated $p$-values or simulated probabilities of the null hypothesis in Expression 12.1 for various choices of a non-inferiority margin, $\delta$.

In all examples (Tables 12.4 through 12.6) the posterior probabilities and 95% credible intervals for the Bayesian methods are respectively similar to $p$-values and the 95% confidence intervals from the large sample normal and Satterthwaite methods. In each case the Bayesian-$T$ and Satterthwaite methods gave quite similar results with the Bayesian-$T$ method, producing slightly wider 95% confidence/

**TABLE 12.5**

Summary of $p$-Values, Posterior Probabilities, and 95% Confidence/Credible Intervals

| Non-Inferiority Margin, $\delta$ | Bayesian-$\gamma$ | Bayesian-$T$ | Large Sample Normal | Satterthwaite | Equal Variance |
|---|---|---|---|---|---|
| 0 | 0.833 | 0.828 | 0.833 | 0.829 | 0.850 |
| 1 | 0.610 | 0.610 | 0.609 | 0.608 | 0.617 |
| 2 | 0.340 | 0.344 | 0.339 | 0.341 | 0.328 |
| 3 | 0.136 | 0.140 | 0.134 | 0.139 | 0.118 |
| 4 | 0.039 | 0.044 | 0.036 | 0.042 | 0.029 |
| 5 | 0.009 | 0.010 | 0.006 | 0.010 | 0.005 |
| 95% CI | (−1.49, 4.32) | (−1.59, 4.41) | (−1.44, 4.24) | (−1.57, 4.37) | (−1.28, 4.08) |

**TABLE 12.6**

Summary of $p$-Values, Posterior Probabilities, and 95% Confidence/Credible Intervals

| Non-Inferiority Margin, $\delta$ | Bayesian-$\gamma$ | Bayesian-$T$ | Large Sample Normal | Satterthwaite | Equal Variance |
|---|---|---|---|---|---|
| 0 | 0.666 | 0.662 | 0.666 | 0.666 | 0.663 |
| 1 | 0.331 | 0.329 | 0.331 | 0.331 | 0.334 |
| 2 | 0.096 | 0.096 | 0.096 | 0.097 | 0.101 |
| 2.5 | 0.042 | 0.042 | 0.041 | 0.042 | 0.045 |
| 3 | 0.016 | 0.016 | 0.015 | 0.016 | 0.017 |
| 95% CI | (–1.78, 2.77) | (–1.82, 2.79) | (–1.76, 2.75) | (–1.79, 2.78) | (–1.83, 2.82) |

credible intervals. For the Bayesian-$\gamma$ method, when the posterior probability of the null hypothesis was very small, the posterior probability lied between the $p$-values for the large sample normal method and the Satterthwaite method and in each example the upper limit of the 95% credible interval for $\mu_C - \mu_E$ lied between the upper limits of the 95% confidence intervals from the large sample normal and Satterthwaite methods.

## 12.2.5 Sample Size Determination

The selection of sample size for a non-inferiority test of a continuous endpoint might be easier than for a binary endpoint, as discussed in the previous chapter. The sample size depends on the difference in the selected alternative for $\mu_C - \mu_E$ and the hypothesized value for $\mu_C - \mu_E$ in the null hypothesis (i.e., $\delta$). A superiority trial (i.e., $\delta = 0$) where it is believed that $\mu_C - \mu_E = -15$ will require the same sample size for a desired power as a non-inferiority trial where it is believed that $\mu_C - \mu_E = -5$ and $\delta = 10$. In each case the difference between the selected alternative and the hypothesized value in the null hypothesis is –15. Therefore, the sample size for a non-inferiority trial can be determined by using sample size software designed for a superiority trial where the selected superior effect is the difference between the selected non-inferiority alternative and $-\delta$.

It is tempting to assume that there is no difference in the true means for the experimental and control arms when powering/sizing the non-inferiority trial. However, if the control therapy has demonstrated the best efficacy among many available or tried therapies, it may be appropriate to power the trial on the basis of an appropriate difference in favor of the control therapy. It may therefore be optimistic to size the study under the assumption that the experimental therapy is as efficacious as the best therapy among many therapies, leading to an undersized, underpowered trial. In such cases it may be appropriate to assume that the experimental therapy has less efficacy than the control therapy (but still meaningfully efficacious if compared to a placebo).

Alternatively, when the control therapy is not terribly effective, the non-inferiority margin is small. Assuming no difference in the true means for the experimental and control arms not only will lead to a large sample size but will also reflect the belief that the experimental therapy is not terribly effective.

In deriving sample formulas, we borrow from ideas in Kieser and Hauschke.[7] Let $\mu_1$ and $\mu_2$ represent the assumed true means for the experimental and control arms, respectively, where $\mu_2 - \mu_1 < \delta$. Let $\sigma_1$ and $\sigma_2$ denote the respective assumed underlying standard deviations, where $l = \sigma_1/\sigma_2$. Let $k = n_E/n_C$ denote the allocation ratio. On the basis of the assumption of independent normal random samples and using a Satterthwaite-like approximation for the degrees of freedom, the test statistic $Z$ in Expression 12.2 will be modeled as having an approximate noncentral $t$ distribution with noncentrality parameter given by $\dfrac{\mu_2 - \mu_1 - \delta}{\sigma_2\sqrt{(l^2/k+1)/n_C}}$ and degrees of freedom given by

$$\nu = \frac{(1+l^2/k)^2}{1/(n_C-1)+l^4/[k^2(kn_C-1)]}.$$ We will use the approximate relation that the

100$\beta$th percentile of the noncentral $t$ distribution is approximately equal to the noncentrality parameter plus the 100$\beta$th percentile of the $t$ distribution having the same degrees of freedom. Then an iterative sample size formula for $n_C$ ($n_C$ appears on both sides of the equation) is given by

$$n_C = (1+l^2/k)(t_{\alpha/2,\nu}+t_{\beta,\nu})^2(\sigma_2/(\mu_2-\mu_1-\delta))^2 \qquad (12.6)$$

For the case where equal, unknown underlying variances is assumed, Equation 12.6 applies with $\nu = 2n_C - 2$ and $l = 1$. For the large sample normal test, Equation 12.6 can directly be applied (no iterations needed) where the term $t_{\alpha/2,\nu} + t_{\beta,\nu}$ is replaced with the term $z_{\alpha/2} + z_\beta$. The control sample size calculated for the large sample normal test can be used as the starting value for the iterations for the $t$-test procedures.

The overall sample size is $1 + k$ times the solution for $n_C$ in Equation 12.6 or the corresponding analog for one of the other test statistics. The optimal allocation ratio $k$ that minimizes the overall sample size can be found by a "grid search," by determining the overall sample size for many candidates for $k$, or for the large sample normal procedure by taking a derivative of the formula for the overall sample size with respect to $k$ and setting the result equal to zero and then solving for $k$. As stated in Chapter 11, before choosing a particular allocation ratio, its impact on the evaluation of secondary endpoints and/or safety endpoints should be considered.

The above sample size expressions are based on substituting the assumed standard deviations for the random sample standard deviations. The actual power for the determined sample sizes may be slightly less than the desired/targeted power.

For the Bayesian procedures, when a specific alternative is selected, the required sample size should be similar to the calculated required sample size for the frequentist procedures. For the Bayesian-$T$ procedure, the required sample size should be similar to that calculated for the Satterthwaite-like procedure. Alternatively for the frequentist or Bayesian methods, the required sample size can be based on an assumed distribution over all possibilities for $(\mu_1, \mu_2)$.

The use of the sample size formulas will be illustrated in Example 12.5.

## Example 12.5

Consider an endpoint that is the improvement from baseline in some score. The non-inferiority margin for the difference in mean improvement is 10. Both a 1:1 and a 2:1 experimental to control allocation are potentially being considered. The trial will be sized on the basis of the following assumptions: the true mean improvement is 38 and 40 for the experimental and control arms, respectively, and the corresponding underlying standard deviations are 6 and 12, respectively. For the equal variance approach, the common underlying variance will be assumed as 90 (the average of the variances for standard deviations of 6 and 12). Table 12.7 provides the determined sample sizes for the control arm.

For this example, the calculated sample sizes were smaller for the large sample normal and equal variance methods. For the large sample normal method, when the inference is based on the difference in means with equal and known underlying variances, the overall sample size for a 2:1 allocation will be 12.5% greater than the overall sample size for a 1:1 allocation. In this example, for each method, the percentage increase in the calculated value for the overall sample size going from a 1:1 allocation to a 2:1 allocation is greater than 12.5%. This is due to the unequal variances and the iterative nature of the sample size equations for the Satterthwaite and equal variance methods.

For the large sample normal and Satterthwaite methods, the optimal allocation ratio is $l = \sigma_1/\sigma_2$. In this example, for the large sample normal and Satterthwaite methods, $k = 0.5$ is the allocation ratio that minimizes the calculated overall sample size. For the large sample normal method, the corresponding allocation is 16 subjects to the control arm and 8 subjects to the experimental arm. For the Satterthwaite method, the corresponding allocation is 18 subjects to the control arm and 9 subjects to the experimental arm. For the equal variance method, a 1:1 allocation ratio minimizes the calculated overall sample size with 15 subjects allocated to each arm.

**TABLE 12.7**

Sample Sizes for the Control Arm

| Method | Allocation Ratio | | Percentage Change in Overall Sample Size (%)[a] |
| --- | --- | --- | --- |
| | 1:1 | 2:1 | |
| Large sample normal | 14 | 12 | 35.1 |
| Satterthwaite | 15 | 14 | 38.6 |
| Equal variance | 15 | 11 | 15.9 |

[a] Based on the calculated value before rounding up.

## 12.3 Fixed Thresholds on Ratios of Means

In some situations the relative differences between arms is the primary interest. Typically in bioequivalence problems, the bioequivalence of two products is based an observed ratio—for example, the ratio of means. In this section we will discuss non-inferiority analysis methods for the ratio of means.

### 12.3.1 Hypotheses and Issues

A ratio of means only makes sense when the means must be positive. Interpretation of the ratio of means can also be difficult if individual observations can be negative. Without loss of generality, we will consider the situation where large outcomes are more desired than smaller outcomes. We will assume that means for the experimental and control arms are positive (i.e., $\mu_E > 0$ and $\mu_C > 0$). The preselected non-inferiority threshold will be denoted by $\delta$, where $0 < \delta < 1$. The null and alternative hypotheses to be considered are

$$H_o: \mu_E/\mu_C \leq \delta \text{ and } H_a: \mu_E/\mu_C > \delta \tag{12.7}$$

Since $\mu_C > 0$, the hypotheses in Expression 12.7 can be reexpressed as

$$H_o: \mu_E - \delta \mu_C \leq 0 \text{ and } H_a: \mu_E - \delta \mu_C > 0 \tag{12.8}$$

Thus, non-inferiority testing can be based on either $\mu_E/\mu_C$ or $\mu_E - \delta \mu_C$. The advantage of using $\mu_E - \delta\mu_C$ in practice is that smaller sample sizes should be needed for $\bar{Y} - \delta\bar{X}$ to be approximately normally distributed than the required sample sizes needed for $\bar{Y}/\bar{X}$ to be approximately normally distributed. Adaptations of the testing procedures discussed in Section 12.2 can be used in testing the hypotheses in Expression 12.8.

### 12.3.2 Exact and Distribution-Free Methods

Consistent with the null hypothesis in Expression 12.7, a permutation or rerandomization test can be done by assuming that subjects on the experimental therapy will have outcomes that are $100\delta\%$ of the outcomes observed on the control therapy. An appropriate test statistic is chosen. The sampling distribution when $\mu_E/\mu_C = \delta$ is approximated by multiple rerandomizations of therapies to study subjects, using the same allocation mechanism of therapies to subjects as was originally used in the clinical trial (e.g., using permuted blocks of the same block size, with the same stratification, etc.).

An easy implementation of this is to use the actual outcomes and multiply or divide by $\delta$ to the actual observed values as necessary. For example, if a subject was assigned to the control arm in the real study but to the experimental arm in the rerandomization, the subject's actual observed value is multiplied by $\delta$ for the rerandomization calculations; if a subject was assigned to the experimental arm in the real study but to the control arm in the rerandomization, the subject's actual observed value is divided by $\delta$ for the rerandomization calculations; and if a subject receives the same allocation in the rerandomization as in the real trial and, the actual observed value is used. In such an approach where outcomes must be positive, we are assuming the shapes of the distributions for the logarithms of the outcomes are identical between the control and experimental arms. That is, the shapes of the underlying distributions for the outcomes differ by a scale factor. Here the permutation test being valid means that the residuals of the logarithms are exchangeable. A sufficient condition for the permutation test to be valid is that each subject would have an outcome, if assigned to receive the experimental therapy that is exactly $\delta$ times the outcome that the subject would have had if assigned to receive the control therapy.

### 12.3.3 Normalized and Asymptotic Methods

We will discuss the use of normalized methods for testing the hypotheses as expressed in Expression 12.8 and the use of a delta method for testing the hypotheses as expressed in Expression 12.7.

However, as discussed in the previous section, distribution-free methods are not easily applied in each situation. In addition, if the data are normally or approximately normally distributed, tests based on normal theory will be more powerful than tests that ignore the underlying distribution. There are some issues with using a normal or similar model. A normal model will have some probability below zero. Likewise, there will be some uncertainty that the true mean is positive that is inherent in using a normal distribution or $t$ distribution to model the sampling distribution of the sample mean. When the sample mean divided by its estimated standard error is far from zero (e.g., beyond ±4), the uncertainty that the true mean is positive will be quite small, and the issue of having some uncertainty that the true mean is positive that is introduced by the model can be ignored.

#### 12.3.3.1 Test Statistics

Let $X_1, X_2, \ldots, X_{n_C}$ and $Y_1, Y_2, \ldots, Y_{n_E}$ denote independent random samples for the control and experimental arms, respectively. Let $\mu_C$ and $\mu_C$ denote the means of the underlying distributions for the control and experimental arms and let $\sigma_C^2$ and $\sigma_E^2$ denote the respective variances. Let $S_C^2$ and $S_E^2$ denote the respective sample variances given by $S_C^2 = \sum_{i=1}^{n_C} (X_i - \bar{X})^2 / (n_C - 1)$ and $S_E^2 = \sum_{j=1}^{n_E} (Y_j - \bar{Y})^2 / (n_E - 1)$.

We will consider four procedures for testing the hypotheses in Equations 12.7 and 12.8: (1) a large sample normal–based inference, (2) using a Satterthwaite-like degrees of freedom, (3) using a $t$ statistic under the assumption of unknown but equal variances, and (4) using a delta-method approach.

*Large Sample Normal Inference.* For large sample sizes, it follows from the central limit theorem that the test statistic

$$Z = \frac{\bar{Y} - \delta\bar{X}}{\sqrt{S_E^2/n_E + \delta^2 S_C^2/n_C}} \tag{12.9}$$

will have an approximate standard normal distribution when $\mu_E - \delta\mu_C = 0$. The null hypothesis in Expression 12.8 is rejected and non-inferiority is concluded at a one-sided level of $\alpha/2$ when the observed value for $Z$ is greater than $z_{\alpha/2}$. When $\bar{x}/\sqrt{s_C^2/n_C}$ is quite large (e.g., greater than 4), an approximate $100(1 - \alpha)\%$ confidence interval for $\mu_E/\mu_C$ can be found using a Fieller approach given by $\{\lambda: -z_{\alpha/2} < (\bar{y} - \lambda\bar{x})/\sqrt{s_E^2/n_E + \lambda^2 s_C^2/n_C} < z_{\alpha/2}\}$, where $\bar{x}$ and $\bar{y}$ are the respective observed sample means and $s_C^2$ and $s_E^2$ are the respective observed sample variances.

*Using Satterthwaite-like Degrees of Freedom.* The hypotheses in Expression 12.8 are still tested using the test statistic in Equation 12.9. However, the observed value for $Z$ is compared with $-t_{\alpha/2,\nu(\delta)}$ instead of $-z_{\alpha/2}$, where the degrees of freedom $\nu(\delta)$ is the greatest integer less than or equal to

$$\frac{\left(\dfrac{\delta^2 s_C^2}{n_C} + \dfrac{s_E^2}{n_E}\right)^2}{\dfrac{\delta^4 s_C^4}{n_C^2(n_C - 1)} + \dfrac{s_E^4}{n_E^2(n_E - 1)}} \tag{12.10}$$

An approximate $100(1 - \alpha)\%$ confidence interval for $\mu_E/\mu_C$ by a Fieller approach is given by $\{\lambda: -t_{\alpha/2,\nu(\lambda)} < (\bar{y} - \lambda\bar{x})/\sqrt{s_E^2/n_E + \lambda^2 s_C^2/n_C} < t_{\alpha/2,\nu(\lambda)}\}$.

*t Statistic Assuming Unknown but Equal Variances.* The test statistic is based on the assumptions that the underlying distributions are normal distributions with equal variances. The test statistic is given by

$$T = \frac{\bar{Y} - \delta\bar{X}}{\sqrt{S^2(1/n_E + \delta^2/n_C)}} \tag{12.11}$$

where $S^2 = [(n_C - 1)S_C^2 + (n_E - 1)S_E^2]/(n_C + n_E - 2)$ is a pooled estimator of the common underlying variance. The null hypothesis in Expression 12.8 is rejected and non-inferiority is concluded at a one-sided level of $\alpha/2$ when the observed value for $T$ is less than $-t_{\alpha/2,n_C+n_E-2}$. An approximate $100(1 - \alpha)\%$

confidence interval for $\mu_E/\mu_C$ by a Fieller approach is given by $\{\lambda: -t_{\alpha/2, n_C + n_E - 2} <$ $(\bar{y} - \lambda\bar{x})/\sqrt{s^2(1/n_E + \lambda^2/n_C)} < t_{\alpha/2, n_C + n_E - 2}\}$.

*A Delta-Method Approach.* A delta-method approach can be considered in the testing of the hypotheses in Expression 12.7. This can be done using either a test statistic based on the ratio of sample means or a confidence interval for $\mu_E/\mu_C$. Hasselblad and Kong[8] considered a delta-method approach to the retention fraction for relative risks, odds ratios, and hazard ratios. Rothmann and Tsou[9] evaluated the behavior of delta-method confidence interval procedures through the maintenance of a desired 0.025 one-sided type I error rate and the quality of estimated standard error for the ratio of two normally distributed estimators.

The theorem behind the delta method can be found in many sources, such as the book by Bishop, Fienberg, and Holland.[10] For independent sequences of random variables $\{U_n\}$ and $\{V_n\}$, we have from the delta-method theorem that if $\sqrt{n}(U_n - \mu_1) \xrightarrow{d} N(0, \sigma_1^2)$ and $\sqrt{n}(V_n - \mu_2) \xrightarrow{d} N(0, \sigma_2^2)$, then

$$\sqrt{n}\left(\frac{U_n}{V_n} - \frac{\mu_1}{\mu_2}\right) \xrightarrow{d} N\left(0, \frac{\sigma_1^2}{\mu_2^2} + \frac{\mu_1^2 \sigma_2^2}{\mu_2^4}\right)$$ provided $\mu_2 \neq 0$. It follows, as noted

by Rothmann and Tsou,[9] that

$$W_n = \sqrt{n}\left(\frac{U_n}{V_n} - \frac{\mu_1}{\mu_2}\right) \bigg/ \sqrt{\frac{\sigma_1^2}{\mu_2^2} + \frac{\mu_1^2 \sigma_2^2}{\mu_2^4}}$$

$$= \left(\sqrt{n}\left(U_n - V_n \frac{\mu_1}{\mu_2}\right) \bigg/ \sqrt{\sigma_X^2 + \left(\frac{\mu_1}{\mu_2}\right)^2 \sigma_2^2}\right)(\mu_2/V_n) = Z_n \times (\mu_2/V_n)$$

Note that if $U_n$ and $V_n$ are distributed as $N(\mu_1, \sigma_1^2/n)$ and $N(\mu_2, \sigma_2^2/n)$, respectively, with $\mu_2 \neq 0$, then $Z_n$ has a standard normal distribution. How close the distribution of $W_n$ is to a standard normal distribution then depends on how close the distribution of $\mu_2/V_n$ is to the degenerate distribution at 1.

Applying the delta-method to the ratio of sample means and replacing the standard errors with the estimated standard errors leads to the test statistic of the hypotheses in Expression 12.7 of

$$W = \left(\frac{\bar{Y}}{\bar{X}} - \delta\right) \bigg/ \sqrt{\left(\frac{\bar{Y}}{\bar{X}}\right)^2 \left(\frac{S_E^2/n_E}{\bar{Y}^2} + \frac{S_C^2/n_C}{\bar{X}^2}\right)} \tag{12.12}$$

The unrestricted delta-method estimator for the standard deviation of $\bar{Y}/\bar{X}$ is

$\sqrt{\left(\frac{\bar{Y}}{\bar{X}}\right)^2 \left(\frac{S_E^2/n_E}{\bar{Y}^2} + \frac{S_C^2/n_C}{\bar{X}^2}\right)}$. Alternatively, the denominator in the test statistic in Equation 12.12 can be modified to use an estimator of the standard error that is restricted to $\mu_E/\mu_C = \delta$. The delta-method approximate $100(1 - \alpha)\%$ confidence interval for $\mu_E/\mu_C$ is given by

$$\bar{y}/\bar{x} \pm z_{\alpha/2}\sqrt{\left(\bar{y}/\bar{x}\right)^2\left(\frac{s_E^2/n_E}{\bar{y}^2} + \frac{s_C^2/n_C}{\bar{x}^2}\right)}$$

Note that $W$ in Expression 12.12 can also be reexpressed as $W = Z \times R$, where $Z$ is as in Equation 12.9 and $R = \sqrt{S_E^2/n_E + \delta^2 S_C^2/n_C}\,/\sqrt{S_E^2/n_E + (\bar{Y}/\bar{X})^2 S_C^2/n_C}$. If $Z$ has an approximate standard normal distribution, then $W$ will have an approximate standard normal distribution when $R \approx 1$ in distribution.

Further note that when $\delta = 0$, $R$ is necessarily smaller than 1 ($0 < R < 1$), and thus $W$ has a more concentrated distribution around zero than $Z$. When $Z$ has an approximate standard normal distribution, the distribution of $W$ has smaller tail areas than a standard normal distribution. Therefore, when $\delta$ is close to zero and the sample sizes are reasonably large, we would expect that the distribution of $W$ is more concentrated near zero than a standard normal distribution. Likewise when $\mu_E/\mu_C$ is much larger than $\delta$, $R$ will tend to be less than 1, and thus have a distribution more concentrated toward zero than $W$. Thus, there will be less power when using the test statistic in Expression 12.12 than using the test statistic in Expression 12.9.

Rothmann and Tsou[9] argued that for independent and normally distributed random variables $Y$ and $X$, where the mean of $X$ is not zero, $Y/X$ will have an approximate normal distribution, if $X$ behaves as approximately a nonzero constant with respect to the distribution of $Y$. This would occur if the standard deviation for $X$ divided by the mean of $X$ is close to zero. To have a type I error rate close to the desired level, Rothmann and Tsou[9] suggested, based on their simulations, that the absolute value of the ratio of the mean of $X$ to its standard deviation for $X$ be "greater than 7 or 8 or even greater, depending on how close is "close."

When the sample sizes are quite large (so that the sample mean for the control arm divided by its corresponding estimated standard error is large), the results from these four methods should be similar. We will compare the results of these methods and two Bayesian methods when we revisit Examples 12.1 through 12.3 with non-inferiority testing of the ratio of means when $\delta = 0.9$ in Section 12.3.4.

### 12.3.4 Bayesian Methods

In this section we will consider non-inferiority analysis of the ratio of means using Bayesian approaches. As before, conditional on the values of the parameters, we will consider the samples from the experimental and control arms as being independent random samples from normal distributions. The procedures in Section 12.2.4 (e.g., the Bayesian-$\gamma$ and Bayesian-$T$ procedures) for obtaining joint posterior distributions for $\mu_C$ and $\mu_E$ still apply. There are various posterior probabilities that can be determined. These include:

(a) The posterior probability that the experimental therapy is superior to the active control therapy.

(b) The posterior probability that the mean for the experimental therapy is greater than $100\delta\%$ of the mean for the control therapy and the mean for the control therapy is positive.

(c) The posterior probability that the mean for the experimental therapy is greater than $100\delta\%$ of the mean for the control therapy and the mean for the control therapy is positive, or the mean for the experimental therapy is positive and greater than the mean for the control therapy.

Example 12.6 revisits Examples 12.1 through 12.3. Here the non-inferiority inference will be based on the ratio of the means. The posterior probabilities of the null hypothesis or $p$-value when $\delta = 0.9$ and the 95% confidence/credible interval will be determined using each Bayesian procedure and each procedure discussed in Section 12.3.3.1.

## Example 12.6

We revisit Examples 12.1 through 12.3. Table 12.8 provides the $p$-value or the posterior probabilities of the null hypothesis in Expression 12.7 (or Expression 12.8 if appropriate) when $\delta = 0.9$ and the corresponding 95% confidence/credible intervals for the six methods discussed in this section. The $p$-values or posterior probabilities of the null hypothesis less than 0.025 are italicized. For the Bayesian methods, the set of values for $\mu_C$ and $\mu_E$ simulated earlier were used. All simulated values for $\mu_C$ and $\mu_E$ were positive and far from zero.

**TABLE 12.8**

Summary of $p$-Values, Posterior Probabilities, and 95% Confidence Intervals

| Procedures | Example 12.1 | | Example 12.2 | | Example 12.3 | |
|---|---|---|---|---|---|---|
| | $p$-Value | 95% CI | $p$-Value | 95% CI | $p$-Value | 95% CI |
| Large sample normal | *0.0230* | (0.901, 1.030) | *0.0217* | (0.902, 1.038) | *0.0125* | (0.910, 1.061) |
| Satterthwaite-like | 0.0271 | (0.899, 1.033) | 0.0265 | (0.899, 1.042) | *0.0137* | (0.909, 1.062) |
| Equal variance | 0.0294 | (0.898, 1.039) | *0.0206* | (0.903, 1.033) | *0.0135* | (0.909, 1.064) |
| Delta-method | *0.0236* | (0.901, 1.030) | 0.0292 | (0.898, 1.033) | *0.0148* | (0.908, 1.059) |
| Bayesian-$\gamma$[a] | 0.0255 | (0.900, 1.032) | 0.0249 | (0.900, 1.039) | *0.0137* | (0.909, 1.062) |
| Bayesian-$T$[a] | 0.0270 | (0.899, 1.033) | 0.0283 | (0.898, 1.042) | *0.0144* | (0.909, 1.063) |

[a] Posterior probabilities of the null hypothesis are given under the $p$-value column.

Statistical significance at a one-sided level of 0.025 (or the Bayesian equivalent) was reached for the large sample normal and delta-method procedures for revisited Example 12.1, by the large sample normal and equal variance procedures for revisited Example 12.2 and for all procedures for revisited Example 12.3. For revisited Examples 12.1 and 12.2, the lower limits of the 95% confidence/credible intervals varied far less across methods than the respective upper limits. Reversing the sample variances in Example 12.1 for Example 12.2 reduced the estimated standard error of $\bar{Y} - 0.9\bar{X}$ for the large sample normal method, which thereby decreased the corresponding $p$-value but increased the estimated standard deviation of $\bar{Y}/\bar{X} - 0.9$ for the delta method, which increased the corresponding $p$-value.

In all examples, the ratio of the sample mean for the control arm to its estimated standard error was quite large. This satisfies a Rothmann and Tsou[9] requirement for the delta-method to produce type I error rates close to the desired level. This also makes the posterior probability that $\mu_C > 0$ microscopically close to 1 for the Bayesian methods and thus simplifying the probabilities given in (b) and (c). For (a), the simulated probabilities of $\mu_E > \mu_C$ can be found from Tables 12.4 through 12.6 by subtracting from 1 each posterior probability of the null hypothesis when the non-inferiority margin for the difference is zero. In these cases, the simulated probabilities of $\mu_E > \mu_C$ are closer to 0.5 for the Bayesian-$T$ method than for the Bayesian-$\delta$ method, which suggests that the Bayesian-$T$ method imposes more variability on $\mu_C$ and $\mu_E$ than the Bayesian-$\gamma$ method. In all three cases, the standard deviations for the simulated marginal posterior distribution of $\mu_C$ and $\mu_E$ were larger for the Bayesian-$T$ method than for the Bayesian-$\gamma$ method ranging from 1.0% larger to 2.8% larger. In these three examples, relative to the other methods, the Bayesian-$T$ method was fairly conservative, producing generally larger $p$-values with wider 95% confidence/credible intervals.

For the Satterthwaite-like procedure in all three examples, the upper and lower confidence limits were determined from multipliers/table values based on different numbers of degrees of freedom. In revisited Example 12.3, the lower limit of the 95% Fieller confidence interval was based on $t_{0.025,92} = 1.9861$, whereas the upper limit of the 95% Fieller confidence interval was based on $t_{0.025,99} = 1.9842$.

### 12.3.5 Sample Size Determination

Standard software packages for sample size and power calculations are not designed for non-inferiority analyses based on ratios. We again provide caution about sizing a study based on the arbitrary assumption that the control and experimental therapies have identical efficacy.

On the basis of the test statistic in Equation 12.11, Kieser and Hauschke[7] determined iterative formulas for the desired sample size assuming an unknown, common underlying variance. Their arguments can be extended to the case that does not assume equal underlying variances. Let $\mu_1$ and $\mu_2$ represent the assumed true means for the experimental and control arms, respectively, where $\mu_1/\mu_2 > \delta$ and $\mu_2 > 0$. Let $\sigma_1$ and $\sigma_2$ denote the respective assumed underlying standard deviations where $l = \sigma_1/\sigma_2$. Let $k = n_E/n_C$ denote the allocation ratio. On the basis of the assumption of independent normal random samples and using a Satterthwaite-like approximation for the degrees of freedom, the test statistic $Z$ in Equation 12.9 will be modeled as having an approximate noncentral $t$ distribution with noncentrality parameter given by $\dfrac{\mu_1 - \delta\mu_2}{\sigma_2\sqrt{(l^2/k + \delta^2)/n_C}}$ and degrees of freedom given by

$v = \dfrac{(\delta^2 + l^2/k)^2}{\delta^4/(n_C - 1) + l^4/[k^2(kn_C - 1)]}$. We will use the approximate relation that the

100$\beta$th percentile of the noncentral $t$ distribution is approximately equal to the noncentrality parameter plus the 100$\beta$th percentile of the $t$ distribution having the same degrees of freedom. Then an iterative sample size formula for $n_C$ ($n_C$ appears on both sides of the equation) is given by

$$n_C = (l^2/k + \delta^2)(t_{\alpha/2,v} + t_{\beta,v})^2(\sigma_2/(\mu_1 - \delta\mu_2))^2 \qquad (12.13)$$

For the case where equal, unknown underlying variances is assumed, Equation 12.13 applies with $v = 2n_C - 2$ and $l = 1$. For the large sample normal test, Equation 12.13 can directly be applied (no iterations needed) where the term $t_{\alpha/2,v} + t_{\beta,v}$ is replaced with the term $z_{\alpha/2} + z_\beta$. The control sample size calculated for the large sample normal test can be used as the starting value for the iterations for the $t$-test procedures. For the delta-method procedure based on the test statistic in Equation 12.12, replace the term $t_{\alpha/2,v} + t_{\beta,v}$ with the term $z_{\alpha/2} + z_\beta$ and replace the term $l^2/k + \delta^2$ with the term $l^2/k + (\mu_1/\mu_2)^2$ in Equation 12.13. Since $\mu_1/\mu_2 > \delta$, the calculated sample size for the delta-method procedure using the test statistic in Equation 12.12 will be larger than the calculated sample size for the large sample normal procedure. This is due to the test statistic in Equation 12.12 not having the standard error in its denominator restricted to the null assumption that $\mu_E/\mu_C = \delta$. A delta-method test statistic that restricts the standard error in the denominator of the test statistic under the assumption that $\mu_E/\mu_C = \delta$ should have a similar sample size calculation as the large sample normal procedure.

The overall sample size is $1 + k$ times the solution for $n_C$ in Equation 12.13 or the corresponding analog for one of the other test statistics. The optimal allocation ratio $k$ that minimizes the overall sample size can be found by similar means as those provided in Section 12.2.5.

We refer the reader to Section 12.2.5 on the sample size determination for non-inferiority trials based on a difference in means for additional comments that also apply for the ratio of means. The use of the sample size formulas will be illustrated in Example 12.7.

### Example 12.7

Consider an endpoint that is the improvement from baseline in some score. The non-inferiority threshold for the ratio of mean improvement is 0.75. Both a 1:1 and a 2:1 experimental-to-control allocation are potentially being considered. As in Example 12.5, the trial will be sized on the basis of true mean improvement of 38 and 40 for the experimental and control arms, respectively, with corresponding underlying standard deviations of 6 and 12, respectively. For the equal variance

**TABLE 12.9**

Sample Sizes for the Control Arm

| Method | Allocation Ratio | | Percentage Change in Overall Sample Size (%)[a] |
|--------|:---:|:---:|:---:|
| | 1:1 | 2:1 | |
| Large sample normal | 20 | 17 | 26.9 |
| Delta method | 28 | 25 | 33.7 |
| Satterthwaite | 21 | 18 | 30.2 |
| Equal variance | 25 | 17 | 4.2 |

[a] Based on the calculated value before rounding up.

approach, the common underlying variance will be assumed as 90. Table 12.9 provides the determined sample sizes for the control arm.

For this example, the calculated sample sizes were smallest for the large sample normal and equal variance methods. The delta method requires a larger sample size than the large sample normal method since $\mu_1/\mu_2 = 0.95 > 0.75$.

When the inference is based on the difference in means with equal and known underlying variances (i.e., when $l/\delta = 1$), the overall sample size for a 2:1 allocation will be 12.5% greater than the overall sample size for a 1:1 allocation. In this example for these methods, the percentage increase in the overall sample size from a 1:1 allocation to a 2:1 allocation was quite different from 12.5% since $l/\delta$ (or $l\mu_2/\mu_1$ when the delta-method is used) was different from 1. The iterative nature of the Satterthwaite and equal variance methods also influences the particular percentage increase. For each of the large sample normal, Satterthwaite ($l/\delta = 2/3$), and delta methods ($l\mu_2/\mu_1 \approx 0.526$), there was around a 30% increase in the overall sample size from a 1:1 allocation to a 2:1 allocation. For the equal variance method $l/\delta = 4/3$ and instead of a 12.5% increase in the calculated value for the overall sample size going from a 1:1 allocation to a 2:1 allocation, there was a 4.2% increase.

The optimal allocation ratio is $l/\delta$ (or $l\mu_2/\mu_1$ for the delta-method is used). In this example, for the large sample normal and the Satterthwaite methods, $k = 2/3$ is the allocation ratio that minimizes the calculated overall sample size. For the large sample normal method, this corresponds to 22 subjects in the control arm and 15 subjects in the experimental arm. For the Satterthwaite-like method, 25 subjects are allocated to the control arm and 16 subjects are allocated to the experimental arm. For the delta-method, $k \approx 0.526$ is the allocation ratio that minimizes the calculated overall sample size with 33 subjects allocated to the control arm and 17 subjects allocated to the experimental arm. For the equal variance method, $k = 4/3$ is the allocation ratio that minimizes the calculated overall sample size with 21 subjects allocated the control arm and 29 subjects allocated to the experimental arm.

## 12.4 Analyses Involving Medians

A median of a distribution for a random variable X, $\tilde{\mu}$, is any value satisfying $P(X \leq \tilde{\mu}) \geq 0.5$ and $P(X \geq \tilde{\mu}) \geq 0.5$. For a continuous distribution, $\tilde{\mu}$ satisfies

$P(X \leq \tilde{\mu}) = 0.5 = P(X \geq \tilde{\mu})$. We will consider only cases involving continuous distributions having unique medians. We will denote the medians for the underlying distributions for the experimental and control arms as $\tilde{\mu}_E$ and $\tilde{\mu}_C$, respectively, and their difference by $\Delta = \tilde{\mu}_E - \tilde{\mu}_C$.

Medians are often used to describe the central location of a distribution that is skewed. They are less frequently used for comparison purposes. A difference in two means between an experimental and control arm in a randomized trial represents the average benefit across trial subjects from being randomized to the experimental arm instead of the control arm. A difference in the two medians does not have any analogous interpretation unless additional assumptions are made on the underlying distributions (e.g., that the shapes of the underlying distributions are equal). Non-inferiority testing involving a median is rare. There are, however, both distinct properties involving non-inferiority testing of medians than for other metrics and there are some common methodology more easily discussed with medians. For positive-valued variables, the median of the log-values is the log of the median. Therefore, the ratio of the medians can be tested through the difference of the medians of log-values. Means do not have such a property (the mean of the log-values is not the log of the mean).

For a sample, the sample median is the middle ordered value of a sample when the sample size is odd. When the sample size is even, any value between the two middle values is a sample median—however, it is common to use the average of the two middle values as the sample median. We will use this common convention as defining the sample median when the sample size is even.

## 12.4.1 Hypotheses and Issues

The non-inferiority tests involving differences in medians are just shifted versions of the tests for superiority. When there are no restrictions on the distributions, testing on medians usually requires determining an estimate of the variance of the estimated median. The variance for an estimator of the median depends on the density at the true median. Since the value of the median is unknown and it may require a rather large sample size to reliably estimate the density function at any likely possibility for the median, inferences on the median are not generally based on asymptotic normal distributions for data that tends to be seen in a clinical trial. To reduce the complexities in basing an inference on the difference of medians, it is a popular practice to make the overall assumption that the underlying distributions have the same shape (i.e., $F_C(y) = F_E(y - \Delta)$ for all $y$ and some $\Delta$). We will discuss procedures based on the median test, Mathisen's test, and the Mann–Whitney–Wilcoxon test having the underlying assumption of equal-shaped distributions.

For a non-inferiority margin of $\delta$ (for some $\delta > 0$), the hypotheses can be expressed as

$$H_o: \Delta \leq -\delta \text{ and } H_a: \Delta > -\delta$$

For the nonparametric methods that will be discussed, non-inferiority testing can be done by first adding $\delta$ to every outcome from the experimental arm and then testing for superiority against the control arm. Alternatively, non-inferiority is concluded when the associated confidence interval of appropriate level for the difference in medians contains only values greater than $-\delta$.

Two–confidence interval approaches, discussed in Chapter 5, are popular for non-inferiority testing. These approaches combine into an inference the results from an external evaluation of the effect size of the control therapy and the estimated difference between treatment arms in the non-inferiority trial. Some approaches comparing two–medians can easily be expressed as two–confidence interval procedures. Hettmansperger[11] investigated the use and properties of confidence intervals for the difference in medians where the limits are differences of the confidence limits of the individual confidence intervals for the medians. The $100(1 - \alpha)\%$ confidence interval for the difference in medians has the form $(L, U) = (L_E - U_C, U_E - L_C)$, where $(L_E, U_E)$ is a $100(1 - \alpha_E)\%$ confidence interval for the median of the experimental arm and $(L_C, U_C)$ is a $100(1 - \alpha_C)\%$ confidence interval for the median of the control arm. The confidence coefficients for the individual confidence intervals are selected so that when those two intervals are disjoint, $H_o: \Delta = 0$ is rejected at a significance level of $\alpha$. Note that the width of the confidence interval for the difference in medians is the sum of the widths of the individual confidence intervals for the medians $(U - L = (U_E - L_E) + (U_C - L_C))$. As a special case, Hettmansperger[11] showed that the confidence interval for the difference in medians constructed by inverting the median test has such a form.

### 12.4.2 Nonparametric Methods

*The Median Test.* The median test or Mood's test[12] compares the equality of two distributions through the equality of the medians with the overall assumption that the two distributions have the same shape (i.e., $F_C(y) = F_E(y - \Delta)$) for all $y$ and some $\Delta$). Let $X_1, X_2, \ldots, X_{n_C}$ and $Y_1, Y_2, \ldots, Y_{n_E}$ denote independent random samples from distributions having respective distribution functions $F_C$ and $F_E$.

Consider testing under the assumption that the two distributions have the same shape that $H_o: \Delta = 0$ against $H_a: \Delta \neq 0$. The test statistic for the median test is

$$M = \sum_{j=1}^{n_E} I(R(Y_j) > (N+1)/2) \qquad (12.14)$$

where $R(Y_j)$ is the rank of $Y_j$ in the ordering of the combined sample, $I$ is an indicator function, and $N$ is the combined sample size. The test statistic $M$ is the number of observations in the experimental arm that are greater than the median of the combined sample. Under the null hypothesis, $M$ has a hyper-geometric distribution, where $P(M = m) = \binom{N/2}{m}\binom{N/2}{n_E - m} \Big/ \binom{N}{n_E}$ for $m = 0,$

1, 2, ..., $n_E$, where $\binom{a}{b} = 0$ whenever $a < b$. When $\Delta = 0$, the distribution for $M$ is symmetric about the mean $n_E/2$ with variance $n_C n_E/[4(N-1)]$. The test rejects $H_0$: $\Delta = 0$ when $M < d$ or $M > n_E - d + 1$, where $\alpha/2 = P(M < d | \Delta = 0) = P(M > n_E - d + 1 | \Delta = 0)$. For large sample sizes, the value for $d$ can be approximated by the greatest integer less than or equal to $n_E/2 - z_{\alpha/2}\sqrt{n_E n_C/[4(N-1)]}$. When $\Delta = 0$ $(F_C = F_E)$, $\dfrac{M - n_E/2}{\sqrt{\lambda n_E/4}} \xrightarrow{d} N(0,1)$ as $n_C, n_E \to \infty$, and $n_C/N \to \lambda > 0$.

Let $X_{(1)} < X_{(2)} < \dots < X_{(n_C)}$ and $Y_{(1)} < Y_{(2)} < \dots < Y_{(n_E)}$ denote the respective order statistics. Without loss of generality, assume that $N$ is an even number and that $n_C \geq n_E$. The test statistic in Equation 12.14 can be reexpressed as

$$M = \sum_{j=1}^{n_E} I(Y_{(j)} - X_{(N/2-j+1)} > 0) \tag{12.15}$$

(see Pratt[13] or Gastwirth[14]). Therefore, from Equation 12.15, the value of $M$ depends on the two samples through the ordered values of

$$Y_{(1)} - X_{(N/2)} < Y_{(2)} - X_{(N/2-1)} < \dots < Y_{(n_E)} - X_{((n_C-n_E)/2+1)}$$

As provided by Hettmansperger,[11] rejecting $H_0$: $\Delta = 0$ is therefore tantamount to examining whether 0 is in the confidence interval for the difference in medians of $(L, U) = (Y_{(d)} - X_{(N/2-d+1)},\ Y_{(n_E-d+1)} - X_{((n_C-n_E)/2+d)})$. This confidence interval for the difference in medians can be expressed as $(L, U) = (L_E - U_C, U_E - L_C)$, where $(L_E, U_E) = (Y_{(d)},\ Y_{(n_E-d+1)})$ is a confidence interval for the median of the experimental arm and $(L_C, U_C) = (X_{((n_C-n_E)/2+d)},\ X_{(N/2-d+1)}) = (X_{(d_C)},\ X_{(n_C-d_C+1)})$ is a confidence interval for the median of the control arm where $d_C = (n_C - n_E)/2 + d$. The confidence coefficients are determined from respective binomial distributions based on $n_E$ and $n_C$ Bernoulli trials, where all trials have success probability of 0.5.

The confidence coefficient for an individual confidence interval will be larger for the arm that has the smaller sample size. This is easily seen for large sample sizes from the arguments of Hettmansperger,[11] who argued that for large $n_C$ and $n_E$ with $\lambda = n_C/N$,

$$z_{\alpha_E/2} \approx z_{\alpha/2}\sqrt{\lambda} \quad \text{and} \quad z_{\alpha_C/2} \approx z_{\alpha/2}\sqrt{1-\lambda} \tag{12.16}$$

where $1 - \alpha_E$ and $1 - \alpha_C$ are the respective confidence coefficients for the individual confidence interval for the medians of the experimental and control arms.

When $(L_E, U_E) = (Y_{(d_E)}, Y_{(n_E - d_E + 1)})$ and $(L_C, U_C) = (X_{(d_C)}, X_{(n_C - d_C + 1)})$ for some $d_E$ and $d_C$, we have that for large $n_C$ and $n_E$ from Theorem 2.2 of Hettmansperger,[11] the confidence coefficients are related by

$$\sqrt{\lambda} z_{\alpha_E/2} + \sqrt{1 - \lambda} z_{\alpha_C/2} \approx z_{\alpha/2} \tag{12.17}$$

In addition, the asymptotic width of the confidence interval does not depend on the choice of $\alpha_E$ and $\alpha_C$. From Theorem 2.3 of Hettmansperger,[11] with probability of 1

$$\sqrt{N}(U - L) \to z_{\alpha/2} / \left[ \sqrt{\lambda(1 - \lambda)} f_C(0) \right]$$

where $f_C(0)$ is the common density at the median. Thus, there are many pairs of $\alpha_E$ and $\alpha_C$ that lead to a confidence interval for the difference in medians of $(L_E - U_C, U_E - L_C)$ that has confidence coefficient of approximately $1 - \alpha$ having the same asymptotic width/efficiency. Note from Equation 12.17, when it is desired to have $\alpha_E = \alpha_C$, then set

$$z_{\alpha_E/2} = z_{\alpha_C/2} = z_{\alpha/2} / \left( \sqrt{\lambda} + \sqrt{1 - \lambda} \right) \tag{12.18}$$

For allocation ratios ranging from 1 to 3, Hettmansperger[11] compared three methods of confidence intervals for the difference in medians of the form $(L_E - U_C, U_E - L_C)$: the median test-based confidence interval, the confidence interval where $\alpha_E = \alpha_C$, and the confidence interval where $U_E - L_E$ and $U_C - L_C$ are asymptotically equal. In the case of asymptotically equal-length confidence intervals, we have from Hettmansperger[11]

$$z_{\alpha_E/2} \approx z_{\alpha/2} / [2\sqrt{\lambda}] \text{ and } z_{\alpha_C/2} \approx z_{\alpha/2} / [2\sqrt{1 - \lambda}] \tag{12.19}$$

The overall assumption of the two underlying distributions having the same shape (i.e., $F_C(y) = F_E(y - \Delta)$ for all $y$) is necessary for Equations 12.16 through 12.19 and other properties to hold. If the shapes of the two distributions are quite different, then these methods may not produce confidence intervals for the difference in medians of a desired level. Pratt[13] noted that when the true medians are equal with underlying distributions having different shapes, then the two-sided level for the median test is asymptotically equal to $2[1 - \Phi(cz_{\alpha/2})]$, where $c = (1 - \lambda + \lambda\tau)/\sqrt{1 - \lambda + \lambda\tau^2}$, and $\tau$ is the ratio of the underlying densities $(f_C/f_E)$ at the common median. It follows, for large sample sizes, that the desired significance level can be approximately maintained (as noted by Freidlin and Gastwirth[15]) when the underlying assumption of equal-shaped distributions is weakened to having equal density functions in analogous neighborhoods about the medians.

We will consider a variable or endpoint where the larger the value, the better the outcome. For a non-inferiority margin of $\delta$ (for some $\delta > 0$), the hypotheses can be expressed as $H_0: \Delta \leq -\delta$ and $H_a: \Delta > -\delta$. Non-inferiority testing can be done by first adding $\delta$ to every outcome from the experimental arm and then testing for superiority against the control arm. The corresponding test statistic based on the median test is thus

$$M_\delta = \sum_{j=1}^{n_E} I(R(Y_j + \delta) > (N+1)/2)$$

where $R(Y_j + \delta)$ is the rank of $Y_j + \delta$ in the ordering of the combined sample of $Y_j + \delta$'s and $X_i$'s. The null hypothesis is rejected and non-inferiority is concluded whenever $M_\delta > n_E - d + 1$, where $\alpha/2 = P(M_\delta > n_E - d + 1 | \Delta = -\delta)$. Alternatively, non-inferiority is concluded when the appropriate level confidence interval for the difference in medians contains only values greater than $-\delta$.

We will now consider another two confidence interval procedure for the difference in medians based on Mathisen's test.

*Mathisen's Test.* Mathisen's[16] test also compares the equality of two distributions through the equality of the medians with the overall assumption that the two distributions have the same shape (i.e., $F_C(y) = F_E(y - \Delta)$ for all $y$ and some $\Delta$). Again let $X_1, X_2, \ldots, X_{n_C}$ and $Y_1, Y_2, \ldots, Y_{n_E}$ denote independent random samples from distributions having respective distribution functions $F_C$ and $F_E$, and we consider testing under the assumption that the two distributions have the same shape $H_0: \Delta = 0$ against $H_a: \Delta \neq 0$. The test statistic for Mathisen's test is

$$V = \sum_{j=1}^{n_E} I(Y_j > \operatorname*{med}_i X_i) \tag{12.20}$$

where $I$ is an indicator function. The test statistic $V$ is the number of observations in the experimental arm that are greater than the median for the control arm. Under the null hypothesis, $V$ has a symmetric distribution. The test rejects $H_0: \Delta = 0$ when $V < d$ or $V > n_E - d + 1$ where $\alpha/2 = P(V < d | \Delta = 0) = P(V > n_E - d + 1 | \Delta = 0)$. For ease, we will consider the distribution of $V$ only when $n_C$ is odd. When $n_C$ is odd, $P(V = v) = \binom{(n_C - 1)/2 + v}{v} \binom{(n_C - 1)/2 + n_E - v}{n_E - v} \Big/ \binom{N}{n_E}$

for $v = 0, 1, 2, \ldots, n_E$. When $\Delta = 0$, $V$ has a symmetric distribution about $n_E/2$ with mean $n_E/2$ and variance $n_E(N + 1)/(4[n_C + 2])$. For large sample sizes, the value for $d$ can be approximated by the greatest integer less than or equal to

$n_E/2 - z_{\alpha/2}\sqrt{n_E(N+1)/(4[n_C+2])}$. When $\Delta = 0$ $(F_C = F_E)$, $\dfrac{V - n_E/2}{\sqrt{n_E/(4\lambda)}} \xrightarrow{d} N(0,1)$

as $n_C, n_E \to \infty$ and $n_C/N \to \lambda > 0$.

Asymptotic results can also be applied when $n_C$ is even and large. This can be shown from the relation $\sum_{j=1}^{n_E} I(Y_j > X_{(n_C/2+1)}) \le V \le \sum_{j=1}^{n_E} I(Y_j > X_{(n_C/2)})$ and that each bounding sum has the same asymptotic distribution as $V$ when $n_C$ is odd. The probability distributions for these bounding sums are also easily obtained and can be used to obtain approximate critical values. Alternatively, for ease when $n_C$ is even, $\sum_{j=1}^{n_E} I(Y_j > X_{(n_C/2+1)})$ or $\sum_{j=1}^{n_E} I(Y_j > X_{(n_C/2)})$ can be used as the test statistic instead of $V$. Note that neither of these sums have a symmetric distribution when $\Delta = 0$.

The test statistic in Equation 12.20 can be reexpressed as

$$V = \sum_{j=1}^{n_E} I(Y_{(j)} - X_{((n_C+1)/2)} > 0) \tag{12.21}$$

Therefore, from Equation 12.21, the value of $V$ depends on the two samples through the ordered values of

$$Y_{(1)} - X_{((n_C+1)/2)} < Y_{(2)} - X_{((n_C+1)/2)} < \cdots < Y_{(n_E)} - X_{((n_C+1)/2)}$$

Rejecting $H_o$: $\Delta = 0$ is therefore tantamount to examining whether 0 is in the confidence interval for the difference in medians of $(L, U) = (Y_{(d)} - X_{((n_C+1)/2)}, Y_{(n_E-d+1)} - X_{((n_C+1)/2)})$. This confidence interval for the difference in medians can be expressed as $(L, U) = (L_E - U_C, U_E - L_C)$, where $(L_C, U_C) = [X_{((n_C+1)/2)}, X_{((n_C+1)/2)}]$ is a 0% confidence interval for the median of the control arm and $(L_E, U_E) = (Y_{(d)}, Y_{(n_E-d+1)})$ is a confidence interval for the median of the experimental arm with confidence coefficient $1 - \alpha_E$, which satisfies for $n_C$ and $n_E$ by Equation 12.17

$$z_{\alpha_E/2} \approx z_{\alpha/2}/\sqrt{\lambda} \tag{12.22}$$

The decision from Mathisen's test (rejecting or failing to reject $H_o$: $\Delta = 0$) can depend on which sample's median is used. If the roles were switched for the $X$'s and $Y$'s, the decision may change.

We provide a table similar to Table 1 of Hettmansperger.[11] When $\alpha = 0.05$ for each method, Table 12.10 provides the confidence coefficients for the individual intervals for the control and experimental medians for allocation ratios between 1 and 3, and as the allocation ratio goes to infinity. Equations 12.16, 12.18, 12.19, and 12.22 are used to determine the confidence coefficients. Those common entries differ somewhat from those of Hettmansperger.[11] The

**TABLE 12.10**

Confidence Coefficients for the Intervals for Control and Experimental Medians when $\alpha = 0.05$

| $n_C/n_E$ | Mathisen's Test | | Median Test | | Equal Coefficients | Equal Lengths | |
|---|---|---|---|---|---|---|---|
| | $1 - \alpha_C$ | $1 - \alpha_E$ | $1 - \alpha_C$ | $1 - \alpha_E$ | $1 - \alpha_C = 1 - \alpha_E$ | $1 - \alpha_C$ | $1 - \alpha_E$ |
| 1 | 0 | 0.994 | 0.834 | 0.834 | 0.834 | 0.834 | 0.834 |
| 1.5 | 0 | 0.989 | 0.785 | 0.871 | 0.836 | 0.879 | 0.794 |
| 2 | 0 | 0.984 | 0.742 | 0.890 | 0.840 | 0.910 | 0.770 |
| 2.5 | 0 | 0.980 | 0.705 | 0.902 | 0.845 | 0.933 | 0.754 |
| 3 | 0 | 0.976 | 0.673 | 0.910 | 0.849 | 0.950 | 0.742 |
| $\to \infty$ | 0 | 0.95 | 0 | 0.95 | 0.95 | 1 | 0.673 |

methods are arranged so that the confidence coefficient for the control arm increases (experimental arm decreases) going from left to right. For the confidence intervals based on Mathisen's test, as the allocation ratio increased from 1 to 3, the confidence coefficient of the confidence interval for the control median remained at zero and the confidence coefficient of the confidence interval for the experimental median decreased from 0.994 to 0.976. For the remaining three methods, the common confidence coefficient for the individual confidence intervals for the medians was 0.834 when equal sample sizes were used. For the confidence intervals based on the two sample median test, as the allocation ratio increased from 1 to 3, the confidence coefficient of the confidence interval for the control median decreased from 0.834 to 0.673, whereas the confidence coefficient of the confidence interval for the experimental median increased from 0.834 to 0.910. The common confidence coefficient for the equal coefficients case was stable, varying from 0.834 to 0.849 as the allocation ratio ranged from 1 to 3. For the approach of using equal asymptotic length confidence intervals, as the allocation ratio increased from 1 to 3, the confidence coefficient of the confidence interval for the control arm increased from 0.834 to 0.950, whereas the confidence coefficient of the confidence interval for the experimental arm decreased from 0.834 to 0.742. For Mathisen's test, the median test, and the equal coefficients cases, the limiting confidence coefficient for the confidence interval for the experimental median was 0.95 as the allocation ratio approached infinity. For the equal-lengths approach, the limiting confidence coefficient of the confidence interval for the experimental median was 0.673 (i.e., $z_{\alpha_E/2} \to z_{0.025}/2$) as the allocation ratio approached infinity, whereas the limiting confidence coefficient of the confidence interval for the experimental median was 1.

When $n_C/n_E$ is less than 1, except for Mathisen's test, the confidence coefficients can be found by reversing the roles of the control and experimental arms. For Mathisen's test when $\alpha = 0.05$, the confidence coefficient for the individual confidence interval for the experimental median is greater than 0.999 when $n_C/n_E < 0.549$. For Mathisen's test, all the uncertainty in the comparison of medians is reflected in the confidence interval for the experimental median. As the uncertainty in the estimation of the control median becomes larger relative to the uncertainty in the estimation of the experimental median, a greater confidence coefficient is needed for the confidence interval for the experimental median. A two–confidence interval procedure based on Mathisen's test is analogous to using the point estimate of the historical effect of the active control therapy as the true effect of the active control in the non-inferiority trial (and thereby ignoring the uncertainty in the estimate) when the constancy assumption holds.

*Mann–Whitney–Wilcoxon Test.* The Mann–Whitney–Wilcoxon test compares the equality of two distributions. When the assumption is made that the two distributions have the same shape (i.e., $F_C(y) = F_E(y - \Delta)$ for all $y$ and some $\Delta$), inferences can be made on the shift parameter, which equals the difference in the medians. Let $X_1, X_2, \ldots, X_{n_C}$ and $Y_1, Y_2, \ldots, Y_{n_E}$ denote independent random samples from distributions having respective distribution functions $F_C$ and $F_E$.

Consider testing $H_o$: $\Delta = 0$ against $H_a$: $\Delta \neq 0$ under the assumption that the two distributions have the same shape. The Mann–Whitney–Wilcoxon test can be based on the sum of the ranks of the observations in one of the arms among the combined observations (i.e., $\sum_{i=1}^{n_E} R(Y_i)$) or equivalently based on the test statistic

$$W = \sum_{j=1}^{n_E} \sum_{i=1}^{n_C} I(Y_j - X_i > 0)$$

where $I$ is an indicator function. Note that when $\Delta = 0$, $W$ has a symmetric distribution about the mean $n_C n_E/2$ with variance $n_C n_E(n_C + n_E + 1)/12$. The test rejects $H_o$: $\Delta = 0$ when $W < d$ or $W > n_C n_E - d + 1$, where $\alpha/2 = P(W < d|\Delta = 0) = P(W > n_C n_E - d + 1|\Delta = 0)$. The Mann–Whitney–Wilcoxon test is the locally most powerful rank test when $F_C$ has a logistic distribution.

For a non-inferiority margin of $\delta$ (for some $\delta > 0$), the hypotheses can be expressed as $H_o$: $\Delta \leq -\delta$ and $H_a$: $\Delta > -\delta$. The corresponding test statistic is

$$W_\delta = \sum_{j=1}^{n_E} \sum_{i=1}^{n_C} I(Y_j - X_i > -\delta)$$

The null hypothesis is rejected and non-inferiority is concluded whenever $W_\delta > n_C n_E - d + 1$, where $\alpha/2 = P(W_\delta > n_C n_E - d + 1|\Delta = -\delta)$. Alternatively, non-inferiority can be tested by finding the corresponding confidence interval for the difference in medians, $\Delta$, and comparing the interval with $-\delta$.

It can be shown that a $100(1 - \alpha)\%$ confidence interval for $\Delta$ based on the Mann–Whitney–Wilcoxon test is given by

$$(Z_{(d)}, Z_{(n_C n_E - d + 1)})$$

where $Z_{(1)} < \ldots < Z_{(n_C n_E)}$ are the ordered differences of $Y_j - X_i$ for $i = 1, \ldots, n_C$ and $j = 1, \ldots, n_E$. Non-inferiority is concluded when this confidence interval only contains values greater than $-\delta$. For large sample sizes, the value for $d$ can be approximated by the greatest integer less than or equal to

$$n_C n_E / 2 - z_{\alpha/2} \sqrt{n_C n_E (n_C + n_E + 1)/12}$$

When $\Delta = 0$ $(F_C = F_E)$, $W^* = \dfrac{W - n_C n_E/2}{\sqrt{n_C n_E (n_C + n_E + 1)/12}} \xrightarrow{d} N(0,1)$ as $n_C$,

$n_E \to \infty$ and $n_C/N \to \lambda$. The asymptotic behavior of the test can also be

determined when $F_C \neq F_E$. When $F_C \neq F_E$, $W$ has mean $n_C n_E p_1$ and variance $n_C n_E (p_1 - p_1^2) + n_C n_E (n_E - 1)(p_2 - p_1^2) + n_C n_E (n_C - 1)(p_3 - p_1^2)$, where $p_1 = P(Y_1 > X_1)$, $p_2 = P(\min Y_1, Y_2 > X_1)$, and $p_3 = P(Y_1 > \max X_1, X_2)$ (see Theorem 3.5.1 of Hettmansperger[17]). When $F_C \neq F_E$, it follows that as $n_C, n_E \to \infty$ and $n_C/N \to \lambda > 0$, $P(W^* > z_{\alpha/2}) \to 0$ if $p_1 < 0.5$, $P(W^* > z_{\alpha/2}) \to 1 - \Phi(z_{\alpha/2} / \sqrt{12\{(1 - \lambda)(p_2 - 0.25) + \lambda(p_3 - 0.25)\}})$ if $p_1 = 0.5$, and $P(W^* > z_{\alpha/2}) \to 1$ if $p_1 > 0.5$.

Since the test statistics are distribution free when $\Delta = 0$, for small samples $d$ can either be found from mathematical calculations or by simulations. The asymptotic approximate results can be used for large sample sizes.

These methods can be used to find confidence intervals for the ratio of medians when the two underlying distributions are positive-valued and "differ" by a scale factor. In this case we have $F_C(0) = 0$ and $F_C(y) = F_E(\theta y)$ for all $y > 0$ and some $\theta > 0$, which represents the experimental to control ratio of the medians. Then the underlying distributions for the logarithms of the observations have the same shape (i.e., $G_C(y) = G_E(y + \log\theta)$) with the difference in medians of log $\theta$. These methods can be used to find a confidence interval for log $\theta$ that can be converted to a confidence interval for the ratio of medians $\theta$.

### 12.4.3 Asymptotic Methods

In this subsection we discuss some estimators for a median or difference of medians, their asymptotic distributions and variances and their relative efficiencies.

*Sample Median.* Let $X_1, X_2, \ldots, X_n$ be a random sample from a continuous distribution with distribution function $F$ and unique median $\tilde{\mu}$. Let $\tilde{X}$ denote the sample median. Then as $n \to \infty$,

$$\sqrt{n}(\tilde{X} - \tilde{\mu}) \xrightarrow{d} N(0, (4[f(\tilde{\mu})]^2)^{-1})$$

For a normal random sample, this asymptotic variance equals $\pi\sigma^2/(2n)$, whereas the asymptotic variance for the sample mean equals $\sigma^2/n$. For a normal random sample, the efficiency of the sample median relative to the sample mean is $2/\pi \approx 0.637$ (i.e., the sample mean based on 637 observations has approximately equal standard error as the sample median based on 1000 observations). This is also the relative efficiency of the difference in sample medians to the difference in sample means when there are independent normal random samples.

When the variance of the underlying distribution exists with the mean equal to the median, the relative efficiency of the sample median to the sample mean equals $4[f(\tilde{\mu})]^2\sigma^2$. Thus the sample median will have greater limiting efficiency than the sample mean if and only if $2f(\tilde{\mu}) > 1/\sigma$.

The estimate of the median corresponding to a two-sided test is obtained as the value $x$ such that if it is specified as the null value the test result will not favor one side of the alternative hypothesis ($\Delta > x$ or $\Delta < x$) over the other side (e.g., both one-sided $p$-values equal 0.5). For both the median test and Mathisen's test, the corresponding estimator of the difference in medians is given by $\hat{\Delta} = \underset{j}{\mathrm{med}}(Y_{(j)} - X_{(N/2+1-j)}) = \tilde{Y} - \tilde{X}$. As $n_C, n_E \to \infty$, and $n_C/N \to \lambda > 0$,

$$\sqrt{N}(\hat{\Delta} - \Delta) \xrightarrow{d} N(0, (4\lambda(1-\lambda)[f(\tilde{\mu})]^2)^{-1}).$$

When comparing two distributions that have the same shape (i.e., $F_C(y) = F_E(y - \Delta)$ for all $y$ and some $\Delta$) where a common variance exists, $4[f(\tilde{\mu})]^2/\sigma^2$ is also the relative efficiency of the difference in sample medians to the difference in sample means when there are independent random samples. When the variance does not exist, the median or difference in medians is more efficient. For underlying double exponential distributions, the relative efficiency of the difference in sample medians to the difference in sample means is 2.

*Hodges–Lehmann Estimator of the Difference in Medians.* The corresponding estimator of the difference in medians based on the Mann–Whitney–Wilcoxon test is the Hodges–Lehmann estimator of $\hat{\Delta}_{HL} = \underset{i,j}{\mathrm{med}}(Y_j - X_i)$. As $n_C, n_E \to \infty$, and $n_C/N \to \lambda > 0$, $\sqrt{N}(\hat{\Delta}_{HL} - \Delta) \xrightarrow{d} N(0, \tau^{-2})$, where $\tau = \sqrt{12\lambda(1-\lambda)} \int_{-\infty}^{\infty} f_C^2(x)\,dx$.

When the two underlying distributions have the same shape, the efficiency of the Hodges–Lehmann estimator of the difference in medians relative to the difference in sample medians equals $3\left( \int_{-\infty}^{\infty} f_C^2(x)\,dx \right)^2 / f(\tilde{\mu})^2$. For normal random samples with equal underlying variances, this relative efficiency is approximately 1.5 and the relative efficiency of the Hodges–Lehmann estimator to the difference in means is approximately 0.955. For random samples from double exponential distributions, the relative efficiency of the Hodges–Lehmann estimator of the difference in medians to the difference in sample medians is 0.75, whereas the relative efficiency of the Hodges–Lehmann estimator to the difference in means is approximately 1.5.

The asymptotic distributions of these estimators are not easy to use, can be misused, and are not necessary to use when constructing tests or confidence intervals (i.e., from the previous subsection, an expression already exists for determining $d$ that do not require the estimation of $f$). Their use requires estimation of the density function, which is also associated with some uncertainty. Also for those intervals based on the sample median, $f(\tilde{\mu})$ is only the correct term in limit. For determining the lower limit of the confidence interval for a median, $f(\tilde{\mu})$ should be replaced by $f(\xi)$ ($f(\xi)$ is the average over some interval $[\tilde{\mu} - k_1, \tilde{\mu}]$), where $\xi$ is near and below $\tilde{\mu}$. The same is true for the upper limit, but with $\xi$ near and above $\tilde{\mu}$. Let $X_1, X_2, \ldots, X_n$ be a random sample from a continuous distribution with distribution function $F$ and unique median $\tilde{\mu}$. The $100(1 - \alpha)\%$ confidence interval for $\tilde{\mu}$ is given by $(X_{(d)}, X_{(n - d+1)})$, where $d \approx n/2 - z_{\alpha/2}\sqrt{n/4}$ when $n$ is large. Suppose that $n$ is large and odd. Then $X_{((n+1)/2)} - X_{(d)}$ is estimating $F^{-1}(0.5) - F^{-1}(0.5 - z_{\alpha/2}/\sqrt{4n}) = (z_{\alpha/2}/\sqrt{4n})/f(F^{-1}(\eta))$, where $\xi = F^{-1}(\eta)$ is between $F^{-1}(0.5 - z_{\alpha/2}/\sqrt{4n})$ and $\tilde{\mu}$. Similarly, $X_{((n+1)/2)}$ is estimating $(z_{\alpha/2}/\sqrt{4n})/f(F^{-1}(\eta))$, where $\xi = F^{-1}(\eta)$ is between $\tilde{\mu}$ and $F^{-1}(0.5 + z_{\alpha/2}/\sqrt{4n})$. The window about $\tilde{\mu}$ for which $f$ is averaged gets smaller as the sample size gets larger. When $f$ has a local minimum (maximum) at $\tilde{\mu}$, using $f(\tilde{\mu})$ will lead to a wider (narrower) confidence interval with a confidence coefficient greater (less) than $1 - \alpha$.

## 12.5 Ordinal Data

In this section we briefly discuss non-inferiority inference on ordinal data. That is, different outcomes are ordered, but there is no measure for how different the outcomes are. Munzel and Hauschke[18] applied a rank test based on the Mann–Whitney–Wilcoxon statistic for non-inferiority testing of ordinal data. For our purposes (i.e., different from Munzel and Hauschke), larger outcome values will be more favorable outcomes. The procedure is based on $p = P(Y > X) + 0.5 \times P(Y = X)$, where $X$ has the underlying distribution for control arm and $Y$ has the underlying distribution for the experimental arm. This generalizes the Mann–Whitney–Wilcoxon parameter of $P(Y > X)$ by counting a tie as half a "success" and half a "failure." Formulas for one-sided confidence intervals for $p$ and for determining the sample size for non-inferiority and superiority testing are provided in the papers of Munzel and Hauschke[18] and Brunner and Munzel.[19] The hypotheses are expressed as

$$H_o : p \leq 0.5 - \delta \text{ vs. } H_a : p > 0.5 - \delta \qquad (12.23)$$

where $0 \le \delta < 0.5$. The value for $\delta$, the non-inferiority margin should depend on both the effect of the control therapy and the differences in the utility or preferences of the categories. In general, the fewer the number of categories (e.g., only "failure" and "success"), the larger will be the difference in utility between successive categories, and thus the tendency for a smaller margin.

We will describe the test procedure in Munzel and Hauschke[18] for testing the hypotheses in Expression 12.23. Let $X_1, X_2, \ldots, X_{n_C}$ and $Y_1, Y_2, \ldots, Y_{n_E}$ denote independent random samples from distributions having respective distribution functions $F_C$ and $F_E$. An unbiased, consistent estimator for $p$ is given by $\hat{p} = (\bar{R}_E - (n_E + 1)/2)/n_C$, where $\bar{R}_E$ is the arithmetic average of the ranks of the observations in the experimental arm among all observations. That is, $\bar{R}_E = \sum_{j=1}^{n_E} R(Y_j)/n_E$, where $R(Y_j)$ is the rank of $Y_j$ in the ordering of the combined sample. Define also $\bar{R}_C = \sum_{j=1}^{n_C} R(X_j)/n_C$, where $R(X_j)$ is the rank of $X_j$ in the ordering of the combined sample. When ties occur, $R(Y_j)$ $(R(X_j))$ is the midrank. Let $R^{(C)}(X_j)$ denote the rank of $X_j$ among $X_1, \ldots, X_{n_C}$ and let $R^{(E)}(Y_j)$ denote the rank of $Y_j$ among $Y_1, \ldots, Y_{n_E}$. Define $J_C^2 = \sum_{j=1}^{n_C} (R(X_j) - R^{(C)}(X_j) - \bar{R}_C + (n_C + 1)/2)^2/(n_C - 1)$ and $J_E^2 = \sum_{j=1}^{n_E} (R(Y_j) - R^{(E)}(Y_j) - \bar{R}_E + (n_E + 1)/2)^2/(n_E - 1)$. Then for $i = E, C$, define $\hat{u}_i^2 = J_i^2/(N - n_i)^2$, where $N = n_E + n_C$. The test statistic is given by

$$Q = \frac{\hat{p} - (0.5 - \delta)}{\sqrt{\dfrac{\hat{u}_C^2}{n_C} + \dfrac{\hat{u}_E^2}{n_E}}} \tag{12.24}$$

Brunner and Munzel[19] showed that when $N \to \infty$ and $n_C/n_E$ converges to a positive constant, $(\hat{p} - p)/\sqrt{u_C^2/n_C + u_E^2/n_E}$ converges in distribution to a standard normal distribution, where $u_C^2$ and $u_E^2$ are the variances of $F_C(X_1)$ and $F_E(Y_1)$, respectively. Additionally, they showed that $u_C^2$ and $u_E^2$ are consistently estimated by $\hat{u}_C^2$ and $\hat{u}_E^2$, respectively. Therefore, a large sample test of the hypotheses in Expression 12.23 can be performed by comparing $Q$ in Equation 12.24 with the appropriate percentile of a standard normal distribution. Non-inferiority is concluded at an approximate significance level of $\alpha/2$ when $Q > z_{\alpha/2}$. Alternatively, the non-inferiority inference can be based on an approximate $100(1 - \alpha)\%$ confidence interval for $p$ given by $\hat{p} \pm z_{\alpha/2}\sqrt{\dfrac{\hat{u}_C^2}{n_C} + \dfrac{\hat{u}_E^2}{n_E}}$ . Non-inferiority is concluded when every value in the confidence interval exceeds $0.5 - \delta$. For small sample sizes, from simulations, Brunner and Munzel[19] found that the test that rejects $H_0 : p \le 0.5$ when $Q > z_{\alpha/2}$ can be quite liberal or conservative depending on the ratios for the true variances and for the sample sizes. They recommend for per group sample sizes between 15 and 50 using a Satterthwaite-like approximation of the degrees of freedom. The quality of the approximation of the degrees of freedom was dependent on the number of categories and deemed sufficient when there were at least three categories. The value of $t_{\alpha/2,v}$ (where $v$ is the Satterthwaite degrees of freedom) would replace $z_{\alpha/2}$ for the determination of confidence intervals and as a critical value in hypotheses testing.

For sizing a trial, let $p'$ represent the assumed value for $p$. Let $\sigma_1$ and $\sigma_2$ denote the respective assumed underlying standard deviations of $F_C(X_1)$ and $F_E(Y_1)$, where $l = \sigma_1/\sigma_2$. Let $k = n_E/n_C$ denote the allocation ratio. Then the sample size for the control arm is given by

$$n_C = (1 + l^2/k)(z_{\alpha/2} + z_\beta)^2(\sigma_2/(p' - 0.5 + \delta))^2 \qquad (12.25)$$

When the sample sizes are small, the term $z_{\alpha/2} + z_\beta$ in Equation 12.25 can be replaced with $t_{\alpha/2,v} + t_{\beta,v}$ (where $v$ is the Satterthwaite degrees of freedom). This creates an equation where iterations will be needed to determine the sample sizes, since $n_C$ appears on both sides of the equation.

For two noncontinuous distributions, Wellek and Hampel[20] proposed a nonparametric test of equivalence around the parameter $P(Y > X \mid Y \neq X)$. This parameter ignores the ties and the probability of a tie. For both equivalence and non-inferiority testing, a tie is consistent with the alternative hypothesis. Therefore, use of the parameter $P(Y > X \mid Y \neq X)$ will greatly penalize an experimental therapy in situation when a tie is quite likely and lead to conservative testing as noted by Munzel and Hauschke.[18]

For equivalence testing, a constant odds ratio based on the Wilcoxon mid-ranks statistic and derived from the corresponding exact permutation distribution was considered by Mehta, Patel, and Tsiatis.[21]

Often scores are assigned to the ordered categories and the data are treated as continuous. However, it may be difficult to interpret a specific difference between arms in the score, and the scores themselves may be subjective.

---

## References

1. Hollander, M. and Wolfe, D.A., *Nonparametric Statistical Methods*, John Wiley & Sons, New York, 1973.
2. Wiens, B.L., Randomization as a basis for inference in non-inferiority trials, *Pharm. Stat.*, 5, 265–271, 2006.
3. Good, P., *Permutation, Parametric, and Bootstrap Tests of Hypotheses*, Springer, New York, NY, 2005.
4. Satterthwaite, F., An approximate distribution of estimates of variance components, *Biometrics*, 2, 110–114, 1946.

5. Ghosh, P. et al., Assessing non-inferiority in a three-arm trial using the Bayesian approach, Technical report, Memorial Sloan-Kettering Cancer Center, 2010.
6. Gamalo, M. et al., A generalized $p$-value approach for assessing non-inferiority in a three-arm trial, *Stat. Methods Med. Res.*, published online February 7, 2011.
7. Kieser, M. and Hauschke, D., Approximate sample sizes for testing hypotheses about the ratio and difference of two means, *J. Biopharm. Stat.*, 9, 641–650, 1999.
8. Hasselblad, V. and Kong, D.F., Statistical methods for comparison to placebo in active-control trials, *Drug Inf. J.*, 35, 435–449, 2001.
9. Rothmann, M.D. and Tsou, H., On non-inferiority analysis based on delta-method confidence intervals, *J. Biopharm. Stat.*, 13, 565–583, 2003.
10. Bishop, Y.M.M., Fienberg, S.E., and Holland, P.W., *Discrete Multivariate Analysis: Theory and Practice*, MIT Press, Cambridge, MA, 1975, 493.
11. Hettmansperger, T.P., Two-sample inference based on one-sample sign statistics, *Appl. Stat.*, 33, 45–51,1984.
12. Mood, A. M., *Introduction to the Theory of Statistics*, McGraw-Hill, New York, NY, 1950.
13. Pratt, J.W., Robustness of some procedures for the two-sample location problem, *J. Am. Stat. Assoc.*, 59, 665–680, 1964.
14. Gastwirth, J.L., The first-median test: A two-sided version of the control median test, *J. Am. Stat. Assoc.*, 63, 692–706, 1968.
15. Freidlin, B. and Gastwirth, J.L., Should the median test be retired from general use? *Am. Stat.*, 54, 161–164, 2000.
16. Mathisen, H.C., A method of testing the hypothesis that two samples are from the same population, *Ann. Math. Stat.*, 14, 188–194, 1943.
17. Hettmansperger, T.P., *Statistical Inference Based on Ranks*, Wiley, New York, NY, 1984.
18. Munzel, U. and Hauschke, D., A nonparametric test for proving non-inferiority in clinical trials with ordered categorical data, *Pharm. Stat.*, 2, 31–37, 2003.
19. Brunner, E. and Munzel, U., The nonparametric Behrens–Fisher problem: Asymptotic theory and a small sample approximation, *Biom. J.*, 42, 17–25, 2000.
20. Wellek, S. and Hampel, B.A., Distribution-free two-sample equivalence test allowing for tied observations, *Biom. J.*, 41, 171–186, 1999.
21. Mehta, C.R., Patel, N.R., and Tsiatis, A.A, Exact significance testing to establish treatment equivalence with ordered categorical data, *Biometrics*, 40, 819–825, 1984.

# 13

## Inference on Time-to-Event Endpoints

### 13.1 Introduction

Many meaningful clinical endpoints are time-to-event endpoints—for example, overall survival, time to a response, time to a cardiac-related event, and time to progressive disease. When the intention and the outcome is that all subjects are followed until the event is observed (no censoring), time-to-event endpoints can be analyzed as continuous endpoints. The inferences can be based on the mean, median, or some other quantity relevant to continuous endpoints. However, in most practical cases involving a time-to-event endpoint, all subjects are not followed until an event (i.e., some subjects have their times censored). This limits the types of analyses that can be performed. Nonparametric inferences on means and/or medians may not be possible. To base the inference on means or medians may require following subjects for a long and perhaps impractical length of time.

For a time-to-event endpoint, the amount of available information for inferential purposes is tied to the total number of events and increases either by continuing the follow-up on subjects that have not had events or by beginning to follow additional subjects for events. For standard binary or continuous endpoints, the amount of available information increases solely by including the outcomes of additional subjects.

For clinical trials, most time-to-event endpoints are defined as the time from randomization (or enrollment or start of therapy) to the event of interest or the first of many events of interest. Typically, at randomization, the subject does not have the event or any of the events of interest. Starting the time-to-event endpoint at randomization is also important as randomization is fair and it is during randomization that subjects and their prognoses are fairly allocated to treatment arms. In addition, to maintain the integrity of this fairness of randomization, intent-to-treat analyses should be conducted where all subjects are followed, regardless of adherence, to an event or the end of study (i.e., until the data cutoff date or until some prespecified maximum follow-up has been completed). This allows for a valid comparison of the study arms.

When a subject is not followed up to an event or the end of study, it is common practice to censor their time to an event at that the time when their follow-up ended. When a subject's time is censored $x$ months after randomization, their remaining time to an event is represented by those subjects on the same treatment arm still under observation for an event when applying standard nonparametric comparative methods (e.g., a log-rank test or a Cox proportional hazards model with treatment as the sole factor) and noncomparative methods (e.g., Kaplan–Meier estimates). This representation is only valid when random censoring occurs; that is, when the reason for censoring is independent of the subject's prognosis at the time of censoring. *Informative censoring* occurs when the reason for discontinuing follow-up is not independent of the subject's prognosis. Informative censoring introduces bias in the estimation of the univariate distribution and may introduce bias as well as reduce the reliability in the evaluation of the difference in the treatment effects. Premature or administrative censoring reduces the integrity of the randomization and should be avoided, so that the methods of analyses are valid.

The experimental therapy is noninferior to the control therapy if the time-to-event distribution for the experimental arm is better than or not too much worse than that of the control arm. When the "event" is a negative or undesirable outcome, non-inferiority means that the time-to-event distribution for the experimental arm is greater than or not too much less than that for the control arm. When the "event" is a positive or desirable outcome, non-inferiority means that the time-to-event distribution for the experimental arm is less than or not too much greater than that for the control arm. There are many ways to express the difference between the two time-to-event distributions—these include, through a hazard ratio, a cumulative odds ratio, a difference in means when applicable, a difference in medians, or a difference in the probabilities of achieving or not achieving an event by some landmark.

In clinical trials when using a time-to-event endpoint, it is common to use a hazard ratio or a log-hazard ratio to describe the treatment effect or relative treatment effect. A hazard ratio is the ratio of the instantaneous risk of an event between the two arms. Analyses assume that this ratio is constant over time. On its own, a hazard ratio of an experimental group to a reference group (e.g., a placebo group or an active control group) does not provide the benefit or relative benefit of a therapy. The time-to-event distribution for the reference group is also needed to obtain an understanding of the size of the benefit. In practice, an experimental versus placebo survival hazard ratio of 0.75 (the instantaneous risk of death for the experimental group is 75% of that for the placebo group) corresponds to greater benefit when the mean survival in the placebo arm is 24 months than when the mean survival in the placebo arm is 4 months. That is, for an experimental versus placebo survival hazard ratio of 0.75, the experimental therapy extends mean survival by more months when the mean placebo survival is

24 months than when the mean placebo survival is 4 months. The benefit of the experimental therapy on survival is truly defined by the improvement in mean/expected survival. However, usually the inference is not based on the difference in mean/expected time-to-event. Therefore, the results when positive may need to be translated into some form that provides an impression of the clinical benefit of the experimental therapy. Whenever possible, although such instances are few, an inference should be based on a difference of means.

*Composite Endpoints.* Composite time-to-event endpoints are popular. Such an endpoint is the time to the first event in a set of events of interest. Examples include the time to the first event of stroke, myocardial infarction, and death for cardiovascular trials, and the time to the first event of disease progression and death in metastatic or advanced cancer. The use of a composite endpoint may be necessary when the disease can be characterized by many factors. For example, as reported by Chi,[1] a disease may be characterized by its pathophysiology, severity, signs and symptoms, progression, morbidity, and mortality. Since the event or hazard rate for a composite endpoint is greater than that of the individual components, the use of a composite endpoint has the advantage of requiring fewer subjects and having an earlier analysis than a trial designed on an individual component. Many researchers have written on the issues and disadvantages of using composite endpoints.[1-4]

The individual components of a composite endpoint should be relevant and meaningful for subjects and constitute clinical benefit. When the components are equally important, and a new drug demonstrates superior efficacy, the particular distribution of events across the individual components or the differences between arms in the distribution of the events do not contribute any necessary additional information on the new drug's overall benefit. When the importance varies across components, it is more difficult to interpret the endpoint and the corresponding results. For example, a drug that improves a major component (e.g., death) while having an adverse effect on a minor component may be beneficial, but a drug that improves a minor component while having an adverse effect on a major component would not be beneficial. In addition, the severity of a component may change over time owing to improvements in the treatment or management of the component, thus reducing the utility of including that component in the composite endpoint. A composite endpoint may not be sensible if the components have widely different importance.

When the components are not equally important, additional analyses involving subcomposite endpoints may need to be done to assess the effects. The subcomposite time-to-event endpoint should exclude the events of less relative importance. This process of additional analyses by excluding events of less importance may need to be further done until an analysis on a subcomposite time-to-event endpoint that includes only those most important events of equal value is done. To perform valid analyses on these additional subcomposite time-to-event endpoints would require following all subjects until one of the most important events has occurred or until the end of study. In practice, a frequent difficulty in performing a valid analysis on a subcomposite endpoint is the lack of continued follow-up for additional events after the first event for the original composite endpoint has occurred.

The absence of a placebo control, as in a two-arm active controlled non-inferiority trial, can further the difficulty in analyzing and interpreting the results on the basis of a composite endpoint. For a placebo control trial, a superiority analysis on a subcomposite endpoint may lack power, and the estimate of the effect of therapy on a subcomposite endpoint may be unreliable. This would particularly be true when the subcomposite events are rare. When the subcomposite events are rare, it is unlikely that there will be reliable estimates of the effect of the active control on the subcomposite endpoint. Therefore, it will be difficult to evaluate the effect of the experimental therapy on the subcomposite from an active control non-inferiority trial.

When the benefit of therapy has tended to be through the same component, A, at the potential expense of another component, B, there is a potential of a biocreep with respect to effects on component B. This would be a serious problem when component B is a more meaningful event than component A. Also, differences in the frequency of assessments of the different components can affect the comparison of a composite time-to-event endpoint when the benefit of the experimental therapy on one component is countered by its negative effect on another component. If the component for which there is a positive effect is assessed much more frequently, the estimated effect of the experimental therapy will be inflated.

Effective therapies can lower the event rates. This will lead to larger, longer clinical trials or the temptation to include additional components into a revised composite endpoint so as to achieve an event rate consistent with past studies. It is unlikely that the new components will be as meaningful as the original components. Such revisions to a composite endpoint lead to a less meaningful endpoint. Additionally, introducing less meaningful components which the therapies are not expected to affect will make the treatment arms more similar. If this is unaccounted for in the evaluation of the active control effect and the determination of the non-inferiority, the likelihood of concluding that an ineffective therapy is effective will increase.

*Hazard Rate, Hazard Ratio, and Survival Function.* Many inferences involving time-to-event endpoints involve the hazard rates. Let $T$ be a nonnegative continuous random variable representing the time to some event. Let $f$ denote the probability density function for $T$ and $F$ represent the distribution function for $T$. Then for $t \geq 0$, $F(t) = P(T \leq t) = \int_0^t f(x)\,dx$. The survival function $S$ is defined by $S(t) = P(T \geq t) = \int_t^\infty f(x)\,dx$ for $t \geq 0$. The hazard function or hazard rate at time $t$ is defined by

$$h(t) = \lim_{\varepsilon \to 0} \frac{P(t \le T \le t + \varepsilon \mid T \ge t)}{\varepsilon} \tag{13.1}$$

For an individual subject, the hazard rate at time $t$, $h(t)$, represents the instantaneous risk of an event at time $t$ given that the subject has not had an event by time $t$. If the event is death, $h(t)$ represents the instantaneous risk of death at time $t$ for a subject who is alive as time $t$ approaches. Additionally, for a subject who is alive at time $t$ (without an event by time $t$), the probability the subject dies (has an event) during the next $\varepsilon$ of time is approximately $\varepsilon h(t)$ for a small $\varepsilon$. Evaluating the limit in Equation 13.1 gives

$$h(t) = \lim_{\varepsilon \to 0} \frac{P(t \le T \le t + \varepsilon \mid T \ge t)}{\varepsilon} = \frac{f(t)}{S(t)} = -\frac{d}{dt} \log S(t).$$

The cumulative hazard function is given by $H(t) = \int_0^t h(x)\,dx = -\log S(t)$ for $t \ge 0$.

When the hazard functions for the experimental and control arms are proportional, then the common ratio of the hazard functions, called the *hazard ratio*, is often used to measure the difference in the two distributions. The hazard ratio, $\theta$, satisfies $\theta = \dfrac{h_E(t)}{h_C(t)} = \dfrac{H_E(t)}{H_C(t)} = \dfrac{-\log S_E(t)}{-\log S_C(t)}$ and $S_E(t) = [S_C(t)]^\theta$ for all $t \ge 0$.

In this chapter we will discuss the types of censoring, reasons for censoring, and the issue of censoring deaths in Section 13.2. Non-inferiority analyses involving exponential distributions are discussed in Section 13.3. Non-inferiority analysis based on a hazard ratio from a proportional hazards model is discussed in Section 13.4. Non-inferiority analyses either at landmarks or involving medians are discussed in Section 13.5. The extension of the testing problem in Section 13.5 to an inference over a preset interval is discussed in Section 13.6.

## 13.2 Censoring

Throughout this chapter, the only type of censoring that will be considered is right censoring. Right censoring means that a subject's true time is unknown and to the right of (greater than) the censored time. When the term censoring is used, it will refer to right censoring unless otherwise stated.

Informative censoring occurs when the prognosis of a given subject with a censored time is not independent of the censoring. In other words, what to expect for a given subject's ultimate or true time-to-event, which is censored at time $x$, is not represented by the follow-up experience of those subjects in the same group with times that exceed $x$. Whenever a subject is censored because treatment is being withheld because of declining physical condition, the censoring will be informative. Comparative and noncomparative methods commonly used for time-to-event endpoints assume no informative censoring. Therefore, excessive informative censoring invalidates analyzes from such methods.

Valid reasons for censoring a subjects' time-to-event include:

1. The subject did not have an event observed at the time of the data cutoff for the analysis.
2. The subject completed the prespecified required time on study without having an event.

Alternatives to standard censoring at loss to follow-up should be considered when subjects have their follow-up prematurely discontinued. Alternatives would include selecting a different representative group for a subject's remaining time to event, on the basis of having similar prognosis at the time of loss to follow-up to the subject whose follow-up was prematurely discontinued.

*Type I and II Censoring.* There are various types of random censoring. Two types of censoring prevalent in statistical research are defined here. Type I censoring arises when the individual subjects have their own maximum allowed follow-up without having an event. This would occur in a clinical trial if there was a preset study time or calendar time when the analysis will be performed or there is a required preset maximum follow-up for each subject. In the former case, the limit on a given subject's follow-up would equal the time from the subject's start time (i.e., randomization time) to the preset analysis time. For a given arm, the observed time-to-event outcome of a subject is regarded as the minimum of the actual time and an independent censoring time. For the subjects on that arm, the actual times are a random sample. For type I censoring, the number of subjects that have events is random.

For type II censoring, the number of subjects that have events is predetermined. For type II censoring, only the $r$ smallest times are observed among a random sample of $n$ subjects. This would occur if all subjects had the same start time and the stop time was the time of the $r$th event.

The exact distribution for a within-arm or between-arm estimator (e.g., a mean, ratio of means, or hazard ratio) depends on the type of censoring. Type I and II censoring lend themselves better in deriving distributional results for estimators and in proving asymptotic results than other types of censoring. In general, for those quantities of interest (e.g., means, hazard ratios), their asymptotic distribution tends to be approximately the same regardless of how the random censoring is done.

Censoring in clinical trials tends to be neither type I nor type II censoring. In many oncology clinical trials, subjects are accrued over time (and thus have different start times) and followed until death or the end of study, where the end of study is the time of the $r$th death (i.e., for a preset $r$) among all subjects regardless of treatment arm. Because the time for end of study is random, for this type of censoring the random censoring times are not independent of the actual times. Although the censoring times are not independent of the time-to-event endpoints, the censoring is not informative for a given treatment arm provided the actual times to the event are a random sample.

*Inclusion of Death as an Event.* Censoring a subject's time-to-event endpoint because of death or death not related to disease is problematic and creates a hypothetical endpoint. Follow-up ceases at death; there is no remaining time to the event and thus there is no loss to follow-up or missing data. Complete information or follow-up has been done on the subject. When death is not an event of interest for the time-to-event endpoint, there is no actual time-to-event for a subject who dies without being observed for an event of interest. In situations where such censoring occurs, the subjects' time-to-event times in a given arm are not a random sample from that distribution that is being estimated. Such an endpoint is a hypothetical endpoint. That is, for a given arm, what is being estimated is the distribution for the time-to-event, or the time-to-event when death occurs without experiencing the event, and we pretend that the remaining time to an event of dead subjects can be represented by the remaining time to an event of living subjects still under observation of an event. Having living subjects represent dead subjects is beyond informative censoring and common sense. Another setting where this censoring occurs involves analyses that include only disease-related deaths. This not only creates a hypothetical endpoint where living subjects represent dead subjects but also the determination of whether a death was related to the disease, or even to the treatment, may be inexact and subjective.

## 13.3 Exponential Distributions

An exponential distribution is a popular time-to-event distribution. They are often used to model the life length of objects that fail or stop from one random "shock." The instantaneous risk of the "shock" is constant over time (i.e., shocks occur according to a homogeneous Poisson process). Exponential distributions also model the life lengths of objects that have "half-lives." When a time-to-event endpoint has an underlying exponential distribution, the remaining life distributions are identical and equal to the underlying distribution. This property is known as "new as good as used" and corresponds to the concept of no deterioration.

An exponential distribution is unlikely for human time-to-event endpoints. For many clinical time-to-event endpoints, the risk of an event is believed to be increasing over time, which corresponds to having a condition that tends to deteriorate. Because of its simplicity and ease of use, exponential distributions have been assumed when determining the appropriate number of subjects and/or the study time for the analysis for a clinical trial.

We present the case involving exponential distributions because some research has focused on using exponential distributions and to provide a comparison of the results to the semiparametric case using Cox's proportional hazards models.

The probability density function for an exponential distribution having mean $\mu$ is given by $f(t) = (1/\mu)e^{-t/\mu}$ for $t > 0$. For the exponential distribution, the constant hazard rate equals $1/\mu$ and the survival function is given by $S(t) = e^{-t/\mu}$ for $t > 0$, which represents the probability that a random subject has a time to event greater than $t$, or equivalently the probability that a random subject is event-free at time $t$.

*Inferences on One Mean.* Assume that the actual survival times from a given treatment arm is a random sample of size $n$ from an exponential distribution with mean $\mu$. Let $t_1, t_2, \ldots, t_n$ denote the observed times among which $r$ times are uncensored and $n - r$ times are censored. For both type I and type II censoring, the maximum likelihood estimate for $\mu$ is given by $\hat{\mu} = \sum_{i=1}^{n} t_i/r$ (provided $r > 0$).

For type II censoring for a given arm, it can be shown that $2r\hat{\mu}/\mu$ has a $\chi^2$ distribution with $2r$ degrees of freedom. For type I censoring, per Cox,[5] treating $2r\hat{\mu}/\mu$ as having a $\chi^2$ distribution with $2r + 1$ degrees of freedom leads to satisfactory tests and confidence intervals. These results can be used to derive a confidence interval for the mean or ratio of means (i.e., the hazard ratio).

Inferences based on large sample normal approximations can also be done. For type II censoring, it follows that $(\hat{\mu} - \mu)/(\mu/\sqrt{r})$, $(\hat{\mu} - \mu)/(\hat{\mu}/\sqrt{r})$, and $\sqrt{r}(\log \hat{\mu} - \log \mu)$ each converge in distribution to a standard normal distribution as $r \to \infty$. For type I censoring, it follows from the proposal of Cox[5] that treating $(2r\hat{\mu}/(2r+1) - \mu)/(\mu/\sqrt{r+0.5})$, $(2r\hat{\mu}/(2r+1) - \mu)/(\hat{\mu}/\sqrt{r+0.5})$, and $\sqrt{r+0.5}(\log \hat{\mu} - \log \mu + \log(1+0.5/r))$ as having a standard normal distribution for large $r$ may lead to satisfactory tests and confidence intervals. These results or adaptations of such can be used to derive a confidence interval for the mean, difference in means, or ratio of means (via a difference in log means).

The above results apply, or approximately apply, to many other types of random censoring. For most of the discussion on time-to-event endpoints in later sections, asymptotic results based on appropriate random censoring will be used without any distinction on whether the censoring is type I, type II, or neither.

## 13.3.1 Confidence Intervals for a Difference in Means

As before, subscripts E and C will be used to identify quantities that correspond to the experimental and control arms, respectively. We will start with non-inferiority testing based on a difference in means, $\mu_E - \mu_C$. When larger times are better than smaller times (i.e., the event is undesirable), the null and alternative hypotheses to be considered are

$$H_o: \mu_E - \mu_C \leq -\delta \text{ vs. } H_a: \mu_E - \mu_C > -\delta \tag{13.2}$$

That is, the null hypothesis is that the active control is superior to the experimental treatment by at least the prespecified quantity of $\delta \geq 0$. The alternative hypothesis is that the active control is superior by a smaller amount, or the two treatments are identical, or the experimental treatment is superior. When $\delta = 0$, the hypotheses in Expression 13.2 reduce to classical one-sided hypotheses for superiority testing. Rejection of the null hypothesis ($H_o$) leads to the conclusion that the experimental treatment is noninferior to the control treatment. When smaller times are more desired than larger times (i.e., the event is desirable), the roles of $\mu_E$ and $\mu_C$ in the hypotheses in Expression 13.2 would be reversed (i.e., test $H_o: \mu_C - \mu_E \leq -\delta$ vs. $H_a: \mu_C - \mu_E > -\delta$).

Let $r_E$ and $r_C$ denote the number of events observed in the experimental and control arms, respectively. For type II censoring, an approximate $100(1 - \alpha)\%$ confidence interval for the difference in means is given by $\hat{\mu}_E - \hat{\mu}_C \pm z_{\alpha/2}\sqrt{\hat{\mu}_E^2/r_E + \hat{\mu}_C^2/r_C}$ when $r_E$ and $r_C$ are sufficiently large. For type I censoring, an approximate $100(1 - \alpha)\%$ confidence interval based on Cox's proposal is given by $2r_E\hat{\mu}_E/(2r_E + 1) - 2r_C\hat{\mu}_C/(2r_C + 1) \pm z_{\alpha/2}\sqrt{\hat{\mu}_E^2/(r_E + 0.5) + \hat{\mu}_C^2/(r_C + 0.5)}$ when $r_E$ and $r_C$ are sufficiently large.

The following example illustrates the use of these formulas.

## Example 13.1

A random sample of 200 observations is taken from an exponential distribution having mean 100. The corresponding values represent the actual survival times for subjects in the experimental arm. A random sample of 100 observations is taken from an exponential distribution having mean 90. The corresponding values represent the actual survival times for subjects in the control arm. Note that the difference in the true means is 10. The 200th smallest observation is determined for the combined sample as 114.88. All subjects in both arms with times larger than 114.88 will have their times censored at 114.88. This random censoring is neither type I nor type II censoring. We will apply each confidence interval formula for the difference in means conditioning on the number of uncensored observations in each arm. We have $r_E = 137$ uncensored observations in the experimental arm with $\hat{\mu}_E = 105.4$ and $r_C$ 63 uncensored observations in the control arm with $\hat{\mu}_C = 117.4$. As can happen when data are random, the order of the sample means differs from the order of the true means. The corresponding confidence intervals for the hazard ratio are provided in Table 13.1. All confidence intervals contain the true difference in means of 10. For a non-inferiority margin of 15, non-inferiority would fail to be concluded regardless of the method. The confidence interval for the type II censoring methods has a slightly greater width. Because there were fewer events in the control arm, the type I censoring method reduced the estimate of the control mean

**TABLE 13.1**

Calculated 95% Confidence Intervals for the Difference in Means by Method

| Method | Estimate | 95% CI | Width |
|---|---|---|---|
| Type I censoring | −11.5 | (−45.3, 22.4) | 67.7 |
| Type II censoring | −12.0 | (−46.0, 21.9) | 67.9 |

more than the estimate of the experimental mean. This led to a larger estimate of the difference in means (−11.5 vs. −12). Had the values for the experimental and control arms been exchanged for each other, the type II censoring method would have the larger estimate for the difference in means (12 vs. 11.5).

## 13.3.2 Confidence Intervals for the Hazard Ratio (Ratio of Means)

We will now consider non-inferiority testing based on a ratio of means, $\mu_E/\mu_C$. When larger times are more desired than smaller times, the null and alternative hypotheses to be considered are

$$H_o: \mu_E/\mu_C \leq 1 - \delta \quad \text{vs.} \quad H_a: \mu_E/\mu_C > 1 - \delta \tag{13.3}$$

That is, the null hypothesis is that the mean for the active control is superior to that of the experimental treatment by at least $\delta\mu_C$, where $\delta \geq 0$. The alternative hypothesis is that the active control is superior by a smaller amount, or the two treatments are identical, or the experimental treatment is superior. When $\delta = 0$, the hypotheses in Expression 13.3 reduce to classical one-sided hypotheses for a superiority trial. Rejection of the null hypothesis ($H_o$) leads to the conclusion that the experimental treatment is noninferior to the control treatment. When smaller times are more desired than larger times, the roles of $\mu_E$ and $\mu_C$ in the hypotheses in Expression 13.3 would be reversed (i.e., test $H_o: \mu_C/\mu_E \leq 1 - \delta$ vs. $H_a: \mu_C/\mu_E > 1 - \delta$). Note that $\mu_E/\mu_C$ is also the scale factor relating the two exponential distributions and the control versus experimental hazard ratio (the ratio of the instantaneous risk of an event).

For type II censoring within each arm, we have that $V = \dfrac{\hat{\mu}_E/\mu_E}{\hat{\mu}_C/\mu_C}$ has

an $F$ distribution with $2r_E$ and $2r_C$ degrees of freedom in the numerator and denominator, respectively. For any $0 < \gamma < 1$, the corresponding $100(1 - \gamma)\%$ percentile, $F_{\gamma,2r_E,2r_C}$ is defined as that value satisfying $\gamma = P(V > F_{\gamma,2r_E,2r_C})$. For other $F$ distributions, similar notation will be used for the percentiles. Note that $F_{1-\gamma,2r_C,2r_E} = 1/F_{\gamma,2r_E,2r_C}$. For $0 < \alpha < 1$, we have that

$$P\left(F_{1-\alpha/2,2r_E,2r_C} < \frac{\hat{\mu}_E/\mu_E}{\hat{\mu}_C/\mu_C} < F_{\alpha/2,2r_E,2r_C}\right) = P\left((\hat{\mu}_E/\hat{\mu}_C)F_{1-\alpha/2,2r_C,2r_E} < \frac{\mu_E}{\mu_C} < (\hat{\mu}_E/\hat{\mu}_C)F_{\alpha/2,2r_C,2r_E}\right).$$

Thus, a $100(1 - \alpha)\%$ confidence interval for the hazard ratio is given by $\left((\hat{\mu}_E/\hat{\mu}_C)F_{1-\alpha/2,2r_C,2r_E}, (\hat{\mu}_E/\hat{\mu}_C)F_{\alpha/2,2r_C,2r_E}\right).$

For type I censoring from applying the proposal of Cox, we analogously have an approximate $100(1 - \alpha)\%$ confidence interval for the hazard ratio given by

$$\left( \frac{r_E(2r_C + 1)\hat{\mu}_E}{r_C(2r_E + 1)\hat{\mu}_C} F_{1-\alpha/2, 2r_C+1, 2r_E+1}, \frac{r_E(2r_C + 1)\hat{\mu}_E}{r_C(2r_E + 1)\hat{\mu}_C} F_{\alpha/2, 2r_C+1, 2r_E+1} \right).$$

These results may apply approximately for other types of random censoring. For outcomes for which larger values yield better outcomes, non-inferiority is concluded if the confidence interval for the ratio of means (the reciprocal of the hazard ratio) contains only values greater than the non-inferiority threshold.

Another way of determining a confidence interval for the hazard ratio applies the asymptotic distributions for the natural log of the maximum likelihood estimator of $\mu_E$ and $\mu_C$. For type II censoring, an approximate $100(1 - \alpha)\%$ confidence interval for the control versus experimental log hazard ratio is given by $\log \hat{\mu}_E - \log \hat{\mu}_C \pm z_{\alpha/2} \sqrt{1/r_E + 1/r_C}$. For type I censoring from applying the proposal of Cox, we analogously have an approximate $100(1 - \alpha)\%$ confidence interval for the control versus experimental log hazard ratio given by

$$(\log \hat{\mu}_E - \log \hat{\mu}_C) + \log \left( \frac{r_C(2r_E + 1)}{r_E(2r_C + 1)} \right) \pm z_{\alpha/2} \sqrt{1/(r_E + 0.5) + 1/(r_C + 0.5)}.$$ These results

may apply approximately for other types of random censoring.

Example 13.2 illustrates these formulas.

## Example 13.2

We revisit Example 13.1. Note that the true control versus experimental hazard ratio is 10/9 ($\approx$1.111). We will apply each confidence interval formula for a hazard ratio or log-hazard ratio conditioning on the number of uncensored observations in each arm. We again have $r_E = 137$ uncensored observations in the experimental arm with $\hat{\mu}_E = 105.4$ and $r_C = 63$ uncensored observations in the control arm with $\hat{\mu}_C = 117.4$. The corresponding confidence intervals for the hazard ratio are provided in Table 13.2. All confidence intervals contain the true control versus exponential hazard ratio of roughly 1.111. For a non-inferiority threshold of 0.8, non-inferiority would fail to be concluded regardless of the method. The confidence intervals for the type II censoring methods have a slightly greater relative width than the type I censoring methods. Which method is more conservative depends on various factors, including whether the experimental or control arm performed better in the clinical trial. In this example, the control arm performed better (despite having a poorer underlying distribution) and the $F$ distribution methods gave more conservative intervals (i.e., had smaller lower limits) than their normal distribution counterparts. Likewise, the type II censoring methods were more conservative than the type I censoring methods. Had the values for the experimental and control arms been exchanged for each other, the opposite

**TABLE 13.2**

Calculated 95% Confidence Intervals for the Hazard Ratio by Method

| Method | Estimate | 95% CI | Relative Width |
|---|---|---|---|
| Type I censoring F distribution | 0.901 | (0.663, 1.205) | 1.816 |
| Type II censoring F distribution | 0.898 | (0.660, 1.201) | 1.820 |
| Type I censoring Normal distribution | 0.901 | (0.670, 1.214) | 1.813 |
| Type II censoring Normal distribution | 0.898 | (0.666, 1.210) | 1.816 |

relations for conservatism would have held as noted by the order of the upper limits of the 95% confidence intervals.

## 13.4 Nonparametric Inference Based on a Hazard Ratio

Without loss of generality, we will assume that the event or events of interest are undesirable. The hypotheses and the test procedures can be modified when the event or events of interest are desirable. We will see that when the hazard ratio is close to 1, the asymptotic standard error for the log-hazard ratio from a Cox proportional hazards model is approximately the same as the asymptotic standard error for the log-hazard ratio based on assuming underlying exponential distributions. Thus, there may be little gained in assuming exponential distributions when such an assumption is correct. Also, when the underlying distributions have proportional hazards but are not exponential distributions, the log-hazard ratio estimator in Section 13.3 (i.e., $\log \hat{\mu}_E - \log \hat{\mu}_C$) will be biased and the form for its asymptotic standard deviation will not be valid.

Compared with other methods of analysis of time-to-event endpoints, the Cox proportional hazards model has the advantage of allowing for the adjustment of meaningful covariates. Let $\mathbf{x}$ denote a $p \times 1$ vector of the values for $p$ explanatory variables. Let $h(t|\mathbf{x})$ denote the hazard rate at time $t$ given $\mathbf{x}$. Then the proportional hazards regression model has

$$h(t|\mathbf{x}) = h_0(t)\exp(\boldsymbol{\beta}'\mathbf{x}) \text{ for } t \geq 0 \tag{13.4}$$

where $\boldsymbol{\beta}$ is a $p \times 1$ vector of parameters and $h_0$ is a nonnegative function over $[0, \infty)$. For the proportional hazards model, the hazard functions for any two subjects are proportional. The function $h_0$ is known as the baseline hazard function and represents the hazard function for a subject having $\beta'\mathbf{x} = 0$, when such is possible. Per Cox,[6] estimation of $\beta$ through a partial likelihood function does not depend on the function $h_0$.

Suppose there are 10 subjects at risk of an event at time $t$ (i.e., as time approaches $t$), for some $t > 0$, having hazard rates $h_i(t)$ for $i = 1, \ldots, 10$ that are continuous at $t$. Given that an event occurred at time $t$ for exactly 1 of the 10 subjects, the probability that subject $j$ had the event is given by

$$h_j(t) \Big/ \sum_{i=1}^{10} h_i(t) \tag{13.5}$$

The probability would remain the same if each subject's hazard rate was multiplied or divided by some positive constant $c$ (e.g., divided by a baseline hazard rate value denoted by $h_0(t)$). The partial likelihood function for $\beta$ is based on the product of conditional probabilities like that in Expression 13.5.

Let $\mathbf{x}_i$ denote the vector of explanatory variables for the $i$th subject, $i = 1, \ldots, n$. Suppose that among the $n$ subjects, $k$ subjects are each followed to an event where their times to an event are different, whereas $n-k$ subjects have their times to an event censored. Let $t_{(1)} < t_{(2)} < \ldots < t_{(k)}$ denote the ordered times when events occurred. For $i = 1, \ldots, k$ define $R(t_{(i)})$ as the set of indices of those subjects at risk of an event as time $t_{(i)}$ approaches (i.e., consists of the indices of subjects whose time to event, censored or uncensored, is at least $t_{(i)}$) and let $\mathbf{x}_{(i)}$ denote the vector of explanatory variables for the subject that had an event at time $t_{(i)}$. Then applying the multiplication rule of probabilities to the $k$ probabilities of the form of Expression 13.5 at the times of the events leads to the partial likelihood for $\beta$ of

$$L(\beta) = \prod_{i=1}^{k} \exp(\beta'\mathbf{x}_{(i)}) \Big/ \sum_{j \in R(t_{(i)})} \exp(\beta'\mathbf{x}_j) \tag{13.6}$$

An iterative process is used to find/approximate the maximum likelihood estimate based on maximizing the log of this partial likelihood. The partial likelihood in Expression 13.6 can be adapted for situations where there are multiple events at the same time.[6]

From Expression 13.6, it should be noted that the estimate of the vector $\beta$ depends on the observed times and their censoring status only on the respective order of the times, not on the magnitude of the times or their differences. Per Cox,[7] the overall likelihood function for $\beta, h_0(t)$ factors into two functions, one being the partial likelihood given in Expression 13.6. The overall likelihood function has Expression 13.6 as a factor if and only if the censoring mechanisms are independent of the actual times to the event. Formally, if the censoring is not independent, using the partial likelihood as a basis for inference may not be justified. It would be particularly problematic if the amount of informative censoring is substantial.

The censoring mechanism being "random" requires that when an individual censored at an early time would have survived without an event to some later time, $t''$, their hazard rate of an event at time $t''$ would be the same as that of another subject, having the same set of values for the explanatory variables, who survived to time $t''$ without having an event. In essence, censoring and the process of achieving an event are determined by independent mechanisms.

Let $\beta_1$ denote the experimental versus control log-hazard ratio corresponding to the model given in Expression 13.4. Then the experimental versus control hazard ratio is $\theta = \exp(\beta_1)$. For a non-inferiority threshold $\theta_0 \geq 1$, the hypotheses are expressed as

$$H_0: \theta \geq \theta_0 \text{ and } H_a: \theta < \theta_0 \tag{13.7}$$

Let $\hat{\beta}_1$ denote the maximum likelihood Cox estimator (often referred to as a *Wald's estimator*) of $\beta_1$ and let $\mathrm{se}(\hat{\beta}_1)$ denote an estimate of its standard error. We will elaborate on the form for the standard error later. An approximate $100(1 - \alpha)\%$ confidence interval for the experimental versus control hazard ratio, $\theta = \exp(\beta_1)$, is given by

$$\exp(\hat{\beta}_1 \pm z_{\alpha/2}\mathrm{se}(\hat{\beta}_1)) \tag{13.8}$$

The null hypothesis in Expression 13.7 is rejected at an approximate significance level of $\alpha/2$, and non-inferiority is concluded when the upper limit of the two-sided $100(1 - \alpha)\%$ confidence interval is less than $\theta_0$. A test can also be constructed through a test statistic of the form $(\hat{\beta}_1 - \ln\theta_0)/\mathrm{se}(\hat{\beta}_1)$. When the value of the test statistic is greater than $z_{\alpha/2}$, the null hypothesis in Expression 13.7 is rejected at an approximate significance level of $\alpha/2$ and non-inferiority is concluded. The log-rank test can also be adapted to non-inferiority testing.[8]

*Standard Error of $\hat{\beta}_1$.* Suppose that for each treatment arm, the underlying time to an event distribution is an exponential distribution. The relative efficiency of the log-hazard ratio estimator as determined by a Cox proportional hazards model having treatment as the sole factor has been compared to that of the maximum likelihood estimator in Efron's study.[9] For the case of no censoring, the asymptotic standard error for the maximum likelihood estimator of the log-hazard ratio is given by $\sqrt{1/n_E + 1/n_C}$. In the no censoring case, the asymptotic standard error for the Cox estimator of the log-hazard ratio is given by

$$\left( \int_0^1 \frac{n_E n_C}{n_C + n_E \theta u^{(\theta-1)/\theta}} \, du \right)^{-1/2} \tag{13.9}$$

Approximate confidence intervals for the hazard ratio, $\theta$, can be determined with Expression 13.8 using the maximum likelihood Cox estimator and the Efron standard error in Expression 13.9. Since the Cox estimate depends on the observed times only through their respective order, Expression 13.9 provides the asymptotic standard error for the Cox estimator whether or not the underlying distributions are exponential. When the true hazard ratio is 1, the asymptotic standard error is given by $\sqrt{1/[\pi(1-\pi)n]} = \sqrt{1/n_E + 1/n_C}$, where $n = n_E + n_C$ and $\pi$ is the proportion of subjects allocated to the control arm. When there is no censoring and $n_E = n_C$, Table 13.3 provides the asymptotic relative efficiency of the Cox estimator to the maximum likelihood estimator (i.e., the reciprocal of the respective ratio of asymptotic variances) and the ratio of the asymptotic standard errors for various hazard ratios.

When censoring occurs, the determination of a form for the standard error is more complicated. The form can depend on the stopping rule for the analysis and the censoring distribution when noninformative censoring occurs.[9]

**TABLE 13.3**

Asymptotic Relative Efficiencies and Ratios of Asymptotic Standard Errors

| Hazard Ratio or Its Reciprocal | Asymptotic Relative Efficiency[a] | Ratio of the Asymptotic Standard Errors[b] |
|---|---|---|
| 1.00 | 1.0000 | 1.0000 |
| 0.95 | 0.9993 | 0.9997 |
| 0.90 | 0.9972 | 0.9986 |
| 0.85 | 0.9935 | 0.9968 |
| 0.80 | 0.9879 | 0.9939 |
| 0.75 | 0.9803 | 0.9901 |
| 0.70 | 0.9703 | 0.9851 |
| 0.65 | 0.9578 | 0.9787 |
| 0.60 | 0.9424 | 0.9708 |
| 0.55 | 0.9238 | 0.9611 |
| 0.50 | 0.9014 | 0.9494 |
| 0.45 | 0.8747 | 0.9352 |
| 0.40 | 0.8429 | 0.9181 |
| 0.35 | 0.8050 | 0.8972 |
| 0.30 | 0.7596 | 0.8716 |

[a] Cox estimator to the maximum likelihood estimator.
[b] Maximum likelihood estimator to the Cox estimator.

Additionally, in the presence of informative censoring, $\hat{\beta}_1$ will be biased and therefore the standard error will be different from the standard deviation.

For cases involving censoring, let $r_E$ and $r_C$ denote the number of events in the experimental and control arms, respectively. In cases where censoring is present and that censoring is noninformative, the quantity

$$\sqrt{1/r_E + 1/r_C} \tag{13.10}$$

has been a useful estimate of the unrestricted standard error of the log-hazard ratio when determining confidence intervals. An approximate $100(1 - \alpha)\%$ confidence interval for the experimental versus control hazard ratio would then be given by

$$\exp(\hat{\beta}_1 \pm z_{\alpha/2}\sqrt{1/r_E + 1/r_C})$$

It should be noted that the standard error provided from statistical packages is determined under the null hypothesis of no difference (i.e., a hazard ratio of 1). Frequently in practice, the standard error restricted to the hazard ratio equal to 1 will be relatively close to the quantity provided in Expression 13.10 and other estimates of the standard error. Thus, using the standard error from the statistical packages tends to lead to the same conclusion as using some other estimate of the standard error (e.g., unrestricted version or restricted to the non-inferiority null hypothesis). However, caution should be taken in the choice of an estimate of the standard error of the log-hazard ratio. We provide two examples (Examples 13.3 and 13.4) to illustrate the use of Expressions 13.9 and 13.10.

### Example 13.3

Consider a two-arm study of the experimental drug A and the active control drug B where 400 subjects are evenly randomized between the two arms. Suppose all 400 subjects are followed to the event of interest (e.g., death). Consider testing $H_o$: $\theta \geq 1.25$ versus $H_a$: $\theta < 1.25$. The test statistic is

$$\frac{\ln\hat{\theta} - \ln 1.25}{s}$$

where $\hat{\theta}$ is the Wald's estimator from a Cox model with treatment as the sole explanatory variable. The value for $s$ is selected as $(2/\sqrt{400})/0.9939 = 0.10061$, where the value of 0.9939 comes from Table 13.3. For an observed hazard ratio of 0.95, the value for the test statistic above is $-2.7277$, which corresponds to a $p$-value of 0.003. For a one-sided significance level of 0.025, non-inferiority is concluded ($0.003 < 0.025$).

## Example 13.4

Consider again testing $H_o$: $\theta \geq 1.25$ versus $H_a$: $\theta < 1.25$ on the basis of a two-arm study where 1000 subjects are evenly randomized to the experimental and control arms. At the time of analysis, there are 320 and 304 events in the experimental and control arms, respectively, with a corresponding estimate of the hazard ratio of 1.10. Using Expression 13.10 gives an estimate of the standard error of $\sqrt{1/320+1/304} = 0.0801$, from which a corresponding 95% confidence interval of 0.940–1.287 is obtained. Since the upper limit of 1.287 is greater than 1.25, noninferiority cannot be concluded.

*Standard Errors for the Effects of Binary Covariates.* For a binary (0–1) variable, the form for the asymptotic standard error for its corresponding regression parameter (a log-hazard ratio) is similar to that of the treatment effect. One difference is that the standard error for the treatment effect (the treatment log-hazard ratio) can be fairly controlled by knowing ahead of time roughly how many subjects will be in each arm and how many events will be needed for the analysis. For a binary covariate, the number of subjects that will have each value, 0 and 1, is random (not controlled). It is customary to condition on the number of subjects that have each value of the binary covariate (and the number of events observed at each level) when determining the corresponding standard error for its log-hazard ratio estimator. It is that conditioning that justifies using the same formulas for the standard error for the binary covariate as for the treatment effect. For $r_i$ ($i = 0, 1$) equal to the observed number of events among subjects having value $i$, a useful estimate of the asymptotic standard error for the log-hazard ratio of the binary covariate is $\sqrt{1/r_0 + 1/r_1}$.

As with the treatment effect, software packages restrict the standard error to the null hypothesis that the true value of the log-hazard ratio for the binary covariate equals zero. If the true log-hazard ratio is not far from zero, these two estimated standard errors should be relatively the same.

It is important to note that when a variable is prognostic, and is still prognostic given the set of values of any other potential covariates, the proportional hazards assumption cannot simultaneously hold for the model that includes that variable as an explanatory variable and the model that omits it as an explanatory variable.

$P(X > Y)$. For an experimental versus control hazard ratio of $\theta$, it can easily be shown that the probability is $1/(1 + \theta)$ that a random subject, $X$, given the experimental therapy will have a longer time to the event than a random subject, $Y$, given the control therapy. This probability that the random subject in the experimental arm has an event after the random subject in the control arm has an event remains constant even when both random subjects have "survived" for $t$ amount of time without having an event.

For a randomly paired designed, when the hazard rates are proportional, a confidence interval can be found for the hazard ratio by first finding a confidence interval for $1/(1 + \theta)$, the probability that a random subject given the experimental therapy will have a longer time to the event than a random subject given the control therapy, and then converting the interval into a confidence interval for $\theta$. This can be done by simultaneously following subjects within their random pairs and conditioning on the number of pairs, $n$, where at least one subject had an event. On this condition, the number of pairs where the experimental subject had a longer time has a binomial distribution with parameters $n$ and $1/(1 + \theta)$.

*Additional Design Considerations.* The treatment parameter, $\beta_1$, being estimated by a Cox model is dependent on the covariates being included and excluded in the model. Suppose that the baseline prognosis of subjects improves as accrual continues (e.g., owing to differences in a known prognostic factor), that at any time of accrual for a subject accrued at that time the hazard rates for the theoretical distribution of the experimental arm are proportional to the hazard rates of the control arm with a log-hazard ratio of $\beta_1 \neq 0$, and that all subjects are followed to the events or the same data cutoff date. Then the "log-hazard ratio" being estimated by a Cox analysis with treatment as the sole explanatory variable is between zero and $\beta_1$. For $\beta_1$ to be that value being estimated as the treatment effect by a Cox analysis, covariates that collectively completely capture the baseline prognosis of subjects need to be included in the model.

Owing to the covariates included or excluded in a Cox model, it is important to realize that the active control versus placebo treatment parameter may also be different across historical trials (and in the non-inferiority trial). Not adjusting for influential covariates in the active control therapy versus placebo trials will tend to "underestimate" the active control effect relative to when there is adjustment for those covariates. Also, when influential covariates are adjusted for, the treatment parameter being estimated will depend on the prognosis of the subjects. Consideration should be given in the historical and non-inferiority trials to capture and adjust by important covariates.

Group sequential non-inferiority trials can be done on a hazard ratio with a fixed non-inferiority margin.[10] When the hazards are not proportional, having multiple analyses at different study times may be acceptable for a superiority trial. However, nonproportional hazards can be problematic for a non-inferiority analysis based on a hazard ratio that uses the same threshold or margin for both analyses. Since the follow-up or censoring distribution will depend on the time of the analysis, when the hazards are not proportional a different parameter or value (that is called the "hazard ratio" or "average hazard ratio") is being estimated at each analysis. If there have also been nonproportional hazards when comparing the active control with placebo, then the effect of the active control therapy versus placebo as measured by a hazard ratio depends on the follow-up or censoring distribution. In the presence of nonproportional hazards, a non-inferiority criterion should consider the amount of subject follow-up to ensure that the rejection of the null hypothesis will truly mean that the experimental therapy is noninferior to the active control therapy. If the non-inferiority criterion is established to indirectly represent that the experimental therapy has any efficacy (is superior to placebo), the criterion should be such that the rejection of the null hypothesis will truly mean that the experimental therapy is superior to placebo.

Per Kalbfleisch and Prentice,[11] this issue of nonproportional hazards can be addressed by basing the inference on an "average hazard ratio" using the same weights over time in all the historical trials as well as in the non-inferiority trials. The experimental versus control average hazard ratio in the non-inferiority trial would equal $\theta_{A,H} = \int_0^\infty (h_E(t)/h_C(t))\,dK(t)$ for some pre-specified distribution function $K$. The historical estimation of the effect of the control therapy would be based on the analogous placebo versus control therapy average hazard ratio using the same distribution function $K$.

The endpoint monitoring frequency influences the estimate of a hazard ratio and that value that the (log) hazard ratio estimator estimates without bias. Less frequent monitoring tends to lead to estimates of the hazard ratio closer to 1 than the true hazard ratio, had the endpoint been continuously monitored. Therefore, although having less frequent monitoring can reduce the power in a superiority trial, it can increase the probability of concluding non-inferiority when the experimental therapy is truly inferior. The impact of the monitoring frequency on the hazard ratio needs to be considered when evaluating the effect of the control therapy from the historical clinical trials and when choosing the monitoring frequency and the non-inferiority margin for the non-inferiority trial. It is recommended that the non-inferiority trial have a monitoring frequency the same or more frequent than the trials used to estimate the effect of the control therapy.

### 13.4.1 Event and Sample Size Determination

Let the respective sample sizes for the experimental and control arms be denoted by $n_E$ and $n_C$. The sample-size determination for a time-to-event endpoint can be regarded as a two-step process. The standard error of the log-hazard ratio estimator can usually be expressed as approximately $c/\sqrt{r}$ for some constant $c$, where $r$ is the total number of events. Since the standard error depends on the total number of events, so will the power to conclude non-inferiority (or superiority). It has therefore been common for the timing of the analysis of a time-to-event endpoint to be based on a prespecified number of events so as to isolate the power.

The first step in determining the necessary sample size is to determine the necessary total number of events. When the true log-hazard ratio is not terribly far from 0 and there is no informative censoring, then the asymptotic standard error for the log-hazard ratio is $[(1+k)^2/k]/\sqrt{r}$, where $k = n_E/n_C$ is the allocation ratio (see Fleming and Harrington,[12] pp. 394–395, when $k = 1$). Additionally, the number of required events will depend on the non-inferiority threshold and the assumed experimental versus control log-hazard ratio. The number of required events is given by

$$\left( \frac{z_{\alpha/2} + z_{\beta}}{\beta_{1,a} - \beta_{1,o}} \right)^2 (1+k)^2/k \qquad (13.11)$$

where $\beta_{1,a}$ is the assumed experimental versus control log-hazard ratio (or selected alternative), and $\beta_{1,o}$ is the non-inferiority threshold for the log-hazard ratio. When powering for a superiority claim, $\beta_{1,o} = 0$. Expression 13.11 was provided by Fleming[10] for a one-to-one randomization ($k = 1$).

After determining the required number of events, the sample size will further depend on

- The accrual period or the rate of accrual
- The assumed underlying distributions
- The desired study time (time since the study opens) or calendar time when the number of events will be reached

Example 13.5 illustrates the determination of the sample sizes for a time-to-event endpoint.

## Example 13.5

Consider a two-arm non-inferiority trial comparing an experimental therapy to an active control therapy. Suppose it is desired to allocate subjects in a 2-to-1 fashion between the experimental and control arms, respectively. The experimental arm will be regarded as noninferior to the control arm with respect to a specific time-to-event endpoint (e.g., overall survival) if the hazard ratio is less than 1.15 (i.e., $H_a$: $\theta < 1.15$ or $H_a$: $\beta_1 < \ln 1.15$). Since it is believed that the experimental therapy will be slightly better than the control therapy, a hazard ratio of 0.95 is selected to power/size the trial. For a one-sided test at a significance level of 0.025, 90% power at a hazard ratio of 0.95 requires

$$\left( \frac{z_{\alpha/2} + z_{\beta}}{\beta_{1,a} - \beta_{1,o}} \right)^2 (1+k)^2/k = \left( \frac{1.96 + 1.2816}{\ln 0.95 - \ln 1.15} \right)^2 (1+2)^2 / 2 \approx 1296 \text{ events.}$$

In determining the appropriate sample size that achieves 1296 events by some target time, we will assume that the subjects will be accrued over 24 months in a uniform fashion and it is desired to have the analysis 12 months after the end of accrual (at a study time of 36 months). A random accrual time will be modeled as uniformly distributed over the first 24 months. For ease in determining the sample size, we will assume that the underlying distributions for the experimental and control arms are exponential distributions with respective medians of 10 and 9.5 months. Then, the probability that a random subject will have had an event by the study time of 36 months is

$$1 - \int_0^{24} \exp(-(36 - x)/(10/\ln 2)) dx/24 = 0.788 \text{ and}$$

$$1 - \int_0^{24} \exp(-(36 - x)/(9.5/\ln 2)) dx/24 = 0.803$$

for the experimental and control arms, respectively. Therefore, after 36 months, we would expect $(2 \times 0.788 + 0.803)/3 = 0.793$ of the subjects to have had events. This leads to a sample size of $1296/0.793 \approx 1635$ subjects. This sample-size calculation provides the number of subjects needed so the expected number of events at the study time of 36 months is the number required for stopping and performing the analysis. There will be some variability to the timing in study months that the analysis is performed (1296 events are reached). This variability may also be considered when designing the trial.

Sample-size formulas for non-inferiority trials can be easily derived[13] on the basis of a fixed overall study duration and constant hazards for the time-to-event distributions.

We next discuss procedures that are commonly used to assess whether the assumption of proportional hazards is reasonable.

### 13.4.2 Proportional Hazards Assessment Procedures

Statistical graphics and tests are often misused for assessing whether the assumption of proportional hazards is reasonable. The assessment methods tend to ignore how the estimates of the hazard ratio are determined. The nonparametric estimate of the hazard ratio depends solely on the ordered arrangement of the observations in the combined ordering of the times from both treatment arms, which includes for each observation the arm the observation belongs to and its censoring/event status. If the same increasing transformation is applied to all observations (censored or uncensored), without changing the censoring/event status, then the ordered arrangement of the observations and the hazard ratio estimate remain the same. Also, for each arm, the collection of all estimated event-free probabilities remains the same; what changes is the corresponding event times.

It makes sense and would be consistent with inferences on a hazard ratio that the evaluation of whether the hazards are proportional depends solely on this ordered arrangement of the observations. Therefore, the criterion for determining whether the hazards are proportional should also be invariant under increasing transformations. The eyeball test using a standard plot supplied by statistical software and the commonly used time-dependent covariate model used to assess proportional hazards are not invariant under increasing, continuous transformations.

When the hazards are proportional, $S_E(t) = S_C(t)^\theta$ for all $t > 0$ and some $\theta > 0$. This is equivalent to $\ln(-\ln S_E(t)) = \ln(-\ln S_C(t)) + \ln\theta$ for all $t > 0$ and some $\theta > 0$. This corresponds to the plots of $y = \ln(-\ln S_E(x))$ and $y = \ln(-\ln S_C(x))$ being parallel, separated vertically by the constant $\ln\theta$ for all $x > 0$. For the Kaplan–Meier estimates, it is common for the statistical software to plot for both treatment arms the pairs $(x, \ln(-\ln \hat{S}(x)))$ or the pairs $(\ln x, \ln(-\ln \hat{S}(x)))$. This provides an eyeball assessment of proportional hazards by determining whether the curves are nearly parallel. In this context, assessing whether these plots are parallel means the difference, $\ln(-\ln \hat{S}_E(x)) - \ln(-\ln \hat{S}_C(x))$, is nearly constant in $x$. It is not an assessment of whether the various intermediate slopes are different. It can be tempting to compare slopes when two curves or functions deviate greatly from being "parallel." Applying an increasing, continuous transformation to all the censored and uncensored times to the event, without changing the censoring status, can extend or shrink the interval of time or log times where the curves deviate from being parallel, even though it does not influence the estimation of the hazard ratio. Such a graphical display also overemphasizes the places where the survival curves are flat or nearly flat (where the estimated hazard rates are near zero), exactly the intervals of time that are not influential in the comparison of the arms.

One case from the experience of one of the authors involved a software-produced plot of the pairs $(\ln x, \ln(-\ln \hat{S}(x)))$ for each arm, where the observed log times fell into three equally spaced, enumerated intervals. There were fewer than 10 events, about 150 events, and about 40 events, respectively, in the intervals. Since the intervals are of equal length (in log times), the eyeball test gives equal weight to each interval. Within the first interval (the smaller times), which had fewer than 10 events between the arms, the plots of $y = \ln(-\ln \hat{S}_E(x))$ and $y = \ln(-\ln \hat{S}_C(x))$ were far from parallel. This is certainly not abnormal given the small number of events. Within the other two intervals, the plots of $y = \ln(-\ln \hat{S}_E(x))$ and $y = \ln(-\ln \hat{S}_C(x))$ did not greatly deviate from being equidistant or parallel. In assessing for proportional hazards, the eyeball test gives one-third weight to the first interval despite fewer than 5% of all the observed events occurring within that interval.

*Plots Invariant under Increasing Transformations.* In the one treatment group case, graphical displays of the survival functions where the scaling on the abscissa ("$x$-axis") has its $j$th gap length equal to the reciprocal of the number of subjects at risk of an event just before the $j$th event can be used.[14,15] This approach can be applied to the two-sample case in multiple ways. One way is by rescaling the $x$-axis so that the $j$th gap length is equal to the reciprocal of the number of subjects at risk of an event just before the $j$th event. To describe the approach in more detail, first let $u_1, \ldots, u_r$ represent the times events were observed when combining the observations from both arms. Then, for $j = 1, \ldots, r$, let $N_j$ denote the number of subjects at risk of an event just before time $u_j$ and let $v_j = \sum_{i=1}^{j} 1/N_i$. The resulting plot has for each arm $(v_j, \ln(-\ln \hat{S}(u_j)))$ plotted for $j = 1, \ldots, r$. Other approaches that can be used include having $j$th gap length, $v_j - v_{j-1}$, equal to the reciprocal of the harmonic mean of the number of subjects in each arm that are at risk of an event just before the $j$th event (or corrected version should one arm have no subjects at risk) or having $v_j = j$ (i.e., enumerating the event times on the $x$-axis). In all these cases, the resulting plot is invariant under increasing transformations.

*Time-Dependent Covariate Model.* Likewise, tests of proportional hazards involving time or log time as a time-dependent covariate are not invariant when applying an increasing, continuous transformation to all the censored and uncensored times. For example, the $p$-values for testing whether the coefficient for the time-dependent covariate is zero will change after the transformation $h(x) = \exp\{x + \exp\{x\}\}$ is applied to all the observations, even though the ordered arrangement of the observations remains the same.

Analogous to the rescaling that was described above for the graphical displays, the censored and uncensored times can be rescaled on the basis of the sum of the reciprocal of the number of subjects at risk of an event. The rescaled survival time of the $j$th event would be $\sum_{i=1}^{j} 1/N_i$. The censored times would also be rescaled so as not to affect the overall ordered arrangement of the observations.

As with other diagnostic plots used to evaluate an assumption of proportional hazards, the plot of the ratio of the hazard rates will also overemphasize the places where the estimated hazard rates are near zero (i.e., the time intervals that are not influential in a comparison of the arms). In a published study, the authors concluded from one such plot that the hazard ratio was not constant, whereas in a quite different type of plot concluded that an assumption of proportional cumulative odds may be appropriate. The conclusion is unusual as the aspects of the estimated survival distribution were such that similar conclusions should have drawn on the proportionality of the cumulative hazards and the cumulative odds. In the example, the estimated survival probabilities within both arms were greater than 90% over all studied time points. In particular, $-\ln(\hat{S}_i(t))/(1 - \hat{S}_i(t)) \approx 1$ for all studied $t$ and $i = E, C$, and $\hat{S}_C(t)/\hat{S}_E(t)$ appears to only vary between 0.98 and 1 over the studied interval of time. Different conclusions on proportionality were drawn because the assessment of proportional hazards was based on the ratio of estimates of the hazard rates (not the ratio of estimates of the cumulative hazards) where outlying estimates of the ratio were observed when the survival curves were nearly flat and the hazard rates were close to zero. The assessment of the cumulative odds ratio was based on a plot of the cumulative odds ratio over time, which did not vary significantly. A plot of the ratio of the cumulative hazards would also not have significantly varied over time.

## 13.5  Analyses Based on Landmarks and Medians

Time-to-event comparison can also be based on the medians or on an event-free probability at some prespecified time (i.e., at some landmark). The difference in the medians represents a particular horizontal difference in the plot of the survival functions, whereas the difference in the event-free probabilities at a specific time point represents a vertical difference in the plots of the survival functions. In practice, with the presence of censoring, the medians and event-free probabilities are often used to describe the results within a given treatment arm. It is common to include the medians in a product label to describe the typical time to an event. When the median time to an event is not estimable, event-free probabilities are often used to describe the results within a given arm. Although hypothesis testing could be based on the medians or the event-free probabilities at a prespecified time point, in practice hypothesis testing tends to be based on a hazard ratio. We will discuss performing non-inferiority testing on the basis of the median times to event and of the event-free probabilities.

In the absence of censoring, medians can be compared as provided in Section 12.4, and the event-free probabilities can be compared by methods discussed in Chapter 11. In the presence of censoring, the precision in the estimates involves not only the underlying distribution and the sample size but also the amount of censoring and where that censoring occurs.

For the event-free probabilities, we consider inferences based on their difference and their ratio. Analyses are primarily based on the Kaplan–Meier estimated event-free probabilities and the corresponding Greenwood's estimated variance. For the median times to event, the ratio of the medians will primarily be used. When the two underlying distributions differ by a scale factor, the ratio of the medians is equal to the scale factor. When assuming that the two underlying distributions differ by a scale factor, non-inferiority procedures on the ratio of medians that will be discussed include converting standard tests for testing a difference to construct confidence intervals for the ratio of medians, and a generalization of the two confidence interval procedures by Hettmansperger[16] (provided in Section 12.4) to the presence of censoring (see Wang and Hettmansperger[17]). For the case in which it is not assumed that the two underlying distributions differ by a scale factor, a procedure from Su and Wei,[18] based on a minimum dispersion statistic, is also provided. In all instances in this section, the event of interest is regarded as undesirable (i.e., larger times to the event are better).

### 13.5.1  Landmark Analyses

We will next consider comparing between the arms the event-free probabilities after a fixed, prespecified duration. A *landmark* is a prespecified time from randomization (or start of treatment) at which time the absence or presence of an event is noted or a variable of interest is measured. For time-to-event endpoints, the parameter of interest for a landmark analysis at time $t^*$ is the probability a random subject is event free at time $t^*$.

In practice, some interested landmarks arise from the natural history of the treatment of the disease. For example, for a curable disease, there may be a common landmark across therapies or across clinical trials after which the rate of cure is very small or is zero. In the setting of preventing disease recurrence, there likewise may be a common landmark after which recurrence is rather unlikely.

When monitoring for the event is scheduled periodically, for unbiased estimation of the event-free probabilities at the landmark, the selected landmark should be a scheduled time for monitoring the event. If monitoring occurs every 5 months, a landmark analysis at 12 months will yield the same results as a landmark analysis at 10 months. It is important that the necessary evaluations for an event occur at the time of the selected landmark, not before or after. Performing evaluations too early or too late from the selected landmark can lead to bias estimates of the event-free probabilities. Differential monitoring between treatment arms can lead to a bias in the estimation of the difference in the event-free probabilities at a landmark.

The survival functions for the experimental and control arms will be denoted by $S_E$ and $S_C$, respectively. Then, the probability that a random subject on the experimental and control arms will be event-free at time $t^*$ are denoted by $S_E(t^*)$ and $S_C(t^*)$, respectively. For a non-inferiority margin of $\delta$ (for some $0 < \delta < 1$), the hypotheses for the difference in event-free probabilities at time $t^*$ are expressed as

$$H_o: S_E(t^*) - S_C(t^*) \leq -\delta \text{ and } H_a: S_E(t^*) - S_C(t^*) > -\delta \tag{13.12}$$

A confidence interval for $S_E(t^*) - S_C(t^*)$ can be determined using the respective Kaplan–Meier estimates and Greenwood's estimates of the corresponding variance. When the lower limit of the confidence interval for $S_E(t^*) - S_C(t^*)$ is greater than $-\delta$, non-inferiority is concluded.

*Kaplan–Meier Estimation.* In the absence of censoring, the determination of the estimated survival function (i.e., the event-free probabilities) for a given arm is straightforward. For $t > 0$, the survival function is given by $\hat{S}(t) =$ the relative frequency of times in that arm $\leq t$. In the presence of censoring, the most common estimate of the survival function is the Kaplan–Meier estimate.[19] As earlier, let $t_{(1)} < t_{(2)} < \ldots < t_{(k)}$ denote the distinct ordered times when events occurred, and for $i = 1, \ldots, k$ define $R(t_{(i)})$ as the set of indices of those subjects at risk of an event as time $t_{(i)}$ approaches (i.e., consists of the indices of subjects whose time-to-event, censored or uncensored, is at least $t_{(i)}$). Let $n_i$ denote the size of $R(t_{(i)})$—that is, the number of subjects at risk of an event as time $t_{(i)}$ approaches—and let $d_i$ denote the number of subjects that had events at time $t_{(i)}$. For ease, we will define $t_{(0)} = 0$. Then for $i = 1, \ldots, k$, $1 - d_i/n_i$ represents the relative frequency of subjects followed completely from time $t_{(i-1)}$ to time $t_{(i)}$ that did not have an event, and represents an estimate of the conditional probability that a subject will not have an event during the interval from $t_{(i-1)}$ to $t_{(i)}$ given they have not had an event by time $t_{(i-1)}$. For intermediate intervals, $t_{(i-1)}$ to $t$, where $t_{(i-1)} < t < t_{(i)}$, the observed relative frequency of subjects followed completely from time $t_{(i-1)}$ to time $t$ that did not have an event is 1. Thus, 1 is the estimate of the conditional probability that a subject will not have an event during the interval from $t_{(i-1)}$ to $t$ given they have not had an event by time $t_{(i-1)}$. The Kaplan–Meier estimate of the survival function applies the multiplication rule to these estimated conditional probabilities.

For $t_{(i)} \leq t < t_{(i+1)}$ $i = 0, 1, \ldots, k$, the Kaplan–Meier estimate of the survival function is given by

$$\hat{S}(t) = \prod_{j=1}^{i}(1 - d_j/n_j)$$

For $t_{(i)} \leq t < t_{(i+1)}$ $i = 0, 1, \ldots, k$, Greenwood's formula provides an estimate of the variance for $\hat{S}(t)$ of $\text{Var}(\hat{S}(t)) \approx (\hat{S}(t))^2 \sum_{j=1}^{i} \dfrac{d_j}{n_j(n_j - d_j)}$.

*Testing Procedures.* The Kaplan–Meier estimators (estimates) of the survival functions for the experimental and control arms will be denoted by $\hat{S}_E$ and $\hat{S}_C$ ($\hat{s}_E$ and $\hat{s}_C$), respectively. For the landmark of $t^*$, the asymptotic test statistic for

testing the hypotheses in Expression 13.12, is $Z^* = \dfrac{\hat{S}_E(t^*) - \hat{S}_C(t^*) + \delta}{\sqrt{\text{Var}(\hat{S}_E(t^*)) + \text{Var}(\hat{S}_C(t^*))}}$, as

provided by Com-Nougue, Rodary, and Patte.[8] At a significance level of $\alpha/2$, the null hypothesis in Expression 13.12 is rejected and non-inferiority is concluded when $Z^* > z_{\alpha/2}$. An approximate $100(1 - \alpha)\%$ confidence interval for

$S_E(t^*) - S_C(t^*)$ is given by $\hat{s}_E(t^*) - \hat{s}_C(t^*) \pm z_{\alpha/2}\sqrt{\text{Var}(\hat{S}_E(t^*)) + \text{Var}(\hat{S}_C(t^*))}$. When

the lower limit of the confidence interval is greater than $-\delta$ the null hypothesis in Expression 13.12 is rejected and non-inferiority is concluded.

As noted by Com-Nougue, Rodary, and Patte,[8] under the assumption of proportion hazards, the non-inferiority margin for the landmark analysis can be linked to a non-inferiority threshold based on a hazard ratio by $\theta_0 = \ln(S_C(t^*) - \delta)/\ln(S_C(t^*))$. This relation along with a guess of the event-free probability at the landmark for the control arm can guide in translating an historical problem where inference was based on a hazard ratio to a non-inferiority problem involving a difference in event-free probabilities or vice versa. First, the historical control effect would be estimated using one of the metrics, and then an appropriate non-inferiority threshold or margin for that metric. Then with a guess of $S_C(t^*)$, the above relation leads to a non-inferiority margin or threshold for the other metric.

Alternatively, the relationship between the event-free probabilities and the hazard ratio can be used to base the non-inferiority test directly on the event-free probabilities with the non-inferiority threshold for the hazard ratio. When the hazards are proportional, then the experimental versus control hazard ratio, $\theta$, satisfies $\theta = \log S_E(t^*)/\log S_C(t^*)$. Likewise, the estimates of the event-free probabilities at time $t^*$ can be transformed into an estimate of the hazard ratio. When the inference on a hazard ratio is based on the cumulative hazards or event-free probabilities at time $t^*$, the hypotheses are expressed as

$$H_0: \theta = \log S_E(t^*)/\log S_C(t^*) \geq \theta_0 \text{ and } H_a: \theta = \log S_E(t^*)/\log S_C(t^*) < \theta_0$$

where the non-inferiority threshold satisfies $\theta_0 \geq 1$. The hypotheses can be reexpressed as

$$H_0: \log S_E(t^*) - \theta_0 \log S_C(t^*) \leq 0 \text{ and } H_a: \log S_E(t^*) - \theta_0 \log S_C(t^*) > 0$$

The corresponding test statistic is $Z^* = \dfrac{\log \hat{S}_E(t^*) - \theta_0 \log \hat{S}_C(t^*)}{\sqrt{\text{Vâr}(\hat{S}_E(t^*))/\hat{S}_E^2(t^*) + \theta_0^2 \text{Vâr}(\hat{S}_C(t^*))/\hat{S}_C^2(t^*)}}$.

When $Z^* > z_{\alpha/2}$, the null hypothesis is rejected and non-inferiority is concluded. A Fieller approach can be used to determine an approximate $100(1 - \alpha)\%$ confidence interval for $\theta = \log S_E(t^*)/\log S_C(t^*)$. Note that the inference on a hazard ratio is most efficient when using the entire time to the event (censored or uncensored) from each subject. The estimate of a hazard ratio is less reliable when based on a single time point.

Alternatively, the testing can be based on the ratio of the survival probabilities at a landmark. For a landmark of $t^*$ and a non-inferiority threshold of $\gamma_0$ (for some $0 < \gamma_0 \leq 1$), the hypotheses are expressed as

$$H_0: S_E(t^*)/S_C(t^*) \leq \gamma_0 \text{ and } H_a: S_E(t^*)/S_C(t^*) > \gamma_0$$

These hypotheses can be reexpressed as

$$H_0: S_E(t^*) - \gamma_0 S_C(t^*) \leq 0 \text{ and } H_a: S_E(t^*) - \gamma_0 S_C(t^*) > 0$$

The corresponding normalized test statistic is $Z^* = \dfrac{\hat{S}_E(t^*) - \gamma_0 \hat{S}_C(t^*)}{\sqrt{\text{Vâr}(\hat{S}_E(t^*)) + \gamma_0^2 \text{Vâr}(\hat{S}_C(t^*))}}$.

When $Z^* > z_{\alpha/2}$, the null hypothesis is rejected and non-inferiority is concluded. A Fieller approach can be used to determine an approximate $100(1 - \alpha)\%$ confidence interval for $S_E(t^*)/S_C(t^*)$. Other testing procedures can be considered. As done by Thomas and Grunkemeier,[20] the testing can be based on a likelihood ratio test-based confidence interval for the ratio of the survival probabilities at a landmark. Methods can also be adapted when the hypotheses and the corresponding test are based on the ratio of probabilities of an event by time $t^*$ (i.e., based on $(1 - S_E(t^*))/(1 - S_C(t^*))$).

## 13.5.2 Analyses on Medians

The medians for the experimental and control arms are denoted by $\tilde{\mu}_E$ and $\tilde{\mu}_C$, respectively, and their ratio by $\Delta = \tilde{\mu}_E / \tilde{\mu}_C$. For a non-inferiority margin of $\delta$ (for some $0 < \delta < 1$), the hypotheses are expressed as

$$H_o: \Lambda \le \Lambda_o = 1 - \delta \text{ vs. } H_a: \Lambda > \Lambda_o = 1 - \delta \qquad (13.13)$$

For inference on one median, Efron[21] and Reid[22] used bootstrap methods to derive confidence intervals for the median. Such bootstrapping methods can be easily applied to determine confidence intervals for the difference or ratio of two medians. Alternatively, confidence sets for one median can be derived by inverting tests similar to a sign test, as done by Brookmeyer and Crowley[23] and Emerson.[24] The confidence interval for one median consists of all values $t^*$ for which a two-sided test of $H_o: S(t^*) = 0.5$ fails to reject the null hypothesis. The Brookmeyer and Crowley procedure[23] uses the Kaplan–Meier estimated survival probability at $t^*$, $\hat{S}(t^*)$, along with the corresponding Greenwood estimated variance. This estimated variance changes as $t^*$ changes. As noted by Wang and Hettmansperger,[17] the confidence set derived from these methods need not be an interval. For the Brookmeyer and Crowley procedure, this inadequacy can be alleviated by choosing the Greenwood estimate variance of $\hat{S}(\tilde{x})$, where $\tilde{x}$ is the observed median. In generalizing to two medians, this type of estimated variance can be used for both arms in a minimum dispersion test statistic, as in Su and Wei's study.[18]

*General Procedures.* For comparing two medians, we will first discuss two procedures that are not based on the assumption that the two underlying distributions are related by scale factor.

Su and Wei[18] derived confidence intervals for the difference and ratio of two medians based on a quadratic test statistic similar to a minimum dispersion test statistic used by Basawa and Koul[25] for continuous data. We will present the test procedure for a ratio of medians.

Let $\tilde{X}$ and $\tilde{Y}$ denote the sample medians of the control and experimental arms, respectively. The observed sample medians $\tilde{x}$ and $\tilde{y}$ satisfy $\tilde{x} = \min\{t : \hat{S}_C(t) \le 0.5\}$ and $\tilde{y} = \min\{t : \hat{S}_E(t) \le 0.5\}$. For testing $H_o: \Lambda = \Lambda_o$ against $H_a: \Lambda \ne \Lambda_o$ (for some $0 < \Lambda_o \le 1$), the test statistic is $G(\Lambda_o) = \min_{t>0}$

$$\left\{ \frac{(\hat{S}_E(\Lambda_o t) - 0.5)^2}{\hat{\sigma}_E^2} + \frac{(\hat{S}_C(t) - 0.5)^2}{\hat{\sigma}_C^2} \right\}, \text{ where } \hat{\sigma}_E^2 \text{ and } \hat{\sigma}_C^2 \text{ are the Greenwood's esti-}$$

mates of the variances of $\hat{S}_E(\tilde{y})$ and $\hat{S}_C(\tilde{x})$, respectively. From a simulation study of Su and Wei,[18] the upper percentiles of the distribution of $G(\Lambda_o)$ when $\Lambda = \Lambda_o$ can be approximated by the upper percentiles of a $\chi^2$ distribution with 1 degree of freedom. Let $\chi_{1,\alpha}^2$ denote the $100\alpha$th upper percentile of a $\chi^2$ distribution with 1 degree of freedom. Then, an approximate $100(1 - \alpha)\%$ confidence interval for $\tilde{\mu}_E / \tilde{\mu}_C$ consists of those positive values $u$ so that $G(u) < \chi_{1,\alpha}^2$. If the lower bound of the approximate $100(1 - \alpha)\%$ confidence interval for

$\tilde{\mu}_E/\tilde{\mu}_C$ is greater than $\Lambda_o = 1 - \delta$, then the null hypothesis in Expression 13.13 is rejected and non-inferiority is concluded. Alternatively, bootstrapping can be used to obtain a critical value.

The sign-like test of one median can also be generalized to a large sample normal test for the ratio of medians that uses the sample medians and estimates of the corresponding standard errors. The standard error estimates involve density estimation at the sample median. For $\varepsilon > 0$ and small, define $m_C(\varepsilon) = (\hat{S}_C(\tilde{x} - \varepsilon) - \hat{S}_C(\tilde{x} + \varepsilon))/(2\varepsilon)$. If $m_C(\varepsilon)$ is fairly stable for reasonable choices of $\varepsilon$, then the interval provided by

$$\tilde{x} \pm z_{\alpha/2}\hat{\sigma}_C/m_C(\varepsilon)$$

is approximately equal to the approximate $100(1 - \alpha)\%$ confidence interval for $\tilde{\mu}_C$ of $(S_C^{-1}(0.5 - z_{\alpha/2}\hat{\sigma}_C), S_C^{-1}(0.5 + z_{\alpha/2}\hat{\sigma}_C))$. For the experimental arm, $m_E(\varepsilon)$ is analogously defined. The procedures in Chapter 12 used for comparing two means for a positive-valued endpoint can be applied for the time-to-event medians where $\hat{\sigma}_C/m_C(\varepsilon)$ and $\hat{\sigma}_E/m_E(\varepsilon)$ are used as the standard errors for the sample medians $\tilde{X}$ and $\tilde{Y}$, respectively. For example, for testing $H_o: \Lambda \leq \Lambda_o$ against $H_a:$ $\Lambda > \Lambda_o$ for some $0 < \Lambda_o \leq 1$, the hypotheses can be expressed as $H_o: \tilde{\mu}_E - \Lambda_o\tilde{\mu}_C \leq 0$ and $H_a: \tilde{\mu}_E - \Lambda_o\tilde{\mu}_C > 0$. A large sample normal test statistic is given by

$$Z^* = \frac{\tilde{Y} - \Lambda_o\tilde{X}}{\sqrt{\hat{\sigma}_E^2/m_E^2(\varepsilon) + \Lambda_o^2\hat{\sigma}_C^2/m_C^2(\varepsilon)}}$$

When $Z^* > z_{\alpha/2}$, the null hypothesis in Expression 13.13 is rejected and non-inferiority is concluded. A Fieller approach can also be used to determine an approximate $100(1 - \alpha)\%$ confidence interval for $\tilde{\mu}_E/\tilde{\mu}_C$.

The remaining procedures that will be discussed are based on the overall assumption that the two underlying distributions differ by a scale factor.

*Adapting Standard Time-to-Event Tests.* Let $X_1, X_2, \ldots, X_{n_C}$ and $Y_1, Y_2, \ldots, Y_{n_E}$ denote independent random samples from distributions having respective distribution functions $F_C$ and $F_E$. The $X$'s and the $Y$'s represent the actual, uncensored times to the event of the control and experimental arms, respectively. The underlying assumption is that the two distributions are related through a scale factor $\Lambda$ (i.e., $F_E(y) = F_C(y/\Lambda)$ for all $y$ and some $\Lambda$). For the control arm, the independent censoring variables are denoted as $A_1, A_2, \ldots,$ $A_{n_C}$, which are assumed to be a random sample having common distribution function $H$. For each subject in the control arm, the variable $X_i^* = \min\{X_i, A_i\}$ and the event status $I(X_i = X_i^*)$ are observed. For the experimental arm, the independent censoring variables are denoted as $B_1, B_2, \ldots, B_{n_E}$, which are assumed to be a random sample having a common distribution function $K$. For each subject in the experimental arm, the variable $Y_i^* = \min\{Y_i, B_i\}$ and the event status $I(Y_i = Y_i^*)$ are observed.

Confidence intervals for $\Lambda$ can be obtained through a test statistic for time-to-event endpoints by altering the values in one of the arms and then testing for the equality of the underlying distributions. For any positive number $c$, replace $X_1, X_2, \ldots, X_{n_C}$ with $cX_1, cX_2, \ldots, cX_{n_C}$ and replace $A_1, A_2, \ldots, A_{n_C}$ with $cA_1, cA_2, \ldots, cA_{n_C}$. For the observations in the control arm, the analysis multiplies each observed censored or uncensored time-to-event by $c$ without changing the event status for those observations. If the null hypothesis of equal medians (i.e., equal underlying distributions for $Y_i$ and $cX_j$) is not rejected at a two-sided significance level $\alpha$, then $\Lambda_o$ is in the (approximate) $100(1 - \alpha)\%$ confidence interval for $\Lambda$. If the lower bound of this approximate $100(1 - \alpha)\%$ confidence interval for the scale factor (i.e., also for $\tilde{\mu}_E/\tilde{\mu}_C$) is greater than $\Lambda_o = 1 - \delta$, then the null hypothesis in Expression 13.13 is rejected and non-inferiority is concluded. This procedure for determining a confidence interval for the scale factor or ratio of medians can be applied to the log-rank test (where the parameter of interest tends to be a hazard ratio, not a scale factor) or any Wilcoxon-like test. This procedure for obtaining a confidence interval for a scale factor is analogous to manipulating the Mann–Whitney–Wilcoxon procedure for deriving a confidence interval for the shift in the distributions (i.e., the difference in medians) provided in Section 12.4.

It is a matter of debate whether there may be some crudeness to this procedure. For the event status to remain the same, the corresponding censoring variables would also need to be multiplied by $c$. In a properly conducted clinical trial, the censoring distributions should be the same across arms. When assumptions that are made on the underlying distribution do not hold, comparisons involving quite different censoring distributions can be difficult to interpret when there is a moderate or large amount of censoring.

*Two Confidence Interval Procedures.* For time-to-event data in the presence of censoring, the use and properties of confidence intervals for the difference in medians where the limits are differences of the confidence limits of the individual confidence intervals for the medians were investigated by Wang and Hettmansperger.[17] Several cases are considered, including the case where the two underlying time-to-event distributions are assumed to have the same shape. The results are fairly analogous to those by Hettmansperger[16] for determining confidence intervals for the difference in medians for continuous data, which are summarized in Section 12.4.

It is, however, unlikely that two time-to-event distributions differ by a shift. If the two underlying distributions differ by a scale factor, which is often assumed when comparing time-to-event distributions, then the distributions for the log times will have the same shape (i.e., differ by a shift). The results of Wang and Hettmansperger[17] can be applied to testing the ratio of the underlying medians when assuming equal shapes for the distribution of the log times. For ease in both presentation and in comparing the results to those of Hettmansperger[16] in Section 12.4, the results of Wang and Hettmansperger[17] will be presented for a difference in medians for the log times. The medians for the log times are denoted by $\tilde{\mu}_{\log,E}$ and $\tilde{\mu}_{\log,C}$ for the experimental and control arms, respectively. Their difference is denoted by $\Delta = \tilde{\mu}_{\log,E} - \tilde{\mu}_{\log,C}$. For a non-inferiority margin of $\delta$ (for some $\delta > 0$), the hypotheses are expressed as

$$H_o: \Delta \le -\delta \text{ and } H_a: \Delta > -\delta. \tag{13.14}$$

As in Section 12.4, the $100(1 - \alpha)\%$ confidence interval for the difference in medians of the log times has the form $(L, U) = (L_E - U_C, U_E - L_C)$, where $(L_E, U_E)$ is a $100(1 - \alpha_E)\%$ confidence interval for the median log time of the experimental arm and $(L_C, U_C)$ is a $100(1 - \alpha_C)\%$ confidence interval for the median log time of the control arm. The confidence coefficients for the individual confidence intervals are selected so that when those two intervals are disjoint, $H_o: \Delta = 0$ is rejected at a significance level of $\alpha$ in favor of the two-sided alternative $H_a: \Delta \ne 0$. The null hypothesis in Expression 13.14 is rejected at a significance level of $\alpha/2$, and non-inferiority is concluded if $L > -\delta$.

The previous notation for the time-to-event and censoring variables will apply here to the log times. Let $X_1, X_2, \ldots, X_{n_C}$ and $Y_1, Y_2, \ldots, Y_{n_E}$ denote independent random samples from distributions having respective distribution functions $F_C$ and $F_E$. The $X$'s and the $Y$'s represent the actual, uncensored log times to the event of the control and experimental arms, respectively. For the control arm, the independent censoring variables for the log times are denoted as $A_1, A_2, \ldots, A_{n_C}$ (i.e., $\exp(A_1), \ldots, \exp(A_{n_C})$ are the censoring variables for $\exp(X_1), \exp(X_2), \ldots, \exp(X_{n_C})$), which are assumed to be a random sample having common distribution function $H$. For each subject in the control arm, the variable $X_i^* = \min\{X_i, A_i\}$ and the event status $I(X_i = X_i^*)$ are observed. The common distribution function for $X_i^*$ is given by $F_C^*(t) = 1 - (1 - F_C(t))(1 - H(t))$. For the experimental arm, the independent censoring variables for the log times are denoted as $B_1, B_2, \ldots, B_{n_E}$, which are assumed to be a random sample having common distribution function $K$. For each subject in the experimental arm, the variable $Y_i^* = \min\{Y_i, B_i\}$ and the event status $I(Y_i = Y_i^*)$ are observed. The common distribution function for $Y_i^*$ is given by $F_E^*(t) = 1 - (1 - F_E(t))(1 - K(t))$.

The left continuous inverse of a distribution $F$ is defined by $F^{-1}$, where $F^{-1}(p) = \inf\{t: F(t) \ge p\}$ for $0 < p < 1$. Let $\hat{F}_C$ and $\hat{F}_E$ denote the Kaplan–Meier estimators of $F_C$ and $F_E$, respectively. As in Section 12.4, dE and dC will denote the "depths" of the interval confidence intervals for the experimental and control arms, respectively. For $i = E, C$, the $100(1 - \alpha_i)\%$ confidence interval for $\tilde{\mu}_i$ is of the form $(L_i, U_i) = (\hat{F}_i^{-1}(d_i/n_i), \hat{F}_i^{-1}(1 - (d_i - 1)/n_i))$, where $d_i$ is asymptotically approximated by $n_i/2 + 0.5 - (\sqrt{n_i}/2)Z_i$ for some appropriately selected multiplier $Z_i$. For each of $i = E, C$, the follow-up needs to be such that $\hat{F}_i^{-1}(d_i/n_i)$ and $\hat{F}_i^{-1}(1 - (d_i - 1)/n_i)$ exists. That is, the Kaplan–Meier curves are not suspended at heights near or above 0.5.

For $i = E, C$, define $\tau_i = \int_{-\infty}^{\tilde{\mu}_i} (1 - F_i^*(t))^{-2} \, dG_i^*(t)$, where $G_C^*(t) = P(X_i^* \le t, X_i^* = X_i)$ and $G_E^*(t) = P(Y_i^* \le t, Y_i^* = Y_i)$. Per Wang and Hettmansperger,[17] $\tau_i$ is the asymptotic variance of $2\sqrt{n_i}(\hat{F}_i(\tilde{\mu}_i) - F_i(\tilde{\mu}_i))$. Asymptotically, the multipliers (i.e., $Z_C$ and $Z_E$) and the confidence coefficients (i.e., $1 - \alpha_C$ and $1 - \alpha_E$) will depend on the values for $\tau_C$ and $\tau_E$ and the value of $\alpha$. Wang and Hettmansperger[17] recommend using Greenwood's formula to estimate $\tau_C$ and $\tau_E$. For $i = E, C$, the proposed estimator $\hat{\tau}_i$ equals $4n_i$ multiplied by the Greenwood estimate of the variance at time $t_i^*$, where $t_i^*$ is the Kaplan–Meier estimated median (i.e., $t_C^* = \tilde{x}$ and $t_E^* = \tilde{y}$).

For $\lambda = n_C/N$ from Theorem 2 of Wang and Hettmansperger,[17] the multipliers are asymptotically related by

$$\sqrt{\lambda}Z_E + \sqrt{1-\lambda}Z_C \approx z_{\alpha/2}\sqrt{\lambda\tau_E + (1-\lambda)\tau_C} \qquad (13.15)$$

In addition, the asymptotic width of the confidence interval does not depend on the choice of $Z_E$ and $Z_C$ that satisfies Equation 13.15. From Theorem 3 of Wang and Hettmansperger,[17]

$$\sqrt{N}(U - L) \rightarrow z_{\alpha/2}\sqrt{\lambda\tau_E + (1-\lambda)\tau_C} / [\sqrt{\lambda(1-\lambda)}f_C(0)]$$

in probability, where $f_C(0)$ is the common density at the median.

According to Wang and Hettmansperger,[17] when it is desired to have equal confidence coefficients for the individual intervals (i.e., $\alpha_E = \alpha_C$), then asymptotically the appropriate multipliers satisfy

$$Z_E = \sqrt{\tau_E/\tau_C}Z_C = z_{\alpha/2}\sqrt{\tau_E}\sqrt{\lambda\tau_E + (1-\lambda)\tau_C}/(\sqrt{\lambda\tau_E} + \sqrt{(1-\lambda)\tau_C})$$

In the case of asymptotically equal-length confidence intervals, we have from Wang and Hettmansperger[17]

$$Z_E = \sqrt{(1-\lambda)/\lambda}Z_C = z_{\alpha/2}\sqrt{\lambda\tau_E + (1-\lambda)\tau_C}/[2\sqrt{\lambda}]$$

The authors also provided formulas for the multipliers in the equal-depth case where $d_E = d_C$.[17]

For the equal confidence coefficient procedure, the common confidence coefficient ranged from 0.83 to 0.88 in the cases studied,[17] where the allocation ratios ranged from 1 to 3 and various relative frequencies of censoring were assumed. We refer the reader to Wang and Hettmansperger's paper[17] for analogously determined confidence coefficient intervals for the equal length and equal depth procedures.

Additionally, Wang and Hettmansperger[17] modified the two confidence intervals procedures for the equal-shape case to obtain procedures for the

Behrens–Fisher problem and for when proportional hazards are assumed. These additional cases require the estimates of the densities at the true medians. There are other approaches that do not require density estimation. An estimator and confidence interval for the ratio of medians or scale factor was based on the Hodges–Lehmann approach in Wei and Gail's paper.[26]

## 13.6 Comparisons over Preset Intervals

Instead of comparing treatment arms at a specific landmark or a specific percentile, the event-free probabilities of the treatment arms can be compared over a prespecified time interval or the quantiles of the treatment arms can be compared over an interval of the corresponding levels of percentiles as in a comparison of trimmed means. There should be greater precision when the inference is over an interval than at a particular point. As with inferences at landmarks or based on medians, there is also the advantage of not having to rely on assumptions such as the proportional hazards assumption.

For non-inferiority testing over a preset interval of time on the difference in the survival functions or on a cumulative odds ratio, test procedures based on simultaneous pointwise confidence intervals and based on a confidence interval for the supremum difference between the survival curves were evaluated by Freitag, Lange, and Munk.[27] Let $[\tau_0, \tau_1]$ denote the interval of interest for some $0 < \tau_0 < \tau_1$. For the difference in the survival functions and a non-inferiority margin of $\delta$ (for some $0 < \delta < 1$), the hypotheses for the difference in event-free probabilities over the interval $[\tau_0, \tau_1]$ are expressed as

$$H_o: S_E(t^*) - S_C(t^*) \leq -\delta \text{ for some } t^* \in [\tau_0, \tau_1] \text{ and}$$

$$H_a: S_E(t^*) - S_C(t^*) > -\delta \text{ for all } t^* \in [\tau_0, \tau_1] \tag{13.16}$$

For the cumulative odds ratio, $\psi(t) = [F_E(t)/(1 - F_E(t))]/[F_C(t)/(1 - F_C(t))]$, the hypotheses are expressed as

$$H_o: \psi(t) \geq \psi_o \text{ for some } t^* \in [\tau_0, \tau_1] \text{ and } H_a: \psi(t) < \psi_o \text{ for all } t^* \in [\tau_0, \tau_1] \tag{13.17}$$

Alternatively, the hypotheses can also be expressed involving the ratio of the cumulative hazards $(\ln S_E(t)/\ln S_C(t))$, the ratio of event-free probabilities $(S_C(t)/S_E(t))$, or the relative risk of an event $((1 - S_E(t))/(1 - S_C(t)))$.

*Pointwise Approach.* Concluding the alternative hypothesis in Expression 13.16 requires that for every $t^* \in [\tau_0, \tau_1]$, the null hypothesis $H_o: S_E(t^*) - S_C(t^*) \leq -\delta$ is rejected in favor of $H_a: S_E(t^*) - S_C(t^*) > -\delta$. The landmark testing procedures in Section 13.5 can be used for testing these hypotheses for each given $t^* \in [\tau_0, \tau_1]$. Freitag, Lange, and Munk[27] used bootstrapping to the construct pointwise confidence intervals at the individual landmarks for testing of the hypotheses in Expressions 13.16 and 13.17. This type of test procedure is conservative and necessarily maintains the desired type I error, provided the desired type I error rate is maintained by the testing of the hypotheses at the individual landmarks in $[\tau_0, \tau_1]$.

*Supremum Approach.* For testing based on the supremum difference in the survival functions, the hypotheses in Expression 13.16 can be reexpressed as

$$H_0: \sup_{t\in[\tau_0,\tau_1]}[S_C(t) - S_E(t)] \geq \delta \text{ and } H_a: \sup_{t\in[\tau_0,\tau_1]}[S_C(t) - S_E(t)] < \delta \qquad (13.18)$$

The hypotheses involving the cumulative odds ratio, ratio of the cumulative hazards, ratio of event-free probabilities, and relative risk of an event, would involve the following supremums being compared to the appropriate non-inferiority margin or threshold: $\sup_{t\in[\tau_0,\tau_1]}\psi(t)$, $\sup_{t\in[\tau_0,\tau_1]}[\ln S_E(t)/\ln S_C(t)]$, $\sup_{t\in[\tau_0,\tau_1]}[S_C(t)/S_E(t)]$, and $\sup_{t\in[\tau_0,\tau_1]}[(1 - S_E(t))/(1 - S_C(t))]$, respectively. Freitag, Lange, and Munk[27] used a hybrid bootstrap-based procedure based on that used by Shao and Tu[28] to construct a confidence interval for $\sup_{t\in[\tau_0,\tau_1]}[S_C(t) - S_E(t)]$ for testing the hypotheses in Expression 13.18. When the upper limit of the confidence interval is less than the non-inferiority margin/threshold, the null hypothesis is rejected and non-inferiority is concluded. This procedure maintains the desired type I error rate and the supremum approach has more power than the pointwise approach.

Comparing the event-free probabilities over an interval makes use of more information in the data than a landmark analysis. As with landmark analyses, it is not necessary to assume proportional hazards, the existence of a scale factor, or proportional cumulative odds. When such an assumption holds (or approximately holds), it is more efficient to base the inference on a procedure that is designed for such an assumption than to restrict the inference to some prespecified interval.

The selected non-inferiority margin or threshold here may represent the maximal allowed difference across $[\tau_0, \tau_1]$, which may be a larger allowed difference than for a landmark analysis at a specific time. When the same margin or threshold is used for the non-inferiority analysis over $[\tau_0, \tau_1]$ as for a non-inferiority landmark analysis at landmark $t^* \in [\tau_0, \tau_1]$, rejecting the null hypothesis in Expressions 13.16 or 13.18 for the interval analysis implies that the null hypothesis in Expression 13.12 is rejected for the landmark analysis.

Besides the need to choose a larger margin than for a landmark analysis, it can be very tricky in using the historical results to determine the effect of the control therapy. The specific information on the estimates of the event-free probabilities over $[\tau_0, \tau_1]$ and their corresponding standard errors for the control therapy and the placebo may not be readily available from some or all of the historical trials. If such information was not readily available but it is still desired to base the non-inferiority inference over $[\tau_0,\tau_1]$, it is likely that the non-inferiority margin would be conservatively chosen.

It may also be difficult (as with landmark analyses) in determining how to incorporate differences in estimated $S_C(t)$ across trials to determine the appropriate/best interval $[\tau_0,\tau_1]$ to consider and to set the non-inferiority margin/threshold. There are analogous concerns and issues when the non-inferiority inference over $[\tau_0,\tau_1]$ is based on the cumulative odds ratio, ratio of the cumulative hazards, ratio of event-free probabilities, or relative risk of an event.

When the assumptions do not hold for proportional hazards, the existence of a scale factor, or a constant cumulative odds ratio, the corresponding sample estimator unbiasedly estimates some quantity that depends on the amount of follow-up (i.e., the censoring distributions) for that trial. Thus, the underlying time-to-event distributions for the control therapy and the placebo can remain constant across trials (across historical trials and the non-inferiority trial), thereby having a constant true effect of the control therapy across trials; however, because the assumption relating the underlying distributions is not true (e.g., the hazards are not proportional) and the amount of follow-up differs across trials, the value the selected estimator (e.g., the hazard ratio estimator) is unbiasedly estimating varies across the trials. Landmark analyses and analyses over an interval would not be affected by differences across trials in the amount of follow-up, although they have their own issues.

*Horizontal Differences in the Survival Functions.* The difference in medians is one specific horizontal difference in the experimental and control survival functions (i.e., $S_E^{-1}(0.5) - S_C^{-1}(0.5)$). For continuous time-to-event distributions, the difference in the means is the average of the horizontal differences in the experimental and control survival functions over all percentiles (i.e., the average of $S_E^{-1}(p) - S_C^{-1}(p)$ for $0 < p < 1$), or simply the mean difference in percentiles. Thus, the difference in means (when the means exist) is given by

$$\mu_E - \mu_C = \int_0^1 (S_E^{-1}(y) - S_C^{-1}(y))dy \qquad (13.19)$$

Graphically, the difference in means is equal to the area between the two survival functions, which is usually represented by

$$\mu_E - \mu_C = \int_0^\infty (S_E(x) - S_C(x))dx \qquad (13.20)$$

When the means exist, the expression in Equation 13.19 can be extended to continuous real-valued distributions to $\mu_E - \mu_C = \int_0^1 (F_C^{-1}(y) - F_E^{-1}(y))dy$. The assumption of continuous distributions is necessary for Equation 13.19 to hold. Therefore, Kaplan–Meier estimates of $S_E$ and $S_C$, which are discrete distributions, cannot directly be substituted into Equation 13.19 to produce an estimate of $\mu_E - \mu_C$. When the Kaplan–Meier estimates of $S_E(t)$ and $S_C(t)$ both approach zero as $t$ increases, the Kaplan–Meier estimates can be substituted into Equation 13.20 to produce an estimate of $\mu_E - \mu_C$.

Alternatively, the same percentiles can be defined by $\hat{S}_i^{-1}(p) = \min\{t : \hat{S}_C(t) \le 1-p\}$ for $i$ = E, C or in some other fashion, and substituted into Equation 13.19 to obtain an estimate of $\mu_E - \mu_C$ when the Kaplan–Meier estimates of the survival functions both approach zero as time increases. In any case, an estimate of $\mu_E - \mu_C$ involves an average of the difference in the $100p$-th sample percentiles for the experimental and control arms, where $p$ is taken uniformly over (0, 1).

Because of the presence of censoring, it is rare that both (or either) Kaplan–Meier survival functions will approach zero as time increases, but rather the Kaplan–Meier survival function will be suspended at some height. Thus, $\mu_E - \mu_C$ will not be estimable. However, a difference in trimmed means may still be estimable. For $i$ = E, C, the $\gamma$-trimmed mean is defined by $\mu_{i,\gamma} = \int_{\gamma}^{1-\gamma} S_i^{-1}(y)\,dy/(1-2\gamma)$ for $0 < \gamma < 0.5$. To estimate the difference in the $\gamma$-trimmed means, it is necessary that both Kaplan–Meier survival function eventually fall below $\gamma$. The chosen $\gamma$ and the intended follow-up in the trial must allow for this. Bootstrapping may be able to provide confidence intervals for $\mu_{E,\gamma} - \mu_{C,\gamma}$; however, this requires that every (or almost every) bootstrap replication has both Kaplan–Meier survival functions eventually fall below $\gamma$.

When statistical software is used to plot a Kaplan–Meier survival curve, it is standard that the plot connects with vertical line segments the drops in curve at the event times. This gives the impression of a continuous curve. From such a representation, when the Kaplan–Meier survival curve eventually falls below $\gamma$, the area between the $100\gamma$th and the $100(1 - \gamma)$th percentiles can be found, which when divided by $(1 - 2\gamma)$ yields an estimate of the $\gamma$-trimmed mean. An estimate of the difference in the trimmed means would be an average of the difference in the $100p$-th sample percentiles for the experimental and control arms where $p$ is taken uniformly over $(\gamma, 1 - \gamma)$.

---

## References

1. Chi, G.Y.H., Some issues with composite endpoints in clinical trials, *Fund. Clin. Pharm.*, 19, 609–619, 2005.
2. DeMets, D.L. and Califf, R.M., Lessons learned from recent cardiovascular clinical trials: Part I, *Circulation*, 106, 746–751, 2002.
3. Montori, V.M. et al., Validity of composite endpoints in clinical trials, *Br. Med. J.*, 330, 594–596, 2005.

4. Kleist, P., Composite endpoints: Proceed with caution, *Appl. Clin. Trial*, May 1, 2006, at http://appliedclinicaltrialsonline.findpharma.com/appliedclinicaltrials/article/articleDetail.jsp?id=324331.
5. Cox, D.R., Some simple approximate tests for Poisson variates, *Biometrika*, 40, 354–360, 1953.
6. Cox, D.R., Regression models and life tables, *J. R. Stat. Soc.*, 34, 187–220, 1972.
7. Cox, D.R., Partial likelihood, *Biometrika*, 62, 269–276, 1975.
8. Com-Nougue, C., Rodary, C., and Patte, C., How to establish equivalence when data are censored: A randomized trial of treatments for B non-Hodgkin lymphoma, *Stat. Med.*, 12, 1353–1364, 1993.
9. Efron, B., Efficiency of Cox's likelihood function for censored data, *J. Am. Stat. Assoc.*, 72, 557–565, 1977.
10. Fleming, T.R., Evaluation of active control trials in AIDS, *J. Acq. Immun. Def. Synd.*, 2, S82–S87, 1990.
11. Kalbfleisch, J.D. and Prentice, R.L., Estimation of the average hazard ratio, *Biometrika*, 68, 105–112, 1981.
12. Fleming, T.R. and Harrington, D., *Counting Processes and Survival Analysis*, Wiley, Chichester, 1991.
13. Crisp, A. and Curtis, P., Sample size estimation for non-inferiority trials of time-to-event data, *Pharm. Stat.*, 7, 236–244, 2008.
14. Cox, D.R., A note on the graphical analysis of survival data, *Biometrika*, 66, 188–190, 1979.
15. Nelson, W., Theory and application of hazard plotting for censored failure data, *Technometrics*, 14, 945–966, 1972.
16. Hettmansperger, T.P., Two-sample inference based on one-sample sign statistics, *J. R. Stat. Soc. C Appl.*, 33, 45–51, 1984.
17. Wang, J.-L. and Hettmansperger, T.P., Two-sample inference for median survival times based on one-sample procedures for censored survival data, *J. Am. Stat. Assoc.*, 85, 529–536, 1990.
18. Su, J.Q. and Wei, L.J., Nonparametric estimation for the difference or ratio of median failure times, *Biometrics*, 49, 603–607, 1993.
19. Kaplan, E.L. and Meier, P., Nonparametric estimation from incomplete observations, *J. Am. Stat. Assoc.*, 53, 457–481, 1958.
20. Thomas, D.R. and Grunkemeier, G.L., Confidence interval estimation of survival probabilities for censored data, *J. Am. Stat. Assoc.*, 70, 865–871, 1975.
21. Efron, B., Censored data and the bootstrap, *J. Am. Stat. Assoc.*, 76, 312–319, 1981.
22. Reid, N., Estimating the median survival time, *Biometrika*, 68, 601–608, 1981.
23. Brookmeyer, R. and Crowley, J., A confidence interval for the median survival time, *Biometrics*, 38, 29–41, 1982.
24. Emerson, J.D., Nonparametric confidence intervals for the median in the presence of right censoring, *Biometrics*, 38, 17–27, 1982.
25. Basawa, I.V. and Koul, H.L., Large-sample statistics based on quadratic dispersion, *Int. Stat. Rev.*, 56, 199–219, 1988.
26. Wei, L.J. and Gail, M.H., Nonparametric estimation for a scale-change with censored observations, *J. Am. Stat. Assoc.*, 78, 382–388, 1983.
27. Freitag, G., Lange, S. and Munk, A., Non-parametric assessment of non-inferiority with censored data, *Stat. Med.*, 25, 1201–1217, 2006.
28. Shao, J. and Tu, D., *The Jackknife and Bootstrap*, Springer, New York, NY, 1995.

# Appendix: Statistical Concepts

## A.1 Frequentist Methods

P-values and confidence intervals are used to evaluate the evidence in the data. A $p$-value measures the strength of evidence against a null hypothesis, while a confidence interval is an interval that contains only those possible values for a parameter that have not been statistically ruled out by the data.

### A.1.1 $p$-Values

A $p$-value is the probability of obtaining results as extreme or more extreme (against the null hypothesis) than the observed results, where the probability is determined under the assumption that the null hypothesis is true. For most cases, when the null hypothesis is true, the $p$-value is a completely random value between 0 and 1, its statistical distribution being a uniform distribution over (0,1). As commonly applied, the null hypothesis is rejected if and only if the $p$-value is less than or equal to the significance level. Hence, the $p$-value can be regarded as the smallest significance level for which the null hypothesis is rejected.

A $p$-value measures the strength of evidence against the null hypothesis in the direction or directions of the alternative hypothesis. The smaller a $p$-value, the stronger is the evidence against the null hypothesis, in favor of the alternative hypothesis. A large $p$-value would correspond to little evidence against the null hypothesis. Little or no evidence against the null hypothesis does not mean that there is great evidence for the null hypothesis.

Examples A.1 through A.3 illustrate some properties of $p$-values. These examples involve dichotomous data (coin tosses), continuous data (hemoglobin levels), and time-to-event data (for an undesirable event).

### Example A.1

We will consider a simple experiment of tossing a coin 10 times. Let $p$ denote the probability that any given toss results as a head. The coin is fair if $p = 0.5$. For the null hypothesis of $p = 0.5$, there are three realistic possibilities for the alternative hypothesis: $p < 0.5$, $p > 0.5$, and $p \neq 0.5$. Suppose eight of these tosses result in a head. Table A.1 summarizes the $p$-value in each of three cases. This example helps illustrate the differences among the three cases in the directions of the strength of

**TABLE A.1**

Summary of $p$-Values for Three Cases Involving Dichotomous Data

| Case | Null Hypothesis | Alternative Hypothesis | Result of the Experiment | As Strong or Stronger Evidence in Favor of $H_a$ | $p$-Value[a] |
|------|----------------|------------------------|--------------------------|--------------------------------------------------|--------------|
| 1 | $H_0: p = 0.5$ | $H_a: p < 0.5$ | 8 heads in 10 tosses | 8 or fewer heads in 10 tosses | 0.989 |
| 2 | $H_0: p = 0.5$ | $H_a: p > 0.5$ | 8 heads in 10 tosses | 8 or more heads in 10 tosses | 0.055 |
| 3 | $H_0: p = 0.5$ | $H_a: p \neq 0.5$ ($p < 0.5$ or $p > 0.5$) | 8 heads in 10 tosses | 8 or more heads in 10 tosses, or 2 or fewer heads in 10 tosses | 0.109 |

[a] $p$-Values as fractions are 1013/1024, 56/1024, and 112/1024, respectively.

evidence. In case 1, the smaller the number of heads, the stronger is the evidence against the null hypothesis in favor of the alternative hypothesis. In case 2, the larger the number of heads, the stronger is the evidence against the null hypothesis in favor of the alternative hypothesis. In case 3, the further the number of heads is from five (50% of the number of tosses), the stronger the evidence against the null hypothesis in favor of the alternative hypothesis. Note also that the $p$-value in case 3 is double the $p$-value in case 2 (double the smaller of the $p$-values in cases 1 and 2). In case 3, two or fewer heads among 10 tosses provide as strong or stronger evidence against $p = 0.5$ in favor of $p < 0.5$ as the strength of eight heads among 10 tosses provides against $p = 0.5$ in favor of $p > 0.5$. If $p = 0.5$, for 10 tosses, the probability of getting two or fewer heads equals the probability of getting eight or more heads.

In this coin-tossing example, the test is based on the number of heads in 10 tosses, which is referred to as the *test statistic*. The $p$-values in cases 1 and 2 are referred to as one-sided $p$-values since the respective alternative hypotheses are one-sided. Likewise, since the alternative hypothesis in case 3 is two-sided, the respective $p$-value is referred as a two-sided $p$-value. Note that, here, the sum of the one-sided $p$-values equals 1 plus the probability of getting the observed number of heads if the null hypothesis is true. Whenever the test statistic has a discrete distribution, the sum of the one-sided $p$-values will equal 1 plus the probability of getting the observed value of the test statistic if the null hypothesis is true.

## Example A.2

Suppose it is known that the measured hemoglobin levels of patients having a particular disease are normally distributed with standard deviation of 0.8. Let $\mu$ denote the mean hemoglobin level of the target population. For the null hypothesis of $\mu = 11$, consider each of the possibilities for the alternative hypothesis of $\mu < 11$, $\mu > 11$, and $\mu \neq 11$. Suppose that the sample mean hemoglobin level for a random sample of 4 patients is 10.5. Table A.2 summarizes the $p$-value in each of three cases. There are various similarities between the examples illustrated in Tables A.1 and A.2. One difference is that the test statistic in hemoglobin level example has a continuous distribution. When the test statistic has a continuous distribution, the sum of the one-sided $p$-values equals 1.

**TABLE A.2**

Summary of $p$-Values for Three Cases Involving Continuous Data

| Case | Null Hypothesis | Alternative Hypothesis | Result of the Experiment | As Strong or Stronger Evidence in Favor of $H_a$ | $p$-Value |
|------|-----------------|------------------------|--------------------------|------------------------------------------------|-----------|
| 1 | $H_0: \mu = 11$ | $H_a: \mu < 11$ | Sample mean from 4 patients is 10.5 | Sample mean from 4 patients is 10.5 or less | 0.106 |
| 2 | $H_0: \mu = 11$ | $H_a: \mu > 11$ | Sample mean from 4 patients is 10.5 | Sample mean from 4 patients is 10.5 or more | 0.894 |
| 3 | $H_0: \mu = 11$ | $H_a: \mu \neq 11$ ($\mu < 11$ or $\mu > 11$) | Sample mean from 4 patients is 10.5 | Sample mean from 4 patients is either 10.5 or less, or is 11.5 or more | 0.211 |

Example A.3 will compare and contrast the calculation of a $p$-value for each of four types of comparisons. For this example, an equivalence comparison will be evaluated in Section A.3.

## Example A.3

Let $\theta$ denote the true experimental arm versus control arm hazard ratio of some undesirable event (e.g., death or disease progression). For an observed hazard ratio of 0.91 based on 400 events in a clinical trial that had a one-to-one randomization, Table A.3 summarizes the $p$-value for each comparison type. For the non-inferiority comparison, a hazard ratio threshold of 1.1 is used.

For the inferiority, superiority, and difference comparisons, the orderings of the strength of evidence against the null hypothesis (in favor of the alternative hypothesis) are analogous to cases 1, 2, and 3 in each of Tables A.1 and A.2.

Note that the order of the strength of evidence is the same for a superiority comparison as with a non-inferiority comparison. For each of these comparisons, the smaller the observed hazard ratio, the more favorable is the result for the experimental arm. For these two comparisons, it is the same event (observing a hazard ratio of 0.91 or less) whose probability is the $p$-value. The $p$-values are different because the probabilities are calculated under different assumptions of the truth ($\theta = 1$ and $\theta = 1.1$). In fact, because the "bar is lower" for a non-inferiority comparison than for a superiority comparison between the same two treatment arms, the $p$-value for the non-inferiority comparison will always be smaller than the $p$-value for a superiority comparison.

Note that had the observed hazard ratio equaled 1, the $p$-values for an inferiority comparison and for a superiority comparison would be equal (both $p$-values equaling 0.5). In this example, the $p$-values for an inferiority comparison and for a non-inferiority comparison would be equal (both $p$-values approximately 0.317) if the observed hazard ratio were the square root of 1.1 (the geometric mean of 1 and 1.1). When these $p$-values are equal, the strength of evidence in favor of inferiority equals the strength of evidence in favor of non-inferiority.

**TABLE A.3**

Summary of p-Values for Various Types of Comparisons Involving Time-to-Event Data

| Case | Null Hypothesis | Alternative Hypothesis | Result of the Experiment | As Strong or Stronger Evidence in Favor of $H_a$ | p-Value[a] |
|---|---|---|---|---|---|
| Inferiority | $H_o: \theta = 1$ or $H_o: \theta \leq 1$ | $H_a: \theta > 1$ | Observed hazard ratio of 0.91 based on 400 events | Observed hazard ratio based on 400 events of 0.91 or greater | 0.827 |
| Superiority | $H_o: \theta = 1$ or $H_o: \theta \geq 1$ | $H_a: \theta < 1$ | Observed hazard ratio of 0.91 based on 400 events | Observed hazard ratio based on 400 events of 0.91 or less | 0.173 |
| Difference | $H_o: \theta = 1$ | $H_a: \theta \neq 1$ | Observed hazard ratio of 0.91 based on 400 events | Observed hazard ratio based on 400 events is either 0.91 or less, or is 1/0.91 or greater | 0.346 |
| Non-inferiority | $H_o: \theta = 1.1$ | $H_a: \theta < 1.1$ | Observed hazard ratio of 0.91 based on 400 events | Observed hazard ratio based on 400 events of 0.91 or less | 0.029 |

[a] p-Values are approximated by using the asymptotic distribution of the estimator of the log-hazard ratio for a one-to-one randomization.

There is fairly suggestive but compelling evidence that the experimental arm is noninferior to the control arm with respect to the time-to-event endpoint. For any significance level less than 0.10, whether a one-sided or two-sided significance level, there is not strong enough evidence that the experimental arm is inferior, superior, or different from the control arm with respect to the time-to-event endpoint.

When the correct value for a parameter or effect is the hypothesized value in the null hypothesis and the test statistic has a continuous distribution, the $p$-value is a random value between 0 and 1, its statistical distribution being a uniform distribution over (0,1). For most hypotheses-testing scenarios in practice, when the correct value is among the alternative hypothesis, the distribution for the $p$-value is smaller. Several authors have examined the distribution of the $p$-value when the alternative hypothesis is true. Dempster and Schatzoff[1] and Schatzoff[2] investigated the stochastic nature of the $p$-value and evaluated test procedures based on the expected (mean) $p$-value at a given alternative. Hung et al.[3] determined, for a fixed significance level and a fixed difference between the true value and the hypothesized value in the null hypothesis, that as the sample size increases, the mean, variance, and percentiles for the distribution of the $p$-value decrease toward zero. They also examined the distribution of the $p$-value in certain cases when the effect size is a random variable. Sackrowitz and Samuel-Cahn[4] extended the work of Dempster and Schatzoff,[1] and also related the expected $p$-value to the significance level and power. Joiner[5] introduced the median significance level and the "significance level of the average" (the significance level that corresponds with the mean value of the test statistic) as measures of test efficiency. Bhattacharya and Habtzghi[6] also used the median $p$-value to evaluate the performance of a test. Below, we provide our own analogous derivation for the distribution of the $p$-value.

In general, the distribution of the $p$-value depends on the sample size and the true value of the parameter (or alternatively on the significance level and the true power of the test). For test statistics that are normally distributed, the distribution of the $p$-value depends on the number of standard error difference in the true value and the hypothesized value in the null hypothesis. For a random sample of size $n$ from a normal distribution with mean $\mu_a$ and standard deviation $\sigma$, we will see that the distribution of the $p$-value for testing the null hypothesis $\mu = \mu_0$ against the alternative hypothesis $\mu < \mu_0$ depends on the value of $(\mu_a - \mu_0)/(\sigma/\sqrt{n})$. We can replace $\mu$, $\sigma$, and $n$ by the equivalent when comparing two means or when using a log-hazard ratio to compare two time-to-event distributions. For test statistics that are modeled as having a normal distribution, when the power is $1 - \beta$ with a one-sided significance of $\alpha$, the number of standard error difference in the true value and the hypothesized value in the null hypothesis reduces to $z_\alpha + z_\beta$.

Suppose we are testing $H_0$: $\theta = \theta_0$ versus $H_a$: $\theta < \theta_0$ on the basis of an estimate $\hat{\theta}$, where the true value is $\theta_a$ and $(\hat{\theta} - \theta_a)/\sigma'$ is modeled as having a standard normal distribution. The test statistic is $(\hat{\theta} - \theta_0)/\sigma'$, and thus the $p$-value is the observed value of $\Phi((\hat{\theta}-\theta_0)/\sigma')$. Let $G$ denote the distribution function for the $p$-value. Then for $0 < w < 1$,

$$G(w) = P(\Phi((\hat{\theta}-\theta_0)/\sigma' < w \,|\, \theta = \theta_a)$$
$$= P((\hat{\theta}-\theta_a)/\sigma' < \Phi^{-1}(w) + (\theta_0 - \theta_a)/\sigma' \,|\, \theta = \theta_a)$$
$$= \Phi(\Phi^{-1}(w) + (\theta_0 - \theta_a)/\sigma')$$
$$= \Phi(\Phi^{-1}(w) + (z_\alpha + z_\beta))$$

where $\alpha$ is the significance level and $1 - \beta$ is the power when the true value of $\theta$ is $\theta_a$. For $0 < y < 1$, the quantile function is given by

$$G^{-1}(y) = \Phi(\Phi^{-1}(y) - (z_\alpha + z_\beta))$$

Since $\Phi^{-1}(p) = z_{1-p}$ for $0 < p < 1$, the $100p$-th percentile of the distribution of the $p$-value is given by $\Phi(z_{1-p} - (z_\alpha + z_\beta))$. Note for any significance level $\alpha$, the $100(1 - \alpha)$-th percentile for the $p$-value is $\beta$ (i.e., 1 minus the power). Also, the $100(1 - \beta)$-th percentile for the $p$-value is $\alpha$. For $0 < w < 1$, the density function for the $p$-value is given by

$$g(w) = \phi(\Phi^{-1}(w) + (z_\alpha + z_\beta))/\phi(\Phi^{-1}(w))$$

We note that it can easily be shown that the distribution of the $p$-value becomes larger with respect to a likelihood ratio ordering as $z_\alpha + z_\beta$ becomes smaller. In particular, for a fixed significance level, $\alpha$, the distribution of the $p$-value becomes smaller with respect to a likelihood ratio ordering when the power increases (which can occur by either increasing the sample size or choosing a more favorable alternative). Thus, for a fixed sample size, when comparing two alternatives, the relative likelihood increases in favor of the more favorable alternative as the smaller the observed $p$-value.

Note also that the test statistic has a normal distribution with mean $-(z_\alpha + z_\beta)$ and variance 1.

For cases where the test statistic is normally distributed, Table A.4 provides the median, 5th percentile, and 95th percentile for the distribution of the $p$-value for various combinations for the significance level and power.

Whenever the power at the true effect size is 80% or greater, the median $p$-value is very small and relatively much smaller than the significance level. If a clinical trial is adequately powered at the actual effect size, the $p$-value will typically be very small. An observed $p$-value that is microscopic (e.g., smaller than $10^{-8}$ if the significance level is 0.005) would tend to be indicative of an overpowered study—that is, the study would have had near 100% power at the true effect size.

**TABLE A.4**

Median and Percentiles for *p*-Value Based on Significance Level and Power

| | | Distribution for *p*-Value | | |
|---|---|---|---|---|
| Significance Level[a] | Power | Median | 5th Percentile | 95th Percentile |
| 0.05 | 0.05 | 0.5 | 0.05 | 0.95 |
| | 0.5 | 0.05 | 0.0005 | 0.5 |
| | 0.8 | 0.0064 | 0.00002 | 0.2 |
| | 0.9 | 0.0017 | 0.000002 | 0.1 |
| 0.025 | 0.025 | 0.5 | 0.05 | 0.95 |
| | 0.5 | 0.025 | 0.0002 | 0.376 |
| | 0.8 | 0.0025 | 0.000004 | 0.124 |
| | 0.9 | 0.0006 | 0.0000005 | 0.055 |
| 0.01 | 0.01 | 0.5 | 0.05 | 0.95 |
| | 0.5 | 0.01 | 0.00004 | 0.248 |
| | 0.8 | 0.0008 | 0.0000007 | 0.064 |
| | 0.9 | 0.0002 | 0.00000007 | 0.025 |
| 0.005 | 0.005 | 0.5 | 0.05 | 0.95 |
| | 0.5 | 0.005 | 0.00001 | 0.176 |
| | 0.8 | 0.0003 | 0.0000002 | 0.038 |
| | 0.9 | 0.00006 | 0.00000002 | 0.013 |

[a] One-sided significance level.

Although *p*-values larger than the significance level are not out of the ordinary when the power at the true effect size is 80% or greater, large *p*-values are out of the ordinary. Large *p*-values are indicative of either the alternative hypothesis being false or that the study is not adequately powered at the true effect size (i.e., the assumed effect size is greater than the true effect size). In the latter case, the effect size chosen to design the study ("power the study") was larger than the true effect size.

Note that reporting a *p*-value for non-inferiority testing is rare. This is primarily due to some subjectivity in the determination of the non-inferiority margin.

## A.1.2 Confidence Intervals

The parameter space for a parameter is the set of all values that the parameter could theoretically be. Some of the values in the parameter space may not be practical. A confidence interval for a parameter is an interval based on data that contain only those values in the parameter space for which the data have not statistically ruled out at some significance level, $\alpha$. Since the probability of "statistically ruling out" the correct value of the parameter is at most $\alpha$, we are thus at least $100(1 - \alpha)\%$ confident that the true value of the parameter is contained in the confidence interval. For example, when $\alpha = 0.05$ for each of a large number of experiments (or repetitions of the same experiment), about 95% of the 95% confidence intervals actually capture the correct value of the respective parameter. The value of $(1 - \alpha)$ is called the confidence coefficient.

For each different choice of an alternative hypothesis as presented in Tables A.1 and A.2, there is a different type of confidence interval. For some real-valued parameter $\theta$ and significance level $\alpha$, $\theta_0$ values where $H_0: \theta = \theta_0$ is not rejected in favor of $H_a: \theta < \theta_0$ form a $100(1 - \alpha)\%$ confidence interval for $\theta$ of the form $(-\infty, U)$. The value for $U$ is referred to as the $100(1 - \alpha)\%$ confidence upper bound for $\theta$. Analogously, $\theta_0$ values where $H_0: \theta = \theta_0$ is not rejected in favor of $H_a: \theta > \theta_0$ at a significance level $\alpha$ form a $100(1 - \alpha)\%$ confidence interval for $\theta$ of the form $(L, \infty)$. The value for $L$ is referred to as the $100(1 - \alpha)\%$ confidence lower bound for $\theta$. These two types of confidence intervals are sometimes referred to as "one-sided confidence intervals" since they are based on tests of one-sided alternative hypotheses.

Values for $\theta_0$ where $H_0: \theta = \theta_0$ is not rejected in favor of $H_a: \theta \neq \theta_0$ at a significance level $\alpha$ form a $100(1 - \alpha)\%$ confidence interval for $\theta$ of the form $(L, U)$. Such confidence intervals are sometimes referred to as "two-sided confidence intervals" since they are based on tests of two-sided alternative hypotheses. In clinical trials, the lower limit and upper limits, $L$ and $U$, of the two-sided $100(1 - \alpha)\%$ confidence interval for $\theta$ that tend to be used are the $100(1 - \alpha/2)\%$ confidence lower bound for $\theta$ and the $100(1 - \alpha/2)\%$ confidence upper bound for $\theta$, respectively. This allows the selected two-sided confidence interval to be "error symmetric." That is, before any data occur, the probability that the two-sided $100(1 - \alpha)\%$ confidence interval for $\theta$ will lie entirely above the true value equals the probability that the two-sided $100(1 - \alpha)\%$ confidence interval for $\theta$ will lie entirely below the true value.

The factors that influence the width of a confidence interval depend on the type of parameter of interest. For a mean of some characteristic within a study arm, the width of the confidence interval depends on the sample size, the variability between patients on that characteristic, and the confidence coefficient. For a treatment versus control log-hazard ratio of some time-to-event endpoint, the width of the confidence interval depends on the breakdown of the number of events (or the total number of events and the randomization ratio) and the confidence coefficient. In general terms, the width of a confidence interval will depend on some quantification of the amount of evidence or information gathered and the confidence coefficient. Increasing the amount of evidence gathered on the correct value of a parameter $\theta$ (increasing the sample size or the number of events) reduces the width of the confidence interval. Also, increasing the confidence coefficient increases the width of the confidence interval (e.g., a 90% confidence interval is wider than the corresponding 80% confidence interval).

Table A.5 summarizes various confidence interval formulas for large sample sizes (event sizes). Here, for some quantitative characteristic, $\bar{x}_E$ and $\bar{x}_C$ denote the sample means in the experimental and control arms, respectively; $s_E$ and $s_C$ denote the respective sample standard deviations in the

## TABLE A.5

Formulas for Approximate $100(1 - \alpha)\%$ Confidence Intervals for Particular Parameters of Interest

| Parameter of Interest | Approximate $100(1 - \alpha)\%$ Confidence Interval |
|---|---|
| Single mean ($\mu_E$) | $\bar{x}_E \pm z_{\alpha/2} s_E / \sqrt{n_E}$ |
| Difference in means ($\mu_E - \mu_C$) | $(\bar{x}_E - \bar{x}_C) \pm z_{\alpha/2} \sqrt{\dfrac{s_E^2}{n_E} + \dfrac{s_C^2}{n_C}}$ |
| Single proportion ($p_E$) | $\hat{p}_E \pm z_{\alpha/2} \sqrt{\dfrac{\hat{p}_E(1 - \hat{p}_E)}{n_E}}$ |
| Difference in proportions ($p_E - p_C$) | $(\hat{p}_E - \hat{p}_C) \pm z_{\alpha/2} \sqrt{\dfrac{\hat{p}_E(1 - \hat{p}_E)}{n_E} + \dfrac{\hat{p}_C(1 - \hat{p}_C)}{n_C}}$ |
| Log hazard ratio ($\theta$) | $\hat{\theta} \pm z_{\alpha/2} \sqrt{\dfrac{1}{r_E} + \dfrac{1}{r_C}}$ |

experimental and control arms, respectively; and $\mu_E$ and $\mu_C$ denote the actual or underlying means for the experimental and control arms, respectively.

For a dichotomous characteristic, where the possibilities will be expressed as "success" or "failure," let $\hat{p}_E$ and $\hat{p}_C$ denote the sample proportions of "successes" in the experimental and control arms, respectively, and $p_E$ and $p_C$ denote the actual probability of a "success" for the experimental and control arms, respectively.

For a time-to-event endpoint, let $\theta$ denote the true experimental arm versus control arm log-hazard ratio and let $\hat{\theta}$ denote its estimate based on $r_E$ and $r_C$ events on the experimental and control arms, respectively. Furthermore, let $n_E$ and $n_C$ denote the sample sizes for the experimental and control arms, respectively, and let the $100(1 - \gamma)$-th percentile of a standard normal distribution be denoted by $z_\gamma$.

The confidence intervals in Table A.5 are all of the same form—the estimate plus or minus the corresponding *standard error* for the estimator multiplied by a standard normal value, which represents the number of standard errors that the estimate and the parameter will be within each other $100(1 - \alpha)\%$ of the time. The standard error is the square root of the average squared distance between the estimator of the parameter and the actual value of the parameter.

As can be seen from Table A.5, the confidence interval for a difference in means (or proportions) is not determined by manipulating the individual confidence intervals for each mean (proportion). The use of separate confidence intervals is conservative in determining whether we can rule out that the two true means are equal. Each separate confidence interval reflects, for that arm only, the possibilities for the true mean where it was not out of the ordinary to observe the data that was observed or more extreme data. The confidence interval for the difference reflects possibilities for the difference in means for which it was not out of the ordinary to observe that collective data from both arms or more extreme data.

Suppose the 95% confidence interval for the mean of one arm is (2–8) and the 95% confidence interval for the mean of the other arm is (5–11). The 95% confidence interval for the difference in means is (–0.43, 6.43). For the first (second) arm, it is not out of the ordinary to observe the data for that arm if the true mean was 2.5 (10.5). However, it would be out of the ordinary to observe the collective data of both arms if the difference in the true means was 8.

The standard error of the estimator of the log-hazard ratio depends on the randomization ratio, the total number of events, and the true hazard ratio. For a one-to-one randomization, when the true hazard ratio is not far from 1, the standard error is approximately 2 divided by the square root of the total number of events. For a fixed total number of events, this provides a specific relationship between the upper limit and lower limits of a confidence interval for the hazard ratio and the hazard ratio estimate. For example, for a one-to-one randomization where there are 400 events, the upper limit of the 95% confidence interval for the hazard ratio should be about 22% greater than the estimate of the hazard ratio, which in turn should be about 22% greater than the lower limit of that 95% confidence interval.

A frequent mistake in calculating confidence intervals is applying asymptotic methods when the sample size is not large enough for the assumptions to approximately hold. The confidence intervals will then not have a level approximately equal to the desired level. The delta method has been applied to many functions where it would take rather large sample sizes for the asymptotic results to approximately hold. Care should be taken when applying such methods.

### A.1.3 Comparing and Contrasting Confidence Intervals and *p*-Values

For standard testing of a difference between two means, or proportions (or other measures of location), the $100(1 - \alpha)\%$ confidence interval for the difference will exclude zero exactly when the two-sided *p*-value for testing for a difference is less than $\alpha$. A *p*-value and a confidence interval present different concepts about the strength of evidence in the data. Consider two clinical trials comparing an experimental therapy with a control therapy with respect to a time-to-event endpoint of an undesirable event where each trial has a one-to-one randomization. Suppose the first trial has an experimental versus control hazard ratio of 0.7 based on 500 events and the second trial has an experimental versus control hazard ratio of 0.6 based on 200 events. For the first trial, the two-sided *p*-value for testing for a difference is .00007 and the 95% confidence interval for the true hazard ratio is (0.59, 0.83). For the second trial the two-sided *p*-value for testing for a difference is 0.0003 and the 95% confidence interval for the true hazard ratio is (0.45, 0.79). The first trial provides stronger evidence than the second trial that the experimental arm gives longer times to the event than the control arm. However, the second trial rules out more possibilities for the hazard ratio away from equality (0.79, 1) than the first trial (0.83, 1).

The use of hypotheses tests and $p$-values has been viewed by some as dichotomizing the results as either "successful" or "unsuccessful," and that the role of a clinical trial should be to get precise estimates of the effect of the experimental therapy relative to the control therapy. Precise estimates would require a particular maximum standard error for the estimate and maximum width of the confidence interval. This would require studies of some minimal sample size.

In hypotheses testing, the conclusion of the alternative hypothesis of one trial is reproduced by another trial, if that other trial also reached the conclusion of the same alternative hypothesis. In such a case, for each trial, the $p$-value is smaller than the respective significance level. In practice, for two superiority trials, a one-sided significance level of 0.025 would be used for each trial. For confidence intervals, it may be unclear what is meant by a "reproduced finding." Is a reproduced finding getting similar confidence intervals from each study or confidence intervals that have a great amount of overlap with respect to their widths?

### A.1.4 Analysis Methods

Non-inferiority testing has been understood mostly as a frequentist testing exercise. The Neyman–Pearson framework has been the basis for frequentist testing. In this section we describe the motivation of Neyman–Pearson testing and discuss Fisherian significance testing (randomization-based testing) as well.

#### *A.1.4.1 Exact and Permutation Methods*

We use the term "permutation test" for tests of continuous data that are based on rerandomization procedures and "exact test" for tests of binary or categorical data that are based on procedures that do not require asymptotic distributional assumptions, while acknowledging some overlap in the two sets of tests. Both permutation and exact tests are useful when sample sizes are small (including sparse cells in categorical data) and when there is insufficient reason to believe that assumptions required for normal theory tests are met.

*Permutation Tests for Continuous Data.* Significance testing was developed using permutation methods,[7] but these are not common in testing of non-inferiority. Assuming only that subjects are randomly assigned to treatment, a significance test of the null hypothesis of no effect of treatment (i.e., a null hypothesis from a superiority study) can be performed by permutation methods. By generating alternative randomization lists using the same mechanism as was used for the actual randomization procedure, one can generate alternate possible realizations of the data, which form a sampling distribution: the set of treatment assignments actually used is but one possible allocation of treatments to subjects, and within the framework of the permutation process any allocation is equally likely to have been used. Comparing the observed data with the sampling distribution allows an experimenter to decide whether the observed data are consistent with other possible permutations or whether the observed data are unusual. If the null hypothesis of no difference is true, then the distribution for the observed value of the test statistic can often be determined or approximated. This allows for the simpler calculation of *p*-values and confidence intervals.

The mechanics of permutation tests for testing hypotheses of non-inferiority with continuous data are straightforward. Similar to a test described by Hollander and Wolfe,[8] testing a hypothesis involving the difference in means $H_0$: $\mu_C - \mu_E \geq \delta$ can be accomplished by subtracting $\delta$ from every outcome in the active control group, resulting in $X_C^T = X_C - \delta$. Testing the null hypothesis $H_0$: $\mu_C - \mu_E \geq \delta$ is identical to testing the null hypothesis $H_0$: $\mu_C^T - \mu_E \geq 0$, where $\mu_C^T = \mu_C - \delta$.

One criticism of this non-inferiority permutation test for continuous data is that the resulting test seems to implicitly assume that the effect of the control treatment (under $H_0$) is exactly $\delta$ on each study subject in the population of interest, whereas the interpretation is usually a less stringent requirement of mean effect of $\delta$ on the population. More generally (and more accurately), the residuals must be exchangeable. With residuals defined as $r_C^T = X_C^T - \mu_C^T$ and $r_E = X_E - \mu_E$, the residuals must be identical in distribution to be exchangeable.[9] Equivalently, having $X_C^T$ (a random outcome from the control arm, less $\delta$) and $X_E$ (a random outcome from the experimental arm) equal in distribution (i.e., the distribution of $X_C$ is shifted $\delta$ to the right of the distribution of $X_E$) is a sufficient condition for the permutation test to be valid. The same condition holds for permutation tests of null hypotheses of no difference, but it is easier to comprehend no effect of treatment when an experimental product is being compared to placebo; under the null hypothesis, both treatments are inert and any variability is conceptually random. In a non-inferiority clinical trial, it can be more difficult to conclude that the transformed values have exchangeable residuals. Differences in the shapes of the distributions of $X_C$ and $X_E$ can occur for many reasons: the treatments have different variances, one treatment has a bimodal response whereas the other has a unimodal response, or other differences exist. Importantly, it is often impossible to know with certainty whether the shapes of the distributions are identical when planning a study; thus, proposing a permutation test carries inherent risks. In addition, with a sufficient sample size, tests using normal theory tend to approximate permutation tests,[10] making permutation tests unnecessary for many non-inferiority trials. For these reasons, permutation tests are unusual in non-inferiority trials when the primary outcome variable is continuous. See Wiens' study[11] for more discussion.

*Exact Tests for Binary and Categorical Data.* For binary data, exact methods have been proposed in the literature and are commonly used. Unlike rerandomization methods for continuous data, exact methods are easy to interpret for binary data. Again, as one moves from a null hypothesis of no difference to a null hypothesis of a nonzero difference, complications arise.

The idea of exact tests was first introduced by Fisher[12] while he was developing a conditional exact test for comparing two independent proportions in a 2 × 2 table. Fisher's exact test deals with the classical null hypothesis of no difference between the two proportions conditioning on the observed marginal totals. In this case, the marginal totals form a sufficient statistic (i.e., a sufficient quantity from the data on which inferences can be based) for the nuisance parameter (common proportion, $p$). Conditioning on the marginal totals yields a hypergeometric distribution for the number of successes for the experimental group. However, because of the discreteness of the hypergeometric distribution, Fisher's test tends to be overly conservative. For the same problem, Barnard[13] has proposed exact unconditional tests in which all combinations of the unconditional sampling space are considered in constructing the test. The probability is thus spread over more possibilities, providing a test statistic that is less discrete in nature than the test statistic for Fisher's test. As a result, these exact unconditional tests generally offer better power than Fisher's test, although they are computationally more involved.

For testing the hypothesis of non-inferiority where the null space contains a nonzero difference or a nonunity relative risk, a nuisance parameter arises, making the calculation of the exact $p$-values more complicated. As there is no simple sufficient statistic for the nuisance parameter, the conditioning argument does not solve the problem. Exact unconditional methods for non-inferiority testing have been proposed by Chan[14] and Röhmel and Mansmann,[15] in which the nuisance parameter is eliminated using the maximization principle—that is, the exact $p$-value is taken as the maximum tail probability over the entire null space. Because the maximization involves a large number of iterations in evaluating sums of binomial probabilities, the exact unconditional tests are computationally intensive, particularly with large sample sizes. Some exact test procedures for non-inferiority and the associated confidence intervals are currently available in statistical software.

Although most exact tests have been developed for the non-inferiority testing, they can be easily adapted for equivalence testing by reformulating the equivalence hypothesis as two simultaneous non-inferiority hypotheses of opposite direction. Then, equivalence of the two treatments can be proved if both one-sided hypotheses are rejected.[16,17] This approach is also recommended in regulatory environments as indicated in the International Conference on Harmonization E9 Guideline[18] "Biostatistical Principles for Clinical Trials," which states that "Operationally, this (equivalence test) is equivalent to the method of using two simultaneous one-sided tests to test the (composite) null hypothesis that the treatment difference is outside the equivalence margins versus the (composite) alternative hypothesis that the treatment difference is within the margins."

### A.1.4.2 Asymptotic Methods

It is common in frequentist testing for non-inferiority (as well as superiority) studies to rely on Neyman–Pearson inference.

Neyman and Pearson[19] introduced a testing principle requiring both a null hypothesis ($H_o$) and an alternative hypothesis ($H_a$). The likelihood of observing the data is determined under the assumption that the alternative hypothesis is true is divided by the likelihood of observing the data under the assumption that the null hypothesis is true. When this relative likelihood exceeds some prespecified threshold, the null hypothesis is rejected. When the null hypothesis and the alternative hypothesis consist of one distribution each, such a test is the most powerful test among all tests having the same or smaller type I error rate. With this as a foundation, it has been shown that a most powerful test or uniformly most powerful test (most powerful test at every possible alternative) can be constructed in many other hypotheses-testing scenarios. Many test procedures including a $t$ test for the difference in means are test procedures based on relative likelihood. Like the $t$ test, the form for these test procedures has been simplified to compare a test statistic (often "normalized") to a critical value.

In Neyman–Pearson testing, the null hypothesis is either rejected in favor of the alternative hypothesis, or not rejected in favor of the alternative hypothesis, without the calculation of a $p$-value. In practice, it is common to blend these two philosophies: reporting a $p$-value even while testing a null hypothesis against an alternative hypothesis.

In his seminal paper on non-inferiority testing, Blackwelder[20] explicitly relied on Neyman–Pearson testing for null hypothesis $H_o: \mu_C - \mu_E \geq \delta$ versus the alternative $H_a: \mu_C - \mu_E < \delta$. In practice, assumptions can usually be made on the underlying distribution to allow a valid non-inferiority analysis for small sample sizes. Often for small samples, the samples are assumed as coming from normal distributions. In this situation, a marked violation of the assumed normality may be easy to detect but a minor violation, even if enough to question the results, may not be easy to detect. Often, a $t$ test is used assuming equal variances for the underlying distributions, or the degrees of freedom for the test statistic are approximated on the basis of the observed variances. When the sample sizes are not small, a central limit theorem is applied for testing the difference of means. The assumptions necessary for valid use of the central limit theorem are usually much milder than the assumptions necessary for a $t$ test. For discrete data, asymptotic approximations such as the normal approximation to the binomial distribution have historically been much easier to compute than exact tests, for any sample size. With the advances in computer hardware and software, the results from exact tests can be readily determined.

*Asymptotic Tests for Continuous Data.* When the primary endpoint is continuous, asymptotic methods are commonly used. Consider the null and alternative hypotheses of $H_o: \mu_C - \mu_E \geq \delta$ versus $H_a: \mu_C - \mu_E < \delta$. By assuming that the estimators of $\mu_C$ and $\mu_E$, the sample means $\bar{X}_C$ and $\bar{X}_E$, have approximate normal distributions (based on the underlying normality of the data or on a central limit theorem), the test statistic $Z = \dfrac{\bar{X}_C - \bar{X}_E - \delta}{\text{se}(\bar{X}_C - \bar{X}_E)}$ will have an approximate standard normal distribution, where $\text{se}(\bar{X}_C - \bar{X}_E)$ is the standard error of the difference in sample means. For sample sizes of $n_C$ and $n_E$, respectively, and common standard deviation $\sigma$, the standard error will be $\sigma\sqrt{1/n_C + 1/n_C}$. Except for the subtraction of the nonzero $\delta$ in the numerator, this test statistic $Z$ is identical to the test statistic for a test of superiority. The null hypothesis is rejected if $Z < -z_\alpha$ in a one-sided test at the level $\alpha$, where $z_\alpha$ is the $100 \times (1 - \alpha)$ percentile of the standard normal distribution (e.g., 1.645 for a one-sided test with significance level 0.05). A $p$-value can also be calculated. When $\sigma$ is unknown and the samples are from normal distributions, a $t$ test can be used where $\sigma^2$ is estimated by a pooled variance. With the large sample sizes common in non-inferiority clinical trials, it is not necessary to assume either an equal variance or that the samples are from a normal distribution. Rather, $\sqrt{s_C^2/n_C + s_E^2/n_E}$, where $s_C^2$ and $s_E^2$ are the respective sample variances, can be used to estimate the standard error of the difference in sample means and replaces $\text{se}(\bar{X}_C - \bar{X}_E)$ in $Z$.

In practice, non-inferiority test procedures are often expressed without test statistics, although test statistics can be used. This is due to the subjectivity involved in the choice of $\delta$, when analyzing non-inferiority trials. Rather, a two-sided $100(1 - \alpha)\%$ confidence interval for $\mu_C - \mu_E$ is determined, and the null hypothesis is rejected if the upper bound of the confidence interval is less than $\delta$. If the confidence interval is calculated as $\hat{\mu}_C - \hat{\mu}_E \pm z_{\alpha/2}\text{se}(\hat{\mu}_C - \hat{\mu}_E)$, the confidence interval approach and the test statistic approach are identical. The confidence interval approach conveys those margins that can and cannot be ruled out by the data. Thus, when there are different perspectives on the non-inferiority margin (e.g., different perspectives among regulatory bodies), individual decisions on non-inferiority are based on the same confidence interval. The two-sided confidence interval is preferred to give information about both the "best-case scenario" and "worst-case scenario" for the experimental treatment. A two-sided confidence interval also can be used to simultaneously test for superiority, non-inferiority, inferiority, and equivalence (if equivalence margin are established).

More details on statistical approaches for continuous data are given in Chapter 12.

*Asymptotic Tests for Binary Data.* When the primary endpoint is binary, inference is based on the proportion of subjects in each arm who have a favorable response or a success. For certain combinations of sample size and success rate, the sample proportion has an approximate normal distribution with mean $p$ and variance $np(1 - p)$, where $p$ is the true or theoretical probability of a success and $n$ is the sample size. A general rule of thumb is that at least five successes and at least five failures should be expected in each treatment group, and sometimes a requirement of at least 30 total subjects is added. For a simple analysis of non-inferiority, the null hypothesis of $H_o$: $p_C - p_E \geq \delta$ can be tested versus the alternative $H_a$: $p_C - p_E < \delta$ using the confidence interval methodology as noted above: reject the null hypothesis if the upper bound of a two-sided $100(1 - \alpha)\%$ confidence interval on $p_C - p_E$ is less than $\delta$. The form of this confidence interval was provided in Section A.1.2. Other confidence intervals, having better properties and more suited for non-inferiority testing, are discussed in Chapter 11.

## A.2  Bayesian Methods

For Bayesian analyses, inferences are made on a parameter by treating the parameter as a random variable. The uncertainty about the parameter of interest is based on a distribution called a *posterior distribution* that is dependent on the observed data and an initial distribution of that parameter called a *prior distribution*. This prior distribution tends to be chosen ahead of time and is defined over the parameter space of the parameter. The choice of the prior distribution may reflect a prior belief of where the certainty and uncertainty of the parameter lies in the parameter space, may be based on prior data and chosen for the purposes of integrating findings, or may be selected to have little or no influence on the final statistical inference. *Posterior probabilities* are probabilities involving the parameter that are based on the posterior distribution.

### A.2.1  Posterior Probabilities and Credible Intervals

We will consider again the simple experiment of tossing a coin 10 times where $p$ denotes the probability that any given toss results in a head. For three prior distributions each having mean 0.5, Table A.6 provides the posterior probability that heads are more likely than tails when eight heads have been observed in 10 independent tosses. Among prior beta distributions having mean 0.5, as the variance decreases (i.e., the common value of $\alpha$ and $\beta$ increases), the posterior probability of $p > 0.5$ (or $p < 0.5$) gets closer to 0.5.

In Example A.4, the posterior probabilities of the corresponding alternative hypothesis are determined on the basis of the comparisons and observed hazard ratio in Example A.3.

## TABLE A.6

Summary of Posterior Probabilities of $p > 0.5$ for Different Prior Distributions When Observing 8 Heads in 10 Tosses

| Case | Prior Distribution for $p$ | Posterior Distribution for $p$ | Posterior Probability of $p > 0.5$ |
|---|---|---|---|
| 1 | Beta $\alpha = 1, \beta = 1$ | Beta $\alpha = 9, \beta = 3$ | 0.967 |
| 2 | Beta $\alpha = 3, \beta = 3$ | Beta $\alpha = 11, \beta = 5$ | 0.941 |
| 3 | Beta $\alpha = 5, \beta = 5$ | Beta $\alpha = 13, \beta = 7$ | 0.916 |

## Example A.4

As before, let $\theta$ denote the true experimental versus control hazard ratio of some undesirable event (e.g., death or disease progression). For an observed hazard ratio of 0.91 based on 400 events in a clinical trial that had a one-to-one randomization, Table A.7 summarizes the posterior probability that $\theta$ lies in the alternative hypothesis for each comparison type.

Since superiority ($\theta > 1$) implies non-inferiority ($\theta > 0.9$), the posterior probability for non-inferiority comparison will always be greater than the posterior probability for a superiority comparison of the same two treatment arms. Also, since the parameter has a continuous posterior distribution, the sum of the posterior probabilities for superiority and for inferiority is 1, and the posterior probability for the alternative hypothesis of a difference comparison is always 1 (even if the observed hazard ratio is 1). In practice, since a difference comparison is a two-sided comparison, the posterior probabilities on each side of no difference would be calculated and compared when making an inference about whether a difference (and the direction of that difference) has been demonstrated.

Note that from Tables A.3 and A.7, for inferiority, superiority, and non-inferiority comparisons, the sum of the $p$-value and the posterior probability of the alternative hypothesis equal 1. This will occur for each of these comparisons whenever a normal model with known variance is used for the estimator and a noninformative

## TABLE A.7

Summary of Posterior Probabilities for Various Types of Comparisons for an Observed Experimental versus Control Hazard Ratio of 1.10 Based on 400 Events

| Case | Null Hypothesis | Alternative Hypothesis | Posterior Distribution for the True Log-Hazard Ratio[a] | Posterior Probability of the Alternative Hypothesis |
|---|---|---|---|---|
| Inferiority | $H_0: \theta \leq 1$ | $H_a: \theta > 1$ | $\theta \sim N(\ln(0.91),$ | 0.173 |
| Superiority | $H_0: \theta \geq 1$ | $H_a: \theta < 1$ | $(0.1)^2)$ | 0.827 |
| Difference | $H_0: \theta = 1$ | $H_a: \theta \neq 1$ | | 1 |
| Non-inferiority | $H_0: \theta = 1.1$ | $H_a: \theta < 1.1$ | | 0.971 |

[a] The posterior distribution is approximated for a one-to-one randomization by using the asymptotic distribution of the estimator of the log-hazard ratio and a noninformative prior on the true log-hazard ratio.

prior distribution is selected for the parameter. More on comparing and contrasting a $p$-value and the posterior probability of the alternative hypothesis will be provided in Section A.3.

The point estimate either represents a specific characteristic of the posterior distribution (e.g., the mean or median) or it is the value that minimizes expected loss. For every possible pair of a potential estimate and a possible value for the parameter, a *loss function* provides the corresponding error, cost, or loss that is incurred. Then the estimate is the value that has the smallest expected loss. In many instances, the estimate based on the posterior distribution can be expressed as a weighted average of the corresponding estimate based on the prior distribution and the frequentist estimate based on the empirical distribution of the data. As more and more data are collected, a greater weight is placed on the frequentist estimate.

A *credible interval* is an interval estimate of the parameter that has a prespecified probability of containing the correct value of the parameter. A 100(1 $- \alpha$)% credible interval has a probability of $1 - \alpha$ of containing the correct value of the parameter based on the posterior distribution. As with confidence intervals, credible intervals can be one-sided or two-sided. Whenever any arbitrary value in the credible interval necessarily has greater posterior density than any arbitrary value outside the credible interval, the credible interval is said to be of highest density. The 100(1 $- \alpha$)% credible interval from the 100($\alpha/2$)-th percentile of the posterior distribution to the 100(1 $- \alpha/2$)-th percentile of the posterior distribution is said to be equal tailed. Equal-tailed credible intervals are usually used in practice instead of highest density credible intervals as they mirror error symmetric confidence intervals.

Asymptotic formulas for credible intervals are analogous to those of confidence intervals given in Table A.5. Either the sample means, proportions, and standard deviations from the data can be used or they can be replaced by the estimates based on the respective posterior distribution. The width of a credible interval depends on the variability of the posterior distribution and the probability coverage for the interval. The factors that influence the variability of the posterior distribution include the variability of the prior distribution and the amount of information in the data (i.e., the sample size(s) or the number of events).

The probability distribution for future observations can also be found by integrating over the parameter space the product of the likelihood function and the posterior distribution. Such distributions can be used to make inferences about prediction.

### A.2.2 Prior and Posterior Distributions

Bayesian statistics integrates the prior knowledge or belief on a parameter along with relevant data to make inferences on the parameter. Let $X$ be a random variable that has a distribution that depends on the value of a parameter $\theta$. Let $\Omega$ denote the parameter space for $\theta$, the set of all possible values for $\theta$. We denote the probability density function (or probability mass function) of $X$ by $f(x|\theta)$. Let $h$ denote the probability density function for the prior distribution of $\theta$. Then for the observed values $x_1, \ldots, x_n$, from a random sample $X_1, \ldots, X_n$, the posterior density function for $\theta$ is given by

$$g(\theta \,|\, x_1, x_2, \ldots, x_n) = \frac{f(x_1 \,|\, \theta)f(x_2 \,|\, \theta) \cdots f\left(x_n \,|\, \theta\right)h\left(\theta\right)}{\displaystyle\int_\Omega f(x_1 \,|\, \theta)f(x_2 \,|\, \theta) \cdots f\left(x_n \,|\, \theta\right)h\left(\theta\right)\mathrm{d}\theta}. \tag{A.1}$$

Since the denominator in Equation A.1 does not involve $\theta$ ($\theta$ is integrated out), we have,

$$g(\theta \,|\, x_1, x_2, \ldots, x_n) \propto f(x_1|\theta)f(x_2|\theta)\cdots f(x_n|\theta)h(\theta). \tag{A.2}$$

To illustrate, let $X_1, \ldots, X_{20}$, be a random sample from a Bernoulli distribution having probability of success $p$, where $p$ has a Beta prior distribution with $\alpha = 3$ and $\beta = 5$. We have for $i = 1, \ldots, 20$,

$$f(x_i \,|\, p) = p^{x_i}(1-p)^{1-x_i} \text{ for } x_i = 0 \text{ or } 1 \text{ and } f(x_i \,|\, p) = 0, \text{ otherwise, and}$$

$$h(p) = 105p^2(1-p)^4 \text{ for } 0 < p < 1 \text{ and } h(p) = 0, \text{ otherwise.}$$

If we observe 8 successes and 12 failures (8 $X_i$'s = 1 and 12 $X_i$'s = 0), then from Equation A.2, the density for the posterior distribution of $p$ satisfies $g(p|x_1, x_2, \ldots, x_{20}) \propto p^8(1-p)^{12} \times p^2(1-p)^4 \propto p^{10}(1-p)^{16}$ for $0 < p < 1$, and $g(p|x_1, x_2, \ldots x_{20}) = 0$, otherwise. Since $g$ is proportional to the density for a Beta distribution with $\alpha = 11$ and $\beta = 17$, $p$ has for a Beta distribution with $\alpha = 11$ and $\beta = 17$ as its posterior distribution.

A given type or parametric family of distributions for the observed data is said to have a *conjugate family of prior distributions* (or have a *conjugate prior*), if there exists a parametric family of priors for which the posterior distribution must also necessarily belong. For example, for a random sample from a Bernoulli distribution having probability of success $p$ where $p$ is modeled as having a beta distribution, the posterior distribution for $p$ is also a beta distribution. Thus, a Bernoulli distribution has a beta distribution as a conjugate family of prior distributions. Table A.8 gives a summary of conjugate family of prior distributions for various distributions of data. The posterior distribution and their respective means are provided in the rightmost column.

*Noninformative Prior Distributions.* A noninformative prior is a prior distribution that contains no information about the parameter. Use of a noninformative prior allows for the inferences on the parameter to be based entirely or almost entirely on the data. Additionally, there may be other prior

**TABLE A.8**

Summary of Conjugate Family of Prior Distributions

| Distribution from which $X_1, ..., X_n$ Is Randomly Drawn | Prior Distribution | Posterior Distribution[a] |
|---|---|---|
| Bernoulli ($p$) | $p \sim$ Beta $(\alpha, \beta)$ | $p \sim$ Beta $(\alpha + \Sigma x_i, \beta + n - \Sigma x_i)$ $$\text{mean} = \left( n\bar{x} + (\alpha + \beta)\left( \frac{\alpha}{\alpha + \beta} \right) \right) \Big/ (n + \alpha + \beta)$$ |
| Normal ($\mu, \sigma^2$) $\sigma^2$ known | $\mu \sim$ Normal $(v, \tau^2)$ | $\mu \sim$ Normal with $$\text{mean} = \left( \left( \frac{n}{\sigma^2} \right)\bar{x} + \left( \frac{1}{\tau^2} \right)v \right) \Big/ \left( \left( \frac{n}{\sigma^2} \right) + \left( \frac{1}{\tau^2} \right) \right)$$ and variance $= 1 \Big/ \left( \left( \frac{n}{\sigma^2} \right) + \left( \frac{1}{\tau^2} \right) \right)$ |
| Poisson ($\lambda$) | $\lambda \sim$ Gamma $(\alpha, \beta)$ where the mean of $\lambda$ is $\alpha\beta$ | $\lambda \sim$ Gamma $(\alpha + \Sigma x_i, 1/(n + 1/\beta))$ mean $= (n\bar{x} + (1/\beta)(\alpha\beta)) / (n + 1/\beta)$ |

[a] For the posterior distributions, the observed values are $x_1, \ldots, x_n$ with sample mean $\bar{x}$.

distributions for which the inferences on the parameter are based almost entirely on the data. In some settings, Bayesian inferences based on such prior distributions are completely analogous or identical to inferences based on frequentist methods. For example, consider an experiment where independent samples of size 100 each are taken from normal distributions each having a variance of 25 and respective means of $\mu_1$ and $\mu_2$. The statistical hypotheses that will be tested are $H_o: \mu_1 \leq \mu_2$ and $H_a: \mu_1 > \mu_2$. Let $\bar{x}_1$ and $\bar{x}_2$ denote the observed values of the respective sample means. The $p$-value for the normalized test statistics is $1 - \Phi\left( \frac{\bar{x}_1 - \bar{x}_2}{\sqrt{1/2}} \right)$. For noninformative prior distributions for $\mu_1$ and $\mu_2$ (or equivalently, a noninformative prior distribution on $\theta = \mu_1 - \mu_2$), the posterior probability that $\mu_1 > \mu_2$ equals $\Phi\left( \frac{\bar{x}_1 - \bar{x}_2}{\sqrt{1/2}} \right)$. Thus, for this example, for any $0 < \alpha < 1$, rejecting $H_o$ (and thus concluding $H_a$) whenever the $p$-value is $\leq \alpha$ is equivalent to rejecting $H_o$ whenever the posterior probability of $H_a$ is $\geq 1 - \alpha$.

*Jeffreys Prior Distributions.* In the univariate setting, a *Jeffreys prior* has a density function proportional to the square root of *Fisher's information*. Fisher's information in a single observation is given by $I(\theta) = -E\left( \frac{\partial^2}{\partial\theta^2} \log f(X \mid \theta) \right)$.

The density for the Jeffreys prior then satisfies $h(\theta) \propto \sqrt{I(\theta)}$. When sampling from a Bernoulli distribution, the Jeffreys prior for $p$ is a Beta distribution having $\alpha = \beta = 0.5$. Jeffreys prior distributions have little influence on the statistical inferences.

### A.2.3 Statistical Inference

As mentioned earlier, either a specific characteristic of the posterior distribution (e.g., the median or mean) or a value that minimizes the expected value of a loss function is chosen as the point estimate of a parameter. The means of the posterior distributions are also provided in Table A.8. In each case, the posterior mean is a weighted average of the sample mean and the mean of the prior distribution. When the data come from a normal distribution with a known variance and a normal prior distribution is used for $\mu$, the weights are inversely proportional to the corresponding variance. When the data come from a Bernoulli distribution with a Beta prior distribution for $p$ with parameters $\alpha$ and $\beta$, the mean of the posterior distribution is the quotient of the sum of the number of successes and $\alpha$ to the sum of the number of trials and $\alpha + \beta$. This prior distribution can be interpreted as starting with $\alpha$ successes in $\alpha + \beta$ trials. The mean of the posterior distribution for $p$ is then the pooled proportion of successes. When the data come from a Poisson distribution with a gamma prior distribution for $\lambda$, the mean of the posterior distribution is the quotient of the sum of the number of events and $\alpha$ to the sum of the sample size and $1/\beta$. This prior distribution can be interpreted as starting with $\alpha$ events among $1/\beta$ patients or trials. As the sample size increases, the proportion of weight given the sample mean (the data) increases toward 1. The weight given the prior mean approaches zero as $\alpha \to 0$ and $\beta \to 0$ for the beta prior distribution, as $\tau \to \infty$ for the normal prior distribution, or as $\beta \to \infty$ for the gamma prior distribution. Additionally, the relative weight given the prior mean approaches zero as the sample size grows without bound ($n \to \infty$).

The coefficient for the prior mean is sometimes referred as a *shrinkage factor*.[21] The shrinkage factor is the proportion of distance that the posterior mean is "shrunk back" from the frequentist estimate toward the prior mean. For example, from Table A.8, the shrinkage factor for the normal conjugate family of prior distributions is $1/(1 + n\tau^2/\sigma^2)$. To illustrate, suppose the observed values are modeled as a random sample from a normal distribution with unknown mean, $\mu$, and a variance of 9. A normal distribution with mean 70 and variance 100 is selected as the prior distribution for $\mu$. If the observed sample mean from a random sample of 25 is 72, then we see from Table A.8 that $\mu$ will have a normal posterior distribution with mean $\approx 71.96$ and variance $\approx 0.352$. Here, the shrinkage factor is roughly 0.022.

A function $L$ that assigns a cost or error to every possible pair of an estimate and the actual value of the parameter is called a *loss function*. An estimate of the parameter is often selected as that value that has the smallest expected loss with respect to the posterior distribution of the parameter.

That is, the estimate, $a$, is the value in the parameter space that minimizes
$= \int_{\Omega} L(\theta,a)\, g(\theta\,|\,x_1,\ldots,x_n) d\theta$. For example, consider the squared-error loss
function, $L(\theta,a) = (\theta - a)^2$. The value $a = E(\theta\,|\,x_1,\ldots,x_n)$, the posterior mean of $\theta$,
minimizes $E(\theta - a)^2 = \int_{\Omega} (\theta - a)^2 g(\theta\,|\,x_1,\ldots,x_n) d\theta$. For an absolute loss function
$(L(\theta,a) = |\theta - a|)$, the expected loss is minimized by using the median of the
posterior distribution as the estimate.

Consider a Jeffreys prior for $p$, the probability of a success. An experiment
results in two successes and eight failures among 10 trials. Here, $p$ has a posterior Beta distribution $\alpha = 2.5$ and $\beta = 8.5$. Table A.9 provides the estimates
for three loss functions.

The posterior median and mean for $p$ are approximately 0.210 and 0.227,
respectively. For cubed absolute loss, the value of approximately 0.242 minimizes the expected loss. When the Beta distribution has a mean less than 0.5,
as in this example, the sequence of $a_h$ that minimizes $E(|\theta - a|^h)$ increases to
0.5. If the Beta distribution has a mean greater than 0.5, this same sequence
decreases to 0.5. When the Beta distribution has a mean equal to 0.5, $a_h = 0.5$
minimizes $E(|\theta - a|^h)$ for all $h > 0$.

A frequentist evaluation of a Bayesian method can also be done. For
Bayesian estimators, the sampling distribution can be determined as can
the mean square error (when it exists) of the estimator, and the asymptotic
properties of the estimator. For example, consider making an inference on a
response rate $p$, based on a sample of 20 subjects and a beta prior distribution for $p$ where the value for each parameter is 2. Let $x$ denote the number
among the 20 subjects that responded. We will model $x$ as a random value
from a binomial distribution based on 20 trials with probability of success
$p$. We denote the mean of the posterior distribution by $\hat{p}$. Then the sampling
distribution for $\hat{p}$ is summarized by $P(\hat{p} = (x + 2)/24) = \binom{20}{x} p^x (1-p)^{20-x}$

for $x = 0, \ldots, 20$. The mean squared error for $\hat{p}$ is $(1 + p + p^2)/144$.

*Credible Intervals.* To illustrate an example of an equal-tailed credible interval, consider a randomized, controlled clinical trial where 10 of 15 patients
on the experimental arm and 4 of 15 patients on the control arm responded.
We will use a Jeffreys prior for the prior distribution for the response rate ($p_C$

**TABLE A.9**

Loss Functions and Corresponding Estimates

| Loss Function = $L(\theta,a)$ | Estimate |
|---|---|
| Absolute loss = $|\theta - a|$ | 0.210 |
| Squared-error loss = $(\theta - a)^2$ | 0.227 |
| Cubed absolute loss = $|\theta - a|^3$ | 0.245 |
| $L(\theta,a) = |\theta - a|^h$ as $h \to \infty$ | 0.5 |

**TABLE A.10**

Summary of 95% Credible Intervals

| Parameter | Prior Distribution | Posterior Distribution | 95% Credible Interval |
|---|---|---|---|
| Control response rate $p_C$ | $p_C \sim$ Beta $(0.5, 0.5)$ | $p_C \sim$ Beta $(4.5, 11.5)$ | 0.097, 0.517 |
| Experimental response rate $p_E$ | $p_E \sim$ Beta $(0.5, 0.5)$ | $p_E \sim$ Beta $(10.5, 5.5)$ | 0.416, 0.860 |
| Difference in response rates $p_E - p_C$ | $p_C$ and $p_E$ are assumed to be independent | $p_C$ and $p_E$ are assumed to be independent | 0.046, 0.663 |

and $p_E$) of each arm. Table A.10 summarizes the equal-tailed 95% credible intervals for the response rate of each arm and for the difference in response rates. The joint posterior distribution (with joint density being the product of the posterior densities for $p_C$ and $p_E$) is used to determine the 95% credible interval for the difference.

Note that the 95% exact confidence intervals for $p_C$ and $p_E$ are (0.078, 0.551) and (0.384, 0.882), respectively. Also, the large sample normal approximate 95% confidence interval for $p_E - p_C$ is (0.073, 0.727). The 95% credible intervals are narrower than the respective exact 95% confidence intervals for $p_C$ and $p_E$. Here, for the difference in response rates, the 95% credible interval is narrower than and slightly shifted to the left of the 95% confidence interval. Section A.3 investigates the relationship between credible intervals for a proportion based on a Jeffreys prior with the corresponding exact confidence interval.

Hypotheses testing can be based on a credible interval, on the magnitude of the posterior probability that the alternative hypothesis is true, or on the expected loss/cost for rejecting or failing to reject the null hypothesis. In any case, there will exist a rejection region, a set of possible samples for which the null hypothesis is rejected. The rejection region can be assessed to determine the type I error rate or size of the test, and the power function. For example, suppose a posterior probability for $p > 0.5$ greater than 0.975 is needed to reject the null hypothesis that $p \leq 0.5$ and conclude the alternative hypothesis that $p > 0.5$. For a Jeffreys prior distribution, this would require at least 15 responses among the 20 subjects. The power function for this test is $\sum_{x=15}^{20} \binom{20}{x} p^x (1-p)^{20-x}$ for $0 < p < 1$, and thus the size of the test is approximately 0.021.

*The Likelihood Principle.* The *likelihood principle* states that when two different experiments lead to the same or proportional likelihood functions, inferences about the parameter should be identical for the two experiments. The likelihood principle is preserved with Bayesian inference but may not be preserved with frequentist inference. This is illustrated in the following example.

Example A.5 is similar to one provided by Goodman,[22] which provides two different experiments where the likelihood function is proportional and the frequentist decision may differ.

### Example A.5

Suppose we learn that among seven subjects that were given an investigational drug in a phase I study, one subject experienced the target toxicity. Suppose we are interested in whether the true probability that a subject will experience the target toxicity, $p$, is less than 0.5. Thus, we are interested in testing $H_o: p = 0.5$ against the hypothesis that $H_o: p < 0.5$. Consider the following two possible designs (with corresponding calculations for the $p$-value):

- Design A (binomial design): The design of the study required the toxicity experiences of exactly seven subjects given the proposed dose of the investigational drug. One patient out of the seven experienced the target toxicity. Here, the $p$-value, the probability that zero or one of the seven patients would experience the target toxicity when $p = 0.5$, equals 1/16 (=0.0625). The one-sided lower 95% confidence interval for $p$ is (0, 0.52).
- Design B (negative binomial design): For the study, subjects were to receive the investigational therapy, one at a time, until a subject experiences the target toxicity or until 10 patients have received the investigational therapy. Here, the $p$-value, the probability that at least seven patients will be treated with the investigational drug when $p = 0.5$, equals 1/64 ($\approx$0.0156). The one-sided lower 95% confidence interval for $p$ is (0, 0.39).

The corresponding likelihood functions for $p$ in the two studies are proportional. However, if formal hypotheses testing were done with a significance level of 0.05, the decision on whether the evidence is strong enough to conclude $p < 0.5$ would differ between the study designs. In the Bayesian setting, we have from Equation A.1 that since the likelihood functions for $p$ are proportional, the posterior distribution for $p$ will not depend on whether the design was A or B. For a Jeffreys prior for $p$ (which leads to a Beta posterior distribution with parameter values 1.5 and 6.5), the posterior probability that $p > 0.5$ is 0.025. The one-sided lower 95% credible interval for $p$ is (0, 0.44). For a Beta prior on $p$ with parameters $\alpha$ and $\beta$, the posterior distribution for $p$ will have parameter values $\alpha + 1$ and $\beta + 6$. Sending $\alpha$ and $\beta$ to zero leads to a limiting Beta posterior distribution with parameter values of 1 and 6. On the basis of this Beta posterior distribution, the posterior probability that $p > 0.5$ is 1/64 with a one-sided lower 95% credible interval for $p$ of (0, 0.39). This provides analogous results to the frequentist analysis using a negative binomial design. It can be shown that for alternative hypotheses of the form $p > p_0$ that the $p$-value for observing the $x$th success in the $n$th trial from a negative binomial design equals the probability of $p > p_0$ when $p$ has a Beta distribution with parameter values $x$ and $n-x$. Further comparison of credible intervals and confidence intervals for proportions is provided in Section A.3.

## A.3 Comparison of Methods

In this section we will compare and contrast the results of analyses based on frequentist and Bayesian methods. Particular attention will be given to the similarity and differences of inferences based on $p$-values, confidence intervals, and posterior probabilities.

### A.3.1 Relationship between Frequentist and Bayesian Approaches

As in Section A.2.3, the $p$-value (and also a desired level confidence interval) can be different depending on the design even when the observed data are identical. The Bayesian inference remains the same (but would change as the prior distribution changes). Additionally, we saw in Example A.5, involving 15 subjects in each of two arms, that the 95% credible interval for the response rates and the difference in response rates were broadly similar to the respective 95% confidence intervals.

In practice, for most tests of hypotheses involving one comparison, the inference is identical whether based on a $p$-value or on a confidence interval. The null hypothesis is ruled out by a $100(1 - \alpha)$% confidence interval exactly when the $p$-value is less than $\alpha$. This consistency between the statistical procedures is a desirable property. This relationship between $p$-values and confidence intervals holds for most inferences on one mean, proportion, variance, or quantile, and holds for most inferences on a difference between groups in proportions, means, quantiles, variances, and log hazards. In those cases, one comparison is made—that is, one parameter or a difference in parameters is compared with a constant.

#### A.3.1.1 Exact Confidence Intervals and Credible Intervals Using a Jeffreys Prior

We will compare for one proportion exact confidence intervals with credible intervals based on a Jeffreys prior distribution.

For $0 < \alpha < 1$, we will compare the $100(1 - \alpha)$% equal-tailed credible interval for $p$, the probability of a success or the response rate, based on a Jeffreys prior distribution for $p$ with the exact $100(1 - \alpha)$% confidence interval for $p$.

Suppose $x$ successes or responses have been observed from $n$ subjects according to binomial sampling. The posterior distribution for $p$ based on Jeffreys prior is a Beta distribution with parameter values $x + \frac{1}{2}$ and $n - x + \frac{1}{2}$. Let $(r_L, r_U)$ denote the $100(1 - \alpha)$% equal-tailed credible interval for $p$ (i.e., $r_L$ and $r_U$ are, respectively, the $100\alpha/2$-th and $100(1 - \alpha/2)$-th percentiles of a Beta distribution with parameter values $x + \frac{1}{2}$ and $n - x + \frac{1}{2}$).

The exact $100(1 - \alpha/2)$% confidence lower bound for $p$ is $p_L$, where $p_L$ satisfies

$$\sum_{j=x}^{n} \frac{n!}{j!(n-j)!} p_L^j (1-p_L)^{n-j} = \alpha/2.$$

Similarly, the exact $100(1 - \alpha/2)\%$ confidence upper bound for $p$ is $p_U$, where $p_U$ satisfies

$$\sum_{j=0}^{x} \frac{n!}{j!(n-j)!} p_U^j (1-p_U)^{j-x} = \alpha/2.$$

The exact $100(1 - \alpha)\%$ confidence interval for $p$ is then given by $(p_L, p_U)$.

It can be shown by applying multiple integrations by parts that $p_L$ and $p_U$ satisfy $\sum_{j=x}^{n} \frac{n!}{j!(n-j)!} p_L^j (1-p_L)^{n-j} = \int_0^{p_L} \frac{n!}{(x-1)!(n-x)!} z^{x-1}(1-z)^{n-x} = \alpha/2$

and $\sum_{j=0}^{x} \frac{n!}{j!(n-j)!} p_U^j (1-p_U)^{n-j} = \int_{p_U}^{1} \frac{n!}{x!(n-x-1)!} z^x (1-z)^{n-x-1} = \alpha/2$. Thus, the

$100(1 - \alpha)\%$ confidence interval for $p$ has the $100\alpha/2$-th percentile of a Beta distribution with parameter values $x$ and $n - x + 1$ for its lower limit ($p_L$) and the $100(1 - \alpha/2)$-th percentile of a Beta distribution with parameter values $x + 1$ and $n - x$ for its upper limit ($p_U$).

Let $q_L$ and $q_U$ denote the exact $100\alpha/2\%$ confidence lower bound for $p$ and the exact $100\alpha/2\%$ confidence upper bound for $p$, respectively. Thus, $q_L$ is the $100(1 - \alpha/2)$-th percentile of a Beta distribution with parameter values $x$ and $n - x + 1$ and $q_U$ is the $100\alpha/2$-th percentile of a Beta distribution with parameter values $x + 1$ and $n - x$. Since we are $100\alpha/2\%$ confident that the actual value of $p$ is in $[q_L, 1]$ and also $100\alpha/2\%$ confident that the actual value of $p$ is in $[0, q_U]$, it would seem reasonable that we are $100(1 - \alpha)\%$ confident that the actual value of $p$ is in $(q_L, q_U)$. However, note that a Beta distribution with parameter values $x$ and $n - x + 1$ is stochastically smaller than a Beta distribution with parameter values $x + \frac{1}{2}$ and $n - x + \frac{1}{2}$, which in turn is smaller than a Beta distribution with parameter values $x + 1$ and $n - x$. Thus, $p_L < r_L < q_U$ and $q_L < r_U < p_U$. Hence, $(q_U, q_L)$ is contained in $(r_L, r_U)$, which in turn is contained in $(p_L, p_U)$.

Note that when a frequentist says that a confidence coefficient (regardless whether a one-sided or two-sided confidence interval) is $\gamma$, that means that the confidence interval will capture the correct value of a parameter at least $100\gamma\%$ of the time. Thus, the interval $(q_U, q_L)$ will capture the correct value of a parameter at most (not at least) $100(1 - \alpha)\%$ of the time. In fact, there are few possibilities (if any) for the actual value of $p$ that would be captured exactly $100(1 - \alpha)\%$ of the time by an exact $100(1 - \alpha)\%$ confidence interval for $p$. Thus, invariably, the "probability coverage" of a $100(1 - \alpha)\%$ exact confidence interval for $p$, $(p_L, p_U)$, is greater than $1 - \alpha$. Exact confidence intervals for $p$ have been regarded as conservative.

Suppose subjects are sampled until $x$ successes occur (negative binomial sampling). We observe the $x$th success from the $n$th subject. Similar to before, it can be shown that the limits, $s_L$ and $s_U$, of the equal-tailed $100(1 - \alpha)\%$ confidence interval satisfy $\sum_{j=x}^{n} \frac{n!}{j!(n-j)!} s_L^j (1-s_L)^{n-j} =$

$\int_0^{p_L} \frac{n!}{(x-1)!(n-x)!} z^{x-1}(1-z)^{n-x} = \alpha/2$ and $\sum_{j=0}^{x-1} \frac{(n-1)!}{j!(n-1-j)!} s_U^j (1-s_U)^{n-1-j} =$

$\int_{p_U}^1 \frac{(n-1)!}{(x-1)!(n-x-1)!} z^{x-1}(1-z)^{n-x-1} = \alpha/2$. Thus, the $100(1 - \alpha)\%$ confidence interval for $p$ has the $100\alpha/2$-th percentile of a Beta distribution with parameter values $x$ and $n - x + 1$ for its lower limit ($s_L$) and the $100(1-\alpha/2)$-th percentile of a Beta distribution with parameter values $x$ and $n - x$ for its upper limit ($s_U$). Note that the lower limit of the $100(1 - \alpha)\%$ exact confidence interval for $p$ is the same for binomial sampling as negative binomial sampling (i.e., $s_L = p_L$). Since a Beta distribution with parameter values $x$ and $n - x$ is stochastically smaller than a Beta distribution with parameter values $x$ and $n - x + 1$, we have $s_U < p_U$. Whether there is an ordering between a Beta distribution with parameter values $x$ and $n - x$ and a Beta distribution with parameter values $x + \frac{1}{2}$ and $n - x + \frac{1}{2}$ (or, i.e., the respective order of $s_U$ and $r_U$) depends on the specific values for $x$ and $n$. Note also that a Beta distribution with parameters $x$ and $n - x$ is the limiting posterior distribution from using a Beta prior distribution for $p$ with parameters $\alpha$ and $\beta$, and then sending $\alpha$ and $\beta$ to zero. Thus, $s_U$ is the upper limit of the respective credible interval based on this limiting posterior distribution.

Inference based on a Jeffreys prior distribution can be directly generalized to comparing two proportions. If the samples involving each proportion are independent, the true probabilities of a success will have independent Beta posterior distributions. The posterior distribution for the difference in the true probabilities (or some other function of the true probabilities) can be determined, which can be used to find an estimate and a credible interval for the difference in the probabilities of a success. The specific approach used to find an exact confidence interval for $p$ cannot be directly extended to making inferences about a difference in two probabilities. However, there are exact confidence interval approaches for the difference of two probabilities (see Chapter 11). These approaches require setting an ordering on the possible observations that may or may not be a priori (i.e., different orderings have been used in practice).

### A.3.1.2 Comparison Involving a Retention Fraction

A retention fraction, the fraction of the effect of the control therapy's effect that is retained by the experimental therapy, is only defined when the control therapy has an effect. Dealing with this fact is an added obstacle in making an inference on the retention fraction. We will compare different methods of constructing a 95% confidence interval or 95% credible interval for the retention fraction. Details on the methods used can be found in Chapter 5. Some methods adjust for the possibility that the control therapy is not effective. We borrow ideas from Simon[23] in constructing some of the credible intervals. We will assume that the sample/event size for the non-inferiority trial is independent of results in estimating the effect of the control therapy.

Consider the following hypothetical example for overall survival. The placebo versus control therapy log-hazard ratio is estimated as 0.20, with corresponding standard error of 0.10. A normal distribution is considered for the sampling distribution of the placebo versus control log-hazard ratio estimator. From the non-inferiority trial, the experimental therapy versus control therapy log-hazard ratio is –0.10, with corresponding standard error of 0.08. Table A.11 gives 95% confidence intervals and 95% credible intervals for the retention fraction, $\lambda$, based on various methods. Those methods that are being introduced are described below. For the Bayesian methods, the prior distribution for the placebo versus control therapy log-hazard ratio, $\beta$, is modeled as a normal distribution with mean 0.2 and standard deviation 0.1, and the posterior distribution for the experimental versus control log-hazard ratio, $\eta$, is modeled as a normal distribution with mean –0.10 and standard deviation 0.08.

The intervals using frequentist methods do not adjust for the uncertainty that the control therapy is less effective than placebo. Here the $p$-value for testing whether the control therapy is better than placebo is 0.019. The (Bayesian) probability that the control therapy is less effective than placebo is also 0.019. If we ignore whether the control therapy is more or less effective than placebo and extend the definition of $\lambda$ to include cases where $\beta < 0$ ($\lambda = 1 - \eta/\beta$), then $P(\lambda > 0.239) = P(\lambda < 4.56) = 0.975$. Here the 95% ("equal-tailed") credible interval analog to the 95% confidence interval is (0.239, 4.56).

The other two credible intervals in Table A.11 do not ignore the uncertainty that the control therapy may be less effective than placebo. When the determination of a credible interval for $\lambda$ requires or is restricted to $\beta > 0$, the 95% ("equal-tailed") credible interval for $\lambda$ is (–0.527, 4.51). That is, $P(\lambda > -0.527, \beta > 0) = 0.975$ and $P(\lambda > 4.51, \beta > 0) = 0.025$. For this case, since $P(\beta > 0) \approx 0.977$, an equal-tailed credible interval with coefficient greater than 0.954 cannot be determined. If possibilities where $\eta - \beta < 0$ and $\beta < 0$ (cases where the experimental therapy is better than placebo, which is better than the control therapy) are regarded as

## TABLE A.11

95% Confidence Interval or Credible Interval for Retention Fraction, $\lambda$, Based on Several Methods

| Method | 95% Confidence Interval or Credible Interval for $\lambda$ |
| --- | --- |
| Fieller (based on a normalized test statistic) | 0.640, 26.612 |
| Delta method | 0.575, 2.424 |
| Bayesian ignore whether $\beta < 0$ or $\beta > 0$ | $P(0.239 < \lambda < 4.56) = 0.95$ |
| Bayesian exclude $P(\eta - \beta < 0$ and $\beta < 0)$ | $P(-0.527 < \lambda < 4.51, \beta > 0) = 0.95$ |
| Bayesian include $P(\eta - \beta < 0$ and $\beta < 0)$ | $P(0.614 < \lambda < 9.90, \beta > 0) = 0.95$ |

having greater relative efficacy than any case where $\lambda > 0$ and $\beta > 0$, then the 95% ("equal-tailed") credible interval for $\lambda$ is (0.614–9.90). That is, $P(\{\lambda > 0.614, \beta > 0\}$ or $\{\eta - \beta < 0, \beta < 0\}) = 0.975$ and $P(\{\lambda > 9.90, \beta > 0\}$ or $\{\eta - \beta < 0, \beta < 0\})) = 0.025$. This last method has the advantage of considering both the uncertainty that the control therapy may be less effective than a placebo and also other cases of greater relative efficacy. The interval (0.614, 9.90) may be the most appropriate 95% CI (Confidence interval or credible interval) for the retention fraction.

When $P(\beta > 0)$ is extremely close to 1, the credible intervals from the above three methods will be approximately the same. The 95% confidence interval from the Fieller method will also be similar. For example, if the estimate of the placebo versus control therapy log-hazard ratio is instead 0.4, then the 95% Fieller confidence interval and each of the three 95% credible intervals are approximately (0.851, 1.81). The approximate 95% confidence interval using the delta method is (0.839, 1.66). The confidence interval for the delta method is noticeably different from that using Fieller's method. The estimator of the retention fraction may not have an approximate normal distribution. Rothmann and Tsou[24] examined the actual coverage of delta method confidence intervals for the retention fraction when estimated by a ratio of independent random variables, each having an approximate normal distribution. When the ratio of the mean to the standard deviation is greater than 8 for the estimator of the effect of the control therapy, then (per Rothmann and Tsou[24]) a hypothesis test based on a delta method confidence intervals for the retention fraction will have approximately the desired type I error rate.

For testing for a retention fraction of more than 0.5, Table A.12 summarizes the one-sided $p$-values and the analogous posterior probabilities using these methods for the example given in Table A.11. For this case, the $p$-value or posterior probability was similar for the normalized test statistic, the delta method, and the Bayesian method, which includes possibilities where $\eta - \beta < 0$ and $\beta < 0$ as having the greatest relative efficacy.

### A.3.1.3 Likelihood Function for a Non-Inferiority Trial

We saw in Section A.2.3 that the frequentist analysis (the $p$-value and a desired level confidence interval) can be different for different designs of

**TABLE A.12**

One-Sided $p$-Values and Analogous Posterior Probabilities for Non-Inferiority for Testing for a Retention Fraction of More than 0.5

| Method | $p$-Value/Posterior Probability |
|---|---|
| Normalized test statistic | 0.017 |
| Delta method | 0.017 |
| Bayesian ignore whether $\beta < 0$ or $\beta > 0$ | $P(\lambda < 0.5) = 0.032$ |
| Bayesian exclude $P(\eta - \beta < 0$ and $\beta < 0)$ | $1 - P(0.5 < \lambda, \beta > 0) = 0.036$ |
| Bayesian include $P(\eta - \beta < 0$ and $\beta < 0)$ | $1 - P(0.5 < \lambda, \beta > 0$ or $\eta - \beta < 0$ and $\beta < 0) = 0.019$ |

experiments even though the observed data are identical. The Bayesian analysis remains the same. This is also pertinent to the design and analysis of non-inferiority trials when the analysis includes the estimation of the effect of the control therapy from previous trials. The frequentist interpretation of the results formally depends on whether the design of the non-inferiority trial was independent or dependent of the estimation of the effect of the control therapy. We will illustrate this by considering two designs for comparing the means from two samples. For the purpose of a non-inferiority comparison, $\mu_1$ represents the effect of the control therapy versus placebo that will be estimated by previous trials and $\mu_2$ represents the difference in the effects of the control and experimental therapies that will be estimated from the non-inferiority trial. Consider the following two experiments.

Case 1: A random sample of size 25 is drawn from a normal distribution having an unknown mean $\mu_1$ and a variance equal to 100, and an independent random sample of size 100 is drawn from a normal distribution having an unknown mean $\mu_2$ and a variance equal to 100.

Case 2: A random sample of size 25 is drawn from a normal distribution having an unknown mean $\mu_1$ and a variance equal to 100. The observed sample mean, $\bar{x}_1$, is noted. An independent random sample of size $m(\bar{x}_1)$ is drawn from a normal distribution having an unknown mean $\mu_2$ and a variance equal to 100, for some positive-integer valued function $m$.

Let $\bar{x}_1$ and $\bar{x}_2$ denote the respective sample means.

In case 1, the likelihood function reduces to (is proportional to)

$$L\left(\mu_1,\mu_2;\bar{x}_1,\bar{x}_2\right)= f\left(\bar{x}_1,\bar{x}_2;\mu_1,\mu_2\right)=\frac{1}{4\pi}\exp\left\{-\left[(\bar{x}_1-\mu_1)^2/8+(\bar{x}_2-\mu_2)^2/2\right]\right\}.$$ The

likelihood function factors into the product of separate functions of $\bar{x}_1$ and $\bar{x}_2$, and also factor into the product of separate functions of $\mu_1$ and $\mu_2$. The two random sample means are independent, and if independent noninformative priors are selected for $\mu_1$ and $\mu_2$, $\mu_1$ and $\mu_2$ will be independent at all stages of sampling. In fact, in such a case, certain frequentist and Bayesian inferences will be the same.

In case 2, the likelihood function reduces to (is proportional to)

$$L\left(\mu_1,\mu_2;\bar{x}_1,\bar{x}_2\right)= f\left(\bar{x}_1,\bar{x}_2;\mu_1,\mu_2\right)=\frac{\sqrt{m\left(\bar{x}_1\right)}}{40\pi}\exp\left\{-\left[(\bar{x}_1-\mu_1)^2/8+(\bar{x}_2-\mu_2)^2/\left(200/\right.\right.\right.$$

$\left.\left.\left.m\left(\bar{x}_1\right)\right)\right]\right\}$. This likelihood function will factor into the product of separate

functions of $\mu_1$ and $\mu_2$. Analogous types of Bayesian methods can be applied in case 2 as in case 1, and if $m(\bar{x}_1) = 100$, the likelihood functions and the posterior distributions will be identical, and hence the inferences will be identical. However, in case 2, the likelihood function cannot be expressed as the product of separate functions of $\bar{x}_1$ and $\bar{x}_2$. In fact, if $m$ is not a constant function, then the difference in the random sample means will not have a normal distribution. Suppose $\mu_1 = \mu_2 = 0$ and $m(x) = 1$, if $x < 0$ and $100/m(x) \approx 0$, if $x > 0$. Then, it is easy to see that $P(\bar{X}_1 - \bar{X}_2 > 0) > 0.5$, even though $E\left(\bar{X}_1 - \bar{X}_2\right)= 0$.

Scenarios like these arise when historical trials are used to estimate the effect of the non-inferiority trial's control therapy. Many of the first such non-inferiority analyses that used historical trials to estimate the effect of the control therapy had this estimation occur retrospectively after the results of the non-inferiority trial were known. It is currently common practice to prospectively estimate the effect of the control therapy before conducting the non-inferiority trial. Thus, the non-inferiority criterion and the sizing of the non-inferiority will depend on the results from estimating the effect of the control therapy. For Bayesian analyses, it does not matter whether the estimate of the control therapy's effect and its corresponding variance influences the sizing of the non-inferiority trial. For frequentist analyses, the sampling distribution, for whatever test statistic used, is altered and may not be approximately determined. Rothmann[25] provided a discussion on how the type I error probability changes across the boundary of a non-inferiority null hypothesis and potential ways of addressing this problem when trying to maintain a desired type I error rate.

## A.3.2 Dealing with More than One Comparison

When more than one comparison is made, the relationships (if any) between $p$-values, confidence intervals, and posterior probabilities change. More than one comparison may mean that the same quantity or parameter is compared multiple times (e.g., as in an equivalence trial), the same endpoint is involved in multiple comparisons (e.g., two experimental arms are compared separately to a control arm), or that different endpoints are used in the comparisons. We illustrate some of the differences in the relationship between $p$-values, confidence intervals, and posterior probabilities in the setting of an equivalence comparison in Example A.6. For an equivalence comparison, there are multiple ways of defining a $p$-value. We begin by using the concept directly to the equivalence hypothesis in Section A.1 to define the $p$-value, and then applying the concept to conducting equivalence by performing two simultaneous one-sided tests.

### Example A.6

For an equivalence comparison, a measured difference between arms is compared with two values. This difference, for example, can be a difference in proportions or means, a relative risk, or a hazard ratio. We revisit Examples A.3 and A.4 involving a hazard ratio based on 400 events. The hypotheses of interest for this example will be

$$H_o: \theta \le 0.8 \text{ or } \theta \ge 1.25 \text{ vs. } H_a: 0.8 < \theta < 1.25 \qquad (A.3)$$

where 0.8 and 1.25 are the equivalence limits. As in examples A.3 and A.4, the observed experimental to control hazard ratio is 0.91. We will define the $p$-value consistent with Section A.1 as the (largest) probability of observing a hazard ratio of 0.91 or more extreme (more in favor of the alternative hypothesis) if the null hypothesis was true. It would seem reasonable, at least conceptually, that the closer the observed hazard ratio is to 1 in a relative sense (the closer the observed log-hazard ratio is to 0), the stronger the strength of evidence against the null hypothesis in favor of "equivalence." On the basis of that approach, the $p$-value is the largest probability of getting an observed hazard ratio between 0.91 and 1/0.91 when the null hypothesis is true, which equals 0.098. In practice, a $p$-value is rarely calculated when performing an equivalence test. In general, equivalence is concluded if a confidence interval (usually a 90% confidence interval) contains only possibilities within the equivalence margin. For example, for the alternative hypothesis of equivalence in Equation A.3, equivalence may be concluded if a 90% confidence interval for $\theta$ lies within (0.8, 1.25). As the 90% confidence interval for $\theta$ is (0.772, 1.072), which is not contained in (0.8, 1.25), equivalence cannot be concluded. Here, the $p$-value is less than .10, but the 90% confidence interval contains possibilities in the null hypothesis. There is thus a different relationship between inferences on a $p$-value and inferences based on a confidence interval for equivalence tests than for a superiority test or a test of a difference.

For determining the posterior probability of the alternative hypothesis in Equation A.3, a noninformative prior distribution will be used for the true log-hazard ratio, and the estimated log-hazard ratio will be modeled as having a normal distribution with standard deviation of 0.1. The posterior distribution for the log-hazard ratio is a normal distribution with mean ln(0.91) and standard deviation 0.1. The posterior probability of the alternative hypothesis in Expression A.3 is 0.900. Note that for the equivalence comparison, the posterior probability of the alternative hypothesis did not equal 1 minus the $p$-value. If a 90% posterior probability were required for a conclusion of equivalence, the result would lie on the boundary of statistical significance.

Thus, while it seems that there is 90% confidence that $0.8 < \theta < 1.25$, the 90% confidence interval for $\theta$ does not lie within the interval (0.8, 1.25). We note that these types of equivalence hypotheses tend to be tested using a 90% confidence interval in various settings, including generic drug settings. Schuirmann[17] showed that such a test has a maximum type I error rate of 0.05. This is the result of treating an equivalence test as performing two simultaneous one-sided tests based on one-sided 95% confidence intervals, both of which need statistical significance at a 5% level. The alternative hypotheses for the one-sided tests are $H_a$: $0.8 < \theta$ and $H_a$: $\theta < 1.25$. More commonly, the $p$-value for the equivalence test is alternatively defined as the maximum of the two $p$-values from the two one-sided tests. For each of the one-sided tests, the definition of the $p$-value in Section A.1 is used. In this example, the respective one-sided $p$-values are 0.099 and 0.0008, resulting in a $p$-value of 0.099 for the equivalence test. This $p$-value is compared with 0.05 (the desired type I error rate), not 0.10. Since, $0.099 > 0.05$, equivalence is not demonstrated. Note also that this $p$-value corresponds to the largest-level confidence interval, having a confidence coefficient $1 - 2 \times p$-value, which lies within (0.8, 1.25). Here, the 80.2% confidence interval is (0.8, 1.035).

For this second definition of the $p$-value for the equivalence test, the largest distribution of the $p$-value under $H_0$: $\theta \leq 0.8$ or $\theta \geq 1.25$ is larger than a uniform distribution over (0,1). Thus, this test can be conservative. For the first definition of the $p$-value for an equivalence test in this example, the largest distribution of the $p$-value under $H_0$: $\theta \leq 0.8$ or $\theta \geq 1.25$ is a uniform distribution over (0,1).

---

## A.4 Stratified and Adjusted Analyses

This section will discuss the use of stratification in the randomization and the use of stratified and adjusted analyses.

### A.4.1 Stratification

Clinical trials are commonly randomized using permuted blocks.[26] Furthermore, randomization is commonly stratified by some predefined prognostic factor. With stratification, separate permuted blocks are generated for each level of stratification, and subjects are assigned the next available randomized treatment from the stratum to which they belong. In this way, subjects from each stratum are assigned to the various treatment groups in numbers approximately equal to the desired randomization ratio (exactly equal to the extent that entire blocks are used). This is desired to balance the levels of meaningful prognostic factors between arms. By doing so, any demonstrated difference between the arms can be attributed to the difference in the treatments received instead of one arm being allocated with better subjects than the other arm.

With stratification, treatment arms tend to be more similar in the distribution of the stratification factors. A clinical trial in which treatment arms are not well balanced can be subject to criticism and difficult to interpret, even though the randomization procedure was fair in its assignment of subjects to arms, and the calculation of the $p$-value accounts for any potential imbalance. Without stratification, the treatment arms will be balanced on average; with stratification, the balance will be much closer for the realized allocation as well as for the mean among many theoretical realizations.

A second advantage of using stratification is that it allows for the use of analyses that have greater power. When a stratification factor is used for the randomization process, the analysis may adjust for the stratification factor by either using that factor as a covariate in the analysis or by integrating the results of the comparisons within each level of the factor (known as a "stratified analysis"). For an analysis of covariance, including the factor as a covariate in the model tends to reduce the associated standard error in estimating the difference in means (the treatment effect). The analysis of covariance allows the covariate to explain its contribution to the total variability of the observed outcomes. The remaining variability is now the background variability in estimating the treatment effect. This is true even if the treatment arms are balanced for the stratification factor, provided the stratification factor is correlated with outcome. For a stratified analysis and for Cox proportional hazards models and logistic regression models, the use of a prognostic covariate in the analysis allows for a comparison of "likes" between treatment arms. That is, the comparison is not obscured by arbitrary differences in the covariate between arms. In a stratified analysis, patients with similar prognostics are compared with the results of these comparisons integrated to form the overall comparison. When a treatment is effective and the covariate is quite prognostic, the effect size, as measured by a hazards ratio or an odds ratio, tends to be larger for an adjusted analysis than for an unadjusted analysis.

If a factor is believed to be associated with outcome, it should be considered as a stratification factor. However, having too many stratification factors can be as bad as having none: in the extreme case, each subject represents a separate stratum, which results in a procedure identical to having no stratification factors. The factors related to the outcome of a primary endpoint tend to be more useful than other factors as a stratification factor. In addition, it may be imperative to balance treatments for a particular prognostic factor to ensure that each group will have a sufficient number for subgroup comparisons. Frequently, one or two stratification factors will generally be sufficient. One or two factors with the greatest association with the primary endpoint where imbalance in their allocation can be impactful should be considered as stratification factors for the randomization.

Stratification factors should have few levels. A factor that has many levels can have the number of levels reduced by combining levels. Examples of such stratification factors include diagnosis (e.g., disease variety) and continuous variables (e.g., younger than 65 years and at least 65 years of age in a study in which age is expected to be prognostic). Choosing the factor levels is as important as choosing the stratification factor itself. The impact from level to level on the relationship of the primary endpoint and the relative number of patients entered with such levels need to be considered. An analysis that adjusts for a factor where there is little impact on the relationship with primary endpoint from level to level will provide similar results as the analysis that does not adjust for that factor. Likewise, when almost all the subjects have the same level and there is not a large enough impact on outcome across levels, the adjusted and unadjusted analyses should provide similar results.

Even when several stratification factors are chosen, there may be other factors that are prognostic, either because the relationship was not appreciated when the study was designed or other factors were thought to be more predictive of response. In such a case, it may be possible and plausible to include the factor as a covariate. We caution against post hoc identification of covariates in the primary model, as bias can result. Covariates can be identified a priori for inclusion in the model, even when they were not used as stratification factors; however, every factor used in the stratification process should be included in the analysis model.

One special stratification factor is investigative site. For many therapeutic areas, it is common to include site as a covariate in the model even when it was not used for stratification in the randomization process. This is especially important when outcome is heavily related to investigator skill at diagnosis or treatment (e.g., new surgical technique) or to geography (e.g., bacterial infections in different climates or societies). Including site as a covariate will be problematic when each site enrolls few study subjects, as it creates many strata, some of which may be confounded with treatment when all subjects at a given site are randomized to the same treatment. When it is appropriate, sites can be grouped by geographic region or by some other meaningful criterion (e.g., grouping based on climate or based on the specialization of the principal investigator whenever the climate or specialization has impact).

## A.4.2 Analyses

Members of the population for which a new drug is being developed have characteristics that are quite heterogeneous. For a clinical trial to be applicable to the entire population (have external validity), the study population will also be heterogeneous. When subjects are provided a treatment, this subject variability contributes to the variability in the observed outcomes. Restricting a study to only patients who have similar values for a very influential prognostic factor will lead to less variable outcomes and will require fewer patients for the desired power. However, the results from such a study will only be externally valid for people similar to the study subjects. Variability can be reduced by stratifying the randomization and analysis, or by adjusting the analysis for prognostic factors not used in the randomization process. Stratified and adjusted analyses therefore allow studies to enroll a diverse group of subjects without requiring a dramatically larger number of subjects as a study using a homogeneous group of subjects.[27,28]

When a clinical trial incorporates a randomization process that uses stratification, the analysis commonly incorporates the stratification factor as a covariate. More generally, when a potentially prognostic baseline variable is identified before the study begins, an analysis may use this variable as a covariate in the model whether or not it was used as a stratification factor in the randomization process. Whether these prognostic factors should be included as covariates in the primary analysis model is a matter of some controversy for superiority analyses.[26,27] For non-inferiority analyses, the same controversies exist along with other concerns.

Covariates that can be included in the analysis model are identified in one of three ways: prospectively identified as being prognostic, retrospectively identified as being correlated with response, or recognized as being imbalanced in the clinical trial under consideration. A factor that is prospectively identified is the easiest to assess; a factor that is identified based on the data observed in the study is more difficult, and its inclusion in the model may introduce bias.

In general, prospectively identified factors can be included as covariates in the analysis model without much controversy. If the covariate influences the conclusion, the argument can be made that the inclusion of the factor in the analysis was made before the data were known and, therefore, the conclusion is not biased. One aspect of covariate analysis that may differ in non-inferiority and superiority analyses is the issue of collinearity. Collinearity occurs when two prognostic factors are related to each other and at least one is related to the outcome variable. A model that includes both covariates commonly shows neither covariate has a statistically significant effect on the outcome. Thus, with each collinear variable considered after adjustment for the other, an effect is not identified for either covariate. When one of these two variables is the randomized treatment group (i.e., by not having the covariate balanced between groups by the randomization), this can have the effect of masking a real effect of treatment. For superiority analyses, this has the effect of decreasing the chance of finding a relationship; for non-inferiority analyses, the impact on conclusions is much less well understood. We thus recommend caution in interpreting a non-inferiority clinical trial in which the analysis uses a covariate not included in the randomization process and in which there is considerable imbalance between comparative treatment arms.

The impact of choosing covariates for inclusion in the model on the basis of data observed in the model is even more difficult to defend. Such post hoc model choices are subject to biases in superiority analyses as well as in non-inferiority analyses. Additionally, choosing covariates on the basis of baseline imbalances can have the effect of causing collinearity, obscuring differences in treatment groups, and resulting in a biased estimate of treatment effect.

To illustrate, consider a simple additive analysis of covariance (ANCOVA) model, in which

$$y_{ij} = \alpha + \kappa_i + \gamma_j + \varepsilon$$

where $\alpha$ is the grand mean, $\kappa_i$ is the effect of treatment group $i$, $\gamma_j$ is the effect of categorical covariate $j$ with two levels, $\varepsilon$ is the error, and $y_{ij}$ is the observed value for a subject with associated covariate and treatment assignment. If the covariate is predictive of the outcome as an additive effect to the treatment, as expected by the model, the confidence interval calculated via ANCOVA will tend to be shorter than the confidence interval calculated without consideration of the covariate. The confidence interval for the true difference in means, $\mu_E - \mu_C$, is calculated using the estimated treatment effect and the mean square error, and the null hypothesis is rejected if the lower bound of the confidence interval is greater than $-\delta$.

Despite the advantages of using covariates in the analysis, we caution that all covariates, like other aspects of the analysis, must be prespecified. Including or excluding covariates to obtain a confidence interval that excludes $-\delta$, based on post hoc analyses, is not appropriate as it inflates the chance of falsely concluding that the experimental drug is noninferior to the standard drug.

It is also often important to investigate whether there is any interaction between treatment and a prespecified covariate on the outcome of interest. Such an interaction effect means that the difference in the effects of the experimental and control therapies varies across the levels of the covariate. For superiority testing, when this difference in effect always favors the same therapy, the interaction is regarded as quantitative. If instead this difference in effects sometimes favors the experimental therapy and sometimes favors the control therapy, the interaction is regarded as qualitative. Determination of whether the interaction is quantitative or qualitative involves comparing the difference of effects with zero (zero being the value specified as the difference in effects in the null hypothesis).[29] For non-inferiority testing, the determination of whether the interaction is quantitative or qualitative involves comparing the difference of effects with the non-inferiority margin. In a stratified analysis, an advantage of the control group in one stratum that is larger than the non-inferiority margin can be offset by an advantage of the experimental group in another stratum, a situation akin to that of a qualitative interaction and causing the two treatments, on average, to look similar.[30] In such a situation, a non-inferiority analysis on the overall population may be problematic. Examination of the treatment effect in each stratum to check for consistency of effect will be important (see Chapter 10 for further details).

## References

1. Dempster, A.P. and Schatzoff, M, Expected significance level as a sensibility index for test statistics, *J. Am. Stat. Assoc.*, 60, 420–436, 1965.
2. Schatzoff, M., Sensitivity comparisons among tests of the general linear hypotheses, *J. Am. Stat. Assoc.*, 61, 415–435, 1966.
3. Hung, H.M.J. et al., The behavior of the p-value when the alternative hypothesis is true, *Biometrics*, 53, 11–22, 1997.
4. Sackrowitz, H. and Samuel-Cahn, E., P-values as random variables—expected p-values, *Am. Stat.*, 53, 326–331, 1999.
5. Joiner, B.L., The median significance level and other small sample measures of test efficacy, *J. Am. Stat. Assoc.*, 64, 971–985, 1969.
6. Bhattacharya, B. and Habtzghi, D., Median of the *p*-value under the alternative hypothesis, *Am. Stat.*, 56, 202–206, 2002.
7. Fisher, R.A., *The Design of Experiments*, Oliver and Boyd, Edinburg, 1935.
8. Hollander, M. and Wolfe, D.A., *Nonparametric Statistical Methods*, John Wiley, New York, NY, 1973.
9. Good P., *Permutation, Parametric, and Bootstrap Tests of Hypotheses*, Springer, New York, NY, 2005.
10. Box, G.E.P., Hunter, J.S., and Hunter, W.G., *Statistics for Experimenters: An Introduction to Design, Data Analysis, and Model Building*, John Wiley, New York, NY, 1978.
11. Wiens, B.L., Randomization as a basis for inference in non-inferiority trials, *Pharm. Stat.*, 5, 265–271, 2006.
12. Fisher, R.A., *Statistical Methods for Research Workers*, Oliver and Boyd, Edinburgh, 1925.

13. Barnard, G.A., Significance tests for 2 × 2 tables, *Biometrika*, 34, 123–138, 1947.
14. Chan, I.S.F., Exact tests of equivalence and efficacy with a non-zero lower bound for comparative studies, *Stat. Med.*, 17, 1403–1413, 1998.
15. Röhmel, J. and Mansmann, U., Unconditional non-asymptotic one-sided tests for independent binomial proportions when the interest lies in showing non-inferiority and/or superiority, *Biom. J.*, 41, 149–170, 1990.
16. Dunnet, C.W. and Gent, M., Significance testing to establish equivalence between treatments with special reference to data in the form of 2 × 2 tables, *Biometrics*, 33, 593–602, 1977.
17. Schuirmann, D., A comparison of the two one-sided tests procedure and the power for assessing the equivalence of average bioavailability, *J. Pharmacokinet. Pharm.*, 15, 657–680, 1987.
18. International Conference on Harmonization of Technical Requirements for Registration of Pharmaceuticals for Human Use (ICH), E9: statistical principles for clinical trials, 1998, at http://www.ich.org/cache/compo/475-272-1.html#E4.
19. Neyman, J. and Pearson, E.L., On the problem of the most efficient tests of statistical hypotheses, *Philos. T. R. Soc. Lond.*, 231, 289–337, 1933.
20. Blackwelder, W.C., Proving the null hypothesis in clinical trials, *Control. Clin. Trials*, 3, 345–353, 1982.
21. Carlin, B.P. and Louis, T.A., *Bayes and Empirical Bayes Methods for Data Analysis*, Chapman and Hall, London, 1996.
22. Goodman, S.N., Towards evidence-based medical statistics: The *P*-value fallacy, *Ann. Intern. Med.*, 995–1004, 1999.
23. Simon R., Bayesian design and analysis of active control clinical trials, *Biometrics*, 55, 484–487, 1999.
24. Rothmann, M.D. and Tsou, H., On non-inferiority analysis based on delta-method confidence intervals, *J. Biopharm. Stat.*, 13, 565–583, 2003.
25. Rothmann, M., Type I error probabilities based on design-stage strategies with applications to non-inferiority trials, *J. Biopharm. Stat.*, 15, 109–127, 2005.
26. Senn, S., Added values: Controversies concerning randomization and additivity in clinical trials, *Stat. Med.*, 23, 3729–3753, 2004.
27. Friedman, L.M., Furberg, C.D., and DeMets, D.L., *Fundamentals of Clinical Trials*, 3rd Edition, Springer, New York, NY, 1998.
28. Montgomery, D.C., *Design and Analysis of Experiments*, John Wiley & Sons, New York, NY, 1991.
29. Gail, M. and Simon, R., Testing for qualitative interactions between treatment effects and patient subsets, *Biometrics*, 41,, 361–372, 1985.
30. Wiens, B.L. and Heyse, J.F., Testing for interaction in studies of non-inferiority, *J. Biopharm. Stat.*, 13, 103–115, 2003.

# Index

Printed in the United States
by Baker & Taylor Publisher Services

Statistics

The increased use of non-inferiority analysis has been accompanied by a proliferation of research on the design and analysis of non-inferiority studies. Using examples from real clinical trials, **Design and Analysis of Non-Inferiority Trials** brings together this body of research and confronts the issues involved in the design of a non-inferiority trial. Each chapter begins with a non-technical introduction, making the text easily understood by those without prior knowledge of this type of trial.

### Topics covered include

- A variety of issues of non-inferiority trials, including multiple comparisons, missing data, analysis population, the use of safety margins, the internal consistency of non-inferiority inference, the use of surrogate endpoints, trial monitoring, and equivalence trials
- Specific issues and analysis methods when the data are binary, continuous, and time-to event
- The history of non-inferiority trials and the design and conduct considerations for a non-inferiority trial
- The strength of evidence of an efficacy finding and how to evaluate the effect size of an active control therapy

A comprehensive discussion on the purpose and issues involved with non-inferiority trials, **Design and Analysis of Non-Inferiority Trials** will assist current and future scientists and statisticians on the optimal design of non-inferiority trials and in assessing the quality of non-inferiority comparisons done in practice.

CRC Press
Taylor & Francis Group
an **informa** business

w w w . c r c p r e s s . c o m

ISBN 978-0-367-57691-2

9 780367 576912

# INDEX OF NAMES

*The numbers refer to the pages*

Printed in the United States
By Bookmasters